SAE International's Dictionary for Automotive Engineers

SAE International's Dictionary for Automotive Engineers

BY
JOHN F. KERSHAW EdD

SAE INTERNATIONAL®

Warrendale, Pennsylvania, USA

400 Commonwealth Drive
Warrendale, PA 15096-0001 USA
E-mail: CustomerService@sae.org
Phone: 877-606-7323 (inside USA and Canada)
724-776-4970 (outside USA)
FAX: 724-776-0790

Library of Congress Catalog Number 2022951323
http://dx.doi.org/10.4271/9781468604078

ISBN-Print 978-1-4686-0406-1
ISBN-PDF 978-1-4686-0407-8
ISBN-epub 978-1-4686-0408-5

To purchase bulk quantities, please contact: SAE Customer Service

E-mail: CustomerService@sae.org
Phone: 877-606-7323 (inside USA and Canada)
724-776-4970 (outside USA)
Fax: 724-776-0790

Visit the SAE International Bookstore at books.sae.org

Publisher
Sherry Dickinson Nigam

Development Editor
Publishers Solutions, LLC
Albany, NY

Director of Content Management
Kelli Zilko

Production and Manufacturing Associate
Brandon Joy

Contents

Preface

Without vision you may not succeed, so the vision for *SAE International's Dictionary of Automotive Engineering* is to become the most comprehensive automotive engineering reference for professionals and students alike.

This authoritative reference provides clearly written, easy-to-understand definitions for over 1,800 terms used in automotive engineering worldwide. Unlike a standard dictionary that provides only definitions, the *SAE International's Dictionary for Automotive Engineers* provides a unique level of details including:

- In-depth definitions including formulas and equations where appropriate.
- Over 300 full-color illustrations to provide clarity for a definition, component, or system identification.
- References to relevant SAE Standards to direct the read to additional information beyond a practical definition.
- Coverage of newer technologies such as electric vehicles, automated vehicles, hydrogen fuel.

Organized in alphabetical order, readers will find most acronyms are listed first followed by the term then the definition to mimic conventional usage of acronyms within the industry.

Whether you use the print or eBook addition, *SAE International's Dictionary of Automotive Engineering* exceeds similar resources providing readers with comprehensive view of all SAE offers by providing SAE Standard Identification whenever appropriate.

Acknowledgments

I want to thank my wife Joan for her continued support in all of my projects. Also, I need to acknowledge my good friend and colleague, noted author James Halderman, who has generously supported this project and many others projects along with Tom Richardson, Vice President of Government Affairs with the National Institute for Automotive Service Excellence (ASE).

List of Acronyms

2WD - Two-Wheel Drive
4WD - Four-Wheel Drive
ABS: - Anti-lock Brake System
AC - Alternating Current
ADS - Automated Driving Systems
AGM - Absorbed Glass Mat (Battery)
ALCL - Assembly Line Communication Link
ALDL - Assembly Line Diagnostic Link
ASD - Automatic Shutdown
ASE - Automotive Service Excellence
ASME - American Society of Mechanical Engineers
ASTM - American Society for Testing and Materials
AT - Automatic Transmission
ATA - American Trucking Associations (Connector)
ATC - Automatic Traction Control (Heavy Truck)
ATC - Automatic Transfer Case
ATCM - Automatic Transfer Case Control Module
ATDC - After Top Dead Center
ATF - Automatic Transmission Fluid
AUTO-4WD - Automatic Four-Wheel Drive
AV - Autonomous Vehicle
AWD - All-Wheel Drive
BA - Brake Assist
BAS - Belt Alternator Starter (Low-Cost Hybrids)
BCM - Body Control Module
B-CAN - Body Control Area Network
BDC - Bottom Dead Center
BEV - Battery Electric Vehicle
BHP - Brake Horsepower
BMEP - Brake Mean Effective Pressure
BOV - Blow-Off Valve
BSFC - Brake Specific Fuel Consumption
BTDC - Before Top Dead Center
BTU - British Thermal Unit
CA - Cranking Amps
CAC - Charge Air Cooler
CAFE - Corporate Average Fuel Economy
CAN - Controller Area Network
CCA - Cold Cranking Amps
CCC (C3) - Computer Command Control
CCI - Calculated Cetane Index
CCM - Comprehensive Component Monitor
CDI - Capacitor Discharge Ignition
CDR - Crankcase Depression Regulator
CDR - Crash Data Retrieval
CEMF - Counterelectromotive Force
CFR - Cooperative Fuel Research
CHT - Cylinder Head Temperature
CKP - Crankshaft Position
CM - Continuous Monitor
CMP - Camshaft Position
CN - Cetane Number
CNG - Compressed Natural Gas
CO_2 - Carbon Dioxide
CO - Carbon Monoxide
COE - Cab-Over-Engine
CPI - Continuous Port Fuel Injection
CPU - Central Processing Unit
CRT - Cathode Ray Tube
CV - Constant Velocity
CVT - Continuously Variable Transmission
DC - Direct Current
DEF - Diesel Exhaust Fluid
DI - Direct Ignition
DI - Distributor Ignition
DIC - Driver Information Center
DIS - Distributorless Ignition System
DLC - Data Link Connector
DMM - Digital Multimeter
DMS - Driver Monitoring System
DOC - Dual Overhead Camshaft
DOC - Diesel Oxidation Catalyst
DOT - Department of Transportation
DPF - Diesel Particulate Filter
DSRC - Dedicated Short-Range Communication
DTC - Diagnostic Trouble Code
EAS - Electrically Assisted Steering
EBA - Emergency Brake Assist or Brake Assist (BA/BAS)
EBCM - Electronic Brake Control Module

ECM - Engine or Electronic Control Module
ECU - Engine or Electronic Control Unit
EEPROM - Electronically Erasable Programmable Read-Only Memory (E^2PROM)
EFI - Electronic Fuel Injection
EFT - Engine Fuel Temperature
EGR - Exhaust Gas Recirculation
EMD - Engine Manufacturer's Diagnostic
EMI - Electromagnetic Interference
EMF - Electromagnetic Field
EMF - Electromotive Force
EP - Extreme Pressure (Additive)
EPA - Environmental Protection Agency
EPHS - Electrically Powered Hydraulic Steering
EPROM - Erasable Programmable Read-Only Memory
EPS - Electric Power Steering
ERFS - Electronic Returnless Fuel System
ESC - Electronic Spark Control
ESC - Electronic Stability Control
ESD - Electrostatic Discharge
ESI - Electronic Service Information
EST - Electronic Spark Timing
EST - Electronic Service Tools
ETBE - Ethyl Tertiary Butyl Ether
EUI - Electronic Unit Injector
EV - Electric Vehicle
EVA - Electronic Vibration Analyzer
EVAP - Evaporative Emissions
F-CAN - Fast Control Area Network
FAD - Front Axle Disconnect
FCD - Floating Car Data
FIFFV - Flexible Fuel Vehicles
FIPG - Formed-In-Place Gasket
FM - Frequency Modulation
FMI - Failure Mode Identifier
FMVSS - Federal Motor Vehicle Safety Standard
FTP - Federal Test Procedure
FWD - Front-Wheel Drive
GAWR - Gross Axle Weight Rating
GCWR - Gross Combined Weight Rating
GDI - Gas Direct Injection
GVWR - Gross Vehicle Weight Rating
HD-OBD - Heavy-Duty On-Board Diagnostics
HEUI - Hydraulically Actuated Electronic Unit Injection
HEV - Hybrid Electric Vehicle
HOAT - Hybrid Organic Acid Technology
HPCR - High-Pressure Common Rail
HSD - High-Side Drivers
HUD - Heads-Up Display
IAC - Idle Air Control
IAT - Intake Air Temperature
IAT - Inorganic Additive Technology
IC - Ignition Control
IC - Integrated Circuit
ICM - Ignition Control Module
IDI - Indirect Diesel Injection
IHP - Indicated Horsepower
IMEP - Indicated Mean Effective Pressure
IRS - Independent Rear Suspension
ISO - International Standards Organization
ITS - Intelligent Transportation System
IVHS - Intelligent Vehicle Highway System
IVT - Intelligent Vehicle Technologies
KAM - Keep Alive Memory
KES - Keyless Entry System
LCC - Lane Centering Control
LCD - Liquid Crystal Display
LDW - Lane Departure Warning
LNG - Liquefied Natural Gas
LPG - Liquid Petroleum Gas
LSD - Limited Split Differential
LTFT - Long-Term Fuel Trim
LVD - Low-Voltage Disconnect
MAF - Mass Air Flow
MANET - Mobile Ad Hoc Network
MAP - Manifold Absolute Pressure
MEP - Mean Effective Pressure
MIC - Multiplex Integrated Control
MID - Message Identifier
MMT - Methyl-cyclopentadienyl Manganese Tricarboxylic
MON - Motor Octane Number
MRFS - Mechanical Returnless Fuel System
MTBE - Methyl Tertiary Butyl Ether
Nimh/Ni-MH - Nickel Metal Hydride Battery
NOP - Nominal Opening Pressure
NOx - Oxides of Nitrogen
NTC - Negative Temperature Coefficient
OAT - Organic Acid Technology
OBD - On-Board Diagnostics
OBD II - On-Board Diagnostics Generation II
OEM - Original Equipment Manufacturers
OHV - Overhead Valves
OL DMM - Digital Multimeter Symbol Over Limits or Overload
PCD - Pitch Circle Diameter
PCM - Powertrain Control Module
PDOF - Principle Direction of Force
PHOAT - Phosphate Hybrid Organic Acid Technology
PID - Parameter Identifier
PM - Particulate Matter
PMAC - Permanent Magnet AC
PROM - Programmable Read-Only Memory
PTC - Positive Temperature Coefficient
PTO - Power Takeoff
PWM - Pulse Width Modulation
RAM - Random Access Memory

RBM - Resistance to Bending Moment
RFG - Reformulated Gasoline
RFI - Radio Frequency Interference
ROM - Read-Only Memory
RON - Research Octane Number
RPM - Revolutions Per Minute
RVP - Reid Vapor Pressure
SAE - Society of Automotive Engineers
SAI - Steering Axis Inclination
SCR - Select Catalyst Reduction
SDS - Safety Data Sheets
SFI - Sequential Fuel Injection
SID - Subsystem Identifier
SLA - Short-Arm/Long-Arm (Suspension)
SLA - Sealed Lead-Acid
SOC: - State of Charge
STFT - Short-Term Fuel Trim
TAC - Thermostatic Air Cleaner
TBI - Throttle Body Injection
TCC - Torque Converter Clutch
TCM - Transmission Control Module
TCS - Traction Control System
TDC - Top Dead Center
TOOT - Toe-Out On Turns (Turning Radius)
TPS - Throttle Position Sensor/Switch
TSB - Technical Service Bulletins
UTQGS - Uniform Tire Quality Grading System
V2I - Vehicle-to-Infrastructure
V2V - Vehicle-To-Vehicle Communication
VANET - Vehicular Ad Hoc Network
VCO - Valve Covering Orifice
V/div - Volts Per Division
VI - Viscosity Index
VIN - Vehicle Identification Number
V-ref - Reference Voltage
VRLA - Valve Regulated Lead Acid (Battery)
VSM - Value Stream Map
VSS - Vehicle Speed Sensor
WSS - Wheel Speed Sensor
WVSCT - Wireless Vehicle Safety Communications Telematics

A

3-4-5 Accident Measuring Triangle

The accident measuring triangle is a measuring setup used by accident reconstructionist experts and automotive forensic engineers as a protocol for quantifying vehicle damages resulting from a collision. It is basically the creation of a carpenter's square or right angle. The triangle measurements from a right triangle can be used to form an exterior vehicle profile to assess the principal direction of force (PDOF). Figure A.1 shows the triangle laid against an exact duplicate vehicle to the one whose damage will be measured. You will compare measurements from the damaged vehicle to the duplicate vehicle. The hypotenuse of the triangle is placed at the center of the front wheel. Side b is the 5-foot side of the triangle. Side a, the 3-foot side, extends from the center of the wheel for 3 ft. Side b, the 4-foot side, goes across from the bottom of Side a to meet the bottom of Side b the hypotenuse. Using the Pythagorean theorem formula $a^2 + b^2 = c^2$, $9 + 16 = 25$ [see Equation (A.1)], the forensic engineer will take coordinate measurements like the ones taken at the scene of the accident. These measurements are taken along the base of the triangle and offset measurements inward from the vehicle base every 6 in. along the damaged area. You need to use a 4-foot level that provides a level plane for these measurements. For more information on this, see **J1674_201807** [1] and *Vehicle Accident Analysis and Reconstruction Methods, Third Edition* by Matthew Brach, Raymond M. Brach, and James Mason [2, 3].

$$a^2 + b^2 = c^2 \quad c = \sqrt{a^2 = b^2} \tag{A.1}$$

FIGURE A.1 Measuring triangle.

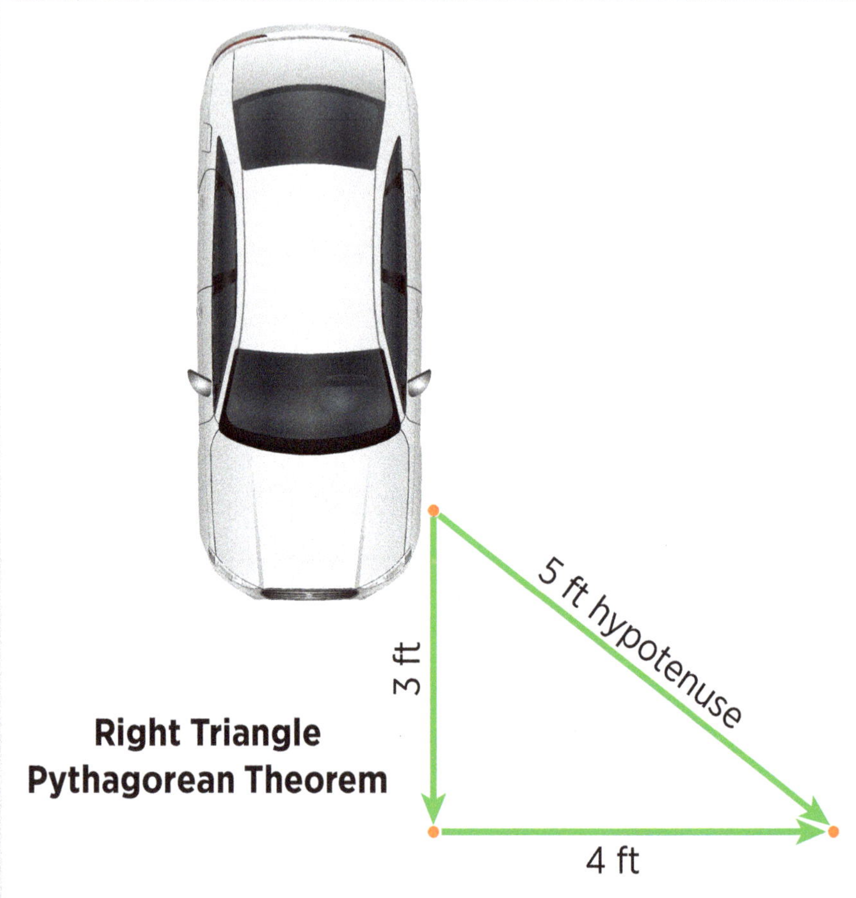

Courtesy of John F. Kershaw Ed.D.

4-Stroke Cycle (Spark Ignition)

The intake stroke begins with the piston at the top spatial relation or top dead center (TDC). A lobe on the camshaft opens the valve either directly or through a disciple, pushrod, and valve rocker assembly. The piston moves down within the bore because of the rotation of the crankshaft. As the piston is moving downward, it pulls outside air through the air filter and into the intake manifold past the open intake valve and into the cylinder. The downward movement of the piston creates a low-pressure area above the piston because as the volume increases, the pressure decreases. Air rushes in to fill the space left by the downward movement of the piston because air pressure is greater than the pressure within the cylinder. The air-fuel mixture is consistent. During the intake stroke, an air-fuel ratio is inducted. The throttle valve controls the atmospheric pressure entering the cylinder. The energy needed to maneuver the piston downward from TDC comes from overlapping power strokes on a multiple-cylinder engine or flywheel inertia. As the piston nears bottom spatial relation, it slows down nearly to a stop. When the piston reaches bottom spatial relation or BDC, the valve closes, sealing the cylinder, and the compression stroke will begin. The crankshaft is still revolving, forcing the piston upward toward TDC. Both intake and exhaust poppet valves are closed. The air-fuel mixture in the cylinder cannot escape, except past the piston rings. The cylinder volume is decreasing because, as the piston rises, the air-fuel gas mixture is compressed. The cylinder gas pressure is inversely proportional to the cylinder gas volume. This purports to Boyle's Gas law, which states that the pressure of a perfect gas varies inversely to its volume at a constant temperature or that the product of the pressure and volume is a constant within the compression of a gas. The volume is decreasing and so the pressure and temperature rise as external work is done on the gas.

Compression ratio is the ratio of the volume at BDC to the TDC clearance volume (compression ratio = volume BDC/volume TDC). A better compression ratio means higher thermal efficiency or that portion of the warmth supplied to the engine became work. When the compression ratio increases, the expansion ratio also increases; thus, thermal efficiency increases. The internal energy of the gas is increased as heat is added to the gas. Near the top of the compression stroke, a sparking plug will ignite the mixture. The power stroke begins right after the air-fuel gas mixture is ignited by a high-voltage (HV) electrical spark and combustion begins. The high gas pressures from combustion developed within the cylinder push the piston downward within the bore, which causes the crankshaft to rotate in the form of translation to rotation. The pressure falls because the volume increases. The temperature falls because the gas does external work. The oxygen and fuel burn and the nitrogen expands, pushing the piston down under power. As the piston continues downward, the combustion gases expand as they offer up their energy.

The power stroke is the only stroke where energy is extracted from the fuel. The cylinder pressure is the highest during the power stroke. As the piston nears the bottom of its movement, the exhaust poppet valve, operated by the camshaft lobe, begins to open. The piston then begins to rise within the cylinder, starting the exhaust stroke. The upward movement of the piston forces the spent gases past the open poppet valve and out the cylinder. The exhaust poppet valve is closed by spring pressure, shortly after the piston begins its downward movement. The exhaust stroke produces no work but uses some of the energy produced during the power stroke to push exhaust gases out of the cylinder. In a spark-ignited gasoline-fueled engine, we have flame speed which increases when engine speed increases. Therefore, the quantity of crank angles occupied by the combustion process is almost independent of engine speed.

FIGURE A.2 Spark-ignition ICS 4-stroke cycle.

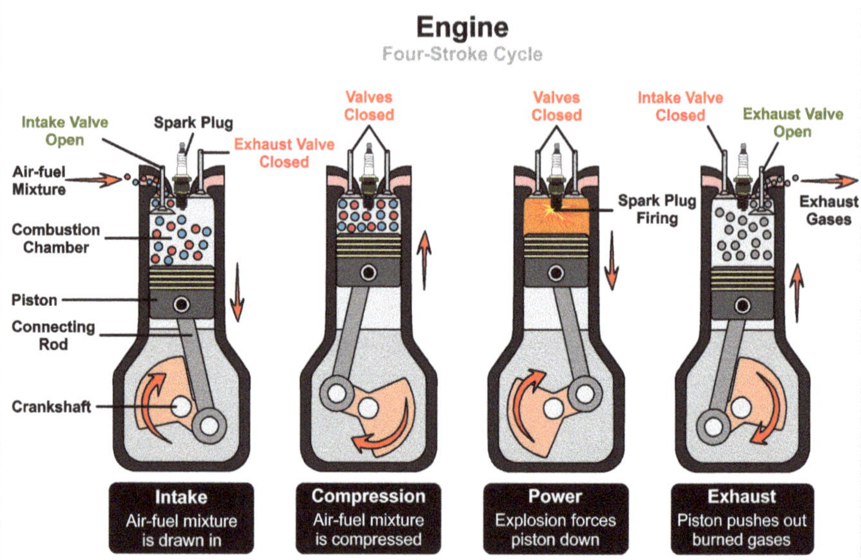

udaix/Shutterstock.com.

4-Stroke Cycle Diesel (Compression Ignition)

The diesel 4-stroke cycle is the same as the Otto 4-stroke cycle. There are differences in combustion, power control, and compression ratio. The 4-stroke cycle consists of intake, compression, power, and exhaust. When all four are completed, the diesel engine has completed the four strokes of one full cycle. The diesel intake stroke starts with the piston at top dead center, or TDC, where the piston is at the top of the cylinder. The camshaft lobe opens the intake poppet valve in the cylinder head. The piston moves down in the bore due to the rotation of the crankshaft. As the piston moves down, it pulls outside air through the air cleaner and into the intake, past the open intake valve and into the cylinder. The downward movement of the piston creates a low-pressure area above the piston (as the volume increases the pressure decreases, which is Boyle's Law). Air rushes in to fill the space left by the downward movement of the piston because atmospheric pressure is greater than the low pressure in the cylinder.

During the diesel intake stroke, only air is inducted; this is because no throttle valve is used, so the cylinder is completely filled with air at the inlet manifold pressure. The air mixes with residual gasses in the cylinder. The energy needed to move the piston from TDC downward comes from either the flywheel or overlapping power strokes from a multiple-cylinder engine. As the piston nears bottom dead center (BDC), it slows down nearly to a stop. When the piston reaches BDC, the intake valve closes, sealing the cylinder filled with air, and the compression stroke begins. The rotating crankshaft forces the piston upward and both intake and exhaust poppet valves are closed so the air cannot exhaust. The volume is decreasing as the piston rises, so the air is compressed. The pressure is inversely proportional to the volume according to Boyle's Law. In the compression of a gas, the volume decreases and the pressure and temperature rise as the gas is pressurized. This causes collisions of the air molecules within the cylinder. For example, if a volume of air is compressed to 1/22 of its original volume as it is in diesel, the open space between the molecules is greatly reduced increasing the number of collisions and the pressure between them. These collisions cause heat due to the kinetic energy of the molecules.

Compression ratio is the ratio of the volume at BDC to the volume at TDC or the cylinder clearance volume. A higher compression ratio means higher thermal efficiency or that portion of the heat supplied to the engine that is converted into work. As the compression ratio increases, the expansion ratio also increases thus thermal efficiency increases. The internal energy of the combustion gas is increased as heat is added to the gas. The high heat generated by greater compression will cause the fuel when injected to atomize or break up into finely divided particles, allowing it to mix easily with the air. In the indirect injection (IDI) engine, mixing is further enhanced by the addition of more heat through the spinning action of a spherical shaped prechamber. Ignition will occur only at the interface where the air and fuel come together, which is called a diffusion flame. The temperature of the compressed air is approximately 1,000°F. The temperature is generally higher than the spontaneous ignition point of diesel fuel, which is 558°F (292°C). Near the end of the compression stroke, fuel will be sprayed into either the prechamber in an IDI engine or the combustion chamber in a direct injected (DI) diesel engine. In the DI engine, the compressed air will swirl in a toroidal piston cavity increasing the friction between air molecules, which will promote mixing with the fuel. The IDI diesel prechamber is joined to the main combustion chamber above the piston by a connecting flow passage. A swirl-precombustion chamber has a spherical shape that mixes the air and fuel by air swirl. This assists in promoting high turbulence by creating a swirling mass of air in the prechamber.

Turbulence is the extreme disturbance of the compressed air in the combustion chamber. It causes the air molecules to move in all directions. The air molecules collide with each other and cause friction and heat, which increases the transfer of heat between the cool liquid-fuel droplets and the hotter air. The power stroke begins after the fuel is sprayed into the direct combustion chamber or prechamber. The fuel nozzle sprays and atomizes the fuel in short bursts forming a fuel-rich core surrounded by air zones. During this period, fuel has entered the main combustion chamber from the prechamber but has not begun to burn. The temperature of the air is much higher than the fuel, so some of it has evaporated, but some has formed into very tiny droplets. It starts to vaporize and mix with the hot compressed air. After about 0.001 sec, any zones that are hot enough and have the correct air-fuel mixture ratio will autoignite. It is important to note that ignition will take place only where air meets fuel. IDI diesel engines begin combustion in the prechamber, and combustion moves into the main chamber as fuel, air, and burned and burning gases while the piston descends on the power stroke. The first fuel burns very rapidly. This rapid burning of the air and fuel creates a sudden pressure causing a highly localized pressure. This localized pressure causes an audible noise, known as diesel knock, whose noise level centers on the pressure rise velocity. The cylinder high pressures force the piston down in the bore, which causes the crankshaft to rotate. The pressure falls as the volume increases. The temperature falls as the gas does external work. The oxygen/fuel burns and the nitrogen expands, pushing the piston down under power. As the piston continues downward, these gases in the cylinder expand and cool as they give up their energy. The power stroke is the only stroke in which energy used from the fuel and cylinder pressure is the highest. As the piston nears the bottom of its travel, the exhaust poppet valve is opened by a camshaft lobe. The piston then begins to rise in the cylinder, beginning the exhaust stroke. The upward movement of the piston forces the burned gases past the open exhaust valve and out of the cylinder. The exhaust valve closes due to spring pressure after the piston begins its downward movement [4].

FIGURE A.3 Diesel 4-stroke cycle in a compression-ignition engine.

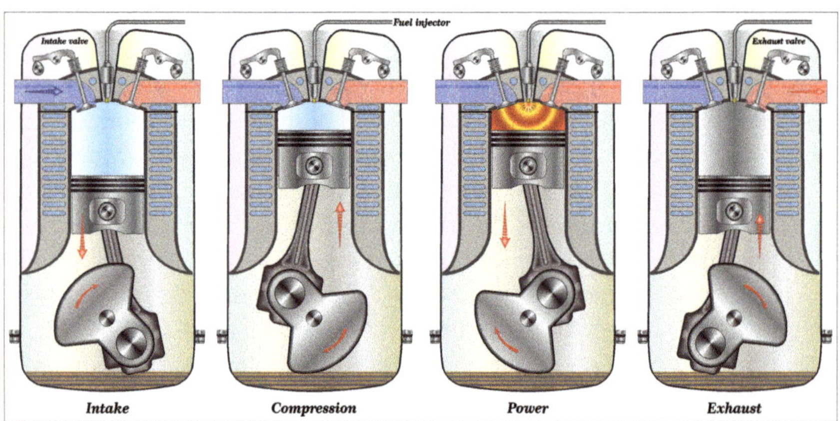

Fouad A. Saad/Shutterstock.com

2WD (Two-Wheel Drive)

The term used to describe a 2WD car or truck. The number "2" indicates the number of wheels that are driven by the transmission/engine. Only the two wheels are propelling this vehicle through a horizontal driveshaft or propeller shaft to a set of bevel gears.

4WD (Four-Wheel Drive)

The term used to describe a 4WD vehicle. The number "4" indicates the number of wheels that are driven by the transmission from the internal combustion engine. Due to open differentials that slip, power goes to the wheel with the least amount of traction. It is also referred to as a 4 × 4.

4WD/AWD Shift Motor

This motor is used to engage the proper clutch to deliver power flow to the front driveshaft on a four-wheel-drive (4WD) vehicle to engage 4WD in high or low. A control module commands the shift motor to rotate to the correct position to apply a clutch. The shift motor turns the control actuator lever shaft and a brake to hold the control actuator shaft in full-clutch position. The control actuator lever shaft moves the clutch lever. The clutch lever pivots on the control lever pivot studs and moves toward the clutch apply plate, to apply the clutch plates. The inner clutch discs are engaged to the clutch hub, and the outer clutch discs are engaged with the clutch housing; the power flow is delivered to the clutch housing. The clutch hub is splined to the rear output shaft and the chain drive sprocket is splined to the clutch housing. Power will flow from the drive sprocket, through the chain, to the chain-driven sprocket to the front output shaft and power is delivered to the front driveshaft. The shift motor rotates the control actuator shaft lever to the 4WD positions. Rotating the control actuator to the various positions changes the clutch torque level. For more information on this, see **J3104_201811** [5].

ABS (Anti-lock Brake System)

A brake control system to prevent brake lockup using a computer that compares the vehicle wheel speeds using wheel speed sensors. ABS can apply the brakes using a modulated pump on the automotive or medium-duty truck hydraulic ABS or release brakes automatically. The system uses a dash indicator (telltale) to indicate failures. If a wheel spins at a faster rate than other wheels, or faster than designed, this indicates a wheel is starting to slip (via that same telltale) and is in danger of losing traction and locking. ABS will reduce the hydraulic pressure to the affected brake, which allows the wheel to speed up so it can regain traction. As traction is recovered, brake pressure is reapplied to again slow the wheel. The opposite is true in that pressure can be applied to a wheel that is faster than the others. SAE Paper 2017-01-1578 cites two control strategies: "Safety preferred control and master cylinder oscillation control, were designed for anti-lock braking on a novel integrated-electro-hydraulic braking system (I-EHB) which has only four solenoid valves in its innovative hydraulic control unit (HCU) instead of eight in a traditional one. The main idea of safety preferred control is to reduce the hydraulic pressure provided by the motor in the master cylinder whenever a wheel tends to be locking even if some of the other wheels may need more braking torque. In contrast, regarding master cylinder oscillation control, a sinusoidal signal is given to the motor making the hydraulic pressure in the master cylinder oscillate in certain frequency and amplitude. Hardware-in-the-loop simulations were conducted to verify the effectiveness of the two control strategies mentioned above and to evaluate them. The simulation platform consists of the I-EHB hardware and software including CarSim and MATLAB/Simulink as well as LabVIEW serving as the communication tool. Conclusions can be reached in the light of testing results that both control strategies were able to achieve anti-lock braking under emergency situations. Compared with safety preferred control, master cylinder oscillation control performed better on the functionality of avoiding braking lock and the reduction of braking distance. Also, it is capable of working with electronic stability control systems (ESC) while safety preferred control cannot." [6].

ABS Truck Hydraulic Control Unit

This unit controls brake fluid pressure to the wheels during an anti-lock stop. It maintains, decreases, or increases the braking force as needed at each affected wheel. It is called a brake modulator on air ABS (anti-lock brake system). For more information on this, see **J2627_200908** [7].

ABS Truck Malfunction System

Medium- and heavy-duty trucks and buses are required to have anti-lock brake system (ABS) under Federal Motor Vehicle Safety Standards 121 and must use ABS malfunction indicator lamps (telltale). These lamps must be yellow and illuminate when the ABS management controller senses a malfunction that affects the generation or transmission of response or control signals from the wheel speed sensors. ABS malfunction indicator lamps do not illuminate for all malfunction codes, only specific safety-related codes. These lamps are required to illuminate for about 5 sec bulb check when the ignition or power switch is engaged. The warning lamps for tractor trailers and dollies are not required to illuminate for a bulb check unless the vehicle is stopped. For more information on this, see **J2627_200908** [7] and FMVSS 101 controls and displays.

Absolute or Dynamic Viscosity

Absolute viscosity is a measure of internal resistance in a liquid lubricant and the tangential force per unit area required to maneuver one horizontal plane with reference to another horizontal plane with the liquid lubricant between them. This takes place at a unit velocity while continuing a unit distance apart within the liquid lubricant.

Absorption

Absorption is the process or action where one unit absorbs or is absorbed by another unit. The unit doing the absorption is called the absorbate and the unit being absorbed is called the absorbent. These units or entities can be a solution of a gas, liquid, or solid in a liquid, attachment of molecules of a gas, vapor, liquid, or dissolved substance to a solid surface by physical forces. The absorption of light at characteristic wavelengths or bands of wavelengths is used to identify the chemical nature of molecules, atoms, or ions and to measure the concentrations of these entities.

AC (Alternating Current)

The electrical potential created by an AC generator is not a steady state, it fluctuates between positive and negative. When such a potential is applied to a circuit, it causes a current that flows first in one direction and then reverses itself and flows in the other direction. In house electrical systems, this direction reversal happens 60 times a second. This current is called alternating current, or AC. Rotating a magnetic field within or around a conductor produces AC in that conductor. When a magnetic field crosses or cuts though a conductor, it induces an AC voltage in that conductor.

AC Coupling

Signal that passes the alternating current (AC) signal component to the meter but blocks the direct current (DC) component. It is an AC signal that rides on a DC signal, like an alternator charging ripple.

AC Induction Motor

This is an electric motor driven by alternating current (AC). It consists of two basic parts, an out of doors stationary stator having coils provided with AC to provide a force field and an internal rotor attached to the output shaft which is given a torque by the rotating field. Nikola Tesla invented the rotating field and pioneered the utilization of a rotary field of force to work machines and designed a two-phase induction motor in 1883. There are two forms of AC motors, depending on the sort of rotor used. The first type is the electric motor, which rotates exactly at the provision frequency or a sub-multiple of the provision frequency. The flux on the rotor is either generated by current delivered through slip rings or by a static magnet. The second type is the asynchronous or induction motor, which is employed in battery electric vehicles. It turns slightly slower than the provision frequency or out of sync as the name suggests. The field of force on the rotor of this motor is made by an induced current.

Accelerate

Accelerate means to increase speed or the rate of change in velocity, which is a vector quantity having both direction and magnitude, where speed is a scalar quantity only having magnitude.

Acceleration

This is defined as a change in velocity (vector quantity has direction and magnitude) as it relates to time. The word comes from the Latin meaning to add speed. It can be either positive, if the vehicle is gaining speed, or negative, if the vehicle is losing speed. Equation (A.2) is used to determine acceleration:

$$a = \frac{V^2}{2 \times D} \tag{A.2}$$

where

a is acceleration
V is velocity
D is the distance

The resultant of this equation can be converted from feet per second or meters per second into g force using Equation (A.3):

$$g = \frac{a}{32\,\text{ft/s or } 9.8\,\text{m/s}} \tag{A.3}$$

FIGURE A.4 Vehicle acceleration change in velocity.

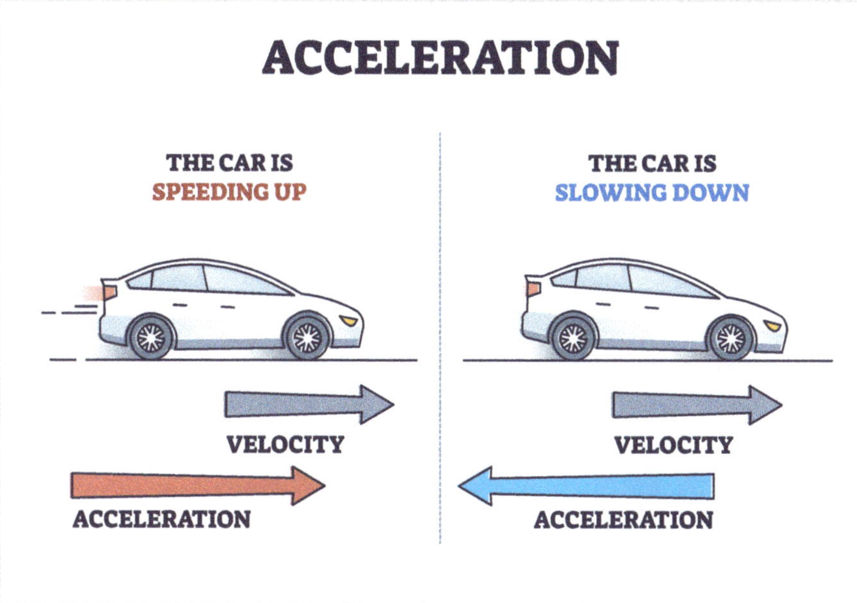

vectormine/Shutterstock.com.

Acceleration Test

This is an internal combustion engine (ICE) ignition system test that measures the spark plug firing voltages when the engine speed is rapidly increased. Also called a load test.

Accelerator Pump

A pumping device used in a carburetor to inject fuel into the boost venturi for additional fuel for combustion to achieve more engine speed during vehicle acceleration.

Accelerometer

A device that measures changes in motion or deceleration. It measures proper acceleration, which is the rate of change of velocity of a body in its own instantaneous rest frame, which is different from coordinate acceleration, which is acceleration in a fixed coordinate system. For example, an accelerometer on Earth's surface will measure an acceleration due to Earth's gravity, straight upward of "g" force, which is 9.81 m/s^2 or 33 ft/s^2. Accelerometers are used in some supplemental restraint systems (SRS).

Accident Reconstruction (Automotive and Motorcycle)

Accident reconstruction is the forensic science of determining how an accident took place while supporting the determination of the accident cause or why a collision or particular event during an accident happened using all the available physical evidence. Scientific methodology is used to determine the circumstances, mechanics, and contributing factors associated with an automotive collision. It requires a working knowledge of the following: physics, vehicle dynamics, mathematics, photogrammetry, and computer applications, simulation or modeling tools, graphics and photo-management software. This would include the use of scan tools and CDR (crash data retrieval) tools and software. Questions needed to be answered, for example, what was the vehicle speed going into the collision, were the brakes applied during ABS braking, or what were the angles of a multiple vehicle collision can be answered by the forensic investigator after a review of all pertinent documents and a physical inspection [1, 2, 3].

Accumulator

A pressure absorbing unit that absorbs and reduces sudden pressure surges in a hydraulic system. Accumulators are used in automatic transmission/transaxle hydraulic systems to cushion and control range shift quality.

Ackermann Principle

This is the correct positioning of the wheels when steering around a corner. It is a geometric alignment of linkages in a vehicle steering such that the wheels on the inside of a turn are able to move in a different circle radius than the wheels on the outside. For more information on this, see **J670_200801** [9].

Ackermann Steering System

A steering system designed by Rudolf Ackermann that used angled steering arms and knuckles. This design allows both front wheels to turn in their own arc. The inner wheel turns at a sharper angle, allowing both wheels to turn from the same pivot point, thus reducing tire wear. This systems has been redesigned over the years. For more information on this, see **J670_200801** [9].

FIGURE A.5 Ackermann steering system that uses angled steering arms to turn a steering knuckle.

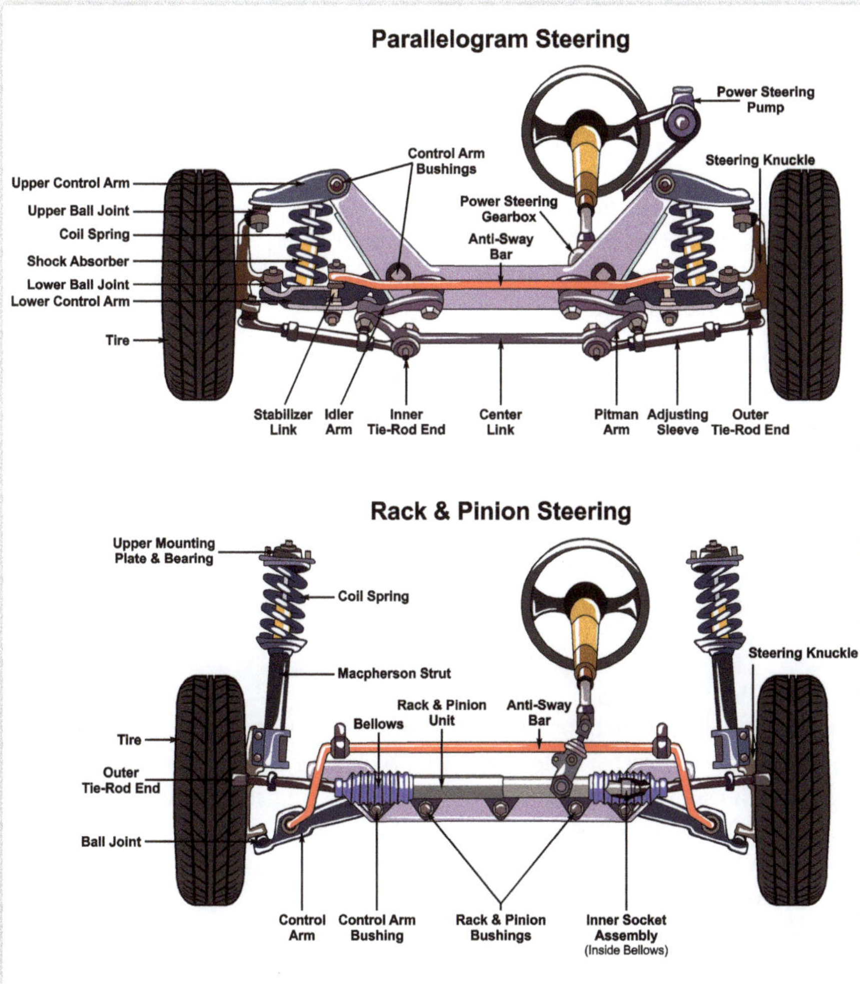

Acrylic Paint

Acrylic enamel paint is a fast-drying paint made of pigment suspended in acrylic polymer emulsion and plasticizers, silicone oils, defoamers, stabilizers, or metal soaps. Acrylic paints are either petrol-based or water-based, but become water-resistant when dry.

Active Restraint System

A restraint system, such as a lap/shoulder belt, that requires the occupants to ready the system through extra effort for them to activate or use. A lap seat belt must be fastened by the passenger to protect them from any crash. This is opposed to a passive restraint system like an SRS (supplemental restraint system) or air bags.

Active Sensor

A sensing device that requires an external source of power to operate; active sensors contrast with passive sensors, which simply detect and respond to some type of input from an outside source.

Actuator

An actuator is any output device like a solenoid or DC motor controlled by a management computer, where the computer operates the device by applying a ground or voltage. Devices would include solenoids or relays. They convert computer output into a mechanical action and perform the physical task dictated by a computer command. Actuators can be used to open and close switches, control the flow of fluid, and control gauges and warning lights.

AD Converter

A solid-state device that converts an analog signal to a digital signal. In electronics, a data converter called an ADC, A/D, or A-to-D converts an analog signal, such as a magnetic pickup used on a rotating shaft or a vehicle speed sensor (VSS) that generates an analog variable voltage signal into a digital signal sent to a management computer. An ADC may additionally provide an isolated measurement like a device that converts an analog input voltage or current to a digital number representing the magnitude of the voltage or current. The digital output may be a 1 and 0 complement binary number that is proportional to the input. Most ADCs are implemented as integrated circuits (ICs). These typically take the shape of metal-oxide semiconductor (MOS) mixed-signal computer circuit chips that integrate both analog and digital circuits. A digital-analog converter (DAC) performs the reverse function; it converts a digital signal into an analog signal. For more information on this, see **J1213/2_198810** [10].

AD (Analog-to-Digital) Converter Circuit

A device circuit used by a digital computer to change continuously variable voltage or current into digital bits.

Adaptive Controller (Radar Cruise Control)

An adaptive cruise controller (ACC) is used to allow the driving force more control of the vehicle by maintaining a clear safe distance, usually 2 to 3 sec, behind the vehicle in its front. If the vehicle at the front slows down, the radar controller of the following vehicle detects the slowing vehicle and automatically reduces its speed to keep a safe distance. Then, if the vehicle at the front increases speed, the radar controller of the following vehicle also allows itself to adjust to the preset speed. For more information on this, see **J3063_202103** [11].

FIGURE A.6 Adaptive or radar controller uses long-range radar to detect faraway objects ahead of the moving vehicle. Some systems use a short-range radar and/or infrared, or optical cameras to detect distances for when the gap between the moving vehicle and another vehicle ahead is reduced.

VectorMine/Shutterstock.com.

Adaptive Learning

The adaptive learning platform delivers learning content that is extremely relevant, adapted to each learner's needs, and is divided into micro-learning elements. To accomplish this, the learning is created through a series of questions and then followed by justifications for why each answer is either correct or incorrect.

Adaptive Memory

Adaptive memory is a feature of computer memory that allows the microprocessor to adjust its memory for computing output response based on changes in vehicle operating conditions.

Adaptive Testing

Adapting testing uses test questions that are developed from key learning points (KLP) focus on important concepts and "need to know" information required to meet the identified performance goal or behavioral change. Each KLP is a single piece of critical information that includes what the learner needs to know/do and why it is important.

ADEMS (Advanced Diesel Engine Management System)

This is a Caterpillar acronym used to describe their heavy-duty diesel engine management electronics using electronic unit injectors (EUI). The ADEMS is a full-authority control system. Every time the EUI camshaft lobe comes near the end of the intake stroke, it begins to depress the injector tappet and plunger. The EUI plunger stroke is cam actuated, so the system control is still restricted to the hard limit window created by the cam profile. In the "normal" state, the poppet valve in the electric solenoid is open. The open valve allows the injector to fill with fuel. and when the injector tappet and plunger are pushed, any pressure developed is bypassed to the fuel return system. When the engine control module (ECM) calls for injection, a 100-V electrical signal is applied to the injector solenoid, and this causes the solenoid to close the bypass poppet valve. With the bypass closed, the moving plunger rapidly develops pressure in the fuel chamber.

When the fuel pressure reaches a moderate level, the injector check valve (lower yellow portion above) opens and fuel injection commences. When the ECM stops injection, it stops sending the 100-V signal to the EUI solenoid, and the solenoid poppet valve opens. Opening the poppet valve instantly bleeds off almost all fuel pressure and fuel is again bypassed to the return system, stopping the fuel injection. With the reduction in internal fuel pressure, the fuel check valve in the lower portion of the EUI closes. The closure of the fuel check valve provides a "clean" injection stop and also prevents combustion pressure or gases from entering the injector. The EUI design permits very rapid starting and stopping of the fuel flow; ECM speed was the limiting factor at first.

In addition to the pilot injection, this high-speed cycling of the injector permits "rate-shaping" during the main-injection burn.

Additive

An additive is used to improve the performance of an engine lubricating oil. Manufacturers blend in chemical additive packages. These packages make up about 25% of the oil. A 5W-30 oil may be 75% oil base stock, 11% viscosity improver, and 14% other additives. The purpose of motor oil additive can be that an additive replaces a property of the oil that was lost during refining, strengthens a natural quality already in the oil, or adds a property that the oil did not naturally have.

Address

This is in reference to a personal computer memory; it is a specific location in the computer memory.

Adhesion

The action of adhering or sticking to a surface or object.

Adhesive

A compound that is designed to seal or stick to a surface, like gluing two items together.

Adiabatic

A thermodynamic term for a low heat rejection heat engine. The adiabatic process, which means impassable, takes place without transferring heat or weight between the thermodynamic system and its environment. In comparison an isothermal thermodynamic process transfers energy to the surroundings only as work. The adiabatic process supports the theory that explains the first law of thermodynamics that energy is neither created nor destroyed, only changed

Adjustable Brake and Accelerator Pedals

Adjustable pedals, pedal height control, or electric adjustable pedals (EAP) place the pedal and so the throttle on motor-operated movable brackets. The height of the brake pedal and the accelerator or treadle pedals are adjusted together and cannot be adjusted individually. Adjustable pedal systems include an adjustable pedal position switch, which allows an actuation to position the pedals, and thus the adjustable pedal assembly that contains the motor, threaded adjustment rods, and a pedal position sensor. The position of the pedals and power seat system are usually included as a component of the memory seat function and can be set for two or more drivers.

ADS (Automated Driving Systems)

An ADS will take over steering, acceleration, and braking in specific scenarios. The ADS must have control systems that are capable of analyzing sensory data to tell apart between different cars on the road. A three-dimensional evaluation model conforming to the functional architecture of autonomous vehicles has to be built, with each dimension representing one among three key functional layers of autonomous vehicle including sensing and perception, decision-making planning, and control and execution. Each dimension contains a set of metrics carefully defined with their weights fairly determined, supporting an entropy weights method. For more information on this, see **J3016_202104** [24].

Aerobic Sealant

A sealant that cures in the presence of air.

Aerodynamic Drag

Aerodynamic drag is the wind resistance of air moving over the size and shape of a vehicle. Air flowing over the body of a moving vehicle is very complex and will require the development of semi-empirical models to represent the resulting effect. The following equation represents aerodynamic drag:

$$D_A = \frac{1}{2} p V^2 C_D A \quad \textbf{(A.4)}$$

where

C_D is the aerodynamic drag coefficient
A is the vehicle frontal area
p is air density

Aerodynamics

Aerodynamics is the study of the motion of air on a moving vehicle produced by drag and viscous friction. This is a sub-field of fluid and gas dynamics. Aerodynamics has focused on issues associated with compressible flow, turbulence, and boundary layers and is computational in nature. The gross flow of air over a vehicle body is controlled by the link between velocity and pressure found in the modified Bernoulli's equation (A.5):

$$P_{\text{static}} + P_{\text{Dynamic}} = P_{\text{Total Air Flow}} \quad \textbf{(A.5)}$$

where

$P_t = P_s + \frac{1}{2} pV^2$
P_s is static air flow
P_t is the total air flow
p is the density of air
V is velocity of air relative to the vehicle

AFR (Air-to-Fuel Ratio)

This is the ratio of the amount of air to the fuel used for combustion. The chemical term stoichiometric is employed to explain the simplest AFR, which is 14.7:1. This number describes the proportions of air at 14.7% to 1 a part of fuel during a combustion reaction, required to complete the consumption of all of the fuel. It is normally expressed by mass or weight.

A-Frame Control Arm

A-frame, or A-arm, is part of the front suspension system found in short- and long-arm (SLA) suspensions and others. A-frame control arms are the primary load-bearing component of the SLA suspension system. The A-frame is also referred to as a wishbone control arm. The arm is a flat triangular component that connects to the frame or subframe at each leg of the "A" with the peak connecting to a ball joint that connects to the steering knuckle that connects to an upper ball joint that connects to the short upper control arm. The arm mounting points are spaced wide apart to prevent forward or backward movement of the steering knuckle. For more information on this, see **J670_200801** [9].

FIGURE A.7 A-frame control arm used in automotive SLA front suspension systems.

Setta Sornnoi/Shutterstock.com.

Afterburner

An afterburner is used as a combustor where heat is added to the working fluid after the last expansion stage. It is a combustor used on some jet engines to increase thrust, usually for supersonic flight and takeoff. For more information on this, see **J604_201108** [23].

Aftercooler (Turbocharger)

A charge air cooler to lower the air temperature to make it cooler and, therefore, denser for more power. You either cool the air before it enters the turbo or after. The aftercooler cools the air after the turbocharger. For more information on this, see **J922_201106** [30].

AGM (Absorbed Glass Mat) Battery

This is a recombinant-type battery or valve regulated lead-acid (VRLA) design. The electrolyte is totally absorbed into the separator, making the battery leak proof and spill proof. The battery is assembled by compressing the cell about 20% and then inserting it into the container. The compressed cell reduces damage caused by vibration and helps keep the acid tightly against the plates. This design uses a pressure release valve in each cell. Most of the hydrogen and oxygen given off during charging remains in the battery. The mat is 90–95% saturated with electrolyte, allowing a portion of the mat to be filled with gas. The gas has spaces that provide channels to allow the hydrogen and oxygen gases to recombine rapidly. Due to the fact that the electrolyte is totally absorbed into the glass mat separator, an AGM battery can be mounted in any direction. For more information on battery definitions used in the automotive industry as they relate to energy storage and batteries for starting, lighting, and ignition applications, as well as for hybrid electric vehicles (HEV) and electric vehicles (EVs), see **J1715/2_202108** [8].

Air ABS (Anti-Lock Brake System)

The air anti-lock braking system (ABS) modulates the atmospheric pressure within the brake chambers to forecast wheel lockup and reduce stopping distances on slippery road surfaces by limiting wheel slip. ABS provides:

- Vehicle stability
- Steerability under emergency braking
- Stable stopping on icy or rain-slicked road surfaces and in curves
- Stopping distance
- Jackknifing or plow out (loss of steering control)
- Tire flat spotting: flat spots caused by abrasion when wheels lockup

Automatic traction control (ATC) systems use ABS components to further enhance the vehicle and trailer traction and improve control. ATC uses sensors and software to scale back wheel slippage and direct power to slipping wheels in traction loss situations like snow or ice. Also, systems used in Europe are emerging within the US market called electronic braking system (EBS) on heavy trucks that has been used on HEV and EVs. EBS is a computer-controlled air brake system where the braking negative feedback circuit is managed electronically and therefore the foot pedal is just an input to an EBS control module. Brake application at the wheels continues to be controlled using air pressure; the difference is that the control is completely electronic. One big difference between hydraulic ABS and air ABS is that, within the mechanism, pressure must be developed by a hydraulic ram, where, within the air ABS, atmospheric pressure is available from the engine-driven compressor. The ABS modulator valve either changes the atmospheric pressure to the brake chamber or holds the prevailing pressure. For more information on this, see **J2627_200908** [7].

Air ABS Blink Codes

ABS blink error codes can be retrieved either with a service tool or by initiating the ABS indicator lamp to show a blink code. Both methods return codes that indicate what system or component caused the error. Blink codes are the easiest and most common method to retrieve errors, but it is important to understand its operating parameters:

- When more than one active fault is stored in the ABS management computer memory call an electronic control unit, or ECU, the first active fault displayed must be corrected before the next fault can be displayed.
- An active fault cannot be erased from the ECU memory until it is fixed.

The blink code feature does not necessarily display the faults in the order they were recorded by the ECU. Always park the vehicle before turning on the ABS blink code lamp. To perform a blink code check, the vehicle voltage system must be between 9.5 and 14 V.

Blink code procedure:

- Turn ignition to the ON position.
- Press and hold the blink code switch for one second and then release.
- Record the number of blink code flashes.
- After recording both numbers of the fault code, turn the ignition switch to the OFF position.
- Identify the recorded blink code.

For more information on this, see **J2627_200908** [7].

Air ABS Brake Modulator

This device regulates the atmospheric pressure to the brakes during ABS action. When not receiving commands from the ABS management computer called an ECU, or electronic control unit, the modulator valve allows air to flow freely and has no effect on the brake pressure. The ECU commands the modulator valve to change the air pressure to the brake chamber or hold the present pressure. It cannot automatically apply the brakes or increase the brake application pressure above the amount applied by the motive force. The modulator valve typically contains two solenoids. The modulator valve and relay valve could also be incorporated into one unit. The modulator valve might also be separate, inserted into the baseline to the brake chamber(s) after any relay valve, or located as close as practicable to the chamber(s) itself. When the modulator valve is separate, it is to manage more air flow and, therefore, includes two larger diaphragm valves which are controlled by the solenoids. It always has three ports. For more information on this, see **J2627_200908** [7].

Air ABS Electronic Control Unit (ECU)

The truck air ABS management computer is called an ECU, or electronic control unit, that monitors the wheel speed sensors to determine the drive axle(s) wheel speed and compare these speeds to the wheels on the steering axle to determine slippage. When the speed of the drive axle(s) exceeds that of the steering wheels by a specified amount, the traction control software within the ABS ECU can command a reduction of engine speed and application of the drive axle brakes on one side of drive axle(s). Engine speed reduction is accomplished by the ABS ECU sending a signal to the engine management computer over the electronic electrical circuit between the two control modules. The communication protocol can follow SAE J1939 controller area network, or CAN. For more information on this see **J2627_200908** [7].

AIR (Air Injection Reactor)

The acronym for air injection reactor emission control systems that pumps fresh air into the exhaust system to provide further reduction in exhaust emissions. It is referred to as secondary air injection in J1930. For details on these terms, see **J3016_202104** [19].

Air Bag

The air bag, as it is called, or the supplemental inflatable restraint (SIR) system includes three important elements. The electrical system includes the impact sensors and therefore the electronic control module. Its main functions are to conduct a system self-check to let the driving force know that it is functioning properly, to detect an effect, and to send a symbol that inflates the restraint. The constraint module is found within the steering wheel for the driving force and within the dash panel for passengers. It contains the restraint and therefore the parts that cause it to inflate. The knee diverter cushions the driving force of the knee from impact and helps prevent the driver from sliding under the bag during a collision. It is located underneath the steering column and behind the steering column trim. Air bags are located in multiple locations in a vehicle: passenger's side, roof curtains, side impact and pillars, etc. For more information on this, see **J1538_201504** [13].

FIGURE A.8 Air bag or supplemental inflation restraint.

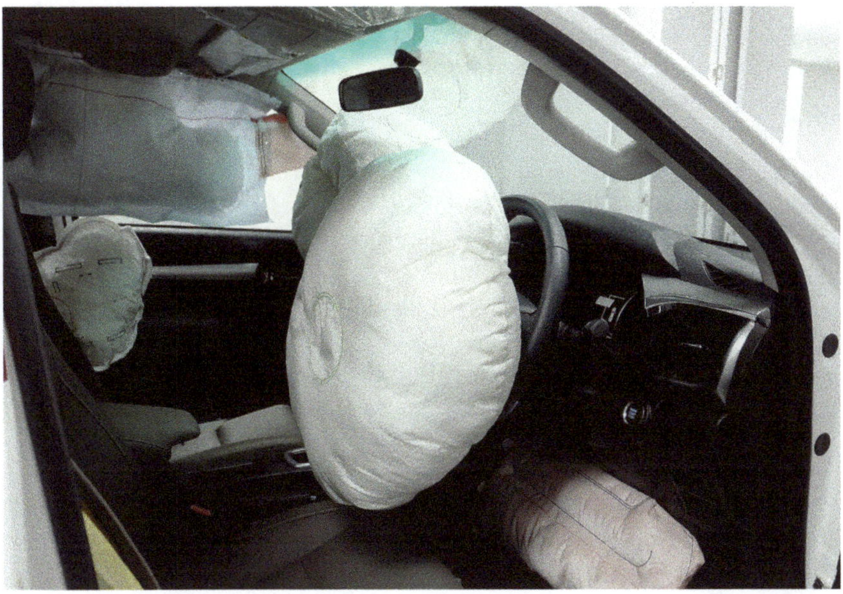

Air Brake Compressor

This compressor provides the compressed air to the truck/bus/train air brake reservoirs, also referred to as air tanks. The compressor is an air pump with pistons, rings, and an inlet and discharge valve to supply a constant pressure to apply and control the air brakes. Air is sucked into the compressor through the inlet valve from a filter, it is compressed when both valve are closed on a compression stroke. The compressor built-up air pressure goes out the discharge valve to the air tanks or reservoirs and ultimately goes to the brake pads or shoes of the foundation brake units. The air pressure is controlled by an unloader piston, which is controlled by air pressure bled off by the governor. The governor is a spring-controlled piston with an inlet and exhaust valve. System air pressure pushes against this adjustable spring, and when it gets to the adjustment valve, it sends air pressure to the bottom of the unloader piston, pushing it up, and opens the inlet valve that unloads the compressor and reduces the air pressure.

Air Brake Double Check Valve

This valve allows control of the trailer service brakes by the use of the trailer hand valve or foot valve. If both systems are applied at the same time, the trailer service brakes will have compound braking. So if the hand valve is applied at 60 psi and therefore the foot pedal at 20 psi, the trailer brakes will be applied at 60 psi. Also it allows blended supply atmospheric pressure taken from the first and secondary reservoirs. If one among the first subsystem loses gas pressure, the two-way check valve will redirect air from the upper pressured secondary system and deliver it to the rear spring brake chambers so that the spring brakes do not apply automatically. Some air brake systems have this valve integrated into the trailer control valve. For more information on this, see **J1409_201607** and **J1410_202012** [15, 16].

Air Brake Dual Air Pressure Gauge

Air pressure gauges monitor the air pressure in air reservoirs and brake lines. Dual air gauge display gas pressure for the front- and rear-service systems on dual air circuit systems using two color-coded needles. The gauge operation is verified by using a master gauge, a good quality, liquid-filled gauge that uses a Bourdon principle of operation. The gauge has two color-coded needles; the colors vary among truck original equipment manufacturers (OEMs). One needle is additionally green to point the front brake reservoir pressure, and thus the opposite could even be red to point the rear service system reservoir gas pressure. Two separate air line connections are located at the rear of the gauge to activate each gauge needle. For more information on this, see **J1409_201607** and **J1410_202012** [15, 16].

FIGURE A.9 Air brake system dash dual air pressure gauge.

Courtesy of John F. Kershaw Ed.D.

Air Brake Pressure Protection Valve

Dual-circuit air brake systems use pressure protection valves, which are normally a closed valve that is used to protect or isolate one air reservoir supply from another by closing automatically at its pressure setting in the event of a system leak. It may also be utilized in rapid buildup air systems by delaying the filling of auxiliary reservoirs until a preset pressure is achieved within the primary reservoir. Some air brake manufacturers discuss using this valve as a charging valve. Compressed gas is delivered to a second air brake reservoir only if the rated pressure for the system within the first reservoir has been reached. If the pressure within the first reservoir falls below that of the second reservoir, there is a feedback supply of air from the second reservoir. Some air brake systems use multi-circuit protection valves. For more information on this, see **J1409_201607** and **J1410_202012** [15, 16].

FIGURE A.10 Air brake pressure protection valve.

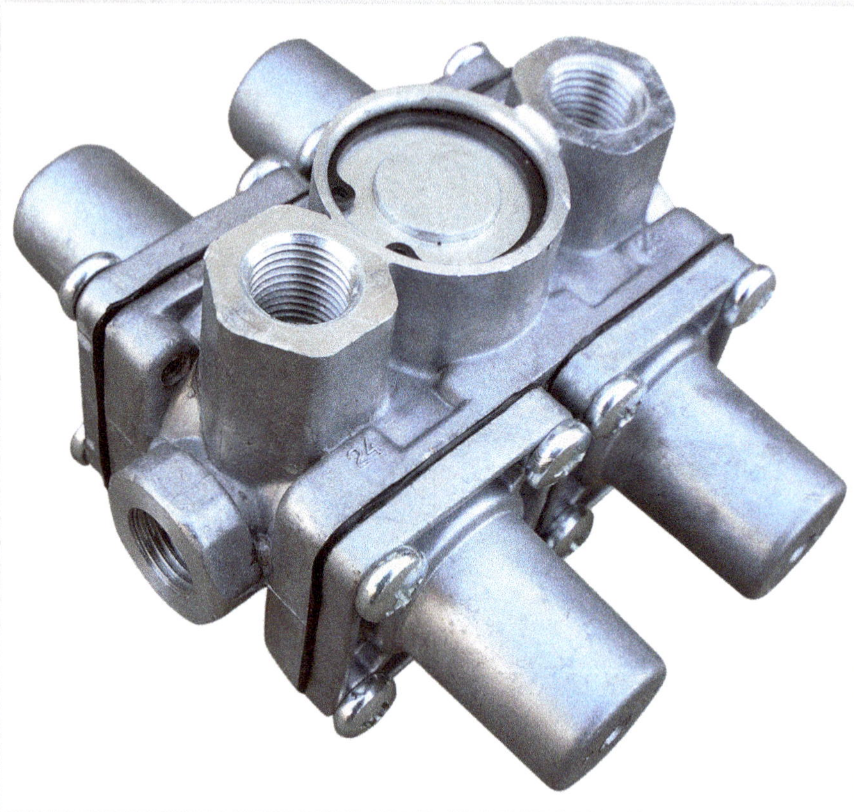

notsuperstar/Shutterstock.com.

Air Brake Reservoir (Air Tank) Drain Valves

A drain valve used to remove any water or condensation from the air brake reservoirs. There are drain valves used for both manual and automatic actuation. The automated drain valve prevents the buildup of water in pipe lines and brake chambers through automatically draining the reservoirs. Air from the auxiliary port on the governor unloader enters the control port and pushes the piston to its lowest position. Water from the reservoir enters a port and passes into a chamber via the undercut diameter on the piston. Water within the control line passes into chamber via a small hole within the piston. When the governor unloader cuts out, the pressure within the control line falls to zero, and therefore the pressure within the reservoir pushes the piston to its uppermost position, and the water is ejected via the undercut diameter. The manual drain valve has a valve on for draining; otherwise, you pass through the stem of the valve to empty out any water. For more information on this, see **J1409_201607** and **J1410_202012** [15, 16].

FIGURE A.11 Truck air brake system showing drain valves, check valves, tanks, relay valve, and trailer brake control valve.

Dr. Norbert Lange/Shutterstock.com.

Air Brake Reservoir Safety Valves

A safety valve used to make sure that the pressure in air reservoirs is not decreased. The compressed air from the air feed line opens a check valve at and reaches the air reservoir, provided its pressure is higher than the pressure in the reservoir. The valve will remain open until the pressures in the feed line and the reservoir are equal. The safety valve prevents the air from returning from the reservoir as, when the pressure in the feed line is reduced, the valve is closed by a spring and the higher reservoir pressure. Air can pass through the check valve only in the direction from the feed line toward the reservoir. For more information on this, see **J1409_201607** and **J1410_202012** [15, 16].

FIGURE A.12 Air brake safety.

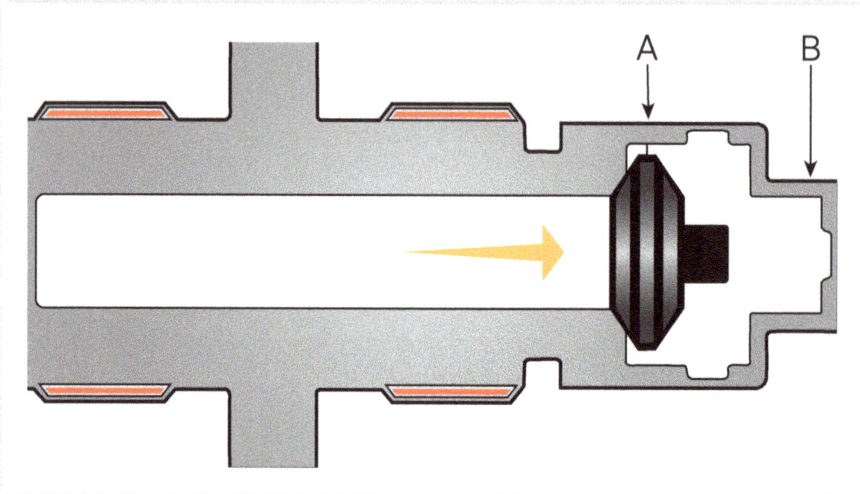

Courtesy of John F. Kershaw Ed.D.

Air Brakes (Truck and Bus)

A brake system used on trucks, buses, and locomotives that uses compressed air–activated service brake chambers to apply force to slack adjusters and caliper pistons to stop the moving wheels of a commercial vehicle. For more information on this, see **J257_201907** and FMVSS 121 Air Brake Systems.

The determining criteria for deciding brake rating power capability are:

- Cold brake stopping ability
- Fade as a result of brake power absorption
- Hot brake stopping ability following brake power absorption
- Brake system stability following brake power absorption
- Functional and structural integrity of brake system [14]

Air Cleaner

This is a paper or oil bath filter that removes dust from the air that passes through it on its way to the throttle valve of a carburetor or throttle body in a fuel injection or diesel system.

Air Conditioning

A finned radiator core is the cooling unit called the evaporator. The air to be cooled passes through the evaporator. The refrigerant boils in the evaporator in boiling the refrigerant absorbs heat and changes into vapor. The vapor is pumped out of the vehicle, and the heat that caused the vapor creation is carried outside the vehicle. Once the vapor is out of the evaporator, the heat it contains must be removed. Since heat expanded the refrigerant from liquid to vapor in the first place, removal of that same heat will let the vapor condense into liquid again. Then the liquid refrigerant can be returned to the evaporator to be used over again. Vapor coming out of the evaporator is very cold and boils at –21.7°F. Heat cannot be removed from subfreezing vapors by "cooling" them in air temperatures, which usually range between 60 and 100°. This is because heat refuses to flow from a cold object toward a warmer object.

The compressor exerts pressure for two reasons: pressure makes the vapor hot enough to cool off in warm air. The compressor raises the refrigerant pressure above the condensation point at the temperature of the surrounding air so it will condense. As the refrigerant leaves the compressor, it is still a vapor and hot. It is ready to give up the heat that was absorbed in the evaporator. In the condenser, refrigerant vapor gives up its heat. This is radiated into the surrounding air through the finned surfaces of the condenser. The vapor in the condenser condenses back into a liquid which collects at the bottom of the condenser. When the refrigerant condenses into liquid refrigerant, it is ready for boiling in the evaporator to remove the heat. The heat-laden vapor is squeezed in a small space. When the vapor is compressed, the heat it contains is concentrated. The pressure and temperature increase. In this way, the vapor gets hotter without adding any heat. So the vapor can cool the warm air through convection. Heat flows from the warm air to the cold evaporator to cool the air.

FIGURE A.13 When the vapor is compressed, the heat it contains is concentrated. The pressure and temperature increase. In this way, the vapor gets hotter without adding any heat. So the vapor in the evaporator can cool the warm air through convection. Heat flows from the warm air to the cold evaporator to cool the air.

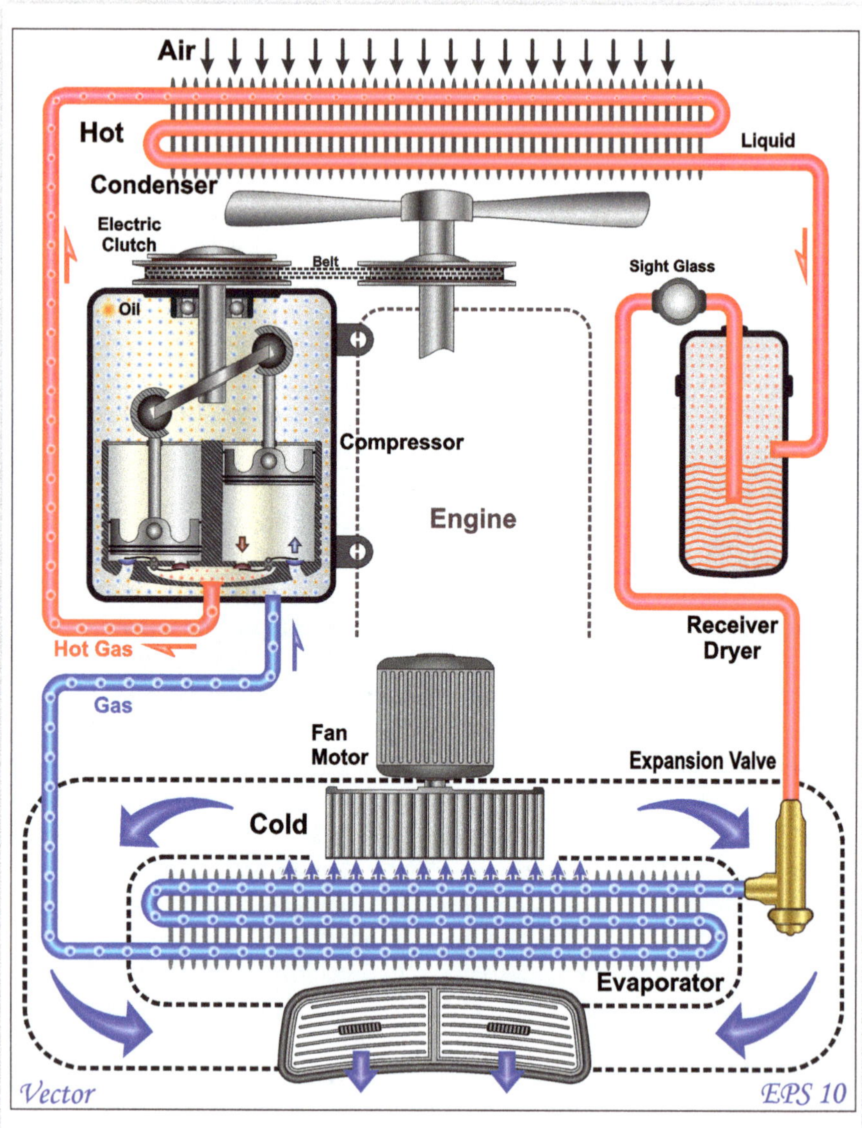

Fouad A. Saad/Shutterstock.com

Air-Cooled Engine

Automotive air-cooled engines require an auxiliary fan to furnish an adequate supply of air for cooling the engine. Forced air engine cooling systems use sheet-metal shrouding to direct air over the heads, cylinders, and oil cooler. They are thermostat-controlled cooling using a bellows-type air flap under each side of the engine. These thermostat flaps open when the temperature of the air under the cylinders reaches 205°F.

FIGURE A.14 Air-cooled boxer engine that is a horizontally opposed engine with finned aluminum cylinder heads and cylinders that are cooled by air flow. It is commonly used in motorcycles, industrial engines, Volkswagens, and Porsche vehicles.

spr/Shutterstock.com.

Air Core Gauge

A measuring gauge that does not operate using a magnetic metal core surrounded by a coil. It is an analog display gauge that enables an indicator to rotate a full 360°. The air core gauge is an air core motor where the sector is made by moving a pivoting static magnet.

Air Dam

Air dams improve air flow at highway speeds because they direct airflow and create a low-pressure area that directs the air through the radiator at highway speeds. Airflow created by vehicle movement is considered ram air because it is forced through the radiator through vehicle movement. Air dams can be attached to the underside of the front of a vehicle or, in the case of a truck, over the cab to improve air flow over the vehicle. They act to improve stability, aerodynamic performance, and engine cooling by redirecting the flow of air. For more information on this, see **J2971_201908** [17] and **J1594_201007** [26].

FIGURE A.15 Air dam on the front of a Dodge Barracuda.

Tony Savino/Shutterstock.com.

FIGURE A.16 Truck air dam.

FotograFFF/Shutterstock.com.

Air Density

A variable used in aerodynamic calculations that depends on temperature, pressure, and humidity conditions. At a standard temperature of 59°F and 29.92 in. of Hg, air density is 0.076 lb/ft^3. In Equation (A.6), the air density is expressed as mass density and is obtained by dividing the acceleration of gravity. The air density equation (A.6) is as follows:

$$p = 0.00236 \frac{P_r}{29.92} \frac{519}{460 + T_r} \quad \text{(A.6)}$$

where

P_r is the atmospheric pressure, in in. of Hg
T_r is air temperature, in °F

Air Disc Brakes (Truck and Bus)

This pneumatic single piston brake is actuated mechanically via a diaphragm brake cylinder or a spring brake actuator, which is fitted directly onto the brake caliper. When the brake diaphragm pushes inward, it moves a crank arm that pushes against the inboard pads that, in turn, pulls in the outward pad, applying equal force on both pads. It is an action-reaction operation. A short outboard friction pad service life is usually an indication of caliper assembly seizure, that is, the sliding action of the brake caliper stops. The caliper slide pins are usually at fault. The rotors used on truck air disc brakes are usually vented to aid in the dissipation of brake heat. In vocational applications which require constant use of the brakes, air disc brake rotors have been vulnerable to heat-related failures such as warpage, which meant that they are not commonly used. For more information on this, see **J2627_200908** [7].

FIGURE A.17 Air screw disc brakes.

Baloncici/Shutterstock.com.

Air Flow

Air flow or airflow is the movement of air from one area to a different area. The first explanation for airflow is the presence of pressure inclines. Air behaves in a fluid manner, meaning particles naturally ensue areas of upper pressure to a lower pressure. Air pressure (14.7 psi) is directly associated with altitude, temperature, and composition. In engineering, airflow may be a measurement of the quantity of air per unit of time that flows through a specific device. The flow of air will be induced through mechanical means (such as by operating an electrical or manual fan) or can occur passively as a function of pressure differentials present within the engine compartment.

Air Flow Meter

An automotive electronic fuel injection air flow meter that is found in air induction system inlet pipe between the air filter and also the throttle body. This device used a measuring plate, compensation plate, return spring, potentiometer, and a bypass passage. The flow meter incorporates the idle mixture adjusting screw, fuel pump switch, and intake air temperature sensor. The intake air volume is a direct measure of the load placed on an engine, and the vane-type air flow meter provides the input to the engine management computer for fuel and spark calculations.

Air Flow Types

Air is a fluid that may exhibit both laminar and flow patterns. Laminar air flow takes place when each air particle follows smooth paths and also the paths never interfere with one another. The air velocity of the fluid is constant at any point within the airflow. The flow is an irregular flow, which is characterized by tiny whirlpool regions. The rate of this fluid is certainly not constant at every point. The flow occurs when there is an irregularity such as a pause in the surface across which the air is flowing that alters the direction of movement. Streamline flow occurs when air can flow smoothly, and exhibits a parabolic velocity profile which may be a velocity profile accustomed to model or measure laminar or smooth airflow. The resistance to flow in air is characterized in terms of the viscosity of the air, if the flow is smooth. In the case of a moving plate in the air, it is found that there is a layer or lamina which moves with the plate and a layer which is actually stationary.

Air-Fuel Charge

The air-fuel charge in the most appropriate proportion of air to fuel that will be induced into the combustion chamber of an ICE. This is typically a stoichiometric measure of 14.7 parts air to 1 part of gasoline in a gasoline-fueled engine. A diesel engine uses a stratified charge that can be 100 parts air to 1 part fuel at idle or 20 parts air to 1 part fuel at full load.

Air-Fuel Ratio

The air-fuel ratio is the amount of air by weight mixed with 1 lb of gasoline, as in 14.7 lb of air to 1 lb of gasoline, or a 14.7 to 1 air-fuel ratio.

Air-Fuel Ratio Sensor

A sensor that monitors air-fuel ratio over a broader range above and below stoichiometric than a conventional oxygen sensor. It can sense air-fuel mixture ratios from 0.9 to 2.2. Also called a wideband oxygen sensor.

Air Gap

Anti-lock brake systems (ABS) that use adjustable wheel speed sensors may require an air gap setting. ABS wheel speed sensors are adjusted by loosening a holding screw and then inserting a nonmagnetic brass or plastic feeler gauge between the tip of the sensor and a high point on the tone ring. The sensor position is adjusted so there is a slight drag on the feeler gauge, and then the screw is tightened.

Air-Over-Hydraulic Brake System

A medium-duty truck brake system where air pressure from the foot brake treadle is used to apply hydraulic pressure to operate wheel cylinders or calipers to apply brakes. The foot brake application valve supplies control line pressure directly to the master cylinder.

Air Spring (Bag) Suspension

Air-ride suspension systems provide enhanced ride characteristics—when ride quality is equally important whether loaded or unloaded. On buses and over the road coaches, the air-ride system provides ride quality and comfort. All the major suspension OEMs offer air suspensions, each using a series of air springs or air bags. These suspensions make it more important to inspect and maintain U-joint angles since air suspensions can actually change their ride height and driveline angles just by adjusting their level using a height control valve. The suspension air control system uses compressed air from the reservoirs to pressurize the air springs. Some tractors are equipped with a dash-mounted control valve, allowing the driver to raise or lower the rear axle suspension when connecting or disconnecting the trailer. For more information on this, see **SAE J511_201604** [18].

FIGURE A.18 Air spring assembly.

Air Standard Dual Cycle

This is the air cycle used in diesel engines. Heat energy is added at constant volume and constant pressure. The constant-volume burning process is carried out until a predetermined pressure is reached, after which any additional burring is carried out at constant pressure.

Air Turbulence

Air turbulence is the extreme disturbance of the compressed air in the combustion chamber. It causes the air molecules to move in all directions. They collide with each other and cause friction and heat. A twisting force like small tornadoes changes the smooth laminar airflow into turbulent flow. This increases the transfer of heat between the cool liquid-fuel droplets and this hotter air.

ALCL (Assembly Line Communications Link)

ALCL was an early acronym for the device now known as a DLC, or data link connector. This is a 6-, 9-, 12-, or 16-pin connector that a diagnostic scan tool or interface with laptop computer is plugged into to extract diagnostic trouble codes (DTC) from the vehicle computer. For more information on this, see **J1930DA_202105** [19].

ALDL (Assembly Line Diagnostic Link)

ALDL is another name for the ALCL (assembly line computer link) and DLC (data link connector), which was defined in SAE J1930 and the most current term. For more information on this, see **J1930DA_202105** [19].

Align

Align means to bring parts of a component into the correct position.

ALLDATA

This is a company that provides an online source for service information that is licensed by the Original Equipment Manufacturers (OEM) to offer this information to the aftermarket.

All-Electric Drive Mode

An all-electric drive mode is a hybrid drive mode in which the hybrid operates like an all-electric vehicle, or EV. In this mode, the high-voltage battery pack provides all of the energy needed to propel the vehicle. This mode is used in reverse on most HEVs. For more information on this, see **J1715/2_202108** [21].

Alloy

A metal that contains one or more elements usually added to change the structure of the metal for increased strength or give the base metal needed properties for the design.

Alphanumeric Code

A code that uses a combination of alphabetical (letters) and numerical characters. They are a group of Latin letters and Arabic digits. It is an identifier composed of alphanumeric characters. An engine RPO (regular production option) code like LT1 is an alphanumeric code.

Alternative Fuels

Alternative fuels are ICE fuels, apart from gasoline, used to fuel an ICE, to scale back exhaust emissions and lessen the dependence on fossil fuels. A number of these fuels are derived from petroleum while others are from non-petroleum sources. Fuels like ethanol or ethyl alcohol or plant product, which are made of grain and located in alcoholic drinks, consists of two carbon atoms and six hydrogen atoms with one added oxygen atom. Propane or LPG and methanol are also used as alternative fuels. For more information on this, see **J1297_201710** [20].

Alternator

An alternator is an AC generator that produces DC voltage and current using electromagnetic induction. The rotor winding generates the magnetic field. The rotor poles may retain some magnetism when not used. Yet residual magnetism is not strong enough to induce voltage across conductors. So excitation winding, or field winding, varies the field current through rotor winding to manage the strength of the magnetic field, and this field current is regulated by a voltage regulator which adds resistance in series to control the field current. The field current is about 1.5–3.0 A. The stator is the conductor in the alternator. The stator windings are laminated to prevent unwanted eddy currents from forming in the iron core. There are three alternator conductors wound onto the cylindrical, laminated core and assembled as a one-piece called the stator. It does not rotate. Each stator winding is molded into a number of coils spaced evenly around the core and there are as many coil conductors as pairs in each of N-S (north-south) poles in the rotor. The brush-type alternator uses carbon brushes that ride on two slip rings. The brushes are connected to a battery suppled current and the circuit carries current to and from the stator. A holder is used to house the brushes.

FIGURE A.19 Automotive alternator or AC generator that is engine driven to provide electrical output to a vehicle that operates on the principle of electromagnetic induction.

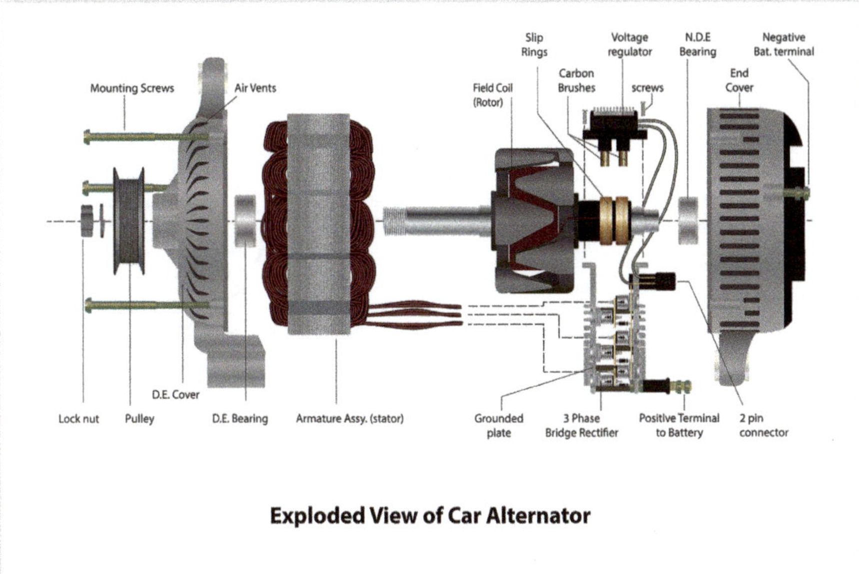

Aluminum Bearing Material

This material uses small quantities of tin and silicone alloyed with it. This makes a stronger but more expensive bearing than either Babbitt or copper-lead alloy. Most of its bearing characteristics are equal to or better than Babbitt and copper lead. These bearings are used for high-speed and load conditions and do not contain lead.

Ambient Air Temperature

Ambient air temperature is the temperature of the air surrounding an object.

Amboid Gearset

The amboid gearset is the opposite of the hypoid gearsets. They have the drive pinion axis plane above ring gear axis plane. They exert thrust on the drive pinion in the direction of power flow.

FIGURE A.20 Hypoid final drive gearset is the opposite of the amboid gearset (see Figure A.21).

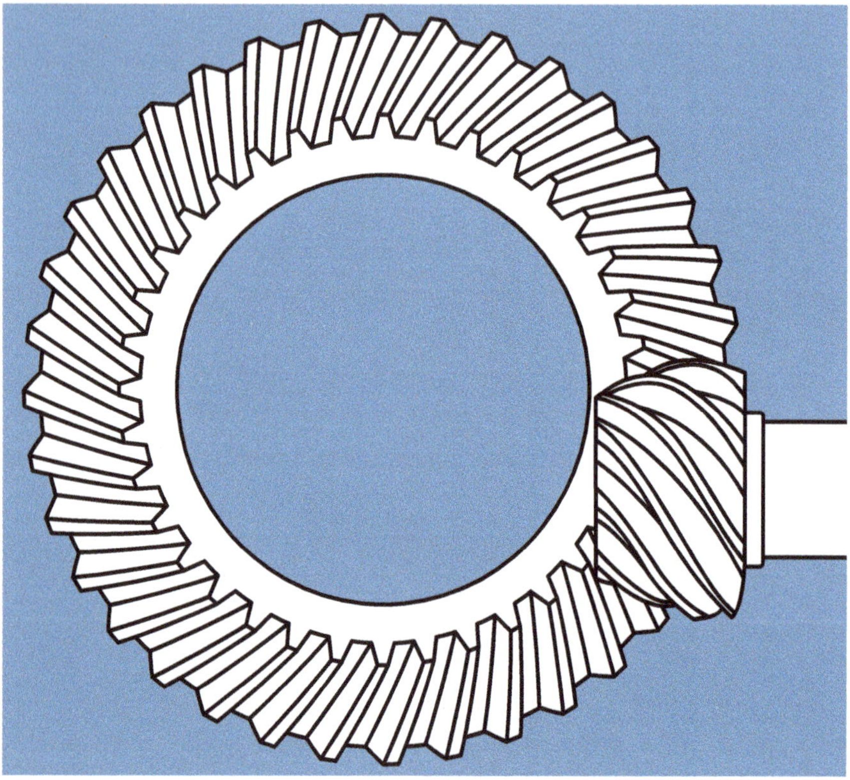

AlexanderZam/Shutterstock.com.

FIGURE A.21 Amboid gearset is the opposite of the hypoid gearset.

Ake Apichai Chumsri/Shutterstock.com.

Amplitude

Amplitude is the maximum value of a periodically varying quantity or waveform. It is the top or apex of the waveform and is used for vehicle vibration diagnostics. It denotes the magnitude of the disturbance. A severe vibration or disturbance would have a high amplitude; a minor disturbance would have a low amplitude. Amplitude is measured by the number of actual movement, or displacement. A radial vibration caused by an out-of-balance wheel moving up and down at 50 mph (80 km/h) would be more severe than a vibration occurring at 25 mph (40 km/h). This takes place because, as the speed increases, the amplitude also increases. For more information on this, see **J3152_202005** [22].

FIGURE A.22 The amplitude waveform.

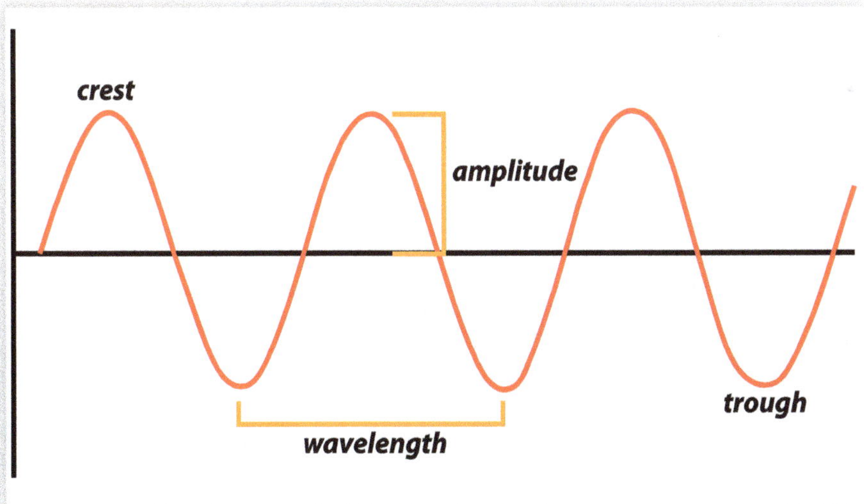

Ammeter

An ammeter measures either direct or alternating electrical current (I) depending on the circuit construction. It is an abbreviation of ampere meter. Electric current is measured in amperes, or amps where you connect the ammeter series with the circuit. Ammeters are no longer stand-alone measuring devices; they are part of a digital multimeter (DMM). The ammeter has low resistance so that it does not result in a significant voltage drop in the circuit being measured.

Ammeter Shunt

An ammeter shunt is a low-resistance conductor like a diode or resistor placed in parallel with the ammeter movement, so most of the current passes through this conductor (shunt) with only a small amount going to the meter. This design extends the usable range of the meter.

Amperage

Amperage is the amount of amperes, or amps, flowing in an electrical circuit or conductor.

Ampere-Hour

This is a current of 1 A flowing for 1 h. If you multiply the current in amperes by the flow time in hours, you will arrive at the total number of ampere-hours.

Ampere-Turn

The measure of the magnetomotive force developed by an electric current in one turn of a coil of wire through which a current of one ampere is flowing.

Amperes

Amperes is the electrical unit for current, which has been shortened to amps. It is also called the rate of flow of electrons. One volt pushed across 1 ohm of resistance causes a current flow of 1 A. A current flow of 1 C/s equals 1 A.

Ampere's Rule

Ampere's rule states that a current flowing in a certain direction is equal to the motion of the positive charges in that specific direction. The magnetic lines of force or flux generated from the current appears as wire circles around a current in a counterclockwise direction when it is approaching the observer.

Amplifier Piston

The amplifier piston is the hydraulically actuated piston that pumps fuel to injection pressure to very high values in the Caterpillar and Navistar HEUI (hydraulically actuated electronic unit injection). It can also be called an intensifier piston.

Anaerobic Sealant

A sealant that will not cure without the presence of a metal along with the lack of oxygen. An anaerobic sealant is used between two closely fitted metal parts or on a nut and bolt to seal in the absence of air. Sometimes called a Superglue.

Analog

Generally referred to as the class of devices or circuits where the output is a continuous function of the input. An analog electrical signal is continuous and variable. A type of dash instrument that indicates values by use of the movement of a needle or similar device.

Analog Multimeter

A meter that uses a needle or pointer to show electrical readings at a constant value that can measure voltage, current, or resistance. The needle moves up and down to show temporary or rapid changes in electrical values.

Analog Signal

A signal that progressively changes in strength which is continuous and variable.

FIGURE A.23 Analog signal waveform on left.

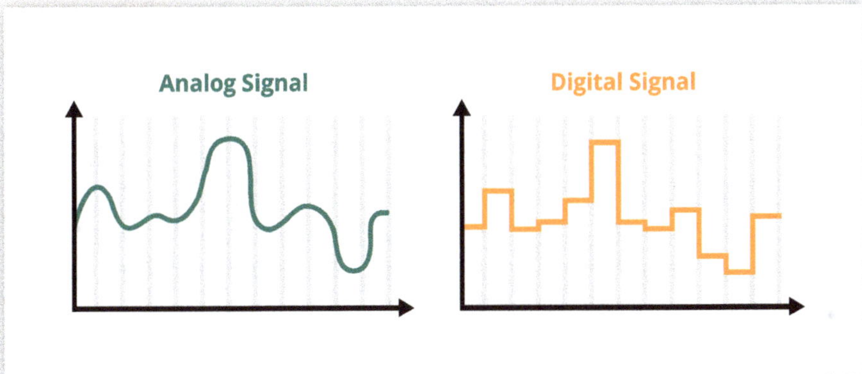

petrroudny43/Shutterstock.com.

Anemometer

An anemometer measures airflow and is sometimes called an airflow meter. This tool may use ultrasound or resistive wire to measure the energy transfer between the measurement device and the passing particles. A hot-wire anemometer, for example, registers decreases in wire temperature, which can be translated into airflow velocity by analyzing the rate of change.

Aneroid

A sealed bellows or capsule that contains a partial vacuum and changes its length or size in response to a change in atmospheric pressure.

Annealing

Annealing is a heat treatment process that changes the physical and chemical properties of a metal to increase its ductility to make it more workable. The metal is heated to above its transition temperature, maintaining a suitable temperature, and then cooled. Annealing can induce ductility, soften material, relieve internal stresses, making it homogeneous, and improve cold-working properties.

Annulus

An annulus is a circular or ring-shaped device that are commonly used in automatic transmissions and transaxles.

Anode

Anode is the positive electrode; the electrode toward which electrons flow.

Anti-dive

A term used to describe the geometry of an automotive suspension that controls the movement of the vehicle during braking. It is normal for a vehicle to dive slightly during braking, and this is designed into most vehicles.

Antifreeze

Antifreeze is a chemical that lowers the freezing point of a water-based liquid and also increases its boiling point. An antifreeze mixture is used to achieve freezing-point depression for cold weather and also achieves boiling-point elevation to allow higher engine temperatures. Freezing and boiling points are properties of a solution, which depend on the concentration of the dissolved substance. Water has good properties as a coolant, water plus antifreeze are used in internal combustion engines. The purpose of antifreeze is to prevent the engine block from bursting due to expansion when water freezes. Commercially, both the additive (pure concentrate) and the mixture (diluted solution) are called antifreeze, depending on the context. Careful selection of an antifreeze can enable a wide temperature range in which the mixture remains in the liquid phase, which is critical to efficient heat transfer and the proper functioning of heat exchangers.

Antioxidants

Antioxidants are additives used to reduce engine varnish by preventing the buildup of acids, destroying chemicals that produce undesirable oxidation by-products, and interrupting the oxidation chain reaction.

Anti-rattle Clips

Anti-rattle clips are one or more metal components designed to keep brake pads from vibrating and rattling.

Anti-roll Bar

See **Stabilizer Bar**.

Anti-squat

A vehicle suspension geometry term that describes the movement of the vehicle body during acceleration. When the body of the vehicle stays level during acceleration, it displays 100% anti-squat. If the level of the vehicle body is less than 100%, this indicates that the body squats down in the rear during acceleration. Anti-squat acts as a lever between the tire and the body, trying to force up the back of the car as acceleration pushes it downward. At 100% anti-squat, the rear of the car will stay level on a launch instead of squatting down (RWD) or raising up (FWD). Anything over 100% anti-squat will raise the body over the tires; any less than 100% will permit the body to lower itself over the tire.

A properly set up suspension will use the axle torque and weight transfer to actually plant the tires and gain traction. This is indicated by the rear body staying level or raising slightly under acceleration, an indication that the tires are being forced downward. When the vehicle body "squats," this is a sign that the axle housing is moving upward. Since it cannot lift off of the ground, the body drops or squats. Rear body squat does not mean that weight is applied to the suspension for better traction, but rather that the suspension is moving up and unloading the tires, therefore minimizing traction. "The anti-squat equation $e/d = h/L$ (e = elevation, L = wheelbase, d = distance from the wheel center to the trailing arm, h = height from the CG to the ground) defines the locus of points extending from the tire contact point on the ground to the height of the center of gravity (CG) over the front axle" [12].

Anti-sway Bar (Sway Bar)

See **Stabilizer Bar**.

A-Pillar

The A-pillar is part of an automotive body that supports the hinges for the front doors and support the windshield. The body B-pillar supports a pin that fortifies the front door locks and the hinges for the rear doors. The C-pillar supports the pin that secures the rear door locks and rear window. The D-pillar supports the rear doors or tailgate of an SUV. Pillars are identified by the side on which they are found (driver side/passenger side) and the pillar location, for example, driver's A-pillar or passenger B-pillar.

FIGURE A.24 Body sections are A-Pillar, B-Pillar, etc.

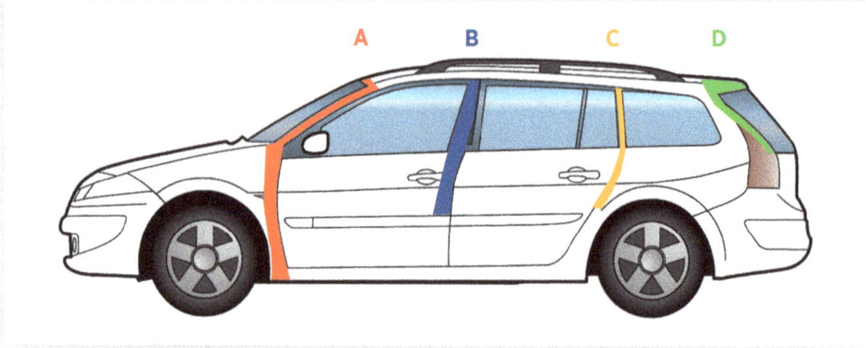

Apex

Apex is the highest point on an object, illustration, or component.

API (American Petroleum Institute) Oil Service Classification

The API classification is a standard that rates engine oils on their ability to lubricate, resist oxidation, prevent high- and low-temperature engine deposits, and protect the engine from rust and corrosion. API has organized a system of letter classifications with two categories, the S category and the C category. API's S (service) category emphasizes oil properties critical to gasoline engines, and the C (commercial) category emphasizes oil properties for diesel engines. To give an oil formulation a particular classification, the oil is run through a series of tests in specific engines.

API Donut

The donut symbol appears as a label printed on the container of oil. The upper half of the symbol displays the API service classifications of the oil, the center of the donut displays the SAE viscosity grade of the oil, and the lower half of the symbol contains the words "Energy Conserving," if the oil is formulated to meet those requirements. When selecting an oil, always make sure the oil quality information on the API donut conforms with the vehicle manufacturer's specifications.

API Service Classification

The API service classification rates engine oils on their ability to lubricate, resist oxidation, prevent high- and low-temperature engine deposits, and protect the engine from rust and corrosion.

API Starburst

The American Petroleum Institute (API) starburst symbol indicates the oil has been certified by the International Lubricant Standardization and Approval Committee (ILSAC) as the correct type for gasoline engines in passenger cars and light trucks. To qualify for the starburst symbol, oils must meet the Energy Conserving II requirements. Only multi-viscosity oils with SAE 0W, 5W, and 10W, such as 5W-30 and 10W-30, will qualify.

FIGURE A.25 ILSAC starburst symbol as an approved engine oil.

petroudny43/Shutterstock.com.

A-Post

See **A-Pillar** (same definition as A-pillar).

APP (Accelerator Pedal Position) Sensor

Mounted on the accelerator pedal assembly, this sensor is two or three individual accelerator pedal position sensors with one housing. Two to three separate circuits for signal, ground, and 5-V reference connect the APP assembly with the engine management computer. It works to provide input to the engine management computer regarding driver requested accelerator pedal position and throttle angle at the throttle body. The APP 1 signal increases as the accelerator pedal is depressed, from 0.67 V at 0% pedal travel (pedal at rest) to 2.51 V at 100% pedal travel (pedal fully depressed). The APP 2 signal decreases from 4.33 V at 0% pedal travel to 2.49 V at 100% pedal travel. The APP 3 signal decreases from 4.0 V at 0% pedal travel to 2.88 V at 100% pedal travel.

Application Valve

The dual air brakes service application valve may be a floor-mounted foot valve, also referred to as a treadle valve. It is two valves in one. The upper portion of the valve is the primary section and the lower portion is the secondary or relay section. Each section features a dedicated feed and its own exhaust port. The upper or primary section of the treadle valve is supplied directly from the first reservoir. The secondary reservoir sends air to the lower or relay section of the valve. The valve is mechanical, that is, it is operated by foot pressure from the driving force. When the driver's foot presses on the applying valve, the first piston is forced downward. This movement first closes the first exhaust port and thus modulates air in proportion to the piston travel to Actuate whatever brakes are connected into the first circuit: this is often usually, but not exclusively, the rear drive axles on a typical tandem drive tractor unit. Actuate the relay or secondary piston, located below the first section in a very dual circuit application valve. Act against the mechanical pressure (foot pressure) to produce brake feel. The secondary or relay section of the twin circuit application valve is actuated pneumatically by the primary circuit air. This section operates similarly to a relay valve, therein an indication pressure value (air from primary section) is employed to displace a relay piston, which then modulates the secondary circuit air to whatever brakes/valves are located within the secondary circuit. In a very typical tractor air brakes, this might normally be the front axle brakes.

Aramid

Aramid is a generic name for aromatic polyamide fibers developed in 1972. Kevlar is the DuPont brand name for aramid.

Armature

The armature is the rotating part of an electric motor that consists of electromagnets, known as field coils, and the segmented contacts, known as the commutator. The armature commutator is an electromagnet created in a loop of wire that is place between two electromagnetic poles. Motors work on principle of magnetic repulsion. This magnetic repulsion takes place when a straight-loop wire conductor made up of an armature, commutator, and brushes is located within a magnetic field and current flows through that wire loop. This situation creates two separate magnetic fields. One produced by the magnet (poles of the magnetic field winding) and another produced by the current flowing through the conductor (armature/commutator/brushes).

Articulated

Articulated means that a vehicle may have one or two sections connected by a flexible joint or connection.

Articulated Bus

A transit bus that has two sections using a flexible joint or fifth wheel so that it can turn corners in the inner city.

Articulation Test

A test specified by some OEMs that tests the amount of force necessary to move the inner tie-rod end in the ball socket assembly. The specification is greater than 1 lb (0.5 kg) and less than 6 lb (2.7 kg) of force.

Asbestosis

Asbestosis is a health condition where asbestos causes scar tissue to form in the lungs causing shortness of breath.

ASD (Automatic Shutdown)

This is the automatic shutdown relay on a Fiat Chrysler system used to power the ignition coil. The ASD relay will not supply voltage to the coil unless the engine is cranking and the computer senses a crankshaft sensor signal. This little fact has fooled many technicians.

ASE (Automotive Service Excellence)

This is the abbreviation for the National Institute for Automotive Service Excellence, a nonprofit organization for the testing and certification of vehicle service technicians. It was originally called NIASE before it was shortened to ASE.

Ash Content

Ash is soot or suspended solids that contain metallic compounds such as sodium. Contained within the fuel oil are ash-forming materials in the form of abrasive solids or soluble metallic soaps. The solids will cause wear of injection components, increase engine deposits, and contribute to engine wear. As content is determined by burning, a given weight of fuel oil in an open container until all the carbon deposits are consumed. The weight of the remaining ash is then expressed as a weight percentage of the original test sample of fuel oil.

ASME (American Society of Mechanical Engineers)

American Society of Mechanical Engineers. ASME is an American association, which promotes the science, and practice of multidisciplinary engineering and allied sciences through: education, training, professional development, standards, research, conferences and publications, and government relations. It is altogether an engineering society, a standards organization, enquiry and development organization, advocacy group, education provider, and a nonprofit organization.

Aspect Ratio (Tire)

Aspect ratio is a comparison of the tires cross-sectional height to cross-sectional width. These two dimensions are found by measuring the widest point and the tallest point of a fully inflated tire. Aspect ratio is determined by the following equation:

$$\text{Aspect ratio} = (\text{Section height} / \text{section width}) \cdot 100 \tag{A.7}$$

- Section height of 130 mm and a section width of 185 mm is described as a 70-series tire.

Assist Hybrid

Assist hybrid is a type of hybrid vehicle that can only move from a standstill when the internal combustion engine is running. A small motor generator assists the gas engine in accelerating and propelling the vehicle from a standstill to about 10 to 20 mph.

Asymmetrical

Asymmetrical means to be different on both sides of center. In an asymmetrical low beam headlamp, the light beam is spread farther to one side of center than to the other.

ATA (American Trucking Associations) Connector

Heavy-duty truck electronic service tools capable of reading data are connected to the on-board electronics by means of a SAE/ATA J1584/J1708 "6" PIN or J1939 "9" PIN Deutsch connector in all current systems. Before heavy-duty on-board diagnostics (HD-OBD), the connector was used by all the truck OEMs and referred to as the ATA connector. All current systems uses the 16-pin HD-OBD DLC, or data link connector. This common connector and the adherence to SAE standards allows the software of one manufacturer to, at least, read the parameters and conditions of their competitors.

FIGURE A.26 6 & 9 pin truck DLC.

Courtesy of John F. Kershaw Ed.D.

ATC (Automatic Traction Control) Heavy Truck

The ATC is a system used on heavy commercial trucks. These systems are integrated with ABS which is now mandatory for all air-braked vehicles and vehicles with hydraulic brakes having a GVWR (gross vehicle weight rating) in excess of 10,000 lb. These systems utilize components of the ABS as well as additional components specific to the ATC. It applies a preset amount of torque to both rear wheels even if one starts slipping. A transmission without ACT will lose traction completely if one rear wheel starts slipping. For more information on this, see **J2627_200908** [7].

ATC (Automatic Transfer Case)

The automatic transfer case can automatically select 2WD and 4WD modes depending on driving conditions. It also uses a driver-controlled shift select switch that allows the driver to manually select 2HI, 4HI, 4LO, AUTO-4WD, and neutral. For more information on this, see **J646_201807** [29].

ATCM (Automatic Transfer Case Control Module)

The ATCM uses speed sensors located in the transfer case to continually monitor front and rear propshaft speeds. If the ATCM detects a difference in propshaft speeds, it reacts to apply the clutch pack. When the clutch pack is applied, torque is directed to the front and rear wheels.

ATDC (After Top Dead Center)

ATDC, or after top center, is an ICE piston position where the piston has moved after it has been at the top of the cylinder.

ATF (Automatic Transmission Fluid)

The ATF is a hydraulic fluid oil employed in vehicles with automatic transmissions that provide automatic up and down gear shifts or changes. It is typically dyed red or green to tell it apart from ICE oil and other vehicle fluids. There are different formulations for this ATF. An ATF was first introduced for Genreal Motors (GM) during the fifties under the brand Dexron and used for all automatic transmissions. The 1940–1949 GM Hydra-Matic Drive used a specialized lubricant called GM Transmission Fluid No. 1.

By using the term "fluid" instead of "oil," this fluid was composed of a base oil and additives to limit oxidation, foaming, rust, corrosion, varnish, and sludge buildup. Dexron is the name used for the technical specifications for the ATF created by GM and a GM registered trademark that licenses the name and specifications to companies which manufacture the fluid. All licensed Dexron fluids have a number that begins with the letters B through J. The first Dexron (B) transmission fluid was introduced on April 1, 1967. Over the years, the first Dexron (B) was supplanted by Dexron-II (C), Dexron-II (D), Dexron-II (E), Dexron-III (F), Dexron-III (G), Dexron-III (H), Dexron-VI (J), Dexron HP, Dexron LV ATF HP, and Dexron ULV, which is the latest fluid used in multiple speed transmissions, HEVs and EVs.

Ford Motor Company created the Mercon ATF technical specification, which is a Ford registered trademark. The first Mercon (M2C185-A) ATF, also called Mercon-type CJ, was introduced in January 1987 and was developed for Ford C-6 transmissions. It was developed to compete with Dexron 2, which had similar specifications. It has been replaced by Mercon "V", Mercon "SP", Mercon LV, and Mercon ULV, which is the latest fluid used in multiple speed and HEV/EV transmissions. Ford has upgraded the Mercon specifications over the years; the newer fluids are not always backward compatible with previous fluids.

Atmospheric Pressure

Atmospheric pressure is the pressure exerted by the atmosphere on all things based on the weight of the air.

Atom

The atom is the smallest particle into which an element can be divided and still have all the characteristics of that element.

Atomization

Atomization is the action of breaking liquid fuel particles into small droplets by pumping it at a high pressure through a miniscule flow area

Atomize

Atomize means to reduce or separate gasoline or diesel fuel into fine or minute particles.

AUTO-4WD (Automatic Four Wheel Drive)

This system uses an automatic transfer case (ATC), which operates like an automatic transmission selecting between two-wheel drive (2WD) and four-wheel drive (4WD). These ATCs come in both single- and two-speed models. The system selects between 2WD, 4WD, and AUTO-4WD, depending on the driving conditions using input sensors and control modules or computers. They use an automatic transfer case control module, which is a computer. For more information on this, see **J646_201807** [29].

Auto Ranging

Auto ranging is an automatic scale used on digital multimeter (DMM) that automatically changes to the correct scale. You would set the rotary switch on the DMM for volts, amps, or ohms.

Automatic Locking Differential

Automatic locking differential has the ability to positively lock the axles to the differential carrier.

Automatic Locking Hubs

Automatic locking hubs connect the wheels to the front axle when 4WD is selected. They connect the wheels to the front axle when 4WD is selected. Torque applied to the front wheels engage a clutch mechanism in the wheel hubs that effectively locks the hub to the axle. The hubs automatically unlock when the vehicle is placed into 2WD. Sometimes it is necessary to drive the vehicle a few feet in reverse to unlock the hubs.

Automatic Transaxle

The automatic transaxle is a transmission that automatically changes forward gear speeds and also contains the final drive unit ring and pinion or another planetary gearset. They use a fluid clutch assembly called a torque converter rather than a mechanical clutch assembly. The torque converter uses ATF (automatic transmission fluid) to provide a fluid link between the engine and transmission. This results in a smooth, cushioned connection between the engine and transmission, and it permits them to remain engaged whether the vehicle is moving or at a standstill. Contained within the torque converter is a lock-up clutch that is also called a torque converter clutch. This clutch directly connects the ICE output to the transmission, bypassing the fluid coupling. These transmissions use 4 to 10 forward speeds, neutral, and reverse. The higher gear ratios are overdrive gear ratio. For more information on this, see **J648_201106, J649_198807**, and **J646_201807** [27, 28, 29].

FIGURE A.27 Automatic transaxle combines transmission with final drive rear axle.

Setta Sornnoi/Shutterstock.com.

Automatic Transmission

An automatic transmission is a transmission that automatically changes forward speed gears and uses a torque convertor, which is a fluid clutch that provides additional torque through three elements: turbine, stator, and pump. This converter uses automatic transmission fluid (ATF) to produce a fluid link between the engine and transmission. This provides a smooth, cushioned connection between the engine and transmission and permits the powertrain to stay engaged whether the vehicle is moving or stationary. The fluid coupling or torque converter enables the vehicle to stop without stalling. Automatic transmissions have four to ten forward speeds, neutral, and reverse. The upper ratios are overdrive gear ratios.

The first automatic transmission was called the automated safety transmission (AST), and it was a four-speed transmission providing power shifting and no clutch. It was called the AST because clutch operation was reduced to one-third to that required by a traditional transmission. The clutch was only used when starting or stopping the vehicle. It was first used by Oldsmobile in 1937. This transmission did not use ATF, but instead used the identical grade of ICE oil for lubrication and hydraulic controls. The AST had two driving ranges: low and high. In the low range, the transmission would shift from first to second gear and hold in second. In the high range, the transmission would start in first gear, go to third gear, and shift to fourth gear. Heavy truck applications now used automatic shifting manual twin-shaft transmissions, which will be stated as automatic transmissions. For more information on this, see **J648_201106**, **J646_201807**, and **J649_198807** [27, 28, 29].

FIGURE A.28 Automatic transmission.

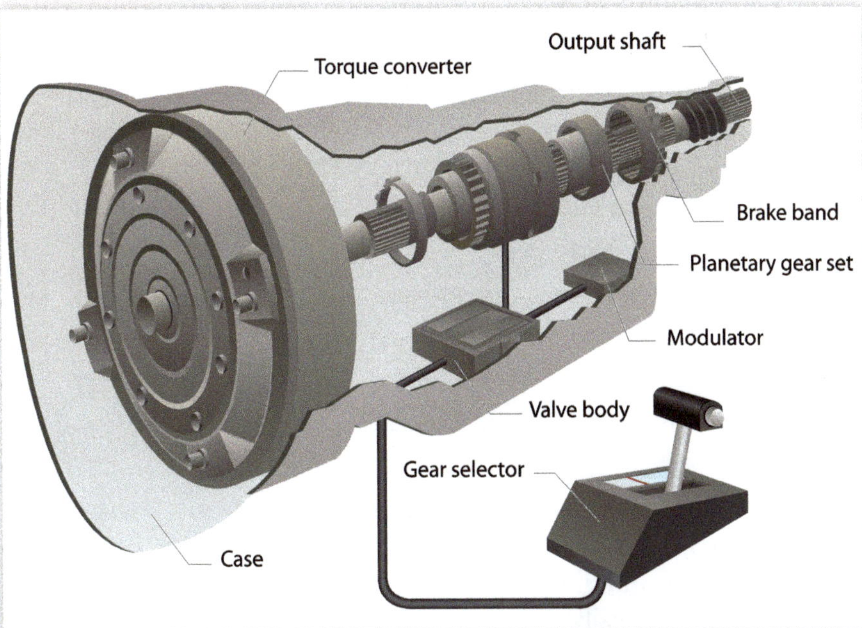

Colin Hayes/Shutterstock.com.

Automotive A/C Servicing—Refrigerant Flushing of a Failed A/C System

"The failure of an A/C system often results in the introduction of contaminants to the A/C system. The sources of the contaminants include debris from damaged components and debris from the surrounding environment. Returning the A/C system to service requires the removal of these contaminants from any reused components. The recommended approach to cleaning contaminated components and systems is to flush with a solvent flushing machine. Previous internal studies have concluded that solvent flushing will remove all contaminants, restoring component and system performance. Many commercial refrigerant recovery and recharge machines include a refrigerant 'flush' feature which can flush oil from the system and components with the systems refrigerant. The effectiveness of using the 'flush' feature of a refrigerant recovery and recharge machine with an added in-line filter to remove contaminants is investigated. One gram of Arizona Test Dust is introduced to the suction line of an automotive A/C system, and the system is run to compressor failure. This introduces debris from the damaged compressor into the condenser and the rest of the system. The condenser is removed, and the refrigerant recovery and recharge machine with the added in-line filter is used in 'flush' mode in an effort to remove the debris from the condenser. The condenser is sectioned and inspected to evaluate the effectiveness of the debris removal. Two machines, made by different manufacturers, were selected to represent machines typically used in A/C service operations. Several compressors were failed in order to create contaminated condenser samples. The condensers were then flushed for varying lengths of time, filled with epoxy, sectioned and photographed under microscope for cleanliness evaluation" [25].

Autonomous Vehicle Levels

The terms related to On-Road Motor Vehicle Automated Driving Systems are listed below; for more information on this, see **J3016_202104**:

- Level 0: No Driving Automation
- Level 1: Driver Assistance
- Level 2: Partial Driving Automation
- Level 3: Conditional Driving Automation
- Level 4: High Driving Automation
- Level 5: Full Driving Automation [24]

AutoShift and AutoShift Generation II Manual Automatic Transmissions

Transmissions that are classified as being semi-automated shift-by-wire models because they automatically select and engage the transmission gears. The driver must, however, use the clutch pedal for both starting and stopping. These transmission are also called three-pedal units because it has an accelerator, brake, and clutch pedal. They require a power-assisted clutch release system on all rear engine coach applications; all other vehicle applications require adjustment-free clutches for all 6-, 7-, 10-, and 18-speed applications. It is a 10-speed base transmission (RTO-XX9 I O-B/C-AS2) with a 5-speed front box and a 2-speed rear box. The 18-speed AutoShift (RTLO-XX918A-AS2) requires a 4-speed rear or back box.

AutoShift Truck Transmissions

Generation II Eaton Fuller Roadranger truck transmissions are classified as being semi-automated shift-by-wire models because they automatically select and engage the transmission gears. The driver must, however, use the clutch pedal for both starting and stopping. So these transmission are also called three-pedal units because it has an accelerator, brake, and clutch pedal. They require a power-assisted clutch release system on all rear engine coach applications; all other vehicle applications require adjustment-free clutches for all 6-, 7-, 10-, and 18-speed applications.

AV (Autonomous Vehicle)

The AV or driverless car, self-driving car, or automated car is an unmanned or occupied ground vehicle that is capable of driving itself by sensing its environment and navigating without human input. Autonomous means self-governing or self-driving. They are possible thanks to the integration of numerous current automotive technologies, such as adaptive cruise control, forward collision warning, lane departure warning, and V2V (vehicle to vehicle) communication being merged into one complex system. These self-driving vehicles generate and maintain an internal map of their surroundings based on a number of input sensors like radar or lidar (light detection and ranging). For example, Uber's self-driving prototypes use 64 different lidar laser beams, along with other sensors, to construct their internal map, and Google's prototypes have used lidar, radar, high-powered cameras, and sonar. For more information on this, see **SAE J3016_201401** [24].

Available Voltage in a Circuit

Sources of low-voltage problems and high-resistance faults are isolated by systematically taking available voltage readings along a circuit while it is carrying current. These available voltage tests provide a quick way to determine overall circuit condition. For example, if an accessory motor is not turning fast enough, available voltage testing can pinpoint the problem area of the circuit. Begin testing at the point of highest potential voltage, or closest to the power source. Then work down the circuit and take readings at each load. The DMM readings can show that there is not enough available voltage to correctly operate the motor. So the problem is either in the switch or the circuit between the switch and the motor.

AWD (All-Wheel Drive)

AWD is a drive system that can drive both the front and rear wheels through all phases of operation. Also called full-time four-wheel drive (4WD). AWD vehicles can be set up on a traditional 4WD vehicle using an inner axle differential to prevent axle windup or with a front-wheel-drive (FWD) vehicle using a transaxle, power transfer unit (PTU), and propeller shaft (driveshaft) sending power to a rear drive unit (RDU). These are FWD-based systems. A single speed PTU can have a differential, a viscous coupling, a center differential, or multi-plate clutch directing power to the rear axles (RDU) so both axles may rotate at the same or different speeds.

FIGURE A.29 A general AWD drivetrain layout for a FWD-based vehicle that would use a front transaxle-based PTU that could connect to differential, viscous coupling or multi-plate clutch to drive the rear wheels.

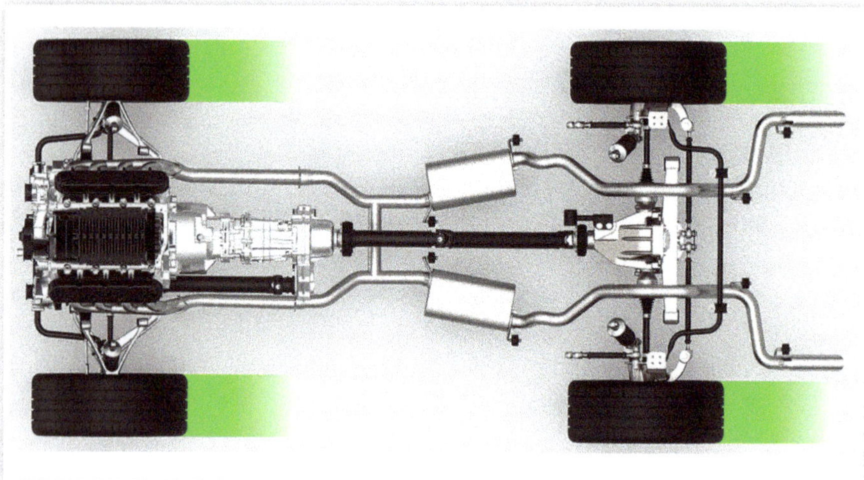

patruflo/Shutterstock.com.

AWG (American Wire Gauge) System

A system designed for solid wire used to indicate the diameter of conductors. A smaller wire gauge number represents a larger conductor diameter, and a larger wire gauge number represents a smaller conductor diameter.

Axial

Axial means to be situated in the line along with the axis or centerline of a part or component. Axial play in a ball joint means looseness in the same axis as the ball joint stud.

Axial Load

Axial load is any line of force applied parallel to the axis.

Axial Motion

Axial motion is the movement along, or parallel to, the centerline or axis of a shaft. When a shaft has revolution, a dynamic seal is required to contain fluids.

Axial Play

Axial play is the movement along, or parallel to, the centerline (axis) of a shaft. It is also called end thrust or endplay. It may also be defined because the gap between the flanks of the threaded spindle and also the threaded nut. The movement of the threaded nut within the axial direction without moving the spindle.

Axis

The axis is the centerline of a rotating part.

Axis System

The axis system is employed to set up a fixed coordinate system within the environment relative to the vehicle because it moves through the air. The axis system is fixed within the body specified, in nominal resting condition on horizontal ground. The X-axis is an imaginary line drawn down the middle of the vehicle from front to back. It is horizontal, parallel to the longitudinal plane of symmetry of the vehicle, and pointing forward. The Y-axis is that the imaginary line across the vehicle from left to right. It is horizontal, perpendicular to the X-axis, and pointing to the left. The Z-axis is that the vertical line that runs through the middle of the vehicle from top to bottom.

Roll is vehicular movement along its X-axis. It is the rolling motion you are feeling when making a pointy corner and is usually what causes rollovers. The Z-axis is vertical, perpendicular to both X and Y axes, and pointing upward. The origin is fixed within the body and chosen because the midpoint of a virtual rear-axle is projected onto the ground. The terms yaw, pitch, and roll describe the movement of a vehicle around the three axes. Pitch is movement round the vehicle Y-axis, which is commonly felt during hard braking or fast acceleration when the front of the vehicle noses down or rises up slightly. Yaw is movement round the Z-axis, which is felt when the vehicle deviates from its straight path, such as when the rear wheels slide out during drifting.

Movement around each of those axes must be controlled during all of the maneuvers of the vehicle for it to be safe. Many safety systems, like electronic stability control, are designed to keep vehicles within the safe limits of those axes. For more information on this, see **J1594_201007** covering axes system, resolving center, forces and moments, attitude angles, force and moment coefficients, vehicle parameters, flow parameters, and yaw-weighted drag coefficient along with **J670_200801** covering axis systems, vehicle bodies, suspension and steering systems, brakes, tires and wheels, operating states and modes, control and disturbance inputs, vehicle responses, and vehicle characterizing descriptors [6, 26].

FIGURE A.30 SAE vehicle axis system coordinates system and sign conventions.

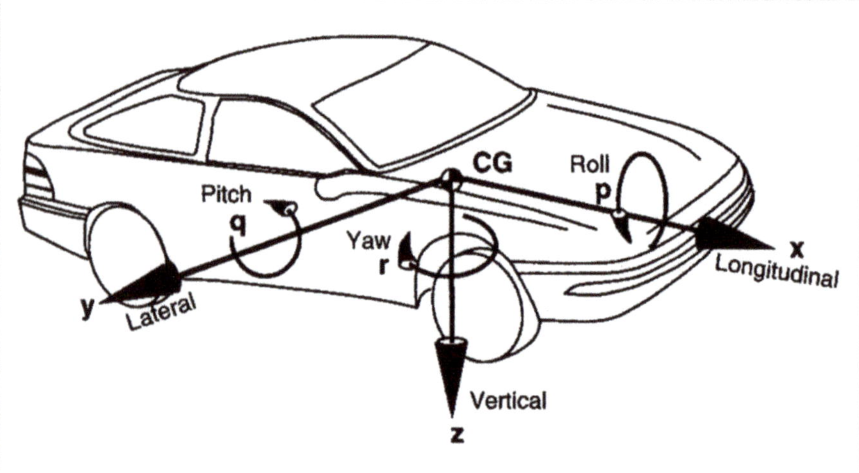

Axle

The axle is a fixed or rotating steel rod or spindle passing through the center of a wheel or group of wheels.

Axle Bearing

The axle bearing is a ball or roller bearing that supports an axle.

Axle Windup

The axle windup is a condition found in solid, driven rear axle configuration commonly found in medium- and heavy-duty trucks. Leaf springs, control arms, pinion snubbers, and torque arms are all means of controlling axle windup. Newton's third law of motion clearly states that for each action there is an equal and opposite reaction. The concept of axle windup is an example of that law. This is because the axle shaft rotates forward to drive the rear wheels, and the motion of the axle shaft rotating forward causes the axle housing to rotate the wrong way causing the axle to twist or lift during hard acceleration.

Axle Yaw

Axle yaw is the feature that uses an open differential to transfer torque to the wheels that have the best grip on the road.

REFERENCES

1. Society of Automotive Engineers, Standard J1674_201807: Early Acquisition and Preservation of Information in a Motor Vehicle Crash, SAE Publishing, Warrendale, PA, 2018.
2. Brach, M., Brach, R.M., and Mason, J., *Vehicle Accident Analysis and Reconstruction Methods*, 3rd ed. (Warrendale, PA: Society of Automotive Engineers, Inc., 2022), Book R-516, ISBN:978-1-4686-0345-3.
3. Society of Automotive Engineers, Standard J3227_202105: Terminology and Measurement Methodology for Determining Vehicle Overlap in Offset-Frontal Crashes, SAE Publishing, Warrendale, PA, 2021.
4. Kershaw, J.F., *GM 6.2/6.5 Liter Diesel Engines* (Forest Lake, MN: CarTech, Inc., 2020), 11-13, ISBN:978-1-61325-560-5
5. Society of Automotive Engineers, Standard J3104_201811: Electric Motors for Driveline Actuation Standard Terminology, Test Parameters, and Equipment Requirements, SAE Publishing, Warrendale, PA, 2009.
6. Liu, T., Yu, Z., Xiong, L., and Han, W., "Anti-Lock Braking System Control Design on an Integrated-Electro-Hydraulic Braking System," *SAE Int. J. Veh. Dyn., Stab., and NVH* 1, no. 2 (2017): 298-306, doi:https://doi.org/10.4271/2017-01-1578.
7. Society of Automotive Engineers, Standard J2627_200908: Braking System Definitions—Truck and Bus, SAE Publishing, Warrendale, PA, 2009.
8. Society of Automotive Engineers, Standard J1715/2_202108: Battery Terminology, Warrendale, PA, SAE Publishing, 2021.
9. Society of Automotive Engineers, Standard J670_200801: Vehicle Dynamics Terminology, SAE Publishing, Warrendale, PA, 2008.
10. Society of Automotive Engineers, Standard J1213/2_198810: Glossary of Reliability Terminology Associated With Automotive Electronics, SAE Publishing, Warrendale, PA, 1988.

11. Society of Automotive Engineers, J3063_202103: Active Safety Systems Terms and Definitions, Warrendale, PA, SAE Publishing, 2021.

12. Gillespie, T.D., *Fundamentals of Vehicle Dynamics* (Warrendale, PA: Society of Automotive Engineers, Inc., 1992)

13. Society of Automotive Engineers, Standard J1538_201504: Glossary of Automotive Inflatable Restraint Systems, SAE Publishing, Warrendale, PA, 2019.

14. Society of Automotive Engineers, Standard J257_201907: Brake Rating Power Requirements - Truck and Bus, SAE Publishing, Warrendale, PA, 2019.

15. Society of Automotive Engineers, Standard J1410_202012: Air Brake Valve - Performance Requirements, SAE Publishing, Warrendale, PA, 2020.

16. Society of Automotive Engineers, Standard SAE J1409_201607: Air Brake Valves Test Procedure, SAE Publishing, Warrendale, PA, 2016.

17. Society of Automotive Engineers, Standard J2971_201908: Truck and Bus Aerodynamic Device and Concept Terminology, SAE Publishing, Warrendale, PA, 2011.

18. Society of Automotive Engineers, Standard J511_201604: Pneumatic Spring Terminology, SAE Publishing, Warrendale, PA, 2011.

19. Society of Automotive Engineers, Standard J1930DA_202105: Electrical/Electronic Systems Diagnostic Terms, Definitions, Abbreviations, and Acronyms Web Tool Spreadsheet, SAE Publishing, Warrendale, PA, 2021.

20. Society of Automotive Engineers, Standard J1297_201710: Alternative Automotive Fuels, SAE Publishing, Warrendale, PA, 2017.

21. Society of Automotive Engineers, Standard J1715/2_202108: Hybrid Electric Vehicle (HEV) and Electric Vehicle (EV) Terminology, SAE Publishing, Warrendale, PA, 2021.

22. Society of Automotive Engineers, Standard J3152_202005: Taxonomy and Guidelines for Design, Fabrication, and Installation of Brake NVH Fixtures for Inertia Dynamometer Testing, SAE Publishing, Warrendale, PA, 2011.

23. Society of Automotive Engineers, Standard J604_201108: Engine Terminology and Nomenclature – General, SAE Publishing, Warrendale, PA, 2011.

24. Society of Automotive Engineers, Standard J3016_202104: Taxonomy and Definitions for Terms Related to Driving Automation Systems for On-Road Motor Vehicles, SAE Publishing, Warrendale, PA, 2021.

25. Lambert, S., Jamo, W., and Kurtz, M., "Automotive A/C Servicing – Refrigerant Flushing of a Failed A/C System," SAE Technical Paper 2017-01-0167, 2017, https://doi.org/10.4271/2017-01-0167.

26. Society of Automotive Engineers, Standard J1594_201007: Vehicle Aerodynamics Terminology, SAE Publishing, Warrendale, PA, 2010.

27. Society of Automotive Engineers, Standard J649_198807: Automatic Transmission Functions – Terminology, SAE Publishing, Warrendale, PA, 1988.

28. Society of Automotive Engineers, Standard J648_201106: Automatic Transmission Hydraulic Control Systems – Terminology, SAE Publishing, Warrendale, PA, 2011.

29. Society of Automotive Engineers, Standard J646_201807: Planetary Gears – Terminology, SAE Publishing, Warrendale, PA, 2011.

30. Society of Automotive Engineers, Standard J922_201106: Turbocharger Nomenclature and Terminology, SAE Publishing, Warrendale, PA, 2011.

B

B20

B20 is a diesel fuel blend consisting of 20% biodiesel and 80% petroleum diesel. It can generally be used in unmodified diesel engines; however, users should consult the engine original equipment manufacturer (OEM) for fuel usage information. For more information on this, see **J1297_201710** *Alternative Automotive Fuels* [1].

Babbitt

An automotive engine bearing material made from a combination of lead, tin, and antimony. Lead and tin are alloyed with small quantities of copper and antimony to provide the needed strength. Babbitt is used in internal combustion engines (ICEs) where a soft material is required for soft shafts running under moderate loads and speeds.

Back Spacing

The distance between the rear wheel rim edge and therefore the center section mounting pad of a wheel. For more information on this, see **J2047_201911** [6].

Back Taper

An angle cut opposite to a chamfer of a gear in order that the spine or tooth narrows just behind the chamfered end.

Backfire

Combustion or an explosion produced by a running ICE that happens in the exhaust, instead of inside the combustion chamber. The fuel burn occurs while the intake valve is open, causing combustion to maneuver backward through the system and out through the intake rather than the exhaust. When the flame moves backward, it is going to even be called a pop-back. A backfire can be caused by either ignition that takes place while the poppet valve is open or unburnt fuel making its way within the hot exhaust. It is possible for a visual flame to momentarily shoot out the throttle valve.

Backing Plate

This device is the mounting for the braking components of a drum brake system. The backing plate holds the hydraulic brakes shoes, springs, wheel cylinder, and other hardware inside the cylinder. The backing plate bolts to the axle housing or spindle, and helps keep dirt and water off the brakes.

Backlash

Clearance between the ring gear and pinion on the final drive. Backlash is determined by mounting a dial indicator to the differential housing and placing the button of the dial indicator gauge against a tooth of the ring gear. Move the ring gear back and forth to indicate movement on the dial indicator.

FIGURE B.1 Final drive ring and pinion backlash.

Courtesy of John F. Kershaw Ed.D.

Backlight

Another name for the rear window glass in a car.

Backpressure

The exhaust system resistance to flow, which can be measured using a gauge installed in the exhaust or a vacuum gauge in the intake manifold on a spark-ignited ICE and measured in pounds per square inch (psi). It is the exhaust gas pressure buildup in the exhaust system, and it can occur before or after the exhaust turbine of the turbocharger.

Backside Spacing (Rear Spacing)

The distance between the back wheel rim edge and the wheel center section mounting pad. This is not the same as offset.

Backup Camera

This camera gives the motive force a view of the realm behind the vehicle when the transmission is placed in reverse. It is utilized in conjunction with a frontway camera that shows the trail forward. The rear vision camera may be a video camera that is produced specifically for the aim of being attached to the rear of a vehicle to assist in backing up and to alleviate the rear blind spot. The backup camera is employed to avoid a collision when stepping into reverse. The rear vision camera provides the driving force with a view of the scene directly behind the vehicle on the infotainment system display (or inside the rearview mirror) to assist in parking and avoid crashing into nearby objects when in reverse. If the vehicle is supplied with a high-definition (HD) rear vision camera, this feature provides the driving force with a high-resolution "digital" view. Analog rear vision cameras are identified by multiple wires at the camera electrical connector. Digital rear vision cameras are identified by a line connection at the camera [2].

Bakelite

Brand name of the Union Carbide Company for phenolformaldehyde resin plastic.

Balance Shaft

A shaft in the engine that is designed so that, as it rotates, it reduces or cancels out any vibration by providing an opposite force to counter the vibrating force.

Ball Bearing

A rotating bearing that can be located between two rotating components like an automotive wheel and a fixed axle, where the rotating part and the stationary part are separated by a ring of small solid metal balls lubricated by grease used to reduce friction.

Ball Joint

Flexible joint having a ball-and-socket kind of construction. Ball joints are typically constructed of a ball and socket. This term is reserved for the primary joints that the steering knuckle pivots on, and tie-rod ends are the term used for the ball-and-socket joints on the steering linkage. They allow the control arms to move up and down with suspension deflection and let the wheel and knuckle assembly rotate for steering. The ball joint in most modern vehicles is a sealed, self-contained unit that is replaced as a unit when it is worn out. It is contained in a pressed-steel housing, fitted with sintered (bonded using pressure and heat) iron seats

FIGURE B.2 Ball joints are swivel connections mounted on the outer ends of the control arms or steering linkage.

Warut Wetsanarut/Shutterstock.com.

Banjo Axle

This axle design uses a hollow central member having a pair of hollow tubes extending outwardly therefrom. Forward and rearward openings are in the central member, each having triangular shaped side portions located adjacent to the tubes.

FIGURE B.3 Heavy truck banjo axle.

Ake Apichai Chumsri/Shutterstock.com.

Bar

Bar is a measure of pressure: 1 bar = 14.5038 psi.

BARO Sensor

A sensor used to measure barometric pressure (BARO) or changes in barometric absolute pressure (atmospheric air pressure). It is vented directly into the atmosphere.

Barometric Manifold Absolute Pressure (BMAP) Sensor

Combination of a BARO and MAP sensor in the same housing.

Barometric Pressure

The measure of atmospheric pressure, in inches of mercury (Hg), which reflects altitude and weather conditions.

Base

The name for the portion of a solid-state silicone transistor that controls the current flow through the transistor.

Base Circle

Lowest spot on an ICE camshaft, which is the area of the camshaft directly opposite the lobe or nose.

FIGURE B.4 The base circle.

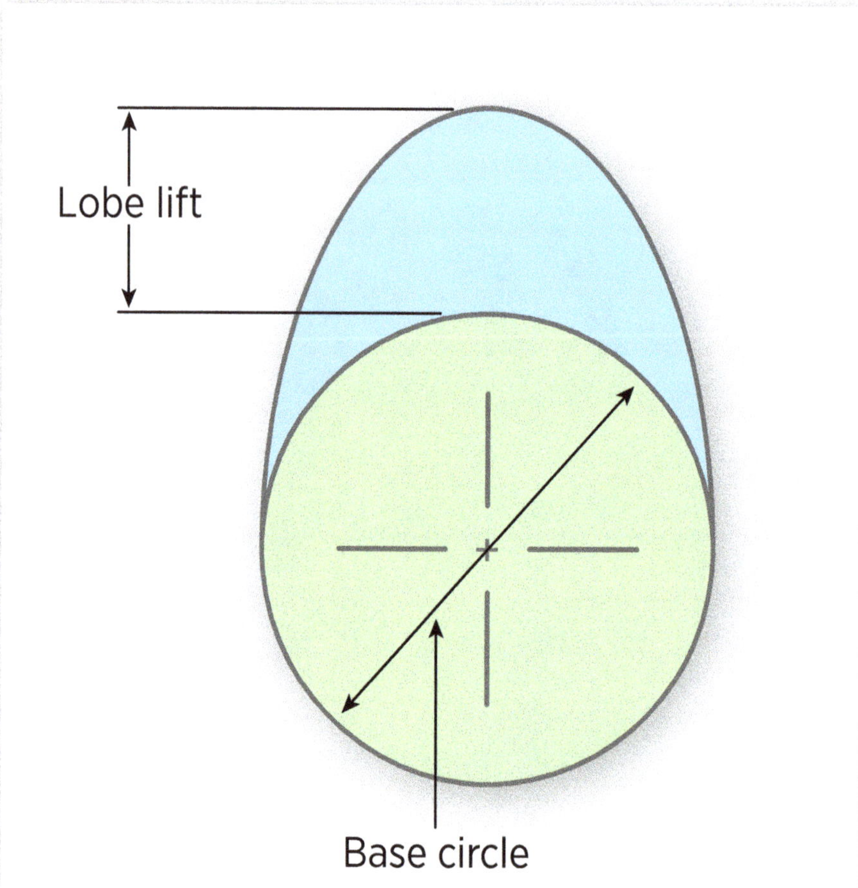

Courtesy of John F. Kershaw Ed.D.

Base Timing

Timing of an ICE ignition spark before the ignition timing is advanced electronically or mechanically.

Basic Mathematics (Addition, Subtraction, Multiplication, and Division)

Mathematics is the manipulation of numbers. Taught at the first or secondary levels. The mastery of mathematics is critical for several professions, including engineering, electronics, carpentry, plumbing, and automobile repair. The order of mathematical actions is Parentheses, Exponents, Multiplication, Division, Addition, and Subtraction. The acronym PEMDAS "Please Excuse My Dear Aunt Sally" is useful in remembering the order of mathematical operations

Battery

A device that uses chemical energy to store and produce electric current and voltage. It changes the electric current generated by the charging system into chemical energy and uses this energy for vehicle systems when the ICE is not running. Chemicals inside the battery store the electrical energy until needed to perform work. It is then changed back into electrical energy and sent through a circuit to the system where it is needed. Electrochemical reaction occurs so that electrons move from one pole to other metals and electrolytes used to control the voltage of the battery. It has one or more cells where chemical energy is converted into electricity and used as a source of power. Automotive batteries use a lead-acid chemical process to store electrical current. The inside of a lead-acid battery is made up of grids, positive plates, negative plates, and separators. The grids form the basic framework of the battery plates. The grids of the positive plates are filled with lead dioxide paste. Negative plates of the battery are coated with sponge lead paste. For more information on this, see **J1715/2_202108** [3].

Battery Case

Positive and negative plates sit in a case or shell of the battery. This case is a strong but lightweight one-piece molded plastic or rubber housing. Inside the battery case is a number of individual cell compartments. For more information on this, see **J1715/2_202108** [3].

Battery Charging

The method in which charging occurs inside of a battery is the reverse of the discharging process. Electricity from the alternator is forced into the battery. This causes the water in the battery to change back to sulfuric acid. This change from water to sulfuric acid causes some of the water to turn into hydrogen and oxygen gas. On vented batteries this gas can vent into the outside air. This causes the electrolyte level inside the battery to drop. For more information on this, see **J1715/2_202108** [3].

Battery Conductance Test Procedure

Using a battery conductance tester:

- Consult the online service information for procedures and guidelines for the battery being tested and tester being used.
- Isolate batteries if they are connected to a bank so that they can be individually tested.
- Identify the type of battery, size, and voltage for input into the tester.
- Turn off all electrical loads.
- Make sure the test leads make good contact with the battery terminals.
- Enter the CCA rating of the battery in the display window if required.
- Press the START button.
- Test results will be displayed in a few seconds.
- Refer to the tester instruction manual for additional information.

Battery Conductance Testing

The number and mobility of free-charge carriers determine the capacity of a private material to conduct electricity. Batteries are designed to store power from chemical action. Conductivity is the ability to transmit heat or electricity. Electrical conductivity is expressed in terms of the present per unit of applied voltage, so conductivity is the reciprocal of resistivity. Resistivity could be a measure of the resistance of a cloth to an electrical phenomenon either through its volume or on a surface. Using the conductance tester you can determine the condition of the battery by using conductance. Conductance may be a measurement of the ability of the battery to provide current. The conductance tester creates a signal impressed through the battery and thus measures a portion of the AC response. Conductance may be a measure of the battery plate surface to work out the quantity of power that a battery can supply. Constant cycling of batteries in heavy-duty service like medium- and heavy-duty trucks will eventually cause subplate growth on the plates and/or shedding of active material. The conductance tester can detect cell defects and short and open circuits that may reduce the ability of batteries to deliver current. The battery to be tested are often less than a 75% state of charge, so you are do not need to wait from 8 to 24 h while the battery is being recharged. For more information on this, see **J1715/2_202108** [3].

FIGURE B.5 Battery conductance tester.

Courtesy of John F. Kershaw Ed.D.

Battery Discharging

The storage of electricity inside a battery depends on the chemical reaction between the active materials on the positive and negative plates. As the battery discharges, the hydrogen and acid in the electrolyte change to water. As the hydrogen and acid converts to water, the specific gravity of the electrolyte will get lower. Measuring the specific gravity of the electrolyte with a hydrometer can be a way to tell the battery charge. For more information on this, see **J1715/2_202108** [3].

Battery Equalizer

A battery equalizer is used to obtain 12-V power in a truck with a dual battery 24-V system. The power harness leads are directly connected to the primary battery. The wiring should never be arranged so that the equalizer becomes a converter, meaning you can never pull 12 V directly from the 12-V post on the equalizer. Battery equalizers are generally original equipment manufacturer (OEM) supplied. For more information on this, see **J1715/2_202108** [3].

FIGURE B.6 Battery equalizer.

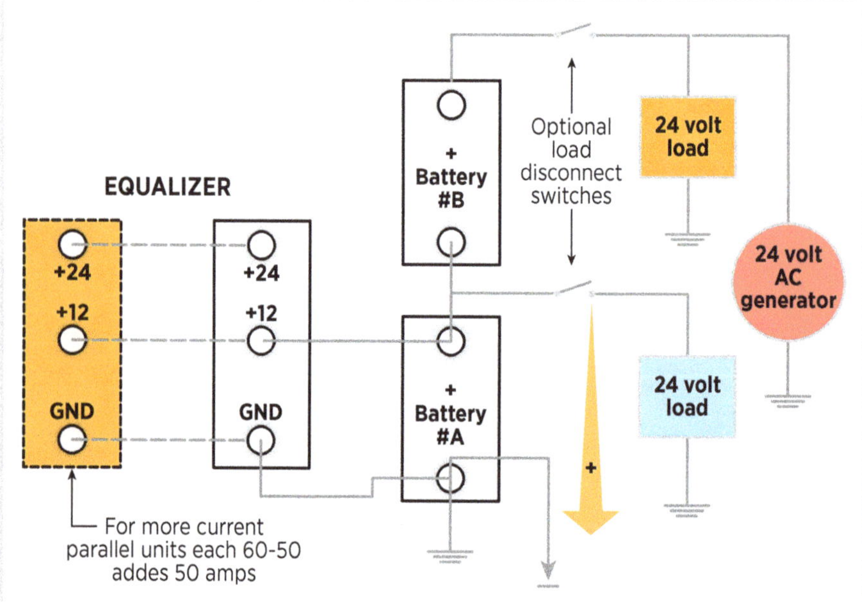

Courtesy of John F. Kershaw Ed.D.

Battery Isolator Systems

Facilitate the charging of an auxiliary battery by the vehicle charging system and electrical separation of the auxiliary battery from the starting circuit when the engine shuts down. Separation of most starting and auxiliary batteries can happen automatically during charging and discharging. Battery isolation systems range from simple, isolating solenoids, or relays, to complex battery management systems that monitor charge rates and voltages for both the starting and auxiliary battery. For more information on this, see **J1715_202105** [4].

Battery Pack SOC (State of Charge)

Share of total charge in a hybrid electric vehicle (HEV) battery pack. For more information on this, see **J1715_202105** [4].

Battery Pack Visual Signal

Lightweight within the dashboard that illuminates if the HV battery pack state of charge is below normal limits. For more information on this, see **J1715_202105** [4].

Battery SOC (State of Charge)

Level of electrical charge in a battery relative to its capacity. The units of SOC are percentage points (0% = empty; 100% = full). It can even be measured in voltage for a sealed lead-acid (SLA) battery at 2.1 V per cell. The SOC for a totally charged battery is 12.6 V. For more information on this, see **J1715/2_202108** [3].

Battery State-of-Charge Indicator

A built-in hydrometer that is installed within the top of an SLA-defined battery. This indicator shows if the electrolyte level has fallen below a minimum level. The indicator could be a plastic rod located at the highest of the battery and into the electrolyte. The Delphi design uses a green plastic ball suspended in a cage from the rod bottom. The precise gravity of the electrolyte determines if the ball floats or sinks within the cage, changing its look from green to dark. If the attention is dark, the battery must be charged. Some designs use a red and a blue ball side by side within the cage. If the precise gravity is high, you may only see the blue ball. Due to the reduced relative density, the blue ball will sink, allowing the red ball to take its place. If the battery is recharged, with increase in relative density, the blue ball will move upward, forcing the red ball back to the side of the cage. For more information on this, see **J1715/2_202108** [3].

Battery Terminals

Automotive batteries use two terminals on the outside of the case: a positive (+) and a negative (–). These terminals can be either two tapered posts or threaded studs on top of the case, or two internally threaded connectors on the front side. These terminals connect to either end of either the positive or negative series of cells inside the battery. Both terminals are marked as positive (+) or negative (–). For more information on this, see **J1715/2_202108** [3].

FIGURE B.7 Tapered type automotive battery terminal.

ratmaner/Shutterstock.com.

Baud Rate

The speed at which bits of computer information are transmitted on a serial data stream; measured in bits per second (bps).

B-CAN (Body Control Area Network)

A group of computers that work together to control relay modules, entry lights, exterior lights, the horn, interlocking system, key-in reminder, keyless entry, power door locks, safety belt reminder, security alarm, turn and hazard signals, wiper-washer system, and other functions. B-CAN communicates at a knowledge rate of about 33 kbps.

BCM (Body Control Module)

The BCM, also called a body computer, is a generic term for an electronic control unit or computer used for monitoring and controlling various electronic accessories in an automotive vehicle. The BCM controls the power windows, power mirrors, air conditioning, immobilizer system, central locking, etc. The BCM communicates with other on-board computers via the car controller area network (CAN) bus, and its main application is controlling load drivers—actuating relays that successively perform actions within the vehicle, like locking the doors, flashing the turn signals (in older cars), or dimming the inside lighting.

BDC (Bottom Dead Center)

The exact bottom of a piston stroke of an ICE.

Bead

"The part of a tire that comes into contact with the rim and is shaped to secure the tire to the rim." For more information on this, see **J2047_201911** [6].

FIGURE B.8 Tire construction (example from SAE J2047_201911).

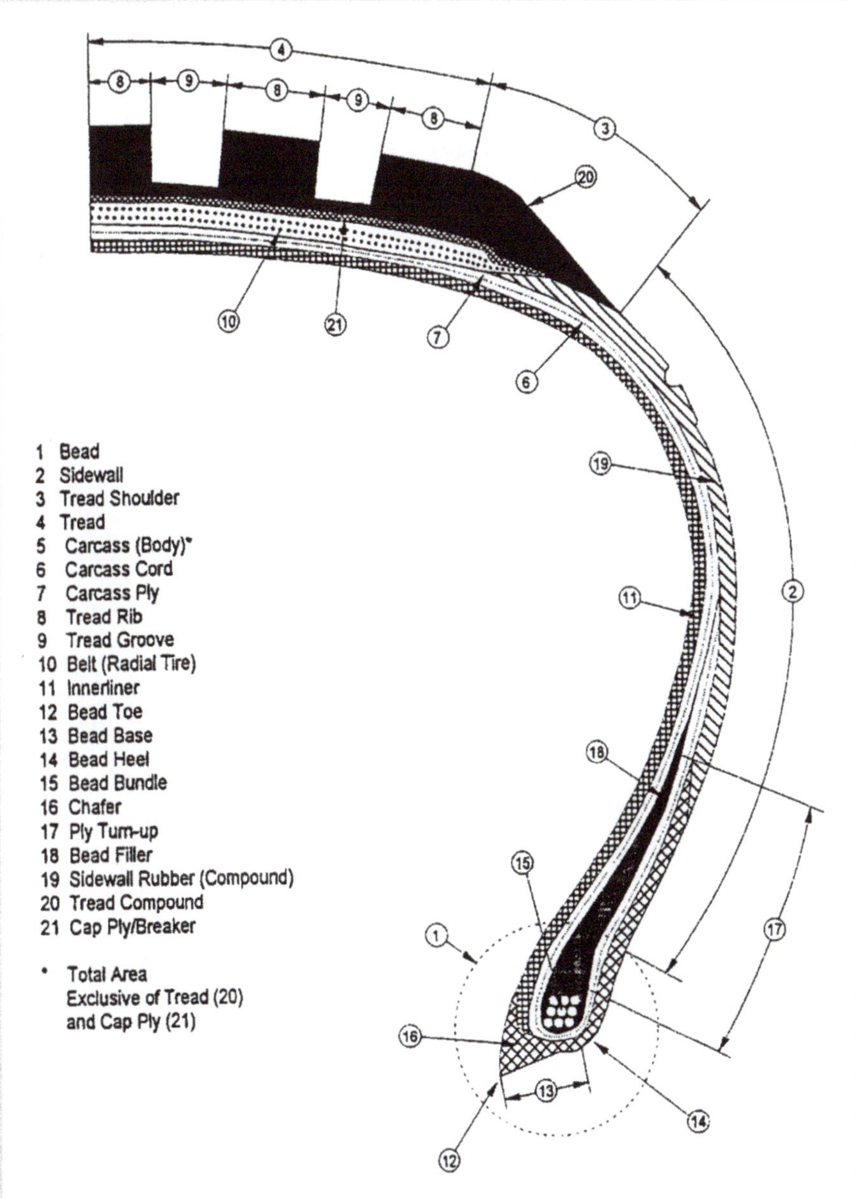

Bead Base

"Inner portion of the bead that is seated on the bead seat." For more information on this, see **J2047_201911** [6].

Bead Bundle (Bead Coils, Bead Cord)

"A circumferentially steel hoop made of steel wires embedded in the bead which resists the inflation pressure generated forces." For more information on this, see **J2047_201911** [6].

Bead Face

"The portion of the bead above the bead heel that interfaces with the rim flange." For more information on this, see **J2047_201911** [6].

Bead Filler (Apex)

"A rubber compound fillet between the bead bundle and adjacent ply cords." For more information on this, see **J2047_201911** [6].

Bead Heel

"Outer portion of the bead base." For more information on this, see **J2047_201911** [6].

Bead Seat

"The part of the wheel rim which provides radial support to the tire and air pressure seal for tubeless tires." For more information on this, see **J2047_201911** [6].

Bead Toe

"Inner portion of the bead base." For more information on this, see **J2047_201911** [6].

Bearing

A support device where the contacting surface of a revolving part rests in order to minimize wear and friction between two surfaces. A friction-reducing device found between two moving parts. Bearings made from Babbitt are used between the connecting rod and cap and the crankshaft and main bearing cap. They are lubricated and cushioned with oil, and the front-wheel bearings must be repacked with grease at regular intervals. Bearings can be ball or roller type.

Bearing (Inside Diameter)

The diameter across the hole for the journal, with the bearing inserts in place.

Bearing Cap

A device that is bolted in place to secure an ICE bearing in place.

The bottom portion of a connecting rod or main ICE bearing that holds the bearing to the shaft.

FIGURE B.9 Bearing caps.

Courtesy of John F. Kershaw Ed.D.

Bearing Cone

The inner race of a tapered roller bearing.

Bearing Crush

When the bearing is installed, each end of the bearing shell is slightly above the parting surface. When the bearing cap is tightened, the ends of the shells are forced together. This force is named bearing crush. Th crush holds the bearing in situ and keeps the bearing from turning when the engine runs. The rush must exert a force of a minimum of 12,000 psi (82,740 kPa) at 250°F (121°C) to carry the bearing securely.

Bearing Lubrication

Lubrication is critical to bearing durability and reliability. The lubricant should be of the NLGI GC-LB type. Wheel bearings should never have differing kinds of grease intermixed as possible incompatibility between lubricants could occur, causing thinning and lack of lubrication. If no lubrication is provided, even for a brief period of time, the controller can become so hot that the rollers or balls are welded to the race. Often, the heat is enough to cause small particles to come back off and weld to a different location. This can be named as "smearing" or "scoring." In tapered roller bearings, the primary signs of inadequate or lack of lubrication appear as scoring on the massive ends of the rollers. This soon progresses into heat discoloration on the massive end of the rollers and races. The big end of the roller is the area of sliding contact within a tapered bearing and is the most difficult area to forestall scoring, especially when lubrication is not readily available. If operation under these conditions continues, the damaged bearings eventually soften the fabric, with the top result being bearing seizure or other component damage. Bearings may also develop "peeling," which supplies the races or rollers a frosty appearance. Peeling may indicate the presence of high operating temperatures or low lubricant viscosity. Bearings must be full of the specified lubricant.

Bearingized

A surface that is exceptionally smooth and durable.

Beating Vibration

Action that takes place when two vibrating forces are adding to each other's amplitude. Yet two vibrating forces may subtract from each other's amplitude. The adding and subtracting of amplitudes in similar frequencies is termed beating. In many cases, eliminating either one of the disturbances can correct the condition. For more information on this, see **J3152_202005** [5].

FIGURE B.10 Two separate disturbances.

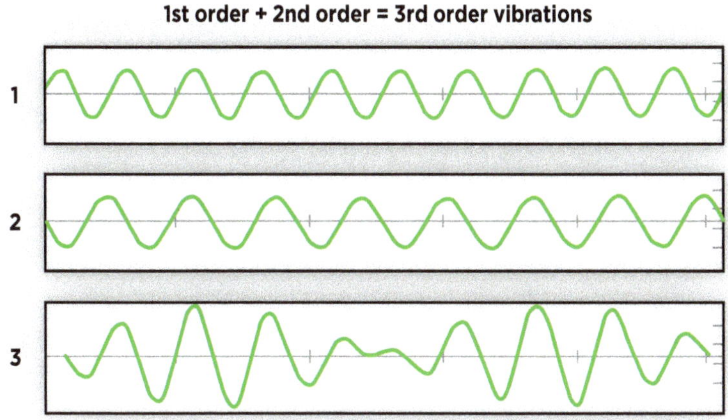

Courtesy of John F. Kershaw Ed.D.

Before Top Dead Center (BTDC)

Position of a piston before it reaches the highest of its stroke and shown in the crankshaft as degrees of rotation.

Bell Housing

A term for the portion of a manual or automatic transmission that covers the flywheel and the clutch or torque converter of the transmission on ICE-powered vehicles.

Belleville Spring

Flat, round piece of spring steel with a hole in the center. Slots are cut from the central hole toward the skin of the spring to create a variety of flexible fingers.

Belt

Fabric or woven steel material over the body plies of a tire, and slightly below the tread area, to assist keep the tire tread from squirming. For more information on this, see **J2047_201911** [6].

Belt Alternator Starter (BAS): Low-Cost Hybrids

BAS is an HEV option. The BAS concept is to interchange the belt-driven alternator with an electrical motor that is a generator and a motor. When the engine is running the motor, and acting as a generator, the system will charge a separate 36-V battery. When the engine must be started, the motor then applies its torque via the accessory belt, and cranks the engine rather than using the starter motor. The motor/generator is made larger than a standard starter motor so more torque can be generated when in the motoring mode. This allows for quicker starts of the engine and makes the start/stop operation possible. Stopping the engine while the vehicle is at idle is a means to conserve fuel. The belt system gets a 2–5% increase in fuel economy. For more information on this, see **J1715_202105** [4].

Belt System

The drive belt system may use one belt or two drive belts. The drive belt is thin so it can bend backward and has several ribs to match the grooves within the pulleys. There also is also a V-belt style belt used in the past to drive certain accessory drive components. The drive belts are a product of differing types of rubbers—Chloroprene or EPDM (Ethylene Propylene Diene Terpolymer), an artificial rubber. They have different layers containing either fiber cloth or cords for reinforcement. Either side of the drive belt can drive the various accessory drive components. When the rear side of the drive belt is employed to drive a pulley, the pulley is smooth.

FIGURE B.11 Typical automotive serpentine drive belt system.

Belt Tension

Tightness of the accessory drive belt.

Belt Tensioner

A device that places tension against a serpentine belt. The crankshaft balancer pulley pulls the drive belt across the accessory drive component pulleys. The spring-loaded drive belt tensioner keeps constant tension on the drive belt to forestall the drive belt from slipping. The drive belt tensioner arm will move when loads are applied to the drive belt by the accessory drive components and also the crankshaft.

FIGURE B.12 Serpentine belt tensioner.

Bendix Drive

A term used for the electric cranking motor clutch that connects the cranking motor drive to the ICE flywheel. It is properly named a Starter Drive. It was named after its inventor, Vincent Hugo Bendix.

FIGURE B.13 ICE cranking motor starter drive clutch.

notsuperstar/Shutterstock.com.

Bevel Gears

Used to alter the direction of a shaft. They are mounted on shafts that are 90° apart and may also work at other angles. The teeth will be straight, spiral bevel, or hypoid. Straight pinion and crown wheel teeth even have the identical noise problem as straight gear wheel teeth. As each tooth engages it impacts the corresponding tooth all without delay. The answer to the current problem is to curve the gear teeth. These spiral teeth engage a bit like helical teeth. The contact starts at one end of the gear and spreads across the entire tooth. On spiral bevel gears, the shafts must be perpendicular to each other and even be within the same plane. If you extended the two shafts past the gears, they'd intersect. The hypoid gearset can engage with the axes in a very different plane.

FIGURE B.14 A bevel gearset using a ring gear and drive pinion.

Sergey Ryzhov/Shutterstock.com.

BHP (Brake Horsepower)

BHP (shaft horsepower) is the power delivered at the engine crankshaft of the engine. The term brake horsepower comes from the method of early engine testing. This consisted of putting a mechanical brake on the engine and measuring the force required to keep the brake from turning. The energy produced was dissipated as heat. Water or air was used to cool the friction surfaces of the brake. Brakes of this type, called Prony brakes, were an early type of absorption dynamometer. The horse terms come from the following: One horsepower would be produced when a horse walked 165 ft in 1 minute pulling a 200-lb weight, or 165 ft × 200 lb = 33,000 ft-lb. BHP may be measured with either a transmission or an absorption dynamometer and is determined by the following formula (see Equation B.1), which is the same as the one listed under horsepower, but stated for BHP:

$$bhp = \frac{2\pi FRN}{33,000} \quad \textbf{(B.1)}$$

where

bhp is brake horsepower

F is the dynamometer load, in lb

R is the radius arm of the dynamometer, in ft

33,000 is the conversion factor, ft-lb/min, to produce 1 hp

N is the engine speed, in RPM

π = 3.1416, a constant

Bias Belted Tire

A type of tire that has the same ply configuration as a bias-ply tire except that the tire has belts added after the plies in the tread area made from fiberglass or steel that is used to prevent punctures in the tread and on top of the cords which form the tire main tire carcass and set diagonally to the center line of the tread. For more information on this, see **J2047_201911** [6].

FIGURE B.15 Comparison between a radial tire and a bias-belted tire.

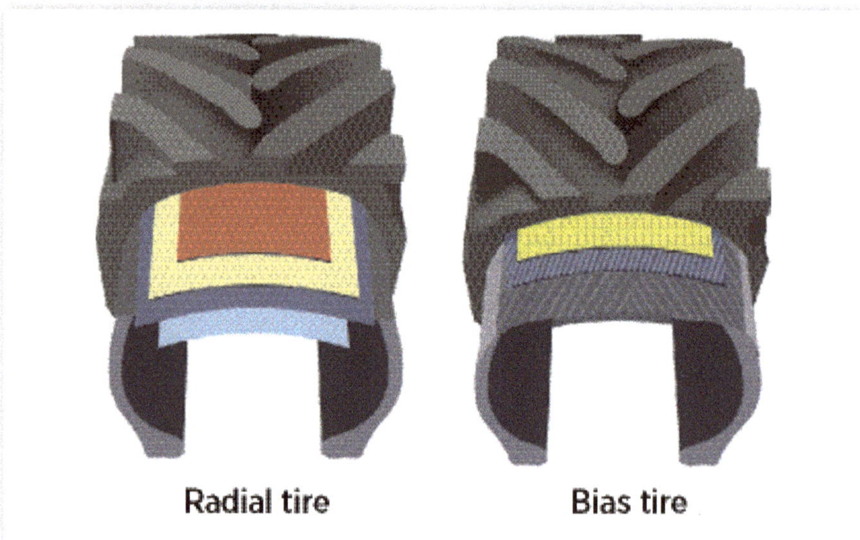

Fehmiu Roffytavare/Shutterstock.com.

Bias Ply Tire

The bias-ply tire construction uses body ply cords that extend and overlap diagonally, on a bias, from bead to bead at 30 to 40° angles with successive plies laid at opposite angles, which result in a crisscross pattern onto which tread is set. This design furnishes a strong, stable casing, but with relatively stiff sidewalls. During cornering, stiff sidewalls can distort the tread and partially lift it off the road surface, which reduces the surface area between the road and the tire. For more information on this, see **J2047_201911** [6].

Bias Voltage

In electrical terms, bias is the voltage applied to a device or component to establish the reference point for operation.

Biasing

Applying current to a junction of semiconductor materials.

Bidirectional Communication

Automotive computer communication that uses serial data as both an input and an output.

Big End

Another term used for the ICE connection rod end that connects to the crankpin of a crankshaft.

Bimetal Gauge

A gauge that uses two kinds of metal bonded to create one strip. The expansion characteristics of every piece of metal are different. Electrical current flows through the bimetal strip causing heat to bend the bimetal strip. This bending movement swings an indicator needle across a gauge face.

Bimetallic

Made or consisting of two metals.

Binary

A mathematical system, consisting of only two digits (0 and 1), that enables an information processing system to read and process input voltage signals.

Biodiesel

Domestically produced, renewable fuel that may be manufactured from vegetable oils, animal fats, or recycled restaurant greases. Biodiesel is safe, biodegradable, and reduces serious air pollutants like particulates, carbon monoxide gas, hydrocarbons, and air toxins. Biodiesel is defined as mono-alkyl esters of long-chain fatty acids derived from vegetable oils or animal fats that conform to ASTM D6751 specifications to be used in diesel engines. Biodiesel refers to the pure fuel before blending with diesel oil.

Bipolar

Having two opposite and opposing forces.

Bit

The individual voltage signal of a serial data stream; also the smallest unit of measurement recognized by a computer.

Black Smoke

Owing to the extremely short time available for mixing, as the fuel-air rate increases beyond a certain value, an appreciable fraction of the fuel fails to find the necessary oxygen for combustion and passes through the cylinder unburned or partially burned black smoke. Fuel burned without air is black smoke. The tendency to produce black smoke is any variable that increases the amount of fuel injected or reduces the amount of air taken into the cylinder. These include the following: any nozzle (injector) malfunctions, turbocharger failures, restricted air flow,

Blink or Flash Codes (OEMs Use Both Terms)

These codes are an on-board means of troubleshooting computer-controlled engine management systems using a dash or computer-mounted electronic malfunction light or CEL (check engine light). Normally, only active codes, i.e., one indicating a malfunction at the time of reading, can be read on the vehicle management computer, usually an engine control module (ECM). Since the adoption of automotive OBD II (On-Board Diagnostics Generation II) and heavy truck HD-OBD (Heavy-Duty On-Board Diagnostics), they have gone away but some OEMs still used them on a limited basis. The following example uses the cruise control function switches of the vehicle to activate the blink code sequence:

1. Turn the ignition key on and wait until the malfunction indicator light (MIL)'s 2-second power-up test is completed. If the lamp remains on after the 2-second power-up test, this indicates an active fault in the memory that can be accessed.

2. Set the cruise control speed control on/off switch to the "off" position and press and hold the Set/Decel switch until the MIL lamp goes out. The lamp will remain off for approximately 1 sec.
3. Immediately after the wait time, the ECM will begin to flash a two-digit blink code on the MIL. The two digits will be separated by an idle time of approximately 1 sec. The on and off time between each flash for each number should be a quarter of a second.
4. Count the number of flashes for the first digit and write the number down.
5. After the 1-sec pause, count the number of flashes for the second digit and record. You now have the two-digit fault code.
6. Cross-reference the blink code with the chart in the service manual to identify the problem area.

Block Diagram

A diagram that uses boxes and symbols, parenthetical components and contours, to represent wires and hoses. Simplifies a system gate so it will be easy to understand how a locality of the system operates and how major components interact.

Block Learn Function

Generically referred to as long-term fuel trim. Fuel delivery adjustments to the injector pulse width supported control system feedback. Short-term fuel trim is predicated on rapidly switching oxygen sensor values, while long-term fuel trim could be a learned value accustomed to make amends for continual deviation of the short-term fuel trim from its central value. For more information on this, see **J1930DA_202105** [7].

Blockchain Framework

Blockchain framework is used in automotive technology to detect unauthorized modifications to vehicle management computers like the engine control module (ECM) or engine control unit (ECU). "A proof of concept blockchain prototype framework is implemented on a set of microprocessors (comparable to those used by simple ECUs) as a means to assess the efficacy of using our blockchain approach to detect unauthorized updates. Traditional cybersecurity protection approaches managed by a centralized authority are not directly applicable to handle automotive cybersecurity given the disparate and distributed nature of the automotive attack vectors. Currently, multiple international, cross-organizational groups are developing guidelines and standards for automotive cybersecurity, but to date, there is no formally adopted or even de facto standard for securing OTA updates or ECU integrity." Refer to SAE Article 11-04-01-0002 for more information [10].

Blocker Rings (Synchronizer)

A speed-operated clutch located within the synchronizer clutch in a manual transmission that synchronizes the speed of the fixed gear shaft with one of the freewheeling speed gears.

FIGURE B.16 Blocker ring.

StockTeamBaku/Shutterstock.com.

Blowby Gases

Combustion gases that leak past the piston rings into the crankcase during the compression and combustion strokes of an internal combustion engine.

Blower Motor

An electric motor and squirrel-cage type of fan that moves air inside the vehicle for heating, cooling, and defrosting.

Blow-Off Valve (BOV)

This is also called a dump valve or vent valve and features an adjustable spring design that keeps the valve closed until a sudden release of the throttle. Allows the manifold pressure to vent to the atmosphere.

Blue Smoke

A blue smoke color comes from engine lubrication oil mixing with the diesel fuel and can be caused by:

- Oil level too high
- Worn valve guides and seals
- Worn piston rings or cylinder walls
- Bad piston oil control rings
- Bad turbocharger seals
- Excessive crankcase pressure

Bluetooth

A radio frequency standard for short-range communications that connects vehicle devices. The range of a typical Bluetooth device is 33 ft (10 m) and operates within the ISM (industrial, scientific, and medical) band between 2.4000 and 2.4835 MHz. It is a wireless standard that works on two levels. It provides physical communication using low power, requiring only about 1 mW (1/1,000 of a watt) of electric power, making it suitable to be used with small handheld or portable devices, like an ear-mounted speaker/microphone. It provides a typical protocol for the way bits of information are sent and received. The Bluetooth standard is wireless. The automotive use of Bluetooth technology is within the operation of a cell phone being connected into the vehicle. The vehicle enables the use of hands-free phone. A vehicle equipped with a Bluetooth phone has the following components:

- Bluetooth receiver may be built into the navigation or existing audio system.
- Microphone allows the driving force to use voice commands moreover as phone conversations from the vehicle to the cell via Bluetooth wireless connections.

All cell phones are equipped with Bluetooth, which allow the caller to use an ear-mounted microphone and speaker. If the vehicle and the cellular phone are equipped with Bluetooth, the speaker and microphone will be used as a hands-free phone when the phone is within the vehicle. The cellular phone will be activated within the vehicle by using voice commands. It was named after Harald Bluetooth, the king of Denmark in the late 900s. The king united Denmark and a part of Norway into one kingdom.

Bluetooth Detection

Bluetooth devices are detected by sensing devices along the road or hand-held cell phones. When these Bluetooth sensors are interconnected they are ready to calculate the time period and supply data for origin and destination matrices. Bluetooth measurement has some differences compared to other traffic measurement technologies. Bluetooth is additionally non-intrusive, which might result in lower-cost installations for both permanent and temporary sites. Bluetooth features are as follows:

- Limited to number of devices broadcasting from a vehicle.
- Systems are generally quick to line up with no calibration needed.
- Since Bluetooth devices become more prevalent on board vehicles and with more portable electronics broadcasting, the number of information collected over time becomes more accurate and valuable for period and estimation purposes

BMEP (Brake Mean Effective Pressure)

A term used for design purposes. BMEP is an academic constant pressure acting on the piston during each power stroke. BMEP produces a power equal to the engine brake horsepower. This term indicates how well an engine uses its displacement to produce torque, and it is a better parameter for comparing engines than horsepower. The BMEP force on the top of the piston is the average pressure exerted on each square inch of the top of the piston, during the power stroke multiplied by the area on top of the piston (Force (f) = area × BMEP). BMEP is determined by the formula:

$$p_b = \frac{BHP \times 33{,}000}{LANK} \qquad \textbf{(B.2)}$$

where

L is the length of stroke, in ft
A is the area of the piston, in in.^{2}
N is the number of power strokes per minute
K is the number of cylinders

BNC Connector

Coaxial-type connector usually used on oscilloscopes. Named after its inventor Bayonet Neil-Concelman (BNC).

Bobtail Proportioning Valve

Senses when the tractor equipped with air brakes is bobtailing and operates when the tractor parking brake is released and the trailer supply valve remains closed. The objective is to reduce the air pressure applied to the tractor drive axle(s) and minimize drive wheel lock-up on slippery pavement. Bobtail proportioning is integral with the relay valve managing the service braking. This valve will function as a relay valve under normal operation. In bobtail mode, the valve ratio is determined by bleeding a percentage of the application air to exhaust.

Body Ground Wire

A wire used on the negative battery cable to confirm that the vehicle body is grounded.

Bogie Axle

A bogie axle may be a modular chassis with a wheel axle attached to it that is attached, or are often detached, as in semi-trailers, and will have its own suspension and brakes.

Boiling Point

Temperature at which a solid or liquid substance turns to vapor. For water at normal water level conditions, the boiling point is 212°F (100°C).

Bolt

Screws, bolts, and studs are the three types of threaded fasteners employed in automotive applications.

Bolt Circle

The diameter of a circle (inches/millimeters) drawn through the middle of the bolt holes in a wheel. For more information on this, see **J2047_201911** [6].

Bolt Pattern

Refers to the amount and spacing of the lug nuts or wheel studs on the wheel hub on the wheel rim. It is also called a bolt circle because the line drawn through the centers of the wheel lugs forms an imaginary circle. Studs are frequently evenly spaced, and the amount of studs determines the pattern. Some smaller vehicles have three studs, some may have four studs, but most passenger cars have five. Pickup trucks and huge SUVs can have as many as six, eight, or ten studs. Vehicles that carry heavier loads generally have more studs. The precise number and pattern of studs vary looking on the vehicle type and OEM design. Some manufacturers also offer a dual bolt pattern, or two possible bolt patterns, on one wheel. For more information on this, see **J2047_201911** [6].

Boom Vibration

A vibration felt in the cabin of a vehicle that travels through the roof caused by a third-order vibration from the driveline and wheels. For more information on this, see **J3152_202005** [5].

Boost

An increase in ICE air pressure above atmospheric at 14.7 psi (pounds per square inch) used when supercharging or turbocharging. Superchargers and turbochargers are designed to provide a pressure greater than atmospheric pressure (14.7 psi) in the intake manifold. This increased pressure forces more air into the combustion chamber over what would normally be forced in by atmospheric pressure to provide a heavier charge with greater mass. This increased charge increases engine power. The higher the boost (pressure), the greater will be the horsepower. Other factors must also be considered when increasing boost pressure: As boost pressure increases, the temperature of the air also increases. As the air temperature increases, so does the combustion temperatures that increases the possibility of detonation.

Boost Control

Both supercharged and turbocharged systems are designed to provide a pressure greater than atmospheric pressure in the intake manifold. This increased pressure forces additional amounts of air into the combustion chamber over what would normally be forced in by atmospheric pressure. The amount of "boost" (or pressure in the intake manifold) is measured in pounds per square inch (psi), in inches of mercury (Hg), bars, or in atmospheres and controlled by several different devices, such as wastegates, boost control valves, and variable position turbine nozzles on turbochargers.

Bore

The diameter or width of an ICE piston or the cylinder diameter. When the bore is larger than the stroke the engine is over squared and most of its power is dependent on the revolutions per minute (RPM) value and generated at higher RPM values. When the stroke is larger than the bore, the engine is undersquare with most of its power dependent on torque development.

Borescope

Optical instrument designed to help visual inspection of narrow, difficult-to-reach cavities, consisting of a rigid or flexible tube with an eyepiece or display on one end, an objective lens or camera on the opposite, linked together by an optical or electrical system in between. The optical system, in some instances, is in the lighting process to boost brightness and contrast. An indoor image of the illuminated object is made by the target lens and magnified by the eyepiece which presents it to the viewer's eye.

Bosch Peak-and-Hold Injector Waveform

The waveform of a peak-and-hold fuel injector is similar to the peak-and-hold waveform except that, instead of lowering the current to hold open the injector, the current is rapidly pulsed on and off.

Bottom End

Industry term for the rotating bottom end of an ICE that includes those made by the crankshaft and the piston and connecting rod assemblies.

Bound Electrons

When there are five or more electrons in the valence ring of an atom, they are held tightly together. This is characteristic of an insulator.

Bowden Cable

A flexible cable that transmits force through the movement of an inner cable relative to a hollow outer cable housing. This cable is used for parking brakes on many vehicles.

Bowed (Buckled) Frame

When one or both frame rails are bent upward in relation to the ends of the rails.

Boxer ICE

A horizontally opposed ICE is a boxer design. It is called a boxer because each set of pistons moves in and out at the same time, like a boxer's gloves. The centrally located crankshaft has the cylinders on each side. The boxer engine design allows a lower center of gravity that lowers the vehicle center of gravity and is more responsive when cornering.

Boyle's Law

The pressure of a perfect gas varies inversely to its volume at a constant temperature, or that the product of the pressure and volume is a constant. The pressure of a perfect gas varies inversely to its volume at a constant temperature, or that the product of the pressure and volume is a constant. Pressure is inversely proportional to the volume occupied by the gas. High pressure equals low volume. At a constant temperature, the pressure of a gas depends on the volume of the vessel holding it. Boyle's Law applies to compression in an ICE. When the piston rises on compression, cylinder volume is reduced and the pressure increases. This exerts pressure on the cylinder walls when the same number of molecules now have less space to move. The general formula is:

$$P \propto \frac{1}{V} PV = \text{Constant} \quad \textbf{(B.3)}$$

Brake Bleeding (Hydraulic Brakes)

An automotive service procedure used to remove trapped air from a hydraulic brake system. Hydraulic brakes need to be bled whenever a hydraulic component of the brake system is replaced, or whenever a diagnosis indicates that air is trapped in the hydraulic system. Methods of bleeding brakes will vary with the type of system. The reason that the brakes are being bled will also influence the method. If a vehicle has a spongy pedal, then bleed the entire brake system. The bleeding procedure for a front/rear split brake system will be different from bleeding a diagonally split brake system. Whenever bleeding an entire brake system, the usual method is to begin by bleeding the master cylinder, then bleed the wheel cylinder farthest from the master cylinder, and continue by bleeding each wheel cylinder from the farthest to the closest in sequence.

There are four methods used: manual, gravity, pressure, and vacuum. Manual or pedal bleeding requires two people to perform the task. One person pumps the brake pedal slowly and firmly to force fluid through the system. After four or five applications of the brake pedal, the pedal is held down. At that time, the second person opens the bleeder valve at that wheel allowing the fluid and air in the line to escape. The pedal sinks to the floor as the fluid and air escapes. The pedal must not be released until the bleeder valve is closed or air will be drawn into the system. The procedure is repeated at each wheel until air bubbles are no longer evident. Gravity bleeding consists of letting the weight of the brake fluid force itself through the system. This is accomplished by simply opening the wheel cylinder bleeder valves one at a time until a solid flow of fluid flows out of the valve.

Pressure bleeding method requires the use of a pressure bleeder, which consists of a tank which is divided into two chambers by a flexible diaphragm. One chamber contains the brake fluid, the other contains air. The air chamber can be pressurized with compressed air from the shop air system. When the air chamber is pressurized, it exerts pressure through the flexible diaphragm against the brake fluid in the fluid chamber. A valve in the fluid side of the tank allows fluid to escape through a hose to an adapter. This adapter fits tightly over the master cylinder reservoir, thus allowing the fluid to enter the brake system under pressure. In this way, the pressure bleeder not only pressurizes the brake system, but it also maintains the fluid level in the master cylinder. Vacuum bleeding is using a Mityvac suction pump attached to the bleeder screw. The pump creates a low-pressure area at the bleeder screw that allows atmospheric pressure to push brake fluid through the system when the bleeder screw is opened, thus removing the air.

Brake Booster (Brake Assist)

Automotive power brakes are used to reduce the force required from the driver when applying the brake pedal. This is done through the use of a brake booster of which there are three types of power assist systems used on most vehicles:

- Vacuum power assist
- Hydraulic boosters
- Anti-lock brake system (ABS) booster

Pneumatic (vacuum) power assist may use vacuum (low pressure) with pressure differential between air pressure and engine vacuum or low to help in pushing the brake cylinder piston. Hydraulic boosters use fluid pressure generated by the power steering pump. Most current vacuum boosters are vacuum-suspended units. The term vacuum-suspended describes the condition of the unit when the car engine is running and therefore the brakes are released. When in this condition, an equal vacuum is present on either side of a diaphragm, thus suspending the diaphragm in a specific vacuum. Although all booster components are combined during a single assembly, a vacuum booster consists of two subassemblies.

Brake Caliper

See **Caliper**.

Brake Chamber

Compressed air from within the brake system is directed through the lines to two or more brake chambers to apply the foundation brakes. Pressurized air from the compressor via the air reservoirs is sent to the application valve. When the driver steps on the treadle, the application valves sends air signals to the quick release valve and relay valve/anti-lock brake system (ABS) modulators. These devices then connect the air supply to the flexible diaphragms of the service brake chambers, which forces a pushrod assembly forward against the force of a coil return spring to move the S-cam to apply the drum brakes. When the air is released from the chamber, the coil return spring pulls the pushrod back to the released position. The number and size of the brake chambers determines the air compressor supply and reservoir requirements. The brake chamber is mounted on a bracket welded to the axle housing assembly as shown in Figure B.17. The chamber pushrod is connected by a yoke and a clevis pin to a slack adjuster. The slack adjuster acts as a lever to multiply the force from the diaphragm and pushrod, which, in turn, rotates the S-cam brake shaft. The S-cam is rotated to permit two cam rollers to be raised. This action, in turn, causes the brake shoes to be forced outward against the spinning brake drum bolted to the wheel. On air disc brakes, as the actuating beam moves the inner brake pad into contact with the rotor, the caliper slides away from the rotor, pulling the outer brake pad into the rotor. For more information on this, see **J1410_202012** and **J2627_202207** [8, 9].

FIGURE B.17 Air brake chambers.

Kevin Norris/Shutterstock.com.

Brake Fade

Brake drums and rotors are forced to soak up a major amount of warmth during braking. Brake fade describes a condition where heat is generated at a faster rate within the brakes than is capable of dissipating the warmth into the surrounding air. During a hard stop the temperature of drums or rotors may rise as high as 100°F in just a few seconds. It should take 30 sec to cool down the temperature of these components before braking. During repeated hard stops, overheating can cause brake inefficiency or a complete failure. Brake fade is the rationale for towing weight limits and trailer brake requirements for vehicles. The added energy resulting from the increased vehicle mass requires added heat conversion capacity when the brakes are applied. There are two types of brake fading caused by heat: mechanical fade and lining fade. Mechanical fade occurs when the drum overheats and expands off from the liner, resulting in increased treadle travel. Rapidly pumping the pedal will help to keep the linings in tune with the drum. Lining fade affects both drum and disc brakes and occurs when the friction material becomes overheated to the point where the coefficient of friction drops. When the coefficient of friction drops, it both reduces friction and also the ability of the brake to convert added heat.

Brake Fluid

Brake fluid is the hydraulic fluid utilized by the automotive hydraulic brakes system to maneuver the pistons within the calipers and wheel cylinder. The US Department of Transportation (DOT) issues specifications for brake fluid. The three main kinds of brake fluid now available are DOT 3, DOT 4, and DOT 5. DOT 3 and DOT 4 are glycol-based fluids, and DOT 5 is silicone-based. The difference is that DOT 3 and DOT 4 is hydroscopic and absorbs water, while DOT 5 does not.

The most important characteristics of brake fluid is its boiling point. Hydraulic systems depend on an incompressible fluid to transmit force. Liquids are generally incompressible while gases are compressible. If the brake fluid boils (becomes a gas), it will lose most of its ability to transmit force. This might partially or completely disable the brakes. The time that drivers are likely to boil the brake fluid is during a period of prolonged braking, such as a driving down a mountain. As a DOT 3 or DOT 4 brake fluid absorbs water, its boiling point decreases. It can absorb water from the air, which is why technicians and drivers should avoid opening the brake fluid reservoir of vehicles. For the same reason, always keep containers of brake fluid tightly sealed.

DOT 5 fluid does not absorb water and is purple in color. This implies the boiling point will remain relatively stable, but it also means any water that does get into the brake will tend to create pure water pockets, which could cause brake corrosion. DOT 5 also does not mix with either DOT 3 or DOT 4. Most OEMs use two kinds of brake fluid: DOT 3 and DOT 5. DOT 3 is that the standard brake fluid used in all cars and lightweight trucks manufactured for the North American market. It is colorless and readily absorbs moisture. For more information on this, see **J1703_201909** [11].

SAE Standard for BHP Rating

- Net horsepower available from the engine with specified induction specifications and engine speed
- Rating guaranteed within 57%

- Corrected to SAE standard ambient conditions
- Air temperature 85°F (29.4°C)
- Elevation 500 ft (159.4 m)
- Dry air density 0.0705 lb/cu ft (11.29 g/m^3)
- May be without the accessories required for the application (compressor, fan, generator, and so on)

Brake Lines

Brake fluid is carried through steel pipes called brake lines. Brake lines are usually constructed of double-walled steel tubing that has double-lap flare or an ISO (International Organization for Standardization) flare on the ends.

Brake Pad

A friction material pad of which there are two pads that grabs the brake rotor like a C-clamp to stop it from turning and that effects braking. A hydraulic caliper piston pushes against the pads to stop the rotor.

FIGURE B.18 Brake pad shown with a caliper and rotor assembly.

LightField Studios/Shutterstock.com.

Brake Rotor

A component used in the disc brake system that is mounted on a front or rear hub at the axle and has a hydraulic caliper mounted above that pushed the two pads to stop the rotor and effect vehicle braking.

FIGURE B.19 Brake rotor.

Setta Sornnoi/Shutterstock.com.

Brake Specific Fuel Consumption (BSFC)

BSFC is the rate of fuel consumption per unit of brake horsepower output. This parameter demonstrates how efficiently the engine converts fuel energy into work. Most OEMs prefer the BSFC to thermal efficiency because it is determined by using the practical units of weight, horsepower, and time. The following formula determines BSFC:

$$\text{BSFC} = \frac{\text{W}}{\text{BHP}} \tag{B.4}$$

where

W is the weight of fuel component, in lb/hr
BHP is the brake, in hp

Engine manufacturers' sales data literature list BSFC, which is shown in either lb/bhp/hr or g/kW-h (grams/kilowatt-hour). One lb/hp-hr is equal to

608.277 g/kW-h. The BSFC for a Caterpillar 3406E engine rated at 475 hp (354 kW) at 1,800 RPM and peak torque of 1,750 lb/ft (2,373 Nm) at 1,200 RPM is approximately 0.316 lb/hp-hr (192 g/kWh) for a fuel rate of 21.3 US gallons (80.8 L/h) when running at 1,800 RPM. At the peak torque rating of 1,200 RPM, the fuel rate is 0.304 lb/hp/hr (185 g/kW-h) for a fuel consumption rate of 16.9 US gallons (64 L/h).

Braking Performance

The automotive braking performance general equation is based upon Newton's Second Law of Motion: Force = Mass times acceleration. The general equation is:

$$Ma_x = \frac{W}{g} D_x = F_{xf} - F_{xr} - D_A - W\sin\theta \quad \textbf{(B.5)}$$

where

- W is the vehicle weight
- g is acceleration due to gravity: g-force
- $D_x = a_x$ is the linear deceleration
- F_{xf} is the front axle braking force
- F_{xr} is the rear axle braking force
- D_A is aerodynamic drag
- θ is the upgrade hill angle

Braking Ratio

Braking ratio is the comparison of the front wheel and rear wheel braking effort. When a vehicle brakes, its weight tends to transfer to the front wheels. This causes the front wheels to be pressed against the road with greater force. At the same time, the rear wheels sometimes lose traction on the road. As a result, the front brakes do more of the braking than the rear.

Braking Stability

Effective braking is also dependent on the pressurization and control of hydraulic fluid. By applying the brake pedal, the hydraulic fluid is pressurized and applies the braking force necessary to stop the vehicle. When braking, the wheels must not lock or the driver will lose steering control (see **ABS**). To prevent wheel lock, the rear wheels must decelerate at a slightly lower rate than the front in order to maintain directional stability. For rear-wheel-drive vehicles, up to 60% of the braking force is applied to the front wheels and 40% is applied to the rear wheels. For front-wheel-drive vehicles, up to 80% of the braking force is applied to the front wheels and 20% is applied to the rear wheels.

Braking System (Automotive)

The automotive brake must be able to rapidly change the state of energy, so the brakes must have more power than the kinetic energy of the vehicle. Brakes in terms of horsepower are 5× that of a standard ICE. A vehicle accelerating from 0 to 60 mph in 10 sec is common. Nominal brakes should stop a vehicle traveling 60 mph within 3–4 sec. The foot lever assembly acts like nothing over a mechanical lever arm to apply force to the brake cylinder piston. When the motive force presses the treadle, the cylinder piston presses against the brake fluid. Brake fluid is not compressible. The force is transmitted as hydraulic pressure. This pressure is sent to the individual wheel cylinders so that the cylinders utilize the friction components of the brake system. All vehicles have a stoplight switch attached to the pedal arm. Depressing the pedal causes the stoplight switch to activate the red stoplights on the rear of the vehicle. The cylinder is actually a foot-operated hydraulic ram that gives the pressure utilized in the braking system. A basic hydraulic brake cylinder consists of a reservoir, reservoir cover with gasket, housing, piston, rubber piston cup, and return spring. The components may vary slightly with different styles or styles of vehicles.

FIGURE B.20 Automotive brake system.

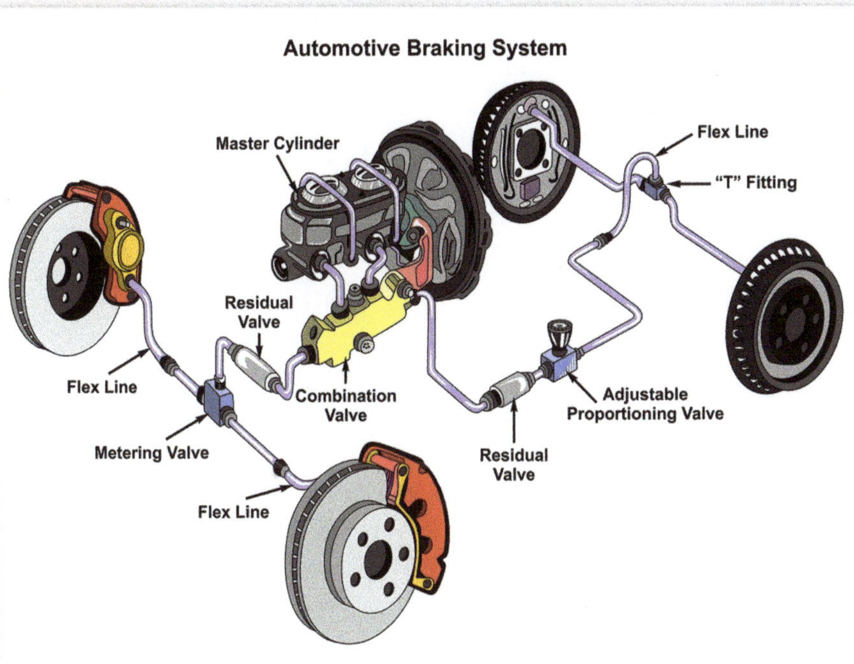

udaix/Shutterstock.com.

Breaker Points

The metal contact points that act as an electrical switch in an ICE ignition distributor to open and close the ignition primary circuit.

Breakout Box

A device that is connected into an electrical circuit or component to check values at specific points in the circuit.

Brinelling

A mechanical failure characteristic of a dent in metal such as what occurs when a shock load is applied to a bearing.

British Thermal Unit (BTU)

The amount of heat required to raise 1 pound of water 1°F at sea level.

Brushes

A copper or carbon conductor used to transfer electrical current from or to a revolving electrical part such as the commutator used in an electrical motor or generator.

BSFC

See **Brake Specific Fuel Consumption**.

Buffer

Circuit or component used to reduce the interaction between two electronic circuits.

Bump Steer

A driving condition that occurs when the front wheels of a front-wheel-drive vehicle steer themselves without input from the steering wheel. This condition is caused by bumps in the track interacting with improper length or angle of your drive axles, suspension, and steering linkages.

Bulkhead

The front partition that separates compartments in a vehicle. It is a metal wall that extends from one side of a vehicle to the other side.

Bus
An electrical conductor, or conductors, serving as a common connection for three or more circuits.

Bus Configuration
A computer network topology in which several of the computers of the vehicle are wired to a single conductor, or bus.

Bushing
A cylindrical metal sleeve that inserts into a machined bore to reduce the effect of friction on moving parts or decrease the diameter of the hole. It is used in joints between moving components and is designed to eliminate the friction that causes premature wear. However, the bushings themselves are subject to wear and may cause play within the connecting points of the component. Bushings are manufactured with bronze or rubber compounds, depending on design requirements. Bushings should be tight and typically do not require lubrication.

Buss Bar
Buss bar is a solid metal strip, or bar, used as a conductor in a multiple fuse panel.

Butt Splice
Wires to be connected together are placed end to end, without overlap.

Butterfly Valve
This is a slang term for the throttle valve. See **Throttle**.

Bypass Hose
A hose that permits coolant to flow through the water pump and back into the engine when the thermostat is closed. The bypass itself may be an external hose or a passage within the thermostat housing or water pump body. It allows for a faster warmup of the ICE.

FIGURE B.21 Bypass hose.

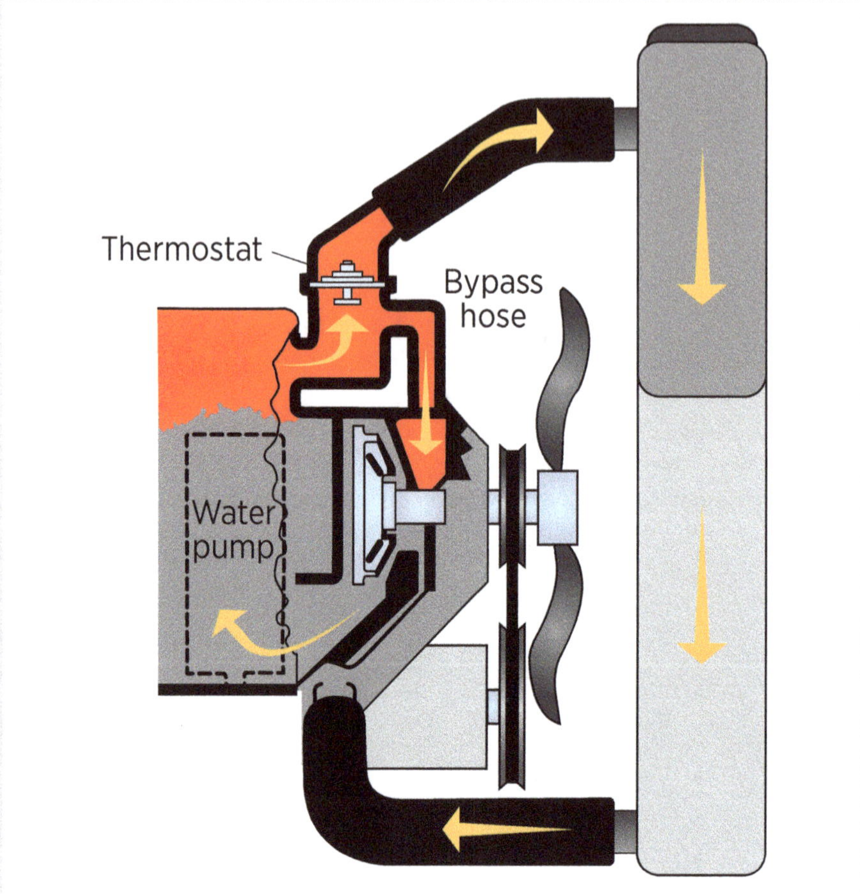

Courtesy of John F. Kershaw Ed.D.

Bypass Valve

See **Blow-Off Valve**.

Byte

A byte is eight bits of computer information processed as a unit and transmitted as a sequence on the serial data stream. Also known as a word.

REFERENCES

1. Society of Automotive Engineers, J1297_201710: Alternative Automotive Fuels, SAE Publishing, Warrendale, PA, 2017.
2. Society of Automotive Engineers, J3063_202103: Active Safety Systems Terms and Definitions, SAE Publishing, Warrendale, PA, 2021.
3. Society of Automotive Engineers, Standard J1715/2_202108: Battery Terminology, SAE Publishing, Warrendale, PA, 2021.
4. Society of Automotive Engineers, Standard J1715_202105: Hybrid Electric Vehicle (HEV) and Electric Vehicle (EV) Terminology, SAE Publishing, Warrendale, PA, 2018.
5. Society of Automotive Engineers, J3152_202005: Taxonomy and Guidelines for Design, Fabrication, and Installation of Brake NVH Fixtures for Inertia Dynamometer Testing, SAE Publishing, Warrendale, PA, 2011.
6. Society of Automotive Engineers, J2047_201911: Tire Performance Terminology, SAE Publishing, Warrendale, PA, 2019.
7. Society of Automotive Engineers, J1930DA_202105: Electrical/Electronic Systems Diagnostic Terms, Definitions, Abbreviations, and Acronyms Web Tool Spreadsheet, SAE Publishing, Warrendale, PA, 2021.
8. Society of Automotive Engineers, Standard J1410_202012: Air Brake Valve - Performance Requirements, SAE Publishing, Warrendale, PA, 2020.
9. Society of Automotive Engineers, Standard J2627_202207: Braking System Definitions—Truck and Bus, SAE Publishing, Warrendale, PA, 2022.
10. Chan, K.H., Pasco, M., and Cheng, B.H., "Towards a Blockchain Framework for Autonomous Vehicle System Integrity," SAE Technical Paper 11-04-01-0002, 2021, https://doi.org/10.4271/11-04-01-0002.
11. Society of Automotive Engineers, Standard J1703_201909: Motor Vehicle Brake Fluid, SAE Publishing, Warrendale, PA, 2019.

C

CA (Cranking Amps)

Same as cold cranking amperes (CCA), except temperature at which the amperes are measured is 32°F and not 0°F. Cranking amperes is an indication of the ability of a battery to crank an engine for 30 s at a temperature of 32°F without battery voltage dropping below 7.2 V.

CAC (Charge Air Cooler)

A heat-exchanging device like an ICE cooling system radiator engine used to cool the air from the turbocharger, making it denser before entering the intake system. This action will increase horsepower. Air passes through the core of the cooler and, through convection, removes the heat from the air and makes the charge dense. It may be in the form of an intercooler, aftercooler, or charge air cooler.

CAFE (Corporate Average Fuel Economy)

Corporate Average Fuel Economy. A system of fuel economy standards developed in the United States (USA) by the Environmental Protection Agency (EPA) under the Clean Air Act.

Cage

Bearing rollers are held in place by a cage between the inner race or cone and the outer race or cup.

Calibration Gas

A mixture of several gases used to calibrate an exhaust emission gas analyzer.

Caliper

A disc brake system brake application device. The hydraulic brake operation begins when fluid pressure from the piston chamber forces the caliper pistons to maneuver in, applying pressure to the pads and forcing them tightly against the rotor to slow or stop the vehicle. The action of a hydraulic brake caliper is compared to the operation of a C-clamp. When applying a C-clamp, the action of turning the handle clockwise moves the inner clamping surface toward the item and the clamp body moves in the other way. This forces the outer clamping surface against the alternative side of the item. A C-clamp may be accustomed easily and uniformly apply high clamping force.

FIGURE C.1 Brake caliper.

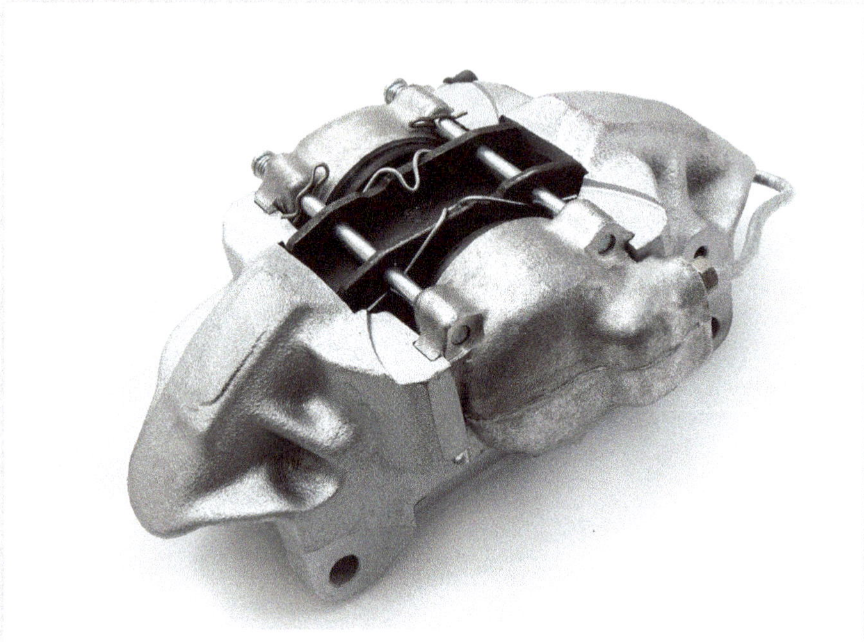

BACHTUB DMITRII/Shutterstock.com.

Calories
A metric unit that expresses the amount of heat needed to raise the temperature of one gram of water by one degree Celsius.

Calorific Value (Fuel)
The heat energy potential of a hydrocarbon fuel measured in BTUs (British thermal units), joules or calories.

Cam Follower
See **Lifter** and **Valve Lifter**.

Camber

Side-to-side vertical tilt of the wheel as viewed from the front of the vehicle and measured in degrees. A positive camber wheel that leans away from the center of the vehicle at the top as viewed from the front of the vehicle. A negative camber wheel leans toward the center of the vehicle. A vehicle with camber pull will pull in the direction of the greatest positive camber. For more information on this, see **J670_200801** [4].

FIGURE C.2 Automotive steering geometry angle called camber, which is the vertical tilt of the wheel as viewed from the front.

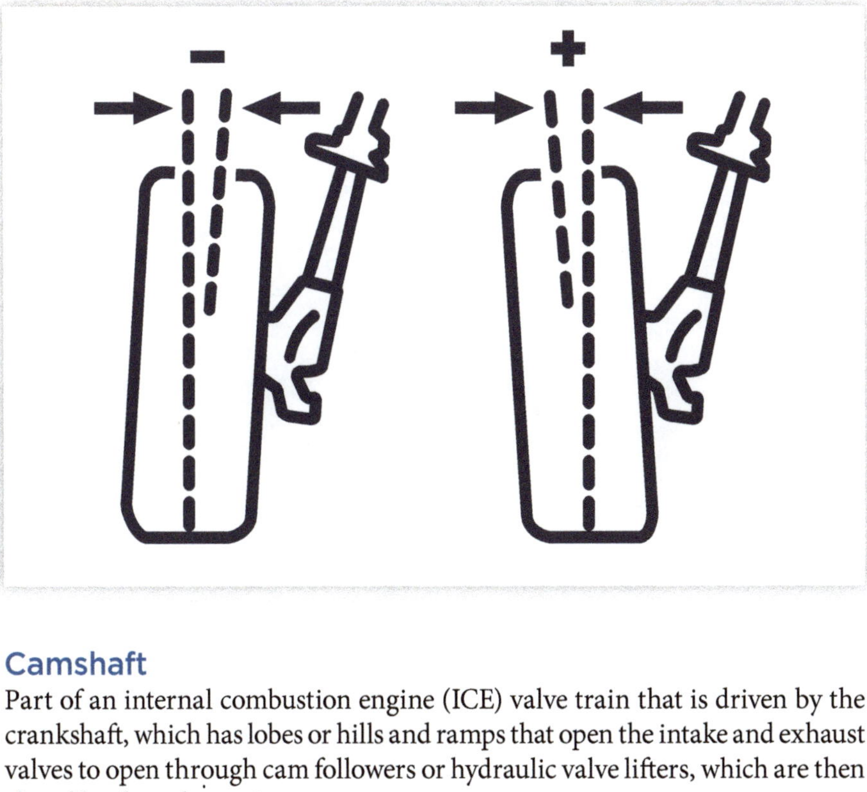

vvushakovv/Shutterstock.com.

Camshaft

Part of an internal combustion engine (ICE) valve train that is driven by the crankshaft, which has lobes or hills and ramps that open the intake and exhaust valves to open through cam followers or hydraulic valve lifters, which are then closed by the valve springs.

Camshaft Gear Backlash

Set the dial indicator up again with the plunger tip resting on the edge of a camshaft gear tooth, in line with (tangential to) the edge of the gear.

Camshaft Gear Runout

To measure the camshaft gear runout, you set up a dial indicator on the engine block with the plunger resting on the face of the camshaft gear just inside the gear teeth and rock it back and forth to measure the clearance between the gears or backlash.

Camshaft Lobe

Peak or high point on the camshaft that will push the cam follower to open the ICE poppet valve. It is opposite the base circle.

FIGURE C.3 Camshaft lobe.

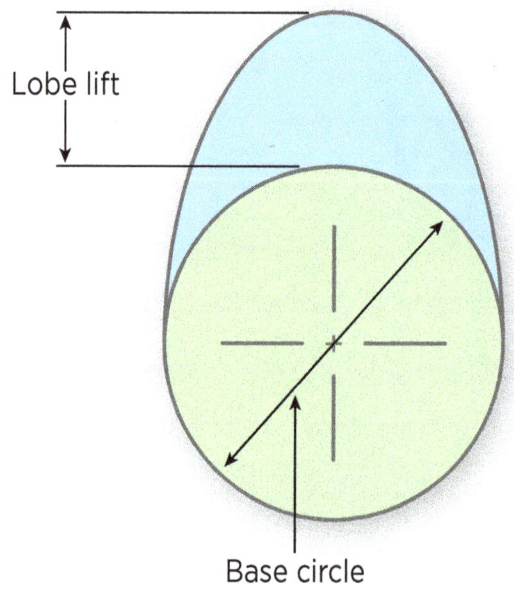

Courtesy of John F. Kershaw Ed.D.

Camshaft Position (CMP) Sensor

A sensor that is mounted in the front cover or block to monitor the rotation position of the engine camshaft. The camshaft turns at one-half crank speed, so the signal generated by this sensor can be used to determine when the number one piston is at top dead center (TDC) on its compression stroke. The computer also uses this signal to determine when a spark should occur at each spark plug. For more information on this, see **J3063_202103** [8].

CAN (Controller Area Network)

"The SAE J1939 documents are intended for light, medium, and heavy-duty vehicles used on or off road, as well as appropriate stationary applications which use vehicle derived components (e.g., generator sets). Vehicles of interest include, but are not limited to, on- and off-highway trucks and their trailers, construction equipment, and agricultural equipment and implements. The purpose of these documents is to provide an open interconnect system for electronic systems. It is the intention of these documents to allow Electronic Control Units to communicate with each other by providing a standard architecture.

This particular document, SAE J1939-21, describes the data link layer using the Classical Extended Frame Format (CEFF) with 29-bit IDs, as defined in ISO 11898-1, December 2015. "For SAE J1939, no alternative data link layers are permitted." The J1939 fault code error message consists of a Source Address (SA) identifying the Electronic Control Module (ECM) sending the DTC (SA0 = Engine Controller #1), a Suspect Parameter Number (SPN) which identifies the parameter sending the fault code error message, and a Failure Mode Identifier (FMI) which identifies the type of fault detected. The SA or Source Address field is the control module sending the message (see Table C.1 in SAE J1939 below) [1].

TABLE C.1 Source address.

SA	Description
1	Engine
3	Transmission
11	ABS

On-Board Diagnostic (OBD) regulations require passenger cars, and light and medium duty trucks, to support communication of a minimum set of diagnostic information to off-board "generic" test equipment. This document supersedes SAE J1979 May 2007, and is technically equivalent to ISO 15031-5 2010, with the addition of new capabilities required by revised regulations from the California Air Resources Board and revised regulations from the European Commission. This document is intended to satisfy the data reporting requirements of On-Board Diagnostic (OBD) regulations in the United States and Europe, and any other region that may adopt similar requirements in the future. This document specifies:

- Message formats for request and response messages,
- Timing requirements between request messages from external test equipment and response messages from vehicles, and between those messages and subsequent request messages,

- Behavior of both the vehicle and external test equipment if data is not available,
- A set of diagnostic services, with corresponding content of request and response messages, to satisfy OBD regulations.

This document includes capabilities required to satisfy OBD requirements for multiple regions, model years, engine types, and vehicle types. Those regulations are not yet final for some regions, and are expected to change in the future. This document makes no attempt to interpret the regulations and does not include applicability of the included diagnostic services and data parameters for various vehicle applications. The user of this document is responsible to verify the applicability of each section of this document for a specific vehicle, engine, model year and region. SAE J1979/ISO 15031-5 specifies diagnostic services and functionally addressed request/response messages required to be supported by motor vehicles and external test equipment for diagnostic purposes which pertain to motor vehicle emission-related data. Any external test equipment meeting the requirements of SAE J1978 is intended to be able to use these messages to retrieve emissions-related information from the vehicle. Each section of this part of SAE J1979/ISO 15031-5 which specifies additional detail to existing sections of ISO 9141-2, ISO 14230-4, SAE J1850, and ISO 15765-4 supersede those specifications. This part of SAE J1979/ISO 15031-5 references the SAE J1979-DA (Digital Annex), which includes all definitions of PIDs, OBDMIDs, TIDs and INFOTYPES. NOTE: SAE J1979/ISO 15031-5 provides the mechanism to satisfy the requirements included in the country-specific regulations and not all capabilities included in this document are required by the country-specific regulations. SAE J1979 is not considered a final authority for interpretation of the regulations, so readers should determine the applicability of capabilities defined in this document for their specific needs. U.S. OBD requirements for 1996 and later model year vehicles. ISO 15031 5 was based on SAE J1979 and was intended to combine the U.S. requirements with European OBD requirements for 2000 and later model year vehicles. Updates for new regulatory requirements from US, China, Europe, and Brazil. This version also has support for the J1979-2 OBDonUDS updates." The SA, or Source Address, field is the control module sending the message (see Table C.1):

- SPN, or suspect parameter number (Table C.2)
- The SPN combines the elements of the J1857 PIDS and SIDS. It is used for multiple diagnosis uses as follows:
  - Identifying the least repairable subsystem that has failed.
  - Identifying subsystems or units that may not have completely failed but may be showing incorrect operation.
  - Identifying a specific event or condition that requires reporting.

- FMI, or failure mode identifier
- The FMI shows the type of failure in the subsystem as identified by the. Conditions can include system events or Status.

TABLE C.2 FMI descriptions.

J1939 Failure Mode Identifiers (FMI) Description	
0	Data valid but above normal operating range
1	Data valid but below normal operating range
2	Data erratic, intermittent or incorrect
3	Voltage above normal or shorted high
4	Voltage below normal or shorted low
5	Current below normal or open circuit
6	Current above normal or grounded circuit
7	Mechanical system not responding properly
8	Abnormal frequency, pulse width, or period
9	Abnormal update rate
10	Abnormal rate of change
11	Failure mode not identifiable
12	Bad intelligent device or component
13	Out of calibration
14	Special instructions
15	Reserved for future use

- PGN, or parameter group number
- The PGN contains both the FMI information, SA, and SPN, which is a larger J1939 message and thus called the PGN. The PGN contains commands, data, requests, acknowledgments, and DTCs. This serial data is sent over the CAN bus. Each PGN will contain a different set of SPNs.

TABLE C.3 SPN and PGN.

SPN	Description	PGN
102	Turbo boost	65270
247	Engine hours	65253
190	RPM	61444
110	Engine coolant temperature	53262
173	Exhaust gas temperature	65270
175	Oil temperature	65262
84	Vehicle speed	65265

- OC, or occurrence count
- The OC, or occurrence count, is the number of times an SPN/FMI has occurred.

For more information on this, see **J1939/21_2021** [1].

Cantilever Axle

A cantilever axle is a rigid unit that extends horizontally and is supported at only one end.

Capacitance

Term used to measure or describe how much charge can be stored in a capacitor (condenser) for a given voltage potential difference and measured in farads or microfarads.

Capacitor

A solid-state device that opposes a change in voltage. The property of opposing voltage change is termed "capacitance," which is additionally accustomed to describe the electron storage capability of a capacitor. A charged capacitor can deliver its stored energy similar to a battery. When capacitors provide electricity, we say they discharge. A capacitor can power a circuit for a brief period of time.

Capacitor Capsule MAP Sensor

MAP sensor that has one or two silicone chips mounted to form a sealed pressure chamber. The chips form diode-resonance circuits that change frequency and capacitance when deflected or moved by pressure changes. As engine intake manifold pressure in the capsule increases or decreases with load, it causes the diode resonance circuits in the sensor to change capacitance in picofarads.

Carbon

A chemical element using the symbol C and the atomic number 6. It contains four electrons available to form covalent chemical bonds and is part of all fossil fuels.

Carbon-Core Wire

Wire that has a constrained amount of resistance designed directly into the metal conductor to reduce radio interference or static from the high-voltage pulses going to the spark plugs. Also called secondary resistance wire. For more information on this, see **J139_202002** [2].

Carbon Dioxide (CO_2)

A colorless gas with a density of approximately 60% higher than dry air. It consists of a carbon atom double bonded to two oxygen atoms. It occurs naturally in Earth's atmosphere as a trace gas. Natural sources include volcanoes, hot springs, and geysers, and it is freed from carbonate rocks by dissolution in water and acids. CO_2 is soluble in water and occurs naturally in groundwater, rivers and lakes, ice caps, glaciers, and seawater. It is present in deposits of petroleum and natural gas. CO_2 is odorless at normally encountered concentrations. CO_2 is produced by animal respiration, which includes humans and all mammals.

Carbon Monoxide (CO)

A colorless, odorless, and tasteless flammable gas that is slightly less dense than air and a by-product of the incomplete combustion of a hydrocarbon such as gasoline or diesel fuel. CO is partially burned diesel fuel that combines with O_2 (oxygen) to form CO_2. CO consists of one carbon atom and one oxygen atom.

Carbon Residue

The amount of carbon residue left within the combustion chamber has a direct bearing on the engine deposits and cleanliness of combustion; therefore, the smaller the amount of carbon residue at the end of the combustion process the longer the engine life will be and the cleaner the exhaust smoke. The amount of carbon in a fuel is determined by burning a given quantity in a sealed container until all that remains is carbon residue. Carbon residue is expressed as a percentage by weight of the original sample of the fuel oil.

Carburetor

The carburetor mixes the right quantity of gasoline with air. If there is not enough fuel mixed with the air, the engine runs lean and hot and could damage the engine. If there is too much fuel mixed with the air, the engine runs rich and either will not run, or will flood in (producing black smoke), which is fuel burned without air. This causes stalling easily, fouled plugs, increased cylinder wall wear, and wasted fuel.

FIGURE C.4 Carburetor.

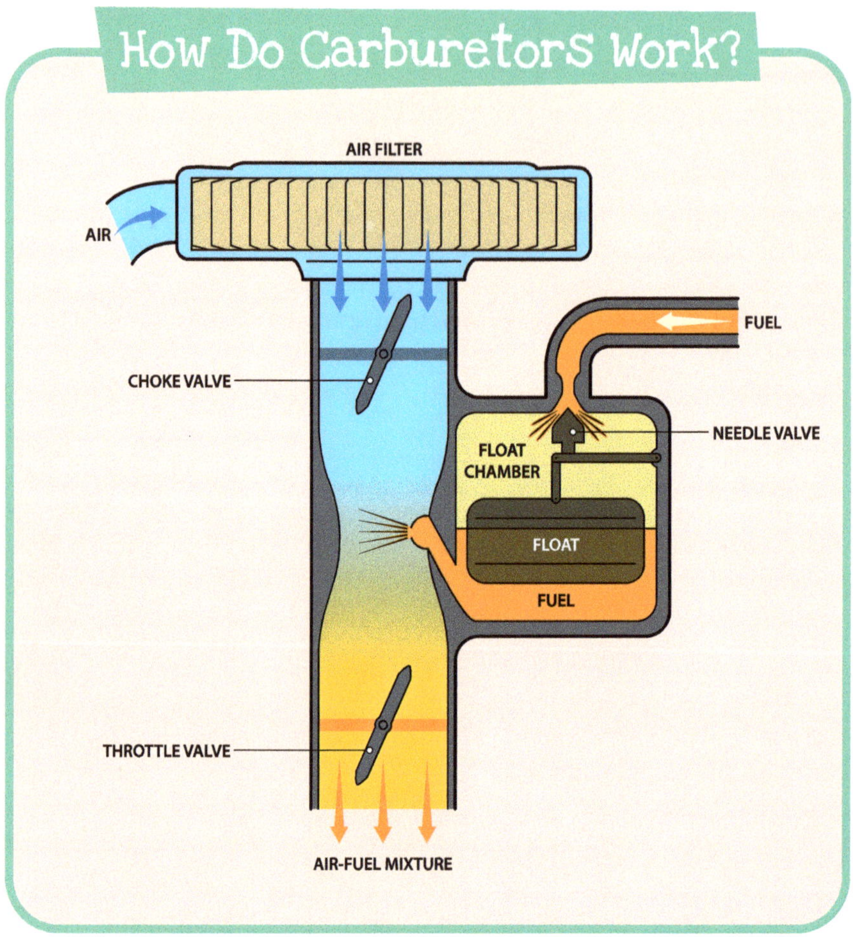

EreborMountain/Shutterstock.com.

Carburizing

A method of case-hardening low-carbon steel in which the metal component is heated above its ferrite-austenite transition in a suitable carbonaceous atmosphere.

Carcass

The main underbody of an automotive tire that the belts and tread are mounted and bonded to. For more information on this, see **J2047_201911** [3].

Carcass Plies

Main framework, or body of an automotive tire. For more information on this, see **J2047_201911** [3].

Case-Hardening

Case-hardening is the hardening of the surface of a metal component while allowing the metal to stay soft underneath, thus forming a skinny layer of harder metal at the surface. When hardening iron or steel with low-carbon content, this process involves infusing additional carbon or nitrogen into the surface layer. This process is completed after the component has been formed into its final shape, but can even be done to extend the hardening element content of bars to be utilized in a pattern welding or similar process.

Case-hardening is desirable for metal components that are subject to sliding contact with hard or abrasive materials because the hardened metal is more immune to surface wear. However, because hardened metal is sometimes more brittle than softer metal, through-hardening is not always the best choice. Case-hardening can produce a component that may not fracture because of the soft core and may absorb stresses without cracking but also provides adequate wear resistance on the hardened surface.

Case/Model-Based Diagnostic Reasoning

Compares system and component performance to expected performance. This is in addition to DTC (diagnostic trouble code) detection and isolation. This form of diagnosis analyzes and categorizes the problem using advanced algorithms. System problems are identified without meddling with system operation by substituting the suspect data from a sensor or component using a default data set.

Caster

The forward or backward tilt of an imaginary line drawn through the steering axis as viewed from the side of the vehicle when compared to the true vertical line. It is this imaginary line that the wheel turns around when the wheel is turned to the side. Caster is positive if the axis is leaning rearward and is negative if the axis is leaning forward. Caster is zero when the steering axis is straight up or down. Caster is measured in degrees. There is a small amount of positive caster designed into all vehicles. For more information on this, see **J670_200801** [4].

FIGURE C.5 Caster is an automotive steering geometry angle that is the forward or rear tilt of the front wheels through an imaginary line drawn through the center of the wheel viewed from the side.

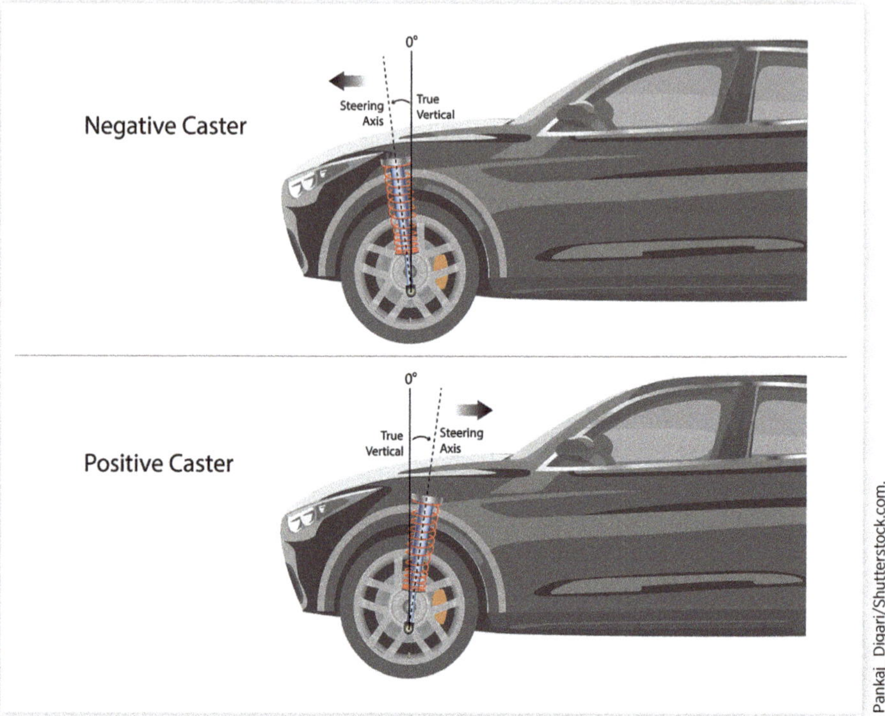

Caster Sweep

A process used to measure caster during a four-wheel alignment adjustment and measuring process where the front wheels are rotated inward about 22° and then outward to make the measurement. For more information on this, see **J670_200801** [4].

Catalyst

A catalyst is a substance that increases the rate of a chemical reaction without itself undergoing a chemical change. Diesel exhaust catalysts speed up the chemical reactions that break down noxious exhaust emissions into harmless by-products and allow reactions to take place at lower temperatures. This is important because diesel engines have low exhaust temperatures and this additional burning helps the conversion of emissions into harmless gases. Diesel catalytic converters can temporarily store emission gases until correct conditions have been met to chemically break them down before going into the diesel particulate filter (DPF). The diesel oxidation catalyst (DOC) has to be near

500°F before it starts to work. Some systems use two DOCs, a close-coupled DOC that is in front of the SCR (selective catalytic reduction) unit and the other DOC is connected to the DPF.

Catalytic Converter

An emission control device located in the exhaust system that changes HC and CO into harmless H_2O and CO_2. In a three-way catalyst, NOx is also divided into harmless nitrogen (N_2) and oxygen (O_2). For more information on this, see **J1145_201109** [10].

Catalytic Substrate

Part of a ceramic converter consists of a ceramic base covered with a wash coat of material containing precious metal (e.g., platinum, palladium). Chemical reactions between the noxious emissions and the metals convert the emissions to nitrogen, water, and carbon dioxide. Metal foil catalysts alloyed with precious earth metals are also used. These catalysts have the capability to be electrically heated.

Cat-Back Exhaust System

A cat-back exhaust system includes a new Y-pipe, intermediate pipe, muffler, and tailpipe.

Catch Tank

An engine coolant tank that uses plumbing that goes into the top of the tank and does not have a hose that goes to the bottom. A catch tank collects or catches expelled coolant to be drained from that tank at a later point. While a recovery, reservoir, expansion, surge, or catch tank holds excess coolant, the recovery, reservoir, expansion, surge tanks will automatically put the coolant back in the system where the catch tank will hold the coolant until emptied.

Cathode

The negative electrode in a solid-state device like a diode.

Cathode Ray Tube (CRT)

An electron beam tube with a cathode at one end and an anode at the screen end. A shot of electrons from the cathode to the anode creates a pattern on the screen.

Cavitation Erosion

Cavitation is the formation of vapor cavities in a liquid, which occurs when the pressure within the water pump drops causing vapor-filled cavities to form. These are bubbles or cavities that are the result of forces acting upon the coolant when it is subjected to rapid changes of pressure. When these vapor-filled bubbles get to areas of higher pressure on their way through the impeller, they collapse or implode with a shock on the impeller vanes. This is followed by a sudden rush of coolant into the voids created by the collapsing bubbles, resulting in mechanical damage. Low pressure can be due to various flow parameters, such as coolant viscosity, temperature, pressure, and flow nature.

FIGURE C.6 Cavitation.

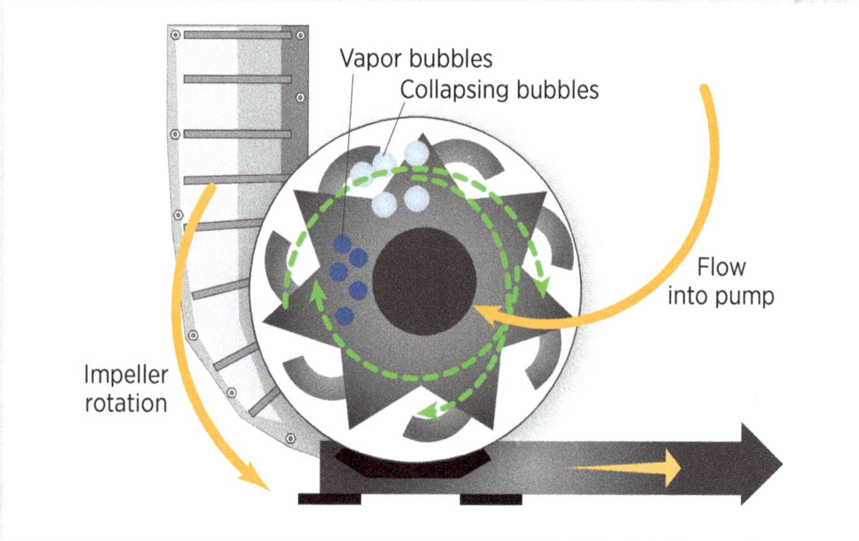

Courtesy of John F. Kershaw Ed.D.

CCA (Cold Cranking Amps)

A rating that indicates how much current the battery can deliver for 30 sec at 0°F (–17.8°C) while maintaining a terminal voltage of 7.2 V (1.2 V per cell).

CCC (C3)

Computer command control (C3) is the name of the General Motors (GM) computer control system that uses an electronically controlled carburetor with a mixture control solenoid using 10% duty cycle.

CCM (Comprehensive Component Monitor)

An internal program used on OBD II (On-Board Diagnostics Generation II) system in the engine management computer that is designed to monitor a failure in any electronic component or circuit (including emission-related and non-emission-related circuits) that provide input or output signals to the engine management computer. For more information on this, see **J1930DA_202105** [6].

CDI (Capacitor Discharge Ignition) System

A solid-state version of the magneto ignition with the breaker points being replaced by analog or solid-state transformer. It is found in some racing vehicles like NASCAR. The CDI system does not have the long charging times found in high-inductance automotive induction ignition coils. So it absolutely is more suitable for small engines and racing engines. Capacitor discharge ignition systems use the discharge current output from a high-voltage capacitor to leap the electrodes in spark plugs. For more information on this, see **J139_202002** [2].

FIGURE C.7 Capacitor DC type.

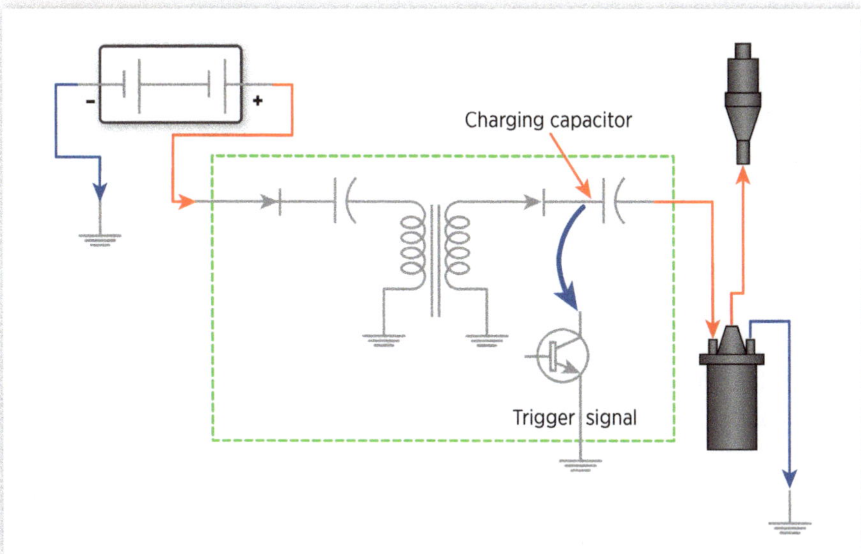

Courtesy of John F. Kershaw Ed.D.

CDR (Crankcase Depression Regulator)

The CDR valve operates by engine vacuum or pressure differential. Inside the CDR valve a spring holds open a valve plate that connects the body with a flexible diaphragm. The valve plate can restrict the outlet passage to the air inlet duct when pressure differential pulls it closed against the spring force. At idle, diesels do not produce significant vacuum (pressure differential) so the airflow past the CDR outlet passage will not be great enough to close the CDR valve, so it is fully open. The CDR valve should be fully open because the crankcase pressure potential at idle is approximately one inch of water and because, at slow speeds there is more time for the air to leak past the rings. At higher engine speeds, the valve closes to provide more restriction. This action prevents the movement of oil vapors into the intake manifold by limiting the crankcase vacuum (measured at 2,000 RPM to be from 2 to 5 in. of water). If you did not restrict the CDR valve, it would attempt to depressurize the crankcase and suck all of the oil out of it.

CEEMAT (Converter Enhanced Electronically Managed Automated Transmission)

Designed originally for the military by Eaton. An electronically controlled manual twin shaft transmission that connect to the ICE with a torque converter and lock up clutch. It is no longer in production.

Cell

A lead-acid battery contains a number of individual cells. These cells are made up of alternating positive and negative plates. Between each plate is a separator that keeps the plates from touching but will let electrolyte (water and sulfuric acid) pass back and forth between them.

CEMF (Counterelectromotive Force)

An induced voltage that opposes the source voltage and any change (increase or decrease). The CEMF is additionally called a back EMF. It is a voltage that opposes the change in current which induced it. It is caused by magnetic induction (see Faraday's law of induction, electromagnetic induction, Lenz's Law).

Center Bearing

Center bearings are used on truck drivelines to support the center portion of the driveline when two or more propeller shafts are used. Driveshafts longer than 65 in. (165 cm) require a center support bearing, or carrier bearing for additional support. The center support bearing is a ball type. The bearing is mounted in a rubber cushion that is attached to a frame crossmember.

FIGURE C.8 Truck driveline center support bearing.

Center Bolt

A bolt used to hold the leaves of a leaf spring together in the center. Also called a centering pin.

Center of Gravity

Single point where all of the vehicle weight is centered. If the vehicle could be balanced from its center of gravity, it would hang from that point in perfect equilibrium. The center of gravity is always above the ground and between the tires. Center of gravity location can be measured in three planes. It can be measured along the lateral axis (the track width), longitudinal axis (wheelbase), and the vertical axis (height above the ground). The center of gravity is always located above the road surface and between the tires. When a vehicle is cornering, this is the point through which all centrifugal force is assumed to act. Its position is determined by the load carried by the front and rear wheels—that is, by how weight is distributed. In a typical rear-wheel-drive vehicle, the weight distribution is approximately 60% fore and 40% aft; 60% of the weight is carried on the front wheels, 40% on the rear, and the center of gravity is closer to the rear than the front. On a typical front-wheel-drive vehicle, the weight distribution is approximately 75% fore and 25% aft. If the lateral weight distribution is unequal, the vehicle can experience unequal handling characteristics when turning in one direction compared to the other direction. For more information on this, see **SAE J1594_201007** [5].

Centerline Steering

A term that describes the position of the steering wheel while driving on a straight, level road. The wheel should be centered or within ±3° as specified by many OEMs.

Centerlink (Relay Rod or Draglink)

The centerlink (relay rod/drag link) moves to the left and right under the vehicle because the steering wheel turns. The centerlink is connected between the pitman arm and also the idler arm. The connections between the pitman arm and idler arm are usually ball-and-socket type; therefore, the centerlink can swivel and move through various angles. Tie rods are attached to the centerlink with similar ball-and-socket-type joints. When the centerlink moves, it causes the tie rods to maneuver the wheel assembly.

Central Processing Unit (CPU)

Microprocessor inside the engine management computer that receives input sensor information, compares the input with information stored in memory, performs calculations, and makes output decisions to the output actuators.

Centrifugal Force

Centrifugal force is the natural tendency of objects, when forced to move in a curved path, to move away from the center of rotation as a result of centripetal force, which is the force directed toward the center. See **Centripetal Force** for formula.

Centrifugal (Mechanical) Advance

Spinning weights that move the distributor cam ahead, advancing the ignition timing in proportion to the engine speed. They react to centrifugal/centripetal force, moving the distributor cam or pole piece.

Centrifugal Oil Filter

An oil-driven rotor inside the oil filter metal housing. The oil flows into the housing, and the oil flow spins both the rotor and the oil. Centrifugal force pushes the heavy contaminants onto the sides of the housing, and the clean oil flows out.

Centrifugal Supercharger

A centrifugal supercharger is similar to a turbocharger but is mechanically driven by the engine instead of being powered by the hot exhaust gases. Centrifugal or spiral-type supercharger has internal compression like a water pump.

Centripetal Force

A force making a body follow a curved path with a direction at a right angle or orthogonal to the body motion and toward the fixed point of the middle of path curvature. This force results in the reactive centrifugal force. Centripetal force is determined by the following formula:

$$F = m\frac{v^2}{r} \quad \text{(C.1)}$$

where

m is the mass
v is velocity
r is radius
F is centripetal force

Cetane

Property of diesel fuel that is a measure of the ignition quality of the fuel relative to a reference fuel mixture composed of cetane and alpha-methylnaphthalene, the percentage, by volume, of cetane in the mixture being the cetane number (CN). CCI stands for calculated cetane index. High CNs specify good ignition quality with a short delay period, and low CNs specify poor ignition quality with a long delay period, white smoke, and misfires. CNs are obtained using a single-cylinder variable-compression-ratio engine similar to determining the octane number of gasoline. The ignition quality of an unknown fuel is matched to a reference fuel. Cetane is a colorless liquid with excellent ignition qualities and is rated at 100. ASTM defines the CN as a percentage by volume of a test fuel containing cetane at a CN of 100 mixed with alpha-methylnaphthalene or heptamethylnonane that has a poor ignition quality CN of 0. The CN is the percentage of cetane and the zero ignition chemical to produce a similar ignition quality measured against the fuel being tested. In other words, a mixture consisting of 45% cetane and 55% alpha-methylnaphthalene has a CN of 45. Numbers 1 and 2 diesel fuel that are used on the highway have a minimum CN of 40.

Cetane Improvers

Fuel additives that will increase the CN value of a fuel. These chemicals should only be used when the OEM (original equipment manufacturer) recommends their use.

CFR (Cooperative Fuel Research) Engine

This is a research ICE with a variable compression ratio that can be adjusted to cause any fuel being tested to detonate, or explode, rather than burn evenly. The fuel tested is run in a CFR engine and the compression ratio is adjusted until the fuel detonates. This engine is used for testing, research, and instruction in the performance of fuels and lubricants for the ICE.

Chain and Sprocket Drive

This type of engine drive mechanism consists of a chain and two sprockets. The sprockets are toothed wheels, over which a single or double roller Morse chain is assembled, that allows the crankshaft of an ICE to turn the camshaft.

FIGURE C.9 ICE chain and sprocket drive.

Courtesy of John F. Kershaw Ed.D.

Chain Tensioner

A spring loaded or hydraulic device that applies a force to create or maintain tension on the camshaft drive chains.

Chamfered Edge

A chamfered edge is another name for beveled edge that allows the easy entry of a shaft into a bore.

Chapman Strut

A type of MacPherson strut independent rear suspension used for sports and racing cars. The name comes from Colin Chapman of Lotus.

Charles's Law

Charles's Law states that a volume of an ideal gas is directly proportional to the absolute temperature at constant pressure. Kelvin temperature and the volume will be in direct proportion when the pressure exerted on a sample of a dry gas is held constant. Charles's Law formula is determined as follows:

$$V \propto T \quad \frac{V}{T} = k \tag{C.2}$$

Chassis

Chassis is the frame, suspension, steering, and machinery of a motor vehicle.

Chassis Ground

A method of grounding a circuit that uses the metal part of a component to attach to ground.

Check Engine Light

A dashboard warning light that is controlled by the vehicle engine or body management computer; since the adaption of SAE standard J-1930, it has been referred to as MIL, or malfunction indicator light. Some vehicle manufacturers refer to this light as the service engine soon (SES) light.

Check Valve

A valve that permits flow in only one direction.

Chlorofluorocarbon (Freon Refrigerant)

Chlorofluorocarbons are fully or partly halogenated paraffin hydrocarbons that contain only carbon, hydrogen, chlorine, and fluorine and commonly known as the DuPont brand name Freon. Freon was formerly used as a refrigerant in air conditioning and invented by Charles Kettering of GM.

Choke

A butterfly valve located in the top of the air horn of a carburetor that when partially closed creates a low-pressure area causing additional fuel to flow out of the main jet to enrichen the fuel mixture for cold starting. This is because cold air does not absorb the moisture in cold air so the additional gasoline will absorb this moisture.

CHT (Cylinder Head Temperature) Sensor

A sensor that measures cylinder head temperature.

Circuit

A circuit is a closed electrical path for a current. For current to flow continuously from a voltage supply, like a battery, there must be a whole circuit or path. Current flows from one end of a circuit to the opposite when the ends are connected to opposite charges (positive and negative). We usually call these ends "power" and "ground." Current flows only in an exceedingly closed or completed circuit. If there is an opportunity somewhere within the circuit, current cannot flow. We usually call a prospect in each circuit as open. Every automotive circuit contains a source of power, conductive material (wires) load, controls, and a ground. These elements are connected to each other with conductors, like a copper wire. The first power source in a car or truck is the battery. As long as there is no external connection between the positive and negative sides, there is no flow of electricity. Once an external connection is connected between them, the free electrons have a path to flow on. The electrical circuit is connected between the positive and negative sides.

FIGURE C.10 Circuits contain a power source, path, control switch, and load.

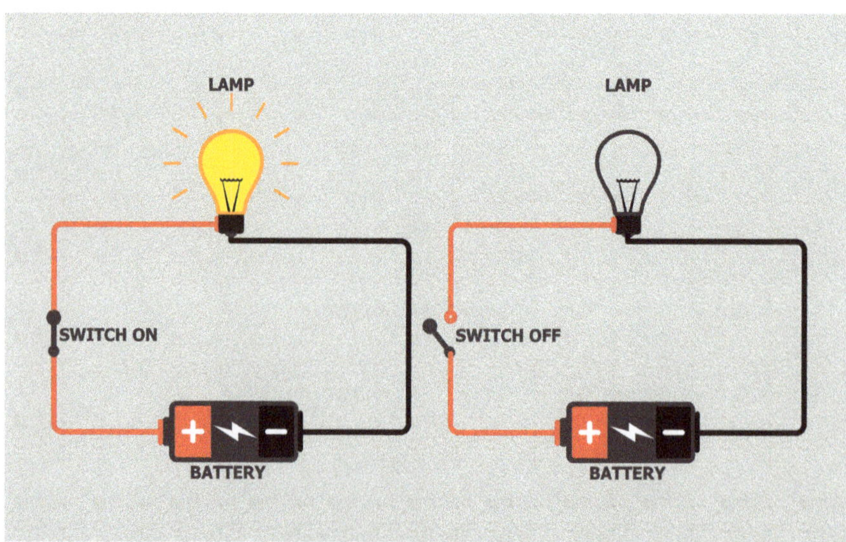

BijanStock/Shutterstock.com.

Circuit Breaker

Circuit protection device consisting of a contact point set attached to a bimetallic strip. The bimetallic strip will heat and bend as current flows. Unlike a fuse, it does not blow out.

Circuit Number

The number, or number and letter, that manufacturers use to identify a specific electrical circuit in a diagram.

Circuit Resistance Tests

Tests used to locate wiring problems in a charging system such as a loose connection, corroded terminal, partially burned wire, or other similar types of troubles.

Circuit Sensing

A system that uses a computer instead of dedicated sensors to monitor components and circuit operation.

Circular Mil

Circular mil equals the area of a circle that has a 0.001-in. diameter.

Circumference

The distance around a closed curved.

CKP (Crankshaft Position) Sensor

Provides a signal to the engine management computer that can be used as a reference to calculate RPM, which tells the computer that the engine is cranking and/or running. It also identifies the cylinder position for sequential fuel injection and ignition and on systems using the sequential COP (coil-on-plug) ignition. It is used in conjunction with a reluctor wheel with a number of teeth with one missing tooth. It has a tooth a specific number of crank degrees. This crank trigger is needed for ignition systems. For more information on this, see **J1930DA_202105** and **J1145_201109** [6, 10].

FIGURE C.11 CKP sensor.

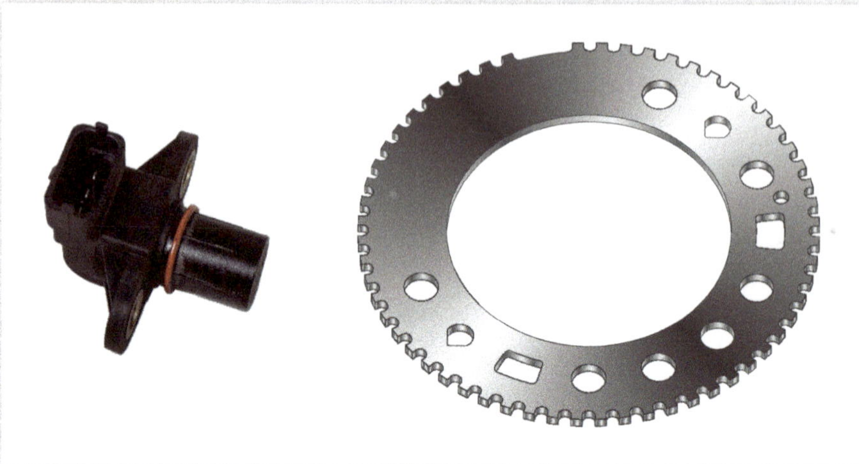

Aleksandr Kondratov/Shutterstock.com.

Fouad A. Saad/Shutterstock.com.

Class 2 Serial Communications

An electronic data transfer medium that toggles the line voltage from 0 to 7 V, with 0 being the rest voltage, and by varying the pulse width.

Class B Data Communications Network Interface

Class B Data Communications Network Interface is used for OBD II communications. "The term 'test tool' is synonymous with OBD II Scan tool. This technical report is being stabilized because it covers technology, products, or processes which are mature and not likely to change in the foreseeable future. This SAE Standard establishes the requirements for a Class B Data Communication Network Interface applicable to all ON/OFF Road Vehicles. It defines a minimum set of data communication requirements such that the resulting network is cost effective for simple applications and flexible enough to use in complex applications. Taken in total, the requirements contained in this document specify a data communications network that satisfies the needs of automotive manufacturers. This specification describes two specific implementations of the network, based on media/Physical Layer differences. One Physical Layer is optimized for a data rate of 10.4 Kbps while the other Physical Layer is optimized for a data rate of 41.6 Kbps.

The Physical Layer parameters are specified as they would be detected on the network media, not within any particular module or integrated circuit implementation. Although devices may be constructed that can be configured to operate in either of the two primary implementations defined herein, it is expected that most manufacturers will focus specifically on either the 10.4 Kbps implementation or the 41.6 Kbps implementation depending on their specific application and corporate philosophy toward network usage. However, low-volume users of network-interface devices are expected to find it more effective to use a generic interface capable of handling either of the primary implementations specified in this document. This SAE document is under the control and maintenance of the Vehicle Networks for Multiplexing and Data Communications (Multiplex) Committee. This committee will periodically review and update this document as needs dictate." For more information on this, see **J1850_201510** [16].

Clearance Volume

Cylinder volume not swept by the piston and enclosed in the copper head gasket. It equals 0.785 times bore diameter squared times the deck clearance. The combustion chamber volume is measured using two different processes to measure the combustion chamber size.

C-Lock Axle

An automotive rear axle that is secured in the differential housing with a C-lock installed at the differential end of the axle shaft. The C-lock is named because it is shaped like the letter "C."

Clock Generator

A quartz oscillator of a computer that generates a gentle stream of one-bit-long voltage pulses. Both the microprocessor and, therefore, the memories monitor the clock pulses while they are communicating. Because they understand how long each voltage pulse should be, they are between a 01 and a 0011. To finish the method, the input and output circuits also watch the clock pulses.

Clock Spring

Electrical current is provided to the supplemental restraint system (SRS), also called an air bag, through the spiral cable, which is also known as a coil or clock spring. This is a contact strip that maintains electrical contact when the steering wheel is turned though it path. For more information on this, see **J1538_201504** [11].

Closed Loop

The OBD fuel system mode in which the engine management computer reads and responds to feedback signals from an exhaust gas oxygen sensor, changing the pulse width or the time of the fuel injectors or mixture control solenoid to keep the fuel ratio stoichiometric. For more information on this, see **J1930DA_202105** [6].

Closing Ramp

The ICE camshaft lobe that opens the intake and exhaust valves has two sides or ramps, a closing and an opening ramp or side. The closing ramp is where the cam follower or lifter rides on to close the intake or exhaust valve.

Cloud Point

A diesel fuel property, that is, the temperature at which the wax crystals in the paraffin base diesel fuel begins to settle out with the result that fuel filter plugging can take place. It is called cloud point because when these wax crystals become big enough, they will cause the fuel to appear cloudy. Cloud point generally occurs 5 to 8°C (9–14°F) above the pour point.

Clutch

A powertrain device used to connect two rotating components. You need a clutch because the engine spins all the time, and the car wheels do not. You need to disengage the transmission from the engine when you stop the vehicle without stopping the engine. A clutch allows you to smoothly engage a running engine to a non-spinning transmission by controlling the slip between the engine and transmission.

The flywheel is connected to the engine, and the clutch plate is connected to the transmission. When your foot is off the pedal, the springs push the pressure plate or clutch cover assembly against the clutch disc, which in turn presses against the flywheel. This locks the engine to the transmission input shaft, causing them to rotate at the same speed. The amount of force the clutch can hold depends on the friction between the clutch plate and flywheel and how much force the spring puts on the pressure plate or clutch cover. There are push-type clutches as shown in C12 and a pull-type, where the release bearing is pulled rather than pushed in C12A. For more information on this, see **J1479_201211** [7].

FIGURE C.12 Push-type clutch (left) and clutch brake (right).

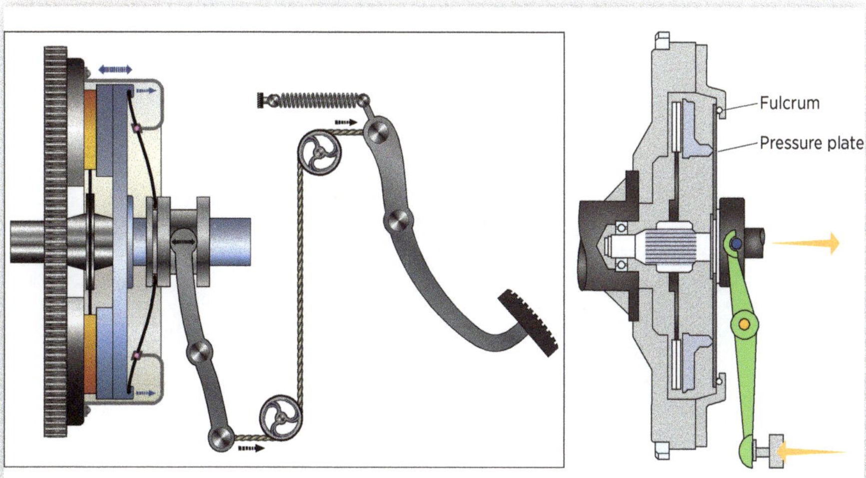

Clutch Brake

Most heavy-duty non-synchronized manual transmissions use a clutch brake that is splined to the transmission input shaft. The clutch brake is a disc, with friction material, that slides on the input shaft. It is squeezed between the release bearing and the transmission front cover during clutch release to stop the rotation of the input shaft. It is used with non-synchronized transmissions to allow shifting into a gear while the vehicle is not moving. It can stop the rotation of the input shaft so that a clash-free shift can be made when the vehicle is at a stop and the engine running. The clutch brake is designed to come into contact with the machined pad at the front of the transmission housing. For more information on this, see **J1479_201211** [7].

FIGURE C.13 Clutch brake.

Dim	MM	IN	Dim.	MM	IN
A	187.5	7.38	X**	5.1	.20
B	349.3	13.75	Y	120.7	4.75
d*	10.3	.406	Z	73	2.90
F	393.7	15.50	A-A	65	2.56
G	63.5	2.50	B-B	147.4	5.81
H	374.7	14.75	C-C**	31.8	1.25
J	74.7	2.94	D-D	124	4.88
L	368.3	14.5	E-E		6
N	12.7	.50	F-F	14.5	.57
P	26.9	1.06	G-G	12.7	.50
R	412.8	16.25	H-H	10.4	.41
T	106.7	4.28	J-J**	4.1	.16
U	58.7	2.31	K-K	19.1	.75
V	.8X45°	.03X45°	L-L	10.03	.395
W	177.8	7.0	M-M**	13.2	.52

Applicable with most
SAE housing size #1 and 2

Housing variations may affect fit.

* For reference only. For exact dimensions consult SAE J1857 flywheel recommended practice. Clutch dimensions should allow for proper allowance.

** Maximum

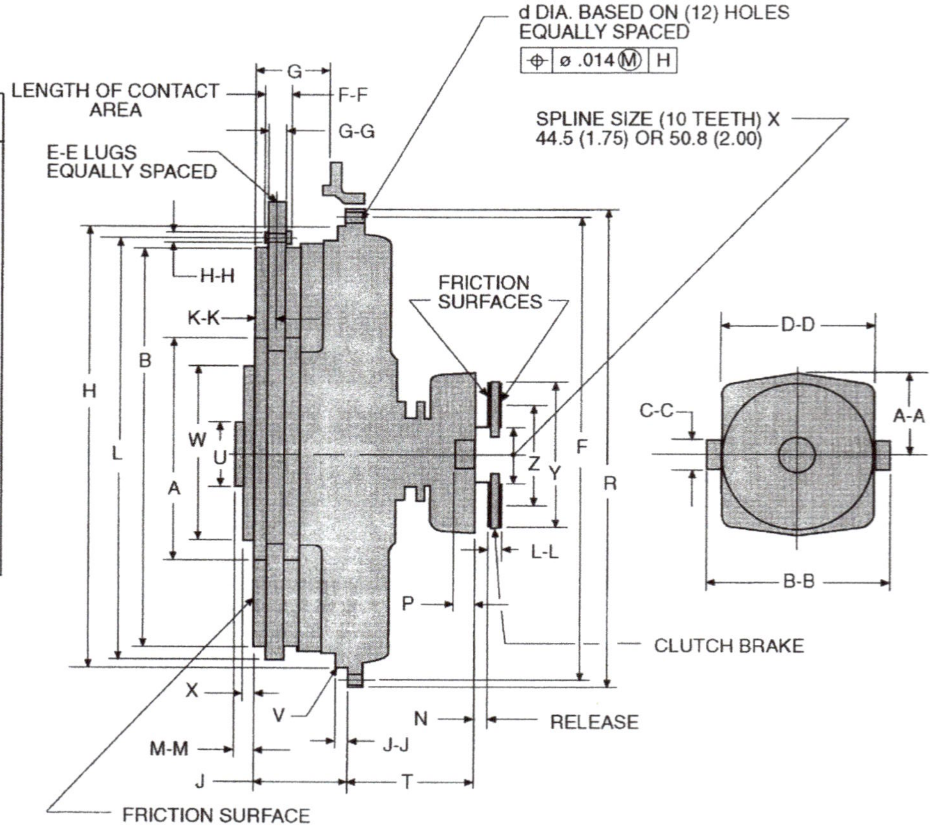

Clutch Disc

The friction disc is a thin piece of steel sandwiched between two layers of friction material. The friction material is the wearable portion of the disc that has a high coefficient of friction and holds onto the flywheel or intermediate plate when clamping force is applied to resist slipping and loss of power. The friction material, either organic or ceramic, is bonded by glue or riveted to the steel. On a riveted disc, the rivet heads must be below the surface of the friction material so that they do not rub on the flywheel or pressure plate. Brass rivets are used because brass is softer than the cast iron flywheel and pressure plate. Even if the friction material wears down far enough to let the rivets touch the flywheel or pressure plate, the rivets wear away instead of damaging the flywheel or intermediate plates. Grooves cut into the surface move dust from worn friction material away from the clutch operating surface. Another design has small pieces of friction material riveted or bonded to cushion springs, also called Marcel springs. Rivets hold the cushion springs to the steel disc. The cushion springs can flex slightly as the clutch engages and disengages.

Breaking the friction material into segments has several advantages. The cushion springs smooth out clutch engagement while the flexing motion helps break any vacuum between disc and flywheel or pressure plate. Small gaps between segments provide better cooling. Clutch friction material can be organic or ceramic facings. Ceramic facings have an increased tolerance to heat and are therefore more commonly found on heavy-duty truck applications. ICE input torque >500 lb/ft (670 Nm) requires the use of a multiple-disc clutch. Multiple-disc clutch systems have two or more driven friction discs that are separated by intermediate driving plates between the pressure plate and flywheel. While single-disc systems work well on cars and light trucks and some medium duty trucks, a single-disc system would fail quickly under the increased load of high-torque diesel engines.

FIGURE C.14 Single clutch (left) and dual disc truck clutch (right).

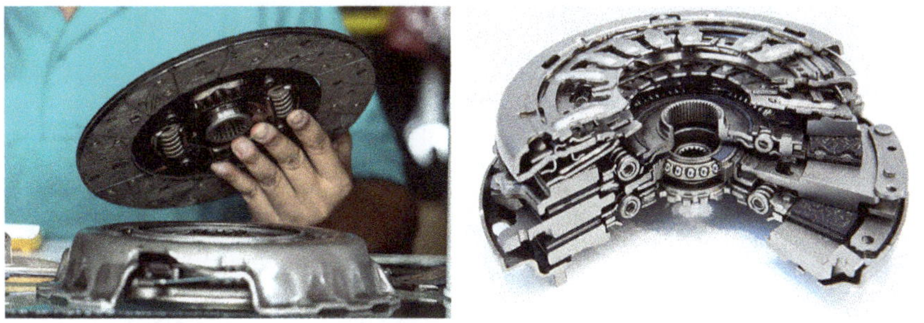

May_Chanikran/Shutterstock.com.

Sergey Ryzhov/Shutterstock.com.

CNG (Compressed Natural Gas)

CNG, or compressed natural gas, is a combustible fuel in a gaseous from made of gasoline which is mainly composed of methane (CH_4), compressed to less than 1% of the volume at standard atmospheric pressure of 14.7 psi. It is stored at a pressure of 20–25 MPa (2,900–3,600 psi).

CO (Carbon Monoxide)

Carbon monoxide, which carries the chemical formula CO, is a colorless, odorless, and tasteless flammable gas that is slightly less dense than air and a by-product of the incomplete combustion of hydrocarbon such as gasoline or diesel fuel. CO is partially burned diesel fuel that combines with O_2 (oxygen) to form CO_2 (carbon dioxide). CO is composed of one carbon atom and one oxygen atom. It is not a major concern in diesel engine emissions.

Cocktail Shaker Piston

Combustion heat transfer that takes place in a reciprocating hollow piston partially filled with engine oil.

COE (Cab-Over-Engine)

Cab-over, or cab-over-engine (COE), cab forward, or forward control (UK), is a truck style with a vertical front or flat face with the cab of the truck sitting above or forward of the engine and front axle.

Coefficient of Friction

The amount of resistance that exists between two surfaces. The formula of friction: F = N (normal force) × μ (coefficient of friction constant). Using the symbol μ, it is a measure of how effortlessly one object moves in relation to another object and the ratio of the force of friction to the normal force found using the formula:

$$\mu = \frac{f}{N} \tag{C.3}$$

where

f is the force of friction
N is normal force

Tire quality has more of an effect on braking. A higher coefficient of friction exists between a given road surface and a tire that has an acceptable amount of tread on it. A lower coefficient exists when a tire with little or no tread.

TABLE C.4 Tire and road conditions.

Road condition	Tire condition	Resultant coefficient of friction
Dry pavement	New tire	1.0 (highest)
Dirt road	New tire	0.9
Dry pavement	Old, worn tire	0.8
Dirt road	Old, worn tire	0.7
Gravel	New tire	0.6
Gravel	Old, worn tire	0.5
Wet road	New tire	0.4
Wet road	Old, worn tire	0.3
Ice	New tire	0.2
Ice	Old, worn tire	0.1 (lowest)

Cogged Belt

A drive belt with cogs or slots that run perpendicular to the belt length. These slots reduce the belt bending resistance.

Coil Spring

A spring steel rod wound in a spiral shape.

Coke

The carbon residue left after ICE combustion.

Collector

The conductive outside layer of semiconductor material in a transistor.

Collision Alert Indicators

A series of red LEDs which will flash when approaching another vehicle too rapidly. The collision alert indicators are located within the electrical device area. The collision alert indicators receive power and ground and are discretely controlled by the instrument cluster through a pair of low control circuits. When requested by the frontview camera controller, the board will pulse the low control circuits, flashing the LEDs as a visible alert that another vehicle is being approached too rapidly. The instrument cluster controls the head-up display via a CAN (controller area network) system. The instrument cluster will command the head-up display to flash the collision alert indicator as a visible alert when approaching another vehicle too rapidly, as requested by the frontview camera module. The EBCM (electronic brake control module) starts out the braking functions of intelligent brake assist and automatic collision preparation. For more information on this, see **J3063_202103** [8].

Combination Valve

Some vehicles combine the pressure differential valve, metering valve, and proportioning valve into a single valve body.

FIGURE C.15 Automotive brake system.

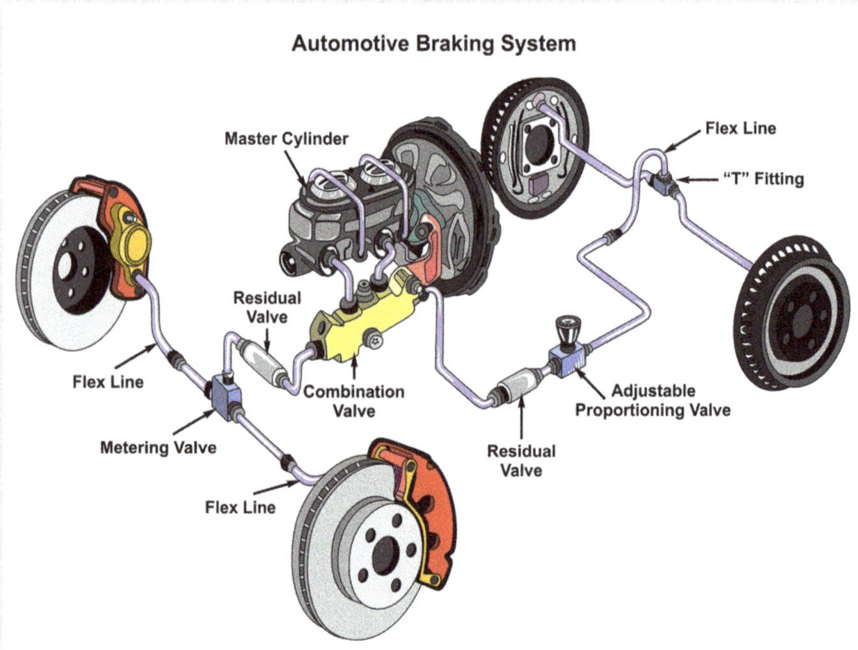

udaix/Shutterstock.com.

Combustion

The high-temperature burning of the fuel and air mix in the combustion chamber of an ICE.

Combustion Chamber

The volume of the space in the cylinder above the piston with the piston at top dead center (TDC) during the compression stroke where the cylinder head and piston crown form the combustion chamber. For more information on this, see **J604_201108** [9].

Comet 5 Piston and Prechamber

The Ricardo Comet 5, or toroidal piston, has a combustion chamber area that is shaped like one-half of a four-leaf clover and designed to overlay a piston displacement (squish) rotary swirl at right angles to the induction-produced swirl around the piston axis. The resulting dual turbulence spirals around the piston axis and resembles a tornado, hence the term toroidal.

FIGURE C.16 Ricardo Comet 5 piston.

Common Rail Fuel Injection

Diesel high-pressure fuel injection systems where fuel pressurization is done independently of engine speed. So no matter what the engine rpm or load is, the fuel is pressurized for injection to the optimal values required. A high-pressure pump driven by the engine pressurizes fuel to the level required for injection for any given engine load and speed condition. Since the quantity of fuel the high-pressure pump can deliver is more than the engine requirements, pressure and volume output of the pump is electronically controlled. The pump supplies fuel at injection pressure to a common fuel rail, which is connected to the electronically operated fuel injectors located in each cylinder. Since all the injectors are connected to the same fuel rail and supplied the same fuel pressure, the system is termed common rail. Injection timing is controlled by an electrical actuator (solenoid) on each injector while the fuel rate is controlled by the on/off times of the actuator and the fuel pressure in the fuel rail.

FIGURE C.17 High-pressure common rail diesel fuel injection.

Mehmet Cetin/Shutterstock.com.

Companion Cylinder (Paired Cylinders)

Used in a waste-spark or distributorless ignition system each end of the secondary winding is connected to a cylinder exactly opposite the other in the firing order.

Companion Flange

Mounted on the differential pinion shaft. The companion flange is a splined flange that transmits power to the pinion gear as discussed in the differential section.

FIGURE C.18 Companion flange being measured for trueness.

Courtesy of John F. Kershaw Ed.D.

Compensating Port (Master Cylinder)

The compensating (vent, fill, inlet) port in front of each piston primary seal is open when the brakes are not applied. The compensating port provides a very small passage between the brake fluid reservoir and the high-pressure chamber of the master cylinder. The compensating port serves two important functions. First, it allows fluid to flow into the master cylinder when the hydraulic system is initially filled, and when closed off, pressure builds. Second, once the system is in service, it allows fluid to flow between the reservoirs and high-pressure chambers to compensate for changes in hydraulic system fluid volume. These changes are caused by variations in fluid temperature or wear of the brake linings.

Compound-Wound Motor

A motor that has both a series and shunt windings.

Comprehensive Component Monitor

See **CCM**.

Compressibility

The ability of a gasket to conform to the surface irregularities.

Compression Brake

A diesel engine does not use a throttle valve except for the regeneration process in a diesel particulate filter, so it cannot use engine compression to slow the vehicle unless the exhaust valves are held open. The high compression ratio of diesel engines results in very high cylinder compressed air pressures to produce the high power and torque at the engine flywheel. This high pressure is used to convert the power-producing engine into a high-pressure air compressor. When activated, the engine compression brake, commonly called a Jake Brake (Jacobs Company), changes the engine exhaust valve opening while consecutively turning off fuel injection into the cylinder to provide a retarding force at the drive wheels. At lower engine speeds, the braking performance is lower. At rated engine speed, the compression braking horsepower will be higher.

Compression Ignition Engine

See **Diesel Engine**.

Compression Ratio

Compression ratio is the ratio of the volume at BDC to the volume at TDC (clearance volume). A higher compression ratio means higher thermal efficiency or that portion of the heat supplied to the engine that is turned into work. As the compression ratio increases, the expansion ratio also increases; thus, thermal efficiency increases. The internal energy of the combustion gas is increased as heat is added to the gas. High heat generated by this greater compression will cause the fuel, upon injection, to atomize or break up into finely divided particles, allowing it to mix easily with the air.

In an ICE, compression ratio is the ratio of the total cylinder volume to the combustion chamber clearance volume. It is the volume at BDC divided by the volume at TDC and calculated in this formula:

$$r = \frac{V_d + V_c}{V_c} \tag{C.4}$$

where

r is the final compression ratio

V_c is volume at TDC

V_d is volume at BDC

For more information on this, see **J604_201108** [9].

Compression Sensing Ignition

Waste-spark ignition systems that use the voltage required to fire the cylinders to see the cylinder position. It requires the next voltage to fire an electrical device under compression than it does when the spark plug is being fired on the exhaust stroke. The electronics within the coil and also the PCM can detect which of the two cylinders that are fired at the same time requires the upper voltage, which indicates the cylinder on the compression stroke. Engines equipped with compression sensing ignition systems, do not require the utilization of a camshaft position sensor to work out the cylinder number.

Condensation Point

Point at which a gas changes into a liquid.

Condenses

The action of making a substance more dense or compact by reducing the volume or extent of the concentrate. The process of applying pressure to a gas, which compresses the expanded gas molecules back into liquid form or cooling the gas so the molecule contracts back into a liquid.

Conditional Driving Automation

SAE Level 3 or Conditional Driving Automation uses various driver assistance systems to make decisions supported changing driving situations round the vehicle. People inside the vehicle donot necessarily supervise the technology, which implies they will engage in other activities. However, a professional driver must be present, alert, and ready to take hold of the vehicle at any time, especially within the case of an emergency because of system failure. You cannnot sleep while sitting in the driver's seat of a Level 3 conditional autonomous vehicle. Audi developed a Level 3 conditional assistance technology for its 2019 A8 sedan, but it never received regulatory approval for the system in Germany and has since been put on hold. Honda became the primary automaker to sell an approved Level 3 conditional assistance system to consumers. Other vehicles equipped with Level 3 driving automation but are looking ahead to regulatory approval include the redesigned 2021 Mercedes-Benz S-Class and therefore the all-new 2022 Mercedes-Benz EQS electric vehicle. The Mercedes technology is termed Drive Pilot. For more information on this, see **J3016_201401** [12].

Conduction

Transfer of heat from molecule to molecule through solids and fluids in close contact at rest. It is heat transfer from one solid to another. No displacement of the heated body takes place during conduction. Heat travels through the rod via conduction from one molecule to another until the end you are holding approaches the temperature of the end in the fire. The action of a solid to conduct heat is called conductivity [13].

FIGURE C.19 Conduction.

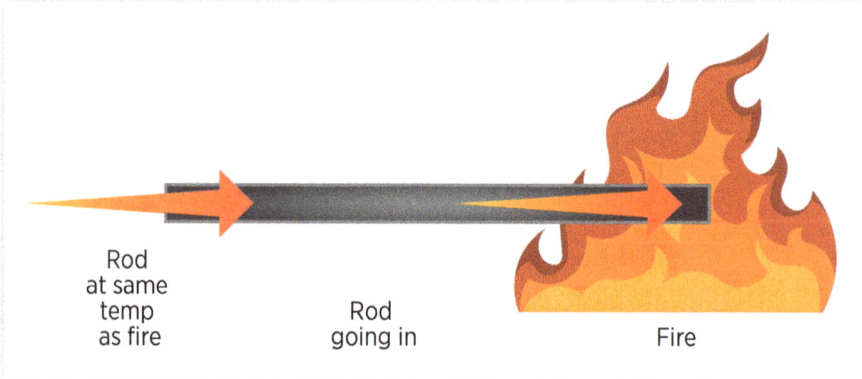

Courtesy of John F. Kershaw Ed.D.

Conductive Material

Conductive materials like a copper wire with an insulator readily permit the flow of electrons and connect the elements of the circuit: power source, load, controls, and ground.

Conductor

A material that gives a path for electricity to flow. Conductors are materials with fewer than four electrons in the outer orbit of their atom. Copper is a wonderful conductor because there is just one electron in its outer orbit. This orbit is far enough from the nucleus of the copper atom that the pull or force holding the outermost electron in orbit is comparatively weak. Copper is the conductor most utilized in vehicles because the cost of copper is affordable compared to the relative cost of other conductors with similar properties.

Cone Clutch

A connection or clutch mechanism used in limited slip differentials that use the principle of a tapered fit to make a solid connection.

Connected Vehicle

A vehicle equipped with different types of telematics to incorporate Internet access, and Wi-Fi, which also stands for wireless local area network (WLAN), to connect the vehicle to the environment outside of the vehicle. Telematics is an interdisciplinary field that encompasses telecommunications, vehicular technologies, road transportation, road safety, engineering (sensors, instrumentation, wireless communications, etc.), and applied science (multimedia, Internet, etc.).

A connected vehicle can share Internet access with other devices both inside as well as outside the vehicle. These vehicles also are equipped with special technologies that tap into the net or WLAN and supply additional benefits to the driving force. For safety-critical applications, it is anticipated that at some point all vehicles are going to be connected using dedicated short-range communication (DSRC) radios, operating within the FCC-granted 5.9 GHz band.

Connecting Rod

Steel rod that connects the piston to the crankshaft and transmits piston up and down motion to rotary motion in the reciprocating piston engine.

Connector

Plastic or rubber device with metal terminals or connection points for two or more wires. Many connectors have locking clips that hold the connectors together, ensuring a good electrical connection.

Contamination

The presence of a minor and undesirable constituent during a material, physique, natural environment, workplace, etc.

Continuity Tester

A piece of equipment used to check a circuit for power or continuity. It uses a probe at one end for touching conductors and a clip for grounding or connecting to the other end of a wire segment. It can also be referred to as a test light.

Continuous Monitor

An OBD II or HD-OBD software monitor that continuously monitors all emission-related controlled systems. There are two types: Misfire type A and Misfire group B, which exceeds EPA federal test procedure (FTP) or cause vehicle to fail inspection and maintenance tailpipe emissions test. The MIL will turn on and a DTC ID stored. For more information on this, see **J1930DA_202105** [6].

Control Arm

A metal strut on the suspension that is found at the highest and bottom of the wheel spindle. The upper and lower control arms allow the front wheels to change direction.

FIGURE C.20 Automotive front suspension control arm.

Greg Brave/Shutterstock.com.

Control Device

Perform many alternative jobs, like turning lights on and off, dimming lights, and controlling the speed of motors. Control devices work by completely stopping current flow or by varying the speed of flow. Controls accustomed stop current flow include switches, relays, and transistors. Controls are used to vary the speed include rheostats, transistors, and other solid-state devices. Control devices may be on the positive or negative side of the circuit.

Control Module

A computer that controls the operation of a system, supported sensor inputs. There are many types: ECM (engine control module), PCM (powertrain control module), BCM (bdy control module), EBCM (electronic bake control module), VCM (vehicle control module), etc.

Controller Area Network (CAN)

See **CAN**.

Convection

Heat transfer by the molecular motion in the heated substance itself and only takes place in liquids and gases. It is heat transfer by circulation though fluids like coolant (or air in air-cooled ICEs) in motion between the fluid and a solid surface in motion like the piston. It is heat that is transferred to the cylinder wall from the heat of combustion where it goes into the coolant and is carried away at the radiator. This form of heat transfer also includes conduction as well as fluid motion. There are two forms of convection heat transfer: natural convection is when the fluid motion is caused by different densities in a gravitational field and forced convection is the method of heat transfer between a fluid and a solid surface in relative motion, when the motion is caused by forces other than gravity. Most of the heat in an ICE flows between the coolant (working fluid) and the engine parts that is transferred via forced convection [13].

FIGURE C.21 Convection.

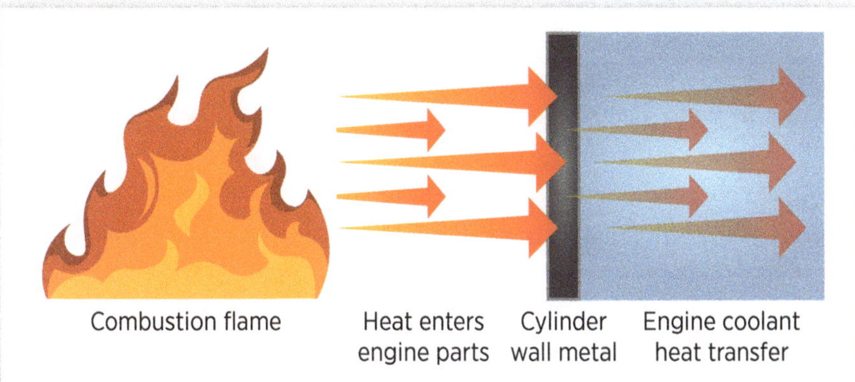

Courtesy of John F. Kershaw Ed.D.

Conventional Current Theory

A theory that says that electrons flow through a circuit from positive (+) to negative (–). Electrons leaving the positive terminal of the battery flow through the circui, and then reenter the battery through the negative terminal.

Coolant

Liquid mixture of antifreeze and water or other chemicals that do not readily freeze in the engine cooling system. Coolant transfers heat from the engine to the radiator, protects the engine and the cooling system from rust and corrosion, and prevents freezing in cold climates. All coolants except the waterless racing type are a mixture of antifreeze and water. Water is best at absorbing more heat per gallon than any other liquid coolant, but it will cause corrosion of the metal in the engine.

Coolant Reservoir Tank

A coolant tank used on older vehicle in the fifties, sixties, and early seventies had a hose attached to the filler neck of the radiator, below the cap, and that hose was vented to the atmosphere by dripping on the ground just below the radiator. When coolant expands from engine temperatures rising, it typically creates steam, and the vented cap allows that steam to vent in order to protect your cooling system. However, as the steam expands, so does the pressure, and often times the vented cap also expels coolant through the hose to the ground. This system was good for engine cooling, but not good for the environment. In the late seventies, the addition of environmental control brought about adding a reservoir to that vent tube allowed the steam to be expelled, but also allowed for the coolant that came with it to be captured in the reservoir tank instead of going to the ground. Capturing that expelled coolant meant that it could then be recovered and reintroduced back into the radiator, hence the term recovery or reservoir tank. Returning coolant back to the radiator is possible because the combination of the reduced steam pressure allows the atmospheric pressure to push coolant from the tank back into the radiator through the vented radiator cap. This adds more coolant to your system and helps to keep the engine cooler. In contrast to the expansion tank, the overflow/recovery/reservoir tank featured a vented cap and was not required to be above the cylinder heads [13].

Cooling Fan

Engine cooling fans are used to draw air through the radiator while the vehicle is at idle or moving at a slow speed. When the vehicle is moving at sufficient speed, air is being forced through the radiator at greater CFM (cubic feet per minute) than the cooling fan is capable of, making the fan somewhat excessive. There are three individual types of engine cooling fans: flex, viscous clutch, and electric [13].

Cooling System

In an ICE, the coolant circulates through the engine to absorb excess heat through the convection process. The major components of a cooling system:

water pump, thermostat, radiator, coolant recovery or expansion tank, radiator fan, hoses, and heater core. When designing an engine cooling system, the selection of these components is very important for engine life and performance. The cooling system can be closed or open. In a closed system the engine coolant has no contact with the outside air. In an open system the coolant has contact with the outside atmosphere. All current systems are a combination of the two, where the coolant, while circulating through the engine, has no contact with the outside air, but at the radiator pressure cap, there is contact through the vacuum or air valve, which allows the entrance of atmospheric pressure [13].

FIGURE C.22 ICE cooling system used to remove the heat of combustion.

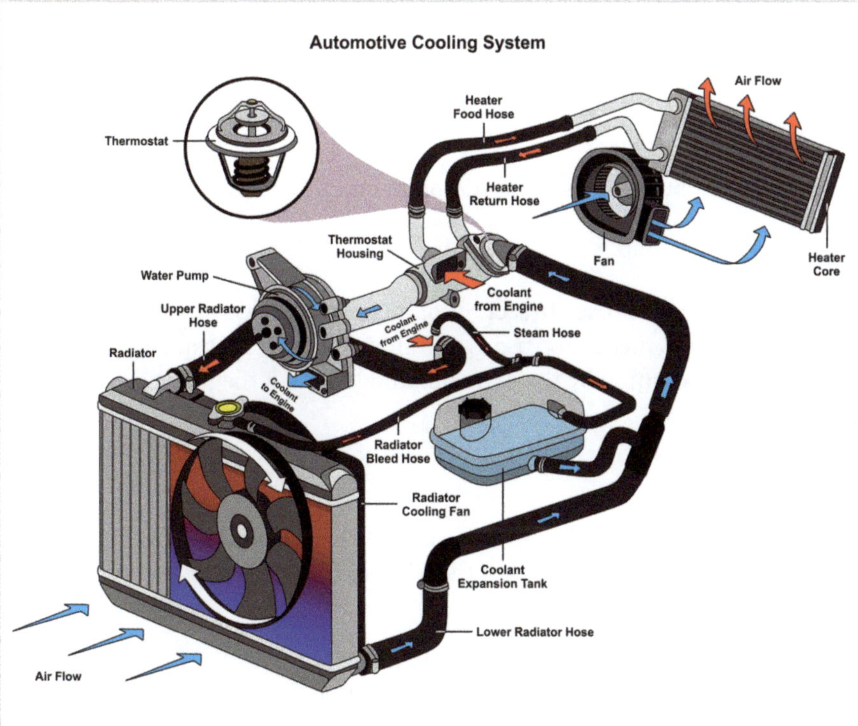

udaix/Shutterstock.com.

COP (Coil-On-Plug) Ignition System

An ignition system that does not use a distributor or a central ignition coil. Each spark plug has its own ignition coil. This system is also called coil-by-plug, coil-near-plug, or coil-over-plug ignition. For more information on this, see **J139_202002** [14].

Core-Out Plug

Seals the holes in the engine block used during the manufacturing process. This component is also referred to as freeze plugs.

Corrosion

Wear by chemical or electrochemical reaction.

Corrosion Inhibitors

Prevents the formation of rust and protect metal parts.

Corrosion Test

Various metals and wheel cylinder cups are submerged in the brake fluid for a specified time.

Corrosive

A material that burns the skin or dissolves metal and other materials.

Cotter Pin

A soft metal pin that is used to secure components in place. Sometimes referred to as cotter keys. A metal loop used to retain castle nuts by installation through a hole. Size is measured by the diameter and length of the pin (e.g., 1/8 in. × 1 1/2 in.).

Co-solvent

Another substance (usually another alcohol) that is soluble in both methanol and gasoline and is used to reduce the tendency of the liquids to separate.

CPI (Continuous Port Fuel Injection)

A form of multipoint fuel injection, or MFI, that uses a central injector that feeds a poppet valve at each cylinder for fuel delivery. The central injector is controlled by the engine management computer.

CPU (Central Processing Unit)

The microprocessor is the central processing unit (CPU) of a computer that does the required mathematical operations and logic decisions that does the processing function. It can be the heart of a computer.

Cracking

Cracking is the crude oil refining process where hydrocarbons with higher boiling points could be broken down (cracked) into lower boiling hydrocarbons by treating them to very high temperatures. This is called thermal cracking. Currently catalytic cracking is used where a catalyst is used to speed up the process without undergoing a permanent chemical change itself. It produces a higher quality gasoline than cracking. For more information on this, see **J312_201902** [15].

Crankcase

The bottom portion of an ICE cylinder block that houses the crankshaft and connecting rods.

Crankcase Ventilation

Blowby gases enter the crankcase from the combustion chambers. Because the combustion chamber gases are under very high pressure, they increase the pressure within the crankcase. If the crankcase is not vented, the pressure will force oil out of the engine. To prevent this, the crankcase is ventilated using the PCV (positive crankcase ventilator) valve on a gasoline fuel engine or a CDR (crankcase depression regulator) valve on a diesel engine.

Cranking Circuit

Electrical components and connections required to crank the engine to start.

Cranking Motor

See **Starter Motor**.

Crankshaft

The rotating member of an ICE. It converts the reciprocating (translating) motion of the engine piston into rotating motion to drive the ICE. Connecting rods are fastened to the crankshaft that connect with the pistons. The crankshafts is additionally either forged or nodular cast iron. Forged crankshafts are stronger than the cast crankshaft and have an oversized separation line. The wide separation line is the result of a grinding process to remove the metal that was extruded from the forging die during the forging process. Most forged crankshafts are made from SAE 4340 or a similar type of steel. The crankshaft is made from a hot steel billet through the use of a series of forging dies. Each die changes slightly the shape of the billet. The crankshaft blank is formed with the last die. The blanks are then machined. Forging makes a dense crankshaft with the metal grain structure running parallel to the principal direction of stress.

Crankshaft configuration may be either cross-plane or flat plane. The flat-plane crankshaft incorporates a 180° angle between crank throws. It is basically two four-cylinder engines connected at the crank. They are utilized in V-engines with eight cylinders. Cadillac first introduced a V8 flat-plane crank engine in 1923. Currently, GM uses a flat-plane crankshaft within the 5.6L V8 engine employed in the mid-engine Chevrolet Corvette.

FIGURE C.23 ICE crankshaft with connecting rods and pistons.

Sashkin/Shutterstock.com.

Crank Web

The part of the crankshaft between the crankpin and the shaft or between adjacent crankpins, which is also called the crank arm or throw.

FIGURE C.24 Crank web.

Crate Engine

A complete replacement ICE that includes all components except accessories. It is a fully assembled, ready-to-run engine delivered in a crate. Most crate engines will include the block, rotating assembly, and cylinder heads.

Cross Camber

Difference in vehicle camber between both sides of a car or truck. Vehicle will pull to the side with the most positive camber. If the left front camber is 1° positive and the right front camber is 1° positive, the effects of pull cancel each other. If the front left camber is 1° negative and the right front 1° positive, the vehicle will pull to the right side.

Cross Caster

Difference in caster between both sides of a vehicle. Caster can cause a vehicle pull, where the vehicle will pull to the side with the most negative caster. When the front left wheel has 2° positive caster and the right front wheel has 2° negative caster, the vehicle pulls to the right. Caster is not a tire wearing angle, so caster is often set 0.5° more negative (less positive) on the left side to create a small pull to the left if a vehicle is driven on highly sloped roads. A difference in caster between one side and the other of more than half a degree may cause a pull or lead toward the side with the least positive caster.

Cross Weight

Weight of the right-rear and left-front and right-front and left-rear wheels. It is used mostly in circle track racing cars. Drag racers are concerned with the front and rear weights, total weight, and the right-rear-wheel weight compared to the left-rear wheel.

Crossfire

Ignition voltage jumping from the distributor rotor on a distributor-type ignition to the wrong spark plug electrode inside the distributor cap or ignition voltage jumping from one spark plug cable to another cable due to defective insulation.

Crossflow Cylinder Head

Cylinder head design with the inlet manifold on one side and the exhaust manifold on the other side of the head so that the inlet and exhaust valves are arranged on opposite sides of the combustion chamber, giving a wider engine but better gas flow.

Cruise Control

A vehicle system that may maintain a preset vehicle speed even on gentle inclines. Some vehicles are programmed to downshift the transmission to maintain vehicle speed on downhill gradients if the speed increases above 5 mph (7 km/h) faster than the preset speed. The typical control system will be set provided that the vehicle speed is 25 mph (40 km/h) or more. Older systems use a throttle actuator to manage the throttle opening, control switches for driver control of controller functions, and electrical brake and clutch (if equipped) pedal-release switches. The actuator uses a stepper motor to maneuver the throttle linkage supported commands from the control module, which is commonly inside the control assembly.

Cruise Control Resume/Increase Switch

The resume/increase switch is used to extend the determined vehicle speed when adaptive control is active. The selected vehicle speed can be increased from the resume/increase switch depending on how long the switch is pressed. The presence of a slower moving vehicle in the path of the adaptive control vehicle will limit the extent to which the determined vehicle speed are often achieved. If there is no preceding vehicle ahead limiting the adaptive control vehicle acceleration, then the vehicle speed that is achieved is the new determined vehicle speed. This determined vehicle speed is displayed by the motive force information center. Acceleration is terminated when the resume/increase switch is released. Momentarily pressing and releasing the resume/increase switch will allow the chosen vehicle to accelerate at 1 mph increments for every time the resume/increase switch is momentarily pressed.

Cruise Control Set/Decrease Switch

The resume/increase switch is used to extend the determined vehicle speed when adaptive control is active. The speed setting can be increased from the resume/increase switch depending on how long the switch is pressed. The presence of a slower moving vehicle in the path of the adaptive controller vehicle will limit the extent to which the determined vehicle speed is achieved. If there is no preceding vehicle before limiting the adaptive control vehicle acceleration, then the vehicle speed that is achieved is the new determined vehicle speed. The present determined vehicle speed is displayed by the driving force information center. Acceleration is terminated when the resume/increase switch is released. Momentarily pressing and releasing the resume/increase switch will allow the chosen vehicle to accelerate at 1 mph increments for every time the resume/increase switch is momentarily pressed.

Cruise Control Switch

A common control switch that is shared between the adaptive cruise control system and the standard cruise control system. The adaptive cruise control system will not operate if one of the cruise switches fails in the circuit. The cruise control switch has the following function switches:

- ON/OFF switch
- Set/decrease switch
- Resume/increase switch

The cruise control system is engaged when the adaptive cruise control ON/OFF switch is turned ON and the set/decrease switch is momentarily pressed and released. When the set/decrease switch is pressed, the selected vehicle speed is set to the current vehicle speed by the engine management computer. The vehicle speed must be near or greater than 25 mph. The selected vehicle speed is displayed by the driver information center. While in the engaged state, the selected vehicle speed and distance can be adjusted. Pressing and holding the set/decrease switch when the adaptive cruise control system is engaged will decrease the selected vehicle speed without deactivating the adaptive cruise control. Momentarily pressing and releasing the set/decrease switch, when the adaptive cruise control is engaged, decreases the selected vehicle speed by 1 mph for each time that the set/decrease switch is pressed.

Curb Weight

The inertia weight (also known as unladen weight or wet weight), which is the total weight of a vehicle with standard equipment, driver, and full tank of fuel, without passengers or cargo.

Current

A movement of electrons in a circuit is the flow of electricity. Another name for the flow of electricity is current. Current flow is measured in amperes (A). This unit expresses the percentage electrons move through a circuit in one second. A current flow of 6.28 pico (billion/billion) (6.28 × 1018) electrons per second is adequate to one ampere.

Current Limiting Hump

Portion of an ICE ignition waveform that indicates the exact point in the dwell when a current limiting device is switched on used with transistor switched ignition coils.

Customary Measuring System

A measuring system established about 1100 AD in then England, now the United Kingdom during the reign of King Henry I. It established the standard measurements such as foot, yard, mile, and the temperature scale using Fahrenheit. Also known as the English system of measurement or FPS, for foot pound, second.

CV (Constant Velocity) Joint

A driveline connection between a driven and drive axle that allows the two shafts to show a near-constant velocity. These joints permit angularity, in and out, and up and down movement. They also permit for unequal axle length, the angles at the CV joints are different, which also leads to the pulling concern and possibility of creating vibration. To combat this condition, many OEMs use an intermediate shaft. The intermediate shaft may be a short section of shaft that typically contains a bearing pressed onto it (similar to a carrier bearing). The intermediate shaft makes it possible that both half-shafts are the identical length from left to right. In some cases the manufacturer uses an extended half-shaft on one side, and a rubber dynamic damper could also be fitted to assist in absorbing vibrations, although this does not reduce torque steer issues. CV joints afford more torque transfer than U-joints because of their larger bearing surfaces and equal angles at both the input and output of the joint. CV joints can operate at greater angles than U-joints. Rzeppa fixed CV joint has an inner race, six steel ball bearings, a sway cage, and an outer race. A sliding spline or a plunge-type joint is used because the inner half-shaft joint accommodates changes in shaft length when traveling over differing kinds of terrain. Plunge-type joints allow smooth power flow while allowing the joint to slip in and out, effectively increasing and decreasing the length of the axle shaft during up and down suspension travel. One kind of plunge CV joint is the tulip tripod joint. The tulip/tripod joint uses three equally spaced fingers shaped like a star. On the ends of the star are three round bearing surfaces that sit on needle bearings on each finger. The round bearing surfaces advance the fingers by needle bearings. The outer race has three straight grooves that run from side to side. This configuration allows in- and-out movement of the shaft while allowing flexing. The fixed-type joint does not slide to permit for shaft lengthening or shortening; it simply allows for angle changes because of the suspension moves. The fixed joint is usually used on the outboard side of the half-shaft.

FIGURE C.25 Constant velocity drive axle joint or connection.

Elena100/Shutterstock.com.

CVT (Continuously Variable Transmission)

An automatic transmission that can change through a series gear ratios continuously using two moving pulleys that move in and out changing the gear ratio.

FIGURE C.26 CVT vector.

fariz guliyev/Shutterstock.com.

Cycle

One complete positive curve and negative curve of a sine wave. The term cycle comes from the word circle that begins and ends at the same point, as does a cycle.

Cylinder Block (CASE)

Central casting of an ICE that houses the crankshaft, connecting rods, pistons, and bearings. In the case of an in-block camshaft, the camshaft is also housed with the cylinder block.

Cylinder Deck

The top of the cylinder case or block that the cylinder head rests on. The first inspection you do on the cylinder deck is with your eyes and hands. You look for excessive scoring, corrosion, erosion, threading pulling up around bolt holes, cracks, dents, and scratches. Run your fingernail across any irregularities you find. If a scratch can catch your fingernail, then it probably is deep enough to be corrected.

Cylinder Head

The cylinder head in an ICE sits on top of the cylinder block above the cylinders and generally forms the combustion chamber. It uses a head gasket to seal the passage between the head and the cylinder block. It also provides passages to feed air and fuel into the cylinder, which allows the exhaust to escape, and contains the valve train and valves, spark plugs, and fuel injectors.

Cylinder Hone

A tool that uses an abrasive tool driven by a drill to smooth out and bring to exact measurement in the ICE cylinder bores.

Cylinder Liner

Diesel engine cylinders use two types of cylinder: dry and wet liners. The dry liner is a thin wall sleeve, while a wet liner is installed in the coolant jacket surrounded by engine coolant. Dry liners are installed into the block bore usually with a marginally loose fit and retained by the cylinder head. The dry sleeve does not transfer heat as well as a wet liner, but they are easily replaced and do not present coolant-sealing problems. Air-cooled engines used finned liners that are cooled using the fins on the liner, like in motorcycle and VW/Porsche boxer engines.

REFERENCES

1. Society of Automotive Engineers, J1939/21_2021: Data Link Layer, SAE Publishing, Warrendale, PA, 2021.
2. Society of Automotive Engineers, J139_202002: Ignition System Nomenclature and Terminology, SAE Publishing, Warrendale, PA, 2020.
3. Society of Automotive Engineers, J2047_201911: Tire Performance Terminology, SAE Publishing, Warrendale, PA, 2019.
4. Society of Automotive Engineers, J670_200801: Vehicle Dynamics Terminology, SAE Publishing, Warrendale, PA, 2008.

5. Society of Automotive Engineers, J1594_201007: Vehicle Aerodynamics Terminology, SAE Publishing, Warrendale, PA, 2010.
6. Society of Automotive Engineers, J1930DA_202105: Electrical/Electronic Systems Diagnostic Terms, Definitions, Abbreviations, and Acronyms Web Tool Spreadsheet, SAE Publishing, Warrendale, PA, 2021.
7. Society of Automotive Engineers, J1479_20121: Automotive Pull-Type Clutch Terminology, SAE Publishing, Warrendale, PA, 2012.
8. Society of Automotive Engineers, J3063_202103: Active Safety Systems Terms and Definitions, SAE Publishing, Warrendale, PA, 2021.
9. Society of Automotive Engineers, J604_201108: Engine Terminology and Nomenclature – General, SAE Publishing, Warrendale, PA, 2011.
10. Society of Automotive Engineers, J1145_201109: Emissions Terminology and Nomenclature, SAE Publishing, Warrendale, PA, 2010.
11. Society of Automotive Engineers, Standard J1538_201504: Glossary of Automotive Inflatable Restraint Systems, SAE Publishing, Warrendale, PA, 2019.
12. Society of Automotive Engineers, J3016_202104: Taxonomy and Definitions for Terms Related to Driving Automation Systems for On-Road Motor Vehicles, SAE Publishing, Warrendale, PA, 2021.
13. Kershaw, J.F., *High-Performance Automotive Cooling Systems* (Forest Lake, MN: CarTech, Inc., 2019), 9-10, ISBN:978-1-61325-504-9
14. Society of Automotive Engineers, J139_202002: Ignition System Nomenclature and Terminology, SAE Publishing, Warrendale, PA, 2020.
15. Society of Automotive Engineers, J312_201902: Automotive Gasolines, SAE Publishing, Warrendale, PA, 2019.
16. Society of Automotive Engineers, J1850_201510: Class B Data Communications Network Interface, SAE Publishing, Warrendale, PA, 2019.

D

Dampening

Reducing movement or vibration such as a shock absorber dampening the oscillation of a spring. For more information on this, see **J3152_202005** [1].

Damper

See **Harmonic Balancer**.

Damping

Ability of an object or material to dissipate or absorb vibration. The automotive shock absorber is a good example. The function of the shock absorber is to absorb or dampen the oscillations of the suspension springs and up and down movement. For more information on this, see **J3152_202005** [1].

FIGURE D.1 Ability of an object or material to dissipate or absorb vibration.

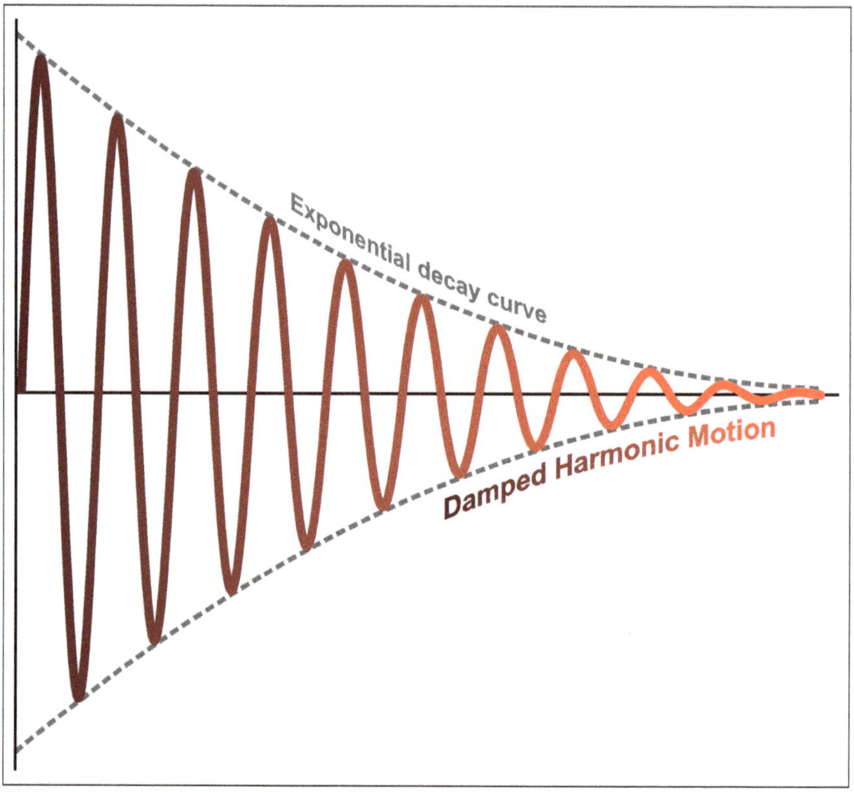

Fouad A. Saad/Shutterstock.com.

Data

Information from a vehicle computer used for mechanical or electronic computation.

Data Bus Communication

The ICE management computer receives information from other vehicle systems, such as the transmission and cruise control, and body computer module and provides important information to other vehicle systems. The computer transmits critical information such as engine speed, throttle position, and engine temperature. The data bus system functions to reduce the amount of wiring needed for the computer to communicate with other vehicle computers. The data bus may be either a single wire or, on older vehicles, a two-wire system. Similar to personal computer systems, the data bus used on today's vehicles becomes faster, uses even less wires, and performs more functions with each passing year. The data bus is also referred to as multiplexing or a controller area network (CAN). For more information on this, see **J1939/21_2021** [2].

Data Stream

The serial data in an automotive engine management computer–controlled vehicle system that can be extracted by a scan tool computer or electronic service tools. For more information on this, see **J1930DA_202105** [3].

Datum Plane

Datum means a basis on which other measurements can be based. The datum plane is the horizontal plane.

DC (Direct Current)

When a steady-state electrical potential is applied to a circuit, the resulting current flows in one direction, which is called direct current, or DC.

DC Generator

A DC generator uses the principle of electromagnetic induction to produce a DC. When the magnetic lines of force cross or cut a conductor (wire loop), it induces a voltage in that conductor. The DC generator uses an iron core or laminated iron sheets to create an electromagnet. When current flows through this coil, a magnetic field or flux is created between the pole pieces. Permanent magnets could be used. A single wire loop is between the north and south poles. When this wire loop turns into a magnetic field, it cuts off the lines of force and induces a voltage. When there is a complete circuit from the wire loop, current will flow.

Decimal

(1) Refers to or denotes a system of numbers and arithmetic that supports quantity in ten, tenth parts, and powers of ten. (2) A fraction whose denominator can be a power of ten and whose numerator is expressed as numbers placed to the right as a percentage point. (3) The amount system that uses the numbers 0, 1, 2, 3, 4, 5, 6, 7, 8, and 9.

Decompressor

A mechanical device used on single- or twin-cylinder air-cooled ICEs that momentarily opens an exhaust valve on the compression stroke of the piston to partly vent the combustion chamber for easy starting.

FIGURE D.2 Decompressor from Honda V2 industrial engine.

Courtesy of John F. Kershaw Ed.D.

De Dion Rear Axle (Dead Axle)

A form of non-independent suspension, different from a swing axle, Hotchkiss drive, or live axle. It does not transmit power to the drive wheels, it is also called a dead axle.

Deep-Cycle Battery

An SLA (sealed lead-acid) battery that provides sustained power over a long period and runs reliably until it is 80% discharged or more, at which point it needs to be recharged. They can be discharged up to 80%, but most battery manufacturers recommend not discharging below 45%.

DEF (Diesel Exhaust Fluid)

DEF (also called reductant and urea) is a reductant. It is composed of urea and purified water and is injected in small amounts into the exhaust where it works with a catalyst to convert NOx into nitrogen and water vapor. DEF is a non-toxic, odorless, and safe solution composed of:

- 67.5% purified water
- 32.5% automotive-grade urea

Urea is produced from natural gas and other sources and is commonly used in fertilizer and industrial applications including emissions control at power plants. It is certified by the API (American Petroleum Institute). Urea is used in Europe where it is known as AdBlue. The following are determined DEF facts:

- DEF freezes at approximately 12°F.
- DEF thaws with no degradation.
- DEF tank and lines are heated for cold weather operation.
- No impact on cold weather engine operation.
- DEF has a 1-year shelf life.
- DEF will degrade at temperatures above 86°F or in direct sunlight for an extended period; it is not an issue if DEF stock is rotated within a year.
- DEF containers have a date stamp.
- On-board warning systems advise the driver when the DEF level is low or of inferior quality.
- The tank includes a filter and screen.
- Anticipated DEF usage of 1–1.25% of diesel fuel usage.

Deflection Rate

Distance an automotive suspension spring bends under a given load or pressure.

Defogger

An automotive device that blows dry air over the front glass and rear backlight to clear away any condensation.

Defroster

A heating grid placed in the rear window glass or backlight in that when heated will defrost the rear window. See **Heated Rear Window Defroster**.

Deglazing

Repair procedure that you use only when both the cylinder walls and the pistons themselves are in good condition. Deglazing roughens the surface of the cylinders without significantly changing their overall diameter.

Delta-Wound Stator

The two basic stator winding designs are delta wound and wye (Y) style. Delta wound are easily identifiable by their shape, as they are triangular.

These windings yield a high current flow at lower RPM. Wye windings produce higher voltage than delta stators at even lower RPM.

Depth

The distance from the top or surface of something to its bottom or the distance from the nearest to the farthest point of something or from the front to the back.

Detergent Additives

These additives fight varnish deposits. Detergents clean piston ring grooves and keep the rings free to seal with maximum effectiveness.

Detonation

An ICE combustion that takes place when an expanding combustion flame front compresses and heats a portion of the air-fuel mixture in a corner of the combustion chamber that is already at a very high temperature. A portion of the air-fuel mixture explodes before the flame front can get to it. This explosion creates a second pressure wave in the combustion chamber that causes the cylinder pressures to rise at an uncontrolled rate, raising the cylinder temperatures above acceptable limits.

DEX-COOL

General Motors (GM) brand name used for organic acid technology (OAT) antifreeze/coolant that does not contain silicates or phosphates. Usually orange in color (may be pink as well), it absolutely was first developed by Havoline and employed in GM vehicles. This extended-life coolant encompasses a longer service interval than the green, ethylene glycol-based coolants.

DI (Direct Ignition)

The DIS, or direct/distributorless ignition system, uses a control module that is part of the coil assembly and connecting wires to replace the ignition distributor. In a waste-spark system, as it is commonly called, a spark plug is attached to each end of the induction coil secondary winding. Each coil of the system fires two plugs at the same time in two companion cylinders. The pistons in these cylinders will reach top dead center (TDC) at the same time. The cylinder at TDC on the compression stroke is the event cylinder, while the cylinder at TDC on the exhaust stroke or TDC overlap is the waste cylinder. When the coil fires, both spark plugs fire at the same time to complete the electric circuit. For more information on this, see **J139_202002** [13].

DI (Direct Injection)

Combustion chamber designs come in two distinct groups: open combustion chamber or DI systems and IDI (indirect Injection) systems. In the DI system, fuel injection and subsequent combustion takes place within the actual working chamber or cylinder of the engine. Fuel is injected directly into the space between the cylinder head and the top of the piston. This is also referred to as an open chamber design. The piston often contains a bowl or has a specially shaped crown to aid in the mixing process for good combustion.

DI (Distributor Ignition)

Distributor ignition (DI) is a term specified by the Society of Automotive Engineers (SAE) J1930 standard for an ignition system that uses a distributor. For more information on this, see **J139_202002** [4].

Diagnostic Chart

A chart that gives logical steps for finding and correcting problems.

Diagnostic Executive

On computer-managed engine management in an OBD-II (On-Board Diagnostics Generation II) systems, the engine management computer uses a special type of software to manage the vehicle system monitors. On Ford and GM systems, this software is called the diagnostic executive. For more information on this, see **J1930DA_202105** [3].

Diagonal Braking System

In a diagonal split brake system, the right-front and left-rear brakes are on the same hydraulic circuit. The left-front and right-rear brakes are on a separate hydraulic circuit. This type of system is found in FWD (front-wheel-drive) and AWD (all-wheel-drive) vehicles. This system offers an advantage of continued performance even when there is a hydraulic failure in half of the system. Brake pressure can be applied to one driven wheel and the opposite rear wheel in order to maintain directional stability.

FIGURE D.3 Automotive diagonal braking system found in FWD and AWD vehicles.

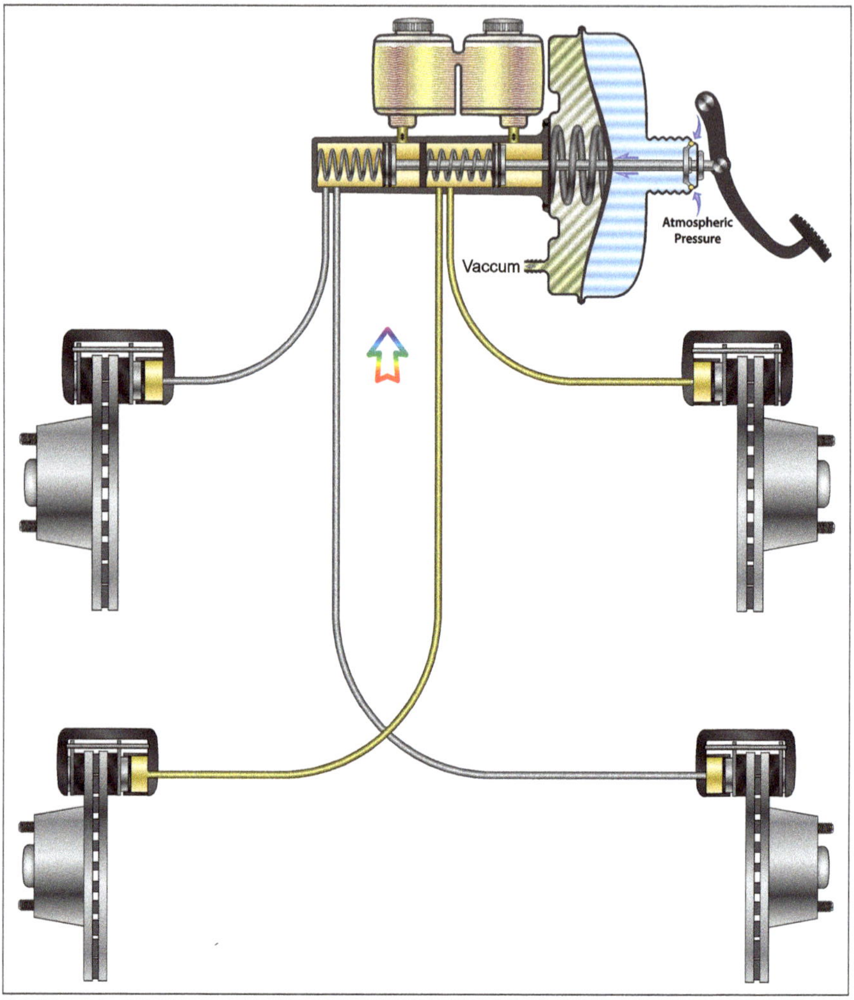

Fouad A. Saad/Shutterstock.com.

Dial Bore Gauge

Special measuring instrument for measuring cylinder and main bearing bores for taper and out of round.

Dial Indicator

The dial indicator is a mechanical device that uses a pin attached to a lever which moves a gear to turn a needle, like a clock, to measure up and down linear travel or rotary movement. It is used to measure endplay, runout, and warpage.

FIGURE D.4 Dial indicator measuring tool.

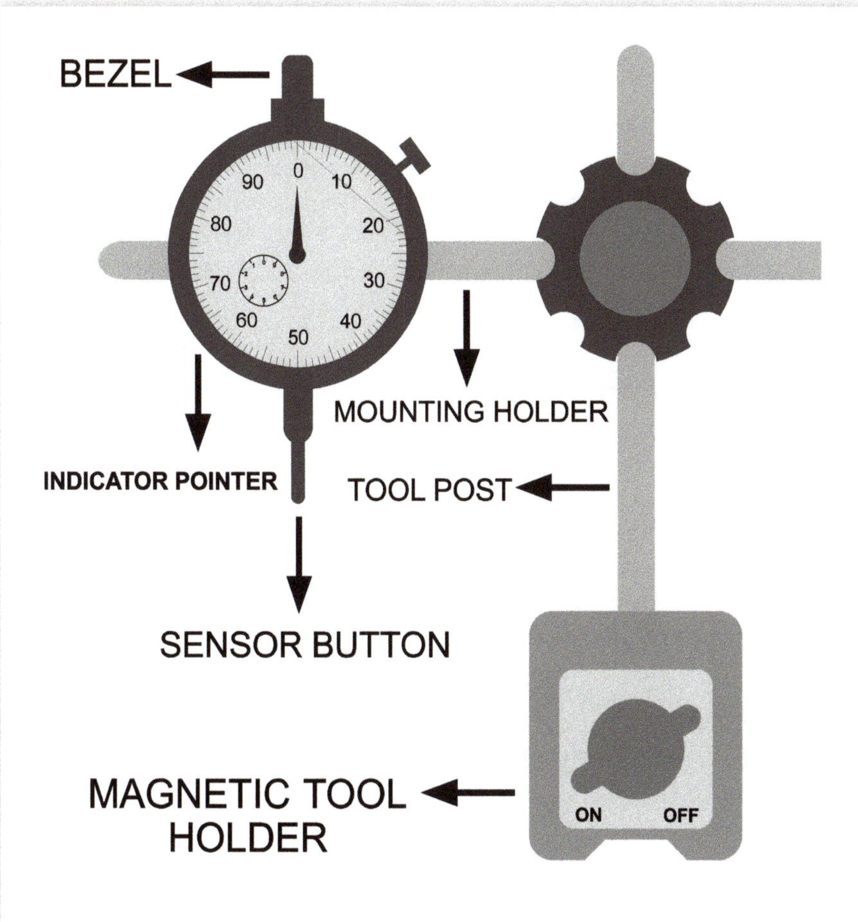

Khoirul Project/Shutterstock.com.

Diameter

The distance between one side of a circular object to the other side straight across and passing through the center.

Diamond-Shaped Frame

Takes place when one frame rail is pushed rearward in relation to the opposite frame rail. Vehicle tracking is affected by this condition. The following causes can result in a diamond: collision, towing, or being towed with the chain attached to only one side of the frame.

Diaphragm

A thin flexible wall separating two cavities, such as the diaphragm in a vacuum advance unit.

Diaphragm Spring

The diaphragm spring is found in automotive clutches. It is a conical or cone-shaped spring instead of a coil spring for producing adequate pressure to engage the clutch. The clutch cover is fastened to the ICE flywheel and the outer rim of the diaphragm spring is in contact with the pressure plate to hold the clutch disc to the ICE flywheel.

DIC (Driver Information Center)

A vehicle communication system that uses messages, audible noises, and haptic impulses to inform the driver of vehicle conditions. The system uses numerous sensor inputs to the engine and powertrain management computers in addition to vehicle mileage to accurately calculate when the oil needs changing. The software uses temperature, highway or city driving, extreme ambient temperatures, and other factors in making the decision. An oil change indicator tells the driver that an oil change is needed.

Die Grinder

A die grinder is a cutting wheel or stone that will easily remove a damaged bolt head.

Dielectric

The insulating material between the two conductive plates of a capacitor.

Dielectric Grease

Special grease that blocks moisture and helps prevent corrosion without affecting electrical conduction.

Dielectric Strength

Dielectric strength is the resistance to electrical penetration.

Diesel Engine

The diesel engine operates on a different principle than the gas engine since fuel is not mixed with air entering the cylinder during the intake stroke. Air alone is compressed during the compression stroke and the diesel fuel is injected or sprayed into the cylinder at the end of the compression stroke. The compression ratio is much higher and provides compressed air temperatures as high as 1,000°F. The temperature at this instant is high enough to ignite spontaneously when the injector sprays or injects fuel into the cylinder. Combustion of the fuel is controlled by the speed at which the fuel is injected into the cylinder. Thus, in the diesel engine, the combustion is not a rapid burning of the fuel already present in the cylinder, as in a gas engine, but a slower burning that produces an even increase in pressure. The diesel operates with a stratified air-fuel mixture in the cylinder, and combustion occurs as the fuel mixes with the air. The inlet throttle becomes unnecessary.

FIGURE D.5 Diesel engine.

Diesel Fuel

A mixture of hydrocarbons (hydrogen and carbon) extracted from fossil fuel through a process called distillation. Diesel is a middle distillate with more energy per unit of volume. It is less complicated to refine than gasoline. Diesel fuel contains higher quantities of sulfur which end in harmful exhaust emissions. The distillation process is primarily accustomed to refine fuel. Fossil crude is subjected to high heat, and its parts are separated out by vaporization as they reach their individual boiling points. The hydrocarbons within the fossil fuel have different boiling points, so if some fossil oil is placed in a very container and heated, the hydrocarbons with very low boiling points are vaporized first, while the others have higher boiling points.

Diesel fuels are processed from that a part of the oil that boils between 350 and 750°F. Automotive/locomotive diesel fuels are derived from refinery products which are commonly observed as middle distillates. Middle distillates represent products which have a better boiling range than gasoline and are obtained from fractionation of the petroleum or from streams from other refining processes. Finished diesel fuel blends represent blends of middle distillates and should contain other blending components of substantially non-petroleum origin, like biodiesel fuel blend stock, and/or middle distillates from non-traditional refining processes, like gas-to-liquid processes. The properties of economic distillate diesel fuels rely upon the refinery practices employed and also the nature of the crude oils from which they are derived. Thus, they will differ both with and within the region in which they are manufactured. Such fuels generally boil, at gas pressure, over a variety of temperature between 130 and 400°C (approximately 270 to 750°F).

Diesel fuel makeup can represent various combinations of volatility, ignition quality, viscosity, sulfur level, density, and other characteristics. Additives are used to impart specific properties to the finished fuel. Diesel fuel accounts for roughly 20% of all crude consumed within the USA. The oil, automobile, and fuel additive industries have worked together for several decades to work out the composition and properties of diesel fuels required to supply satisfactory vehicle operation. This work is ongoing. Since the late 1960s, environmental concerns have led to federal and state regulations to scale back emissions from vehicles and from petroleum storage and transportation facilities. For more information on this see **J313_201706**, which discusses the characteristics of diesel fuels, common standard test methods, specifications developed by ASTM International, and government regulations [5].

Diesel Fuel Heater

Fuel heaters use electric current or warm coolant to heat fuel. Electric inline heaters are placed in line just before fuel enters the filter housing. Other electric heaters use heating elements located within fuel water separators to warm fuel and prevent icing of water contained by fuel. Coolant-type heaters provide greater temperature increases but only after the engine has coolant has warmed up. Some class 8 trucks use a solid-state electric fuel heater that is actually constructed within the fuel line from the fuel tank to the filter assembly.

FIGURE D.6 Diesel low-pressure fuel heater.

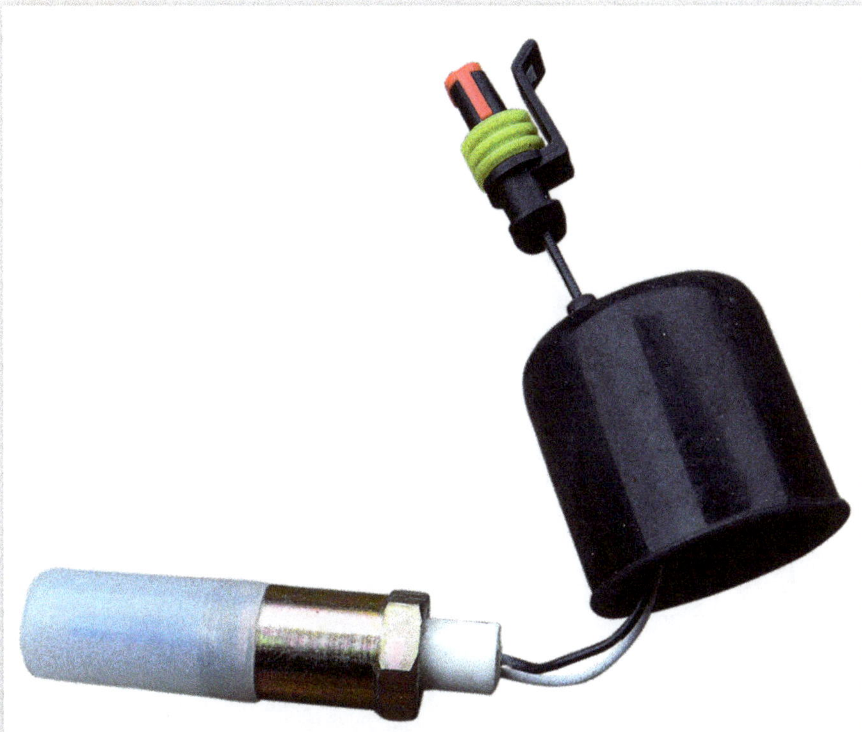

notsuperstar/Shutterstock.com.

Diesel Fuel #1

1D, or #1 diesel, is the most refined diesel fuel among the two primary grades, and it is the most volatile. Number 1 diesel is intended for use in high-RPM engines that experience frequent changes in speed and load.

Diesel Fuel #2

2D, or #2 diesel, is the grade most commonly used in diesel vehicles, especially in warm to moderate climates.

Diesel Knock

In the process of diesel engine combustion, the first fuel burns very rapidly. This rapid burning causes a sudden rise in pressure. This sudden rise causes a highly localized pressure that causes an audible noise known as diesel knock. This diesel knock noise level hinges on pressure rise velocity. This knock occurs in the beginning of the combustion cycle where knock or detonation in a spark-ignited gasoline-fueled engine takes place at the end of the combustion cycle.

Diesel Particulate Filter

See **DPF**.

Diesel Plunger Pump

When individual pumps are contained in an exceedingly single housing with their high-pressure fuel outlets arranged in each line, the assembly is said as a PLN (pump-line-nozzle) system inline fuel injection system pump. They are called PE models (P for "pump" and the E for enclosed camshaft) since they are mounted lengthwise within the bottom of the pump housing and driven from the engine gear train. These pumps may be mounted in one of three ways: base, cradle, or flange. The mounting is set by the drive end of the pump. When an inline pump is flange mounted, a third letter, S, is added to the designation, with the pump designated as a PES unit. Inline pumps are cited by their physical size, which relates to the pumping plunger diameter, the proportion of the fuel they deliver (quantity), and the pressure they' deliver to the nozzle. These pump sizes are M, A, MW, and P.

FIGURE D.7 Bosch line pump.

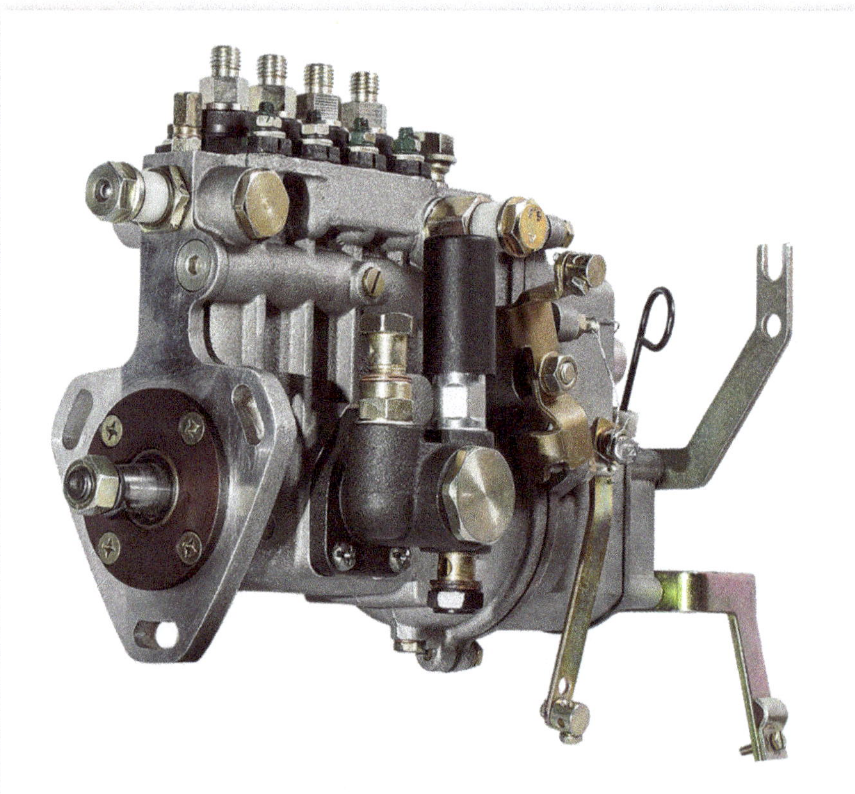

blacklionder/Shutterstock.com

Differential

A gearset within the center of the final drive ring and pinion of a vehicle that permits different wheel speeds at the ultimate drive to permit the vehicle to complete a turn without scuffing the rear tires. The speed of every driven wheel must be allowed to vary slightly as they are going over paved surface irregularities. The open differential consists of several components: differential case, pinion (spider) shaft and gears, and two side (axle end) gears. Torque flows through the ultimate drive to the axles. The method is also lessened into four steps: the drive pinion gear drives the ring gear. A ring gear, bolted to the differential plate, turns the case. The case rotates the differential pinion gear shaft and therefore the pinion gears. The pinion gears drive the side gears, splined to the axle ends, which drive the axles and wheels. For more information on *Nomenclature and Terminology for Truck and Bus Drive Axles*, see **J923_201207** [12].

FIGURE D.8 Automotive final drive differential unit.

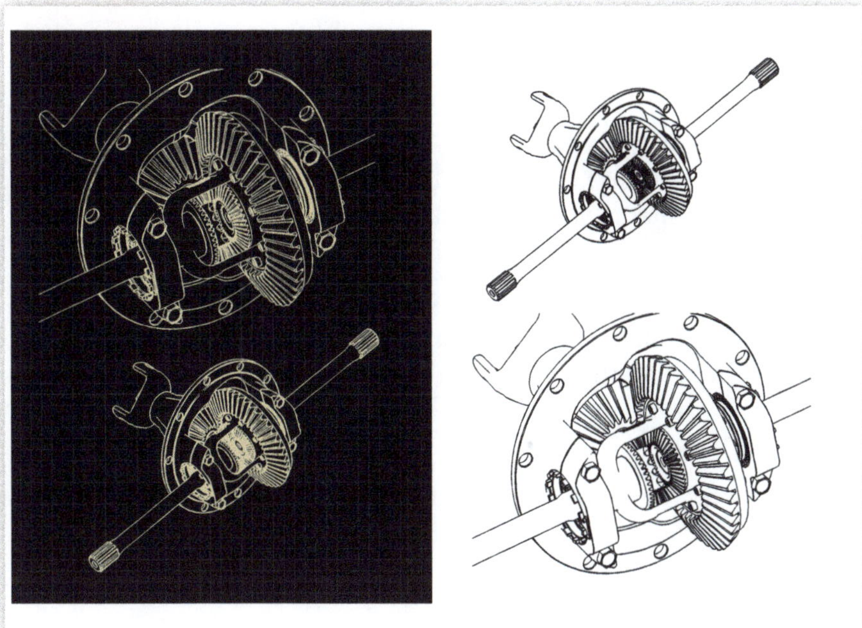

Digital Radio

A radio that uses an electromagnetic signal that broadcasts binary numbers or on/off signals to represent sounds.

Digital Signal

An electronic signal that uses only two voltage levels: on or off.

FIGURE D.9 Analog and digital signal waveform.

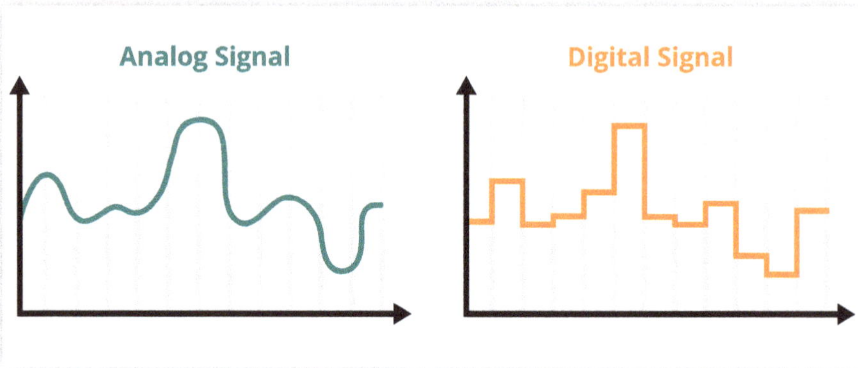

petrroudny43/Shutterstock.com.

Digital Thermometer

An electronic device for measuring temperature. Also called a temperature probe.

Dilution Differential Ratio

The ratio of a specimen to total parts of the solution. A ratio is expressed as parts mixture or specimen to parts diluent.

Diode

A solid-state electronic component that allows current to flow in one direction and is blocked in the other direction.

Diode Trio

A solid-state electronic component that supplies current to the rotor through the electronic regulator.

Dip Stick

A measuring stick to measure the level of oil in an engine or transmission, that is graduated in quarts.

Direct Drive

This action occurs when two meshed gears are the same size and have the same number of teeth, they will turn at the same speed. When the drive gear turns once every revolution of the driven gear, the gear ratio is 1:1, and called direct drive. When a transmission is in direct drive, the engine and transmission turn at the same speed.

FIGURE D.10 Two gears in direct drive with the same number of teeth on each gear.

Syntesi fotografia/Shutterstock.com.

Directional Tires

Tires designed to turn in only one direction to improve wet-road traction by providing more surface contact on the outer edge of the tire with the road.

Directional Wheels

Some vehicles are equipped with directional wheels. The hub location where the wheel nuts are installed shows a casting indicating which location the tire belongs on for some directional wheels. Directional wheels can also be indicated by an arrow around the outer section of the wheel. Directional wheels (with arrows) should always be mounted with the arrow pointing in the direction of rotation. Directional wheels are designed to improve brake cooling with the center section acting as a fan to increase air flow over brake components.

FIGURE D.11 DC Generator.

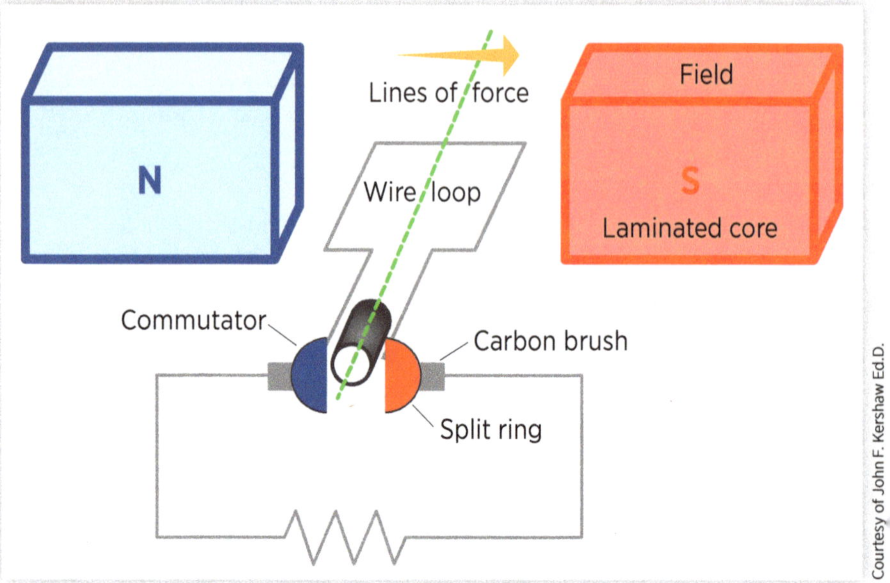

DIS (Distributorless Ignition System)

See **Waste-Spark Ignition**.

Disc Brakes

Disc brake assembly includes the caliper assembly, piston, piston seal, dust seal, bleeder screw, brake pads, and special hardware like anti-rattle springs, and wear indicators. Disc brakes on an automobile add the identical manner. Disc brakes include a rotor between two pads on either side of the rotor. The rotor is attached to the wheel and rotates with it. When the motive force applies the brakes, pressure is applied to the stationary pads; also the pads rub against the edges of the rotor, slowing or stopping its rotation. Disc brakes have advantages over drum brakes:

- Dissipate heat more effectively than drum brakes.
- Apply braking force quicker than drums.
- Disperse water more effectively.

FIGURE D.12 Disc brake assembly on the left and a duo-servo brake assembly on the right.

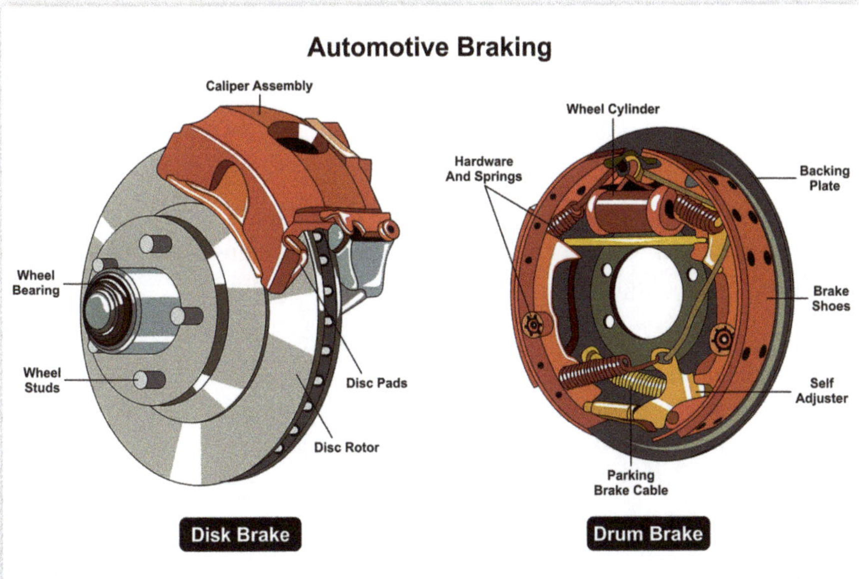

udaix/Shutterstock.com.

Discrete Device

A complete, separately manufactured, individual component with wire leads for connection into a circuit. It is termed discrete because it is not controlled or recognized by the vehicle management computer.

Dispersants

An engine oil additive that reduces sludge formation by keeping the contaminants suspended in the engine oil.

Displacement

Total swept volume of space inside the cylinders of an internal combustion engine (ICE) occupied and then evacuated (swept) by all of the pistons of that ICE. Volume is the measure of how much something holds. The more volume that an engine cylinder has, the more power it can develop. The total piston displacement is found using this formula:

$$V_d = 0.7854d^2 \times s \quad \text{(D.1)}$$

where

d is the piston diameter
s is the stroke
V is cylinder volume

Distributor

Electromechanical component driven by the engine camshaft used to create and distribute the high voltage necessary for spark ignition.

Distributor Indexing

A positive distributor position notch or clamp that permits the distributor to be placed in only one position.

Distributor-Type Diesel Injection Pump

A single, gear-driven pump produces high pressure and a rotating distributor valve routes fuel to the appropriate injection nozzle. The distributor-type fuel injection pump pressurizes and distributes a metered amount of fuel to each cylinder nozzle at the proper time based on the calibrated needs of the engine. This pump uses one pump barrel and a set of plungers to supply all cylinders in a distributing rotation. The pumping element operates eight times because it is used on an eight-cylinder diesel engine and is provided with a distributor or means of connecting the pump delivery to each of the injection nozzles in turn. Some parts have surfaces with machining tolerances measured in microns and require extreme cleanliness during any service. Fuel quantity and timing may be controlled mechanically using a metering valve and speed limiting governor or by an engine management computer.

FIGURE D.13 Distributor-type mechanical diesel fuel injection pump.

Aleksandar Dickov/Shutterstock.com.

Division

A specific segment of an oscilloscope waveform as defined by the grid on the display.

DLC (Data Link Connector)

Connection used in computer-managed systems on cars and trucks to connect the scan tool computer or electronic service tools to the vehicle computers. This device was originally designed for use in a vehicle assembly plant to make sure that all emission-related components were connected and functioning, and was called an ALDL (assembly line diagnostic link) or ALCL (assembly line computer link). For more information on this, see **J1930DA_202105** [3].

DOC (Diesel Oxidation Catalyst)

Causes O_2 (oxygen) to chemically combine with an HC (hydrocarbon) and CO (carbon monoxide) noxious emissions at lower temperatures. The HC and CO are converted to harmless CO_2 (carbon dioxide) and water. Oxidation of these emissions releases heat and takes place without any combustion inside the converter. This design is still used in spark-ignition, gasoline-fueled engines, and they are called two-way converters. Diesel exhaust, oxidation converters can also oxidize black carbon soot to decrease the mass of diesel particulate matter, or PM, emissions to between 5% and 90%.

Dog Clutch

Dog clutches are used where slip is undesirable and/or the clutch does not manage torque. Without slippage, dog clutches are not plagued by wear in the same way that friction clutches are. Dog clutches are used in some manual transmissions and differentials to lock different gears to the rotating input and output shafts. Dog clutches are commonly utilized in racecars.

Dog Tracking

A term accustomed to describe the condition where the rear wheels do not follow directly behind the front wheels; named after the way some dogs run with their rear paws offset toward one side in order that they are going to not hit the front paws while running.

Dolly

Platform on wheels with two axles and four wheels in total, one on each corner. This term is used for a trailer hauling device when a fifth wheel is mounted on a truck axle and uses the fifth wheel to attach to a trailer that is connected to a different trucking rig.

Doppler Effect

The difference between the observed frequency and therefore the emitted frequency of a wave for an observer moving relative to the source of the waves. It is commonly heard when a vehicle sounding a siren approaches, passes, and recedes from an observer. The emitted signal toward the car is reflected back with a variation of frequency that relies on the speed away/toward the radar. This can be only a component of the important speed.

FIGURE D.14 Doppler effect.

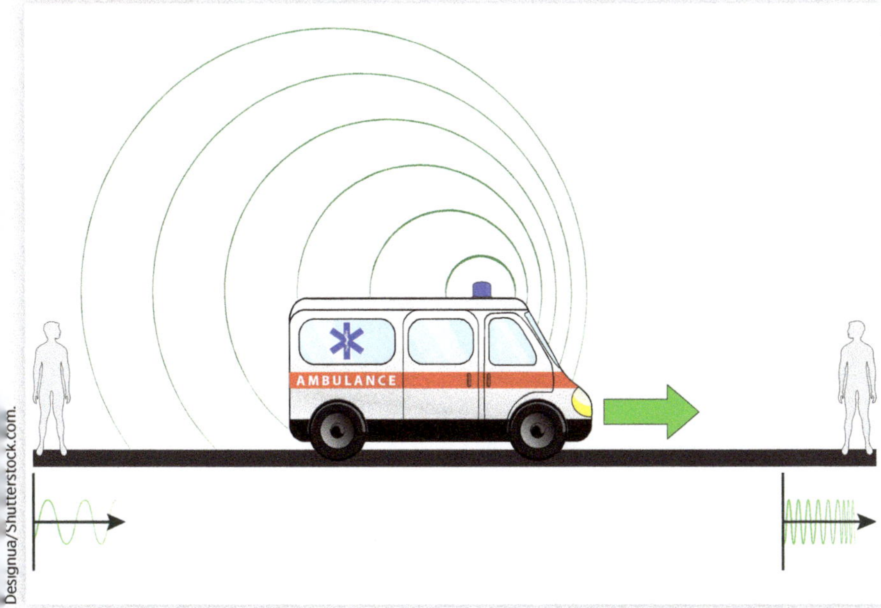

DOT Tire Certification Codes

DOT certification codes are a series of numbers and letters located on the inner sidewall of the tire. It indicates the compliance of a tire and certification with Department of Transportation (DOT) safety standards and may be required for warranty purposes. This label designates the manufacturer, size, and production date of a tire. The production date is a three- or four-digit number; the first two digits signify the week, and the last digit or two digits indicate the year of manufacture. For more information on this, see **J2047_201911** [7].

FIGURE D.15 DOT tire certification codes.

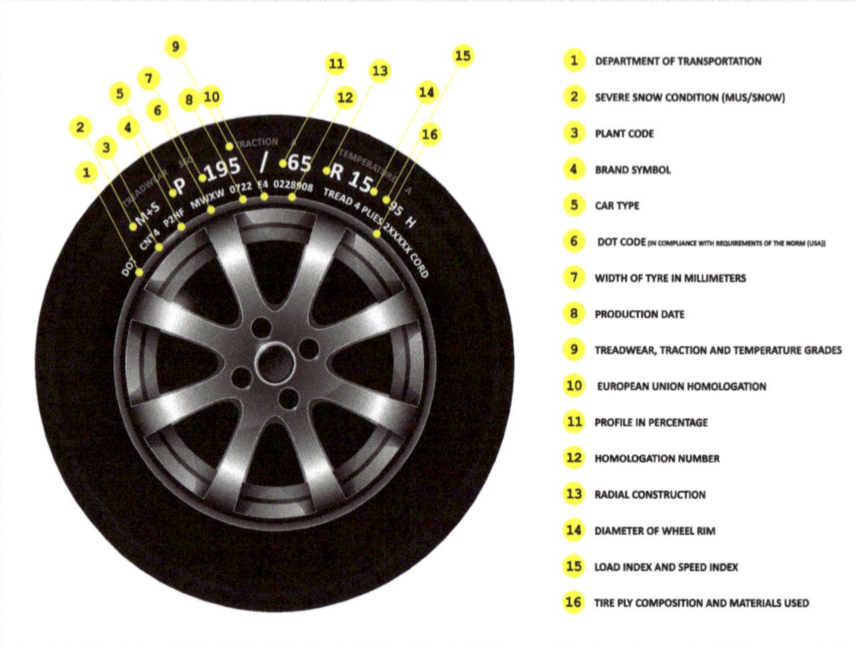

Media Guru/Shutterstock.com.

Double-Wishbone Modified Strut Suspension

The double-wishbone modified strut suspension uses a high-mount upper arm. It uses a coil over the shock absorber (green arrow) damper where the upper end of the stabilizer link (yellow arrow) attaches to an aluminum hub carrier (steering knuckle or upright) for a 1:1 motion ratio. Links at the white arrow connect the lower frame to the stabilizer bar. The 1:1 ratio means this is a direct-acting stabilizer; as the bar offers everything it has, it is often smaller and lighter and the suspension is more responsive.

FIGURE D.16 Double-wishbone modified strut suspension.

patruflo/Shutterstock.com.

Courtesy of John F. Kershaw Ed.D.

Downdraft Carburetor

The carburetor is located above the engine intake and the draft or pull is downward from the low pressure created during the intake stroke. Early systems relied on gravity to deliver fuel to the cylinders.

DMM (Digital Multimeter)

A multimeter that uses a vacuum fluorescent or liquid crystal display to provide a digital readout to measure volts, ohms, frequency, and amperes. For more information on this, see **J1930DA_202105** [3].

DMS (Driver Monitoring System)

This safety system uses an infrared camera mounted on the steering column and infrared illumination provided by infrared light emitting diodes (LEDs) embedded within the wheel to observe the driving force and provides an estimate of the driver's attention state to the Active Safety Control Module. The resultant image of the motive force is not visible to the driving force. The motive force monitoring system is employed as an input to enable/disable the cruise system; therefore, the driver monitoring system should only move when its function is requested by the Active Safety Control Module. The motive force monitoring system infrared LEDs could also be visible as a dim red glow when the system is active.

DPF (Diesel Particulate Filter)

A filter that collects particulate matter (PM) discharged from the engine. When enough PM is accumulated in the filter, regeneration is performed. Passive or automatic regeneration last about 9 to 40 min and occurs in 100–500 miles of driving. After Stage 1, PM consisting of small particles of carbon remaining after combustion are retained from the exhaust gas by the large surface area of the DPF. For more information on this, see **J1145_201109** [6].

FIGURE D.17 Diesel particulate filter (DPF).

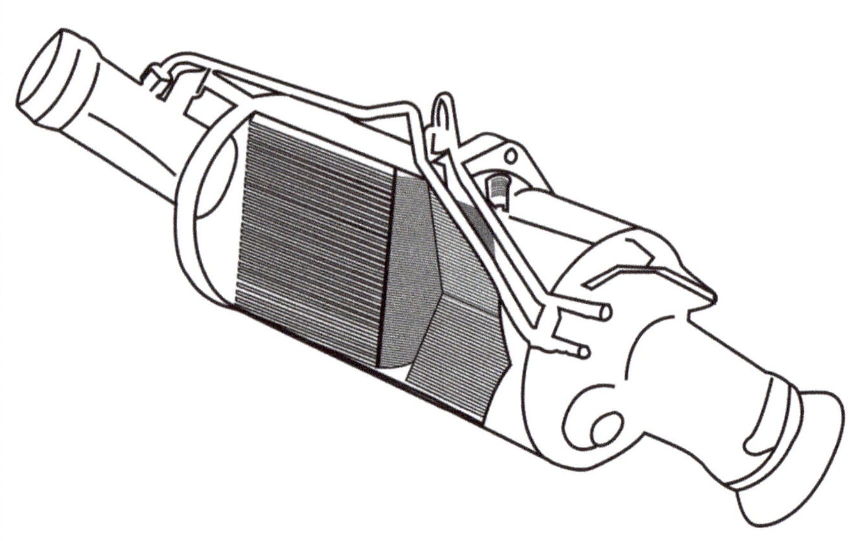

Kanan Shabanov/Shutterstock.com.

Drag

Due to the fact that air flows over the body of a car or truck, it causes a drag on free vehicle movement. A semi-empirical model was developed to represent this drag effect. Aerodynamic drag is demonstrated by the equation:

$$D_A = \frac{1}{2} V^2 C_D A \quad \textbf{(D.2)}$$

where

D_A is the aerodynamic drag
C_D is the aerodynamic drag coefficient
A is the frontal area of the vehicle

Drag Coefficient

Dimensionless quantity accustomed to quantify the drag or resistance of an object in an exceedingly fluid environment, like air or water. It is utilized in the drag equation in which a lower coefficient of drag indicates the article will have less aerodynamic or hydrodynamic drag. The coefficient is usually related to a specific area. For information on this, see **J1594_201007** [8].

Drag Link

A drag link is sometimes used in an automotive steering system to convert rotary motion from a crank arm to a second bell crank. Trucks use a drag link to connect the steering arm that connects to a cross shaft to the opposite steering arm to the pitman arm of the steering gear. Also called a centerlink or relay rod. Drag links are a variation of the centerlink and are used on many modern-day trucks and sport utility vehicles.

FIGURE D.18 Drag link.

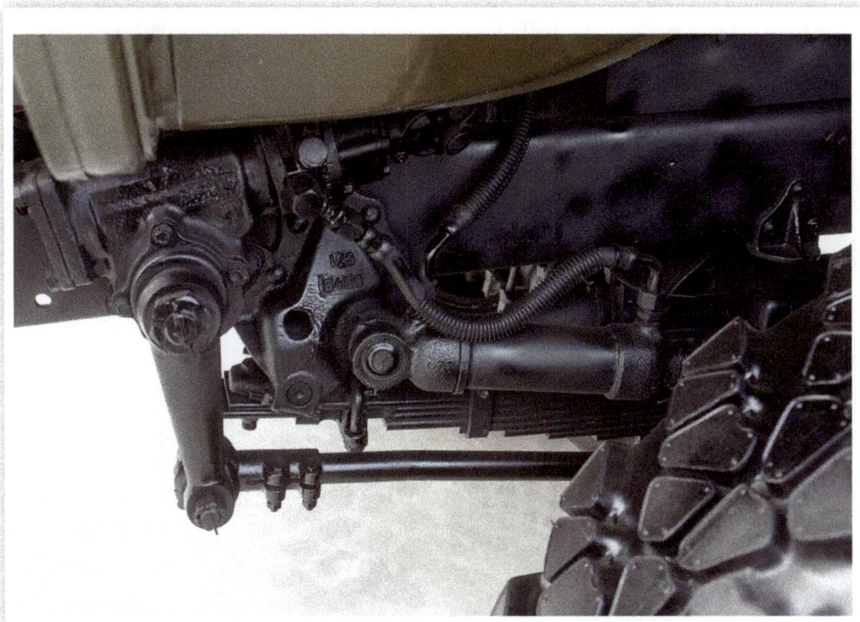

Andrey 69/Shutterstock.com.

Drift

A controlled lateral sliding of a vehicle.

Drivability

The general evaluation of an ICE operating that includes idle smoothness, cold and hot starting, throttle response, and power delivery.

Driveline

Components like universal joint and the propeller shaft or driveshaft that connect the transmission with the driving axles of a vehicle. See **Drivetrain**.

Driveline Angle Phasing

To reduce vibration in a single cardan U-joint, make sure the driving and driven yokes are in phase with one another, which means they are able to cancel out vibration. To check driveline phasing on a vehicle, place the transmission in neutral and raise it on a lift on jack stands. Rotate the rear wheels until the output yoke attached to the transmission output shaft has its ears pointed at the 12 o'clock and 6 o'clock positions. The input yoke attached to the input shaft of the differential should also be in the 12 o'clock and 6 o'clock positions, which is in-phase with the front yoke. Another way to look at this is to observe the yoke ears on the driveline itself. They should be in the same position on both ends of the driveline.

FIGURE D.19 Driveshaft or propeller shaft yokes in phase.

Kudrin Ilia/Shutterstock.com.

Driveshaft (Propeller Shaft)

All RWD (rear-wheel-drive) cars and light-, medium-, and heavy-duty trucks have a driveshaft that is a balanced tubular type. A splined slip joint is provided in each driveline system on a non-tandem axle arrangement. The slip joint is at the forward end of the rear shaft. All driveshafts are designed and built with the yoke lugs (ears) in line with one another. This condition is termed phasing. The shaft will absorb the vibration from the speeding up and slowing down anytime the coupler rotates. Mark the driveshafts when disassembling to assure proper assembly alignment. Some driveshafts use alignment arrows stamped on them at the time of production for correct alignment.

FIGURE D.20 Automotive driveshaft also called a propeller shaft.

notsuperstar/Shutterstock.com.

Driveshaft (Propeller Shaft) Angles Measuring

To measure U-joint and driveshaft angles, the vehicle must be lifted using an axle contact or drive-on lift. You can also lift the vehicle and place the axles on jack stands. This preserves the actual, loaded driveshaft angles. The working angles of the two U-joints on a drive should be within 0.5° of one another to cancel out vibration. To measure the working angle of a U-joint, do the following:

- Place an inclinometer on the front U-joint bearing cap to measure the transmission angle. Let us suppose the transmission angle is 10°.
- Rotate the driveshaft 90° and read the angle of the driveshaft.
- Subtract the larger angle from the smaller angle, which, in this case, is 13° – 10°, which gives you the working angle of the front joint, which is 3°.
- Repeat the same procedure to calculate the working angle of the rear joint, which is shown as 16°.

FIGURE D.21 Measuring driveline angle.

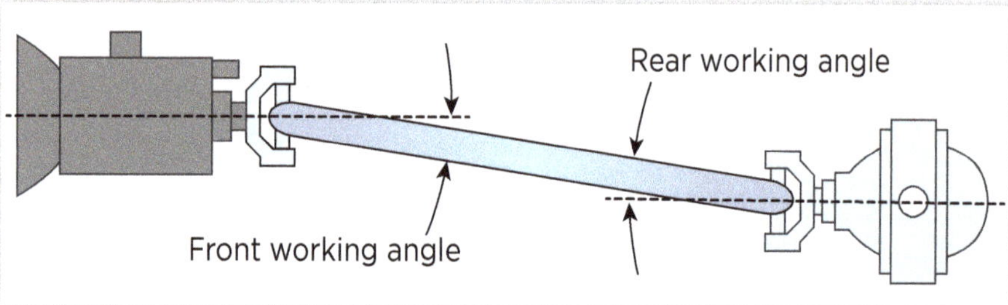

Courtesy of John F. Kershaw Ed.D.

In this example, the working angles are the same, or within 0.5°, and no adjustment is necessary. If the difference between the two working angles is greater than 0.5°; however, shims can be added to bring the working angles closer together. The angle of the rear joint can be adjusted by installing a tapered shim between the leaf spring and the axle. The angle of the front joint can be adjusted by adding or removing shims from the mount under the transmission or by replacing the rear transmission mount.

Driveshaft (Propeller Shaft) Balancing

If the propeller shaft (driveshaft) runout is within specification and a vibration still exists, the balance of the shaft should be checked and fixed. The procedure for testing driveshaft balance using a strobe balancer: Raise the vehicle and mark the propeller shaft with four equally spaced marks around its circumference. Label each mark with the numbers 1, 2, 3, and 4. Attach the strobe balancer sensor to the bottom of the differential housing as close to the companion flange as possible. With the vehicle securely lifted and the drive wheels off the ground, start the engine and put the transmission into gear to allow the drive wheels to rotate. Hold the strobe light close to the marks on the propeller shaft. If the light does not flash, the propeller shaft is balanced and no corrective action is necessary. If the light does flash, observe what number mark is shown by the flashing light. Apply hose clamps so that the screw portion of the clamp(s) is opposite the number seen with the strobe light. The screw portion of the hose clamp is the corrective weight. Remember the strobe light sensor was mounted to the bottom of the differential housing. The strobe light flashes when the heavy part of the propeller shaft is facing downward. If the heavy part of the propeller shaft is down, then corrective weight must be added to the opposite side of the propeller shaft.

Drivetrain

Transfers torque, or a twisting force produced by an engine, to the drive wheels. This system allows the driver to vary the torque and speed of the drive wheels according to driving conditions to disengage engine power from the drive wheels while the engine runs and to select a forward or reverse direction of vehicle travel. In addition, the drivetrain allows the driven wheels to make turns without causing excessive tire and gear wear.

FIGURE D.22 Automotive drivetrain.

patruflo/Shutterstock.com.

Driving Automation—Driver Assistance

At Level 1, the bottom rung of automation, a vehicle has a minimum of one driver web that has steering assistance, or braking and acceleration assistance. The driving force remains liable for driving the vehicle and must be prepared to require control at any time and for any reason. Adaptive controller is an example of Level 1 driver assistance technology. It maintains a secure following distance between your vehicle and traffic ahead with no intervention by the driving force. A steering assistance feature, like lane-centering assistance or lane-following assistance, would also qualify as Level 1 autonomy. However, a vehicle with both of those features working together qualifies as Level 2 driving automation. For more information on this, see **J3016_202104** [9].

Dropping Point

The temperature where grease passes from a semisolid to a liquid state under conditions specified by the ASTM (American Society for Testing and Materials).

Drum Brakes

Drum brakes use a drum that is attached to, and rotates with, the wheel. Inside the drum is a pair of stationary brake shoes that spread apart and rub against the inside of the drum when the brakes are applied. In all current vehicles that use rear wheel drum brakes, they are used only on the rear wheels. There are two basic types of drum brakes: leading-trailing and duo-servo. Drum brakes use the same principles as disc brakes. However, drum brakes have a drum that surrounds the brake shoes and wheel cylinder. The backing plate holds the shoes, springs, and wheel cylinder inside the brake drum. The backing plate bolts to the axle housing or spindle, and also helps keep dirt and water off the brakes. The wheel cylinders use master cylinder pressure to force the brake shoes against the drums. The boots keep water and dirt out of the cylinder. Special rubber seals prevent fluid from leaking past the pistons. Springs hold the rubber cups against the pistons when the pistons are not pressurized. The bleeder screw threads into the wheel cylinder.

Hydraulic pressure can be used to force air and fluid out of the system through the bleeder screw. Return springs pull the brake shoes away from the brake drums, which also forces the wheel cylinder pistons toward the center of their bores. One end of the return spring fits into a hole in the brake shoe. The other end attaches either to the other brake shoe or to an anchor pin on the backing plate. Leading-trailing brakes have the brake wheel cylinder attached to the top of the backing plate and an anchor point at the bottom. The brake shoe at the front is called the "leading" shoe, while the shoe toward the rear of the vehicle is the "trailing" shoe. Duo-servo brakes have a fixed anchor point at the top of the backing plate, above the wheel cylinder. The self-adjuster connects the two brake shoes at the bottom. In the duo-servo system, the front brake shoe is called the primary shoe, while the rear is called the secondary shoe.

When the vehicle is moving forward, and the brake pedal is applied, the primary shoe is forced against the rotating brake drum. Like the leading shoe (leading-trailing type), friction forces the shoe to move more tightly against the drum, causing self-energizing action. There is no fixed anchor point at the bottom of the backing plate, so the primary shoe transfers rotation to the secondary shoe. Since the secondary shoe is held by the anchor point at the top of the backing plate, the bottom of the shoe pivots into the brake drum. This transfers the thrust of the primary shoe to the secondary shoe, causing it to self-energize as well.

FIGURE D.23 Duo-servo drum brake (left) and a leading drum brake (right).

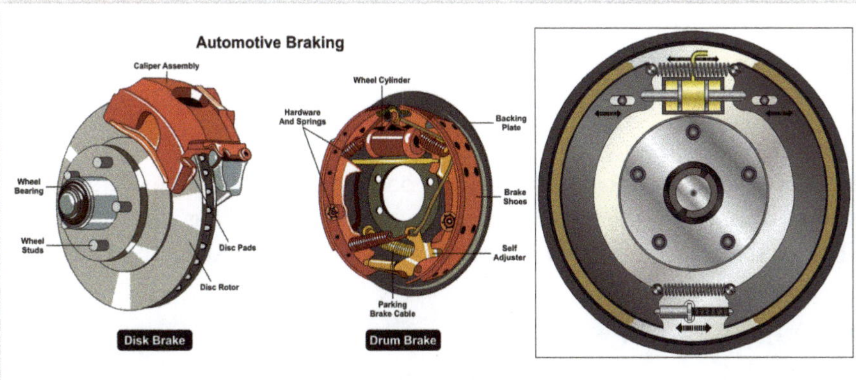

udaix/Shutterstock.com.

Fouad A. Saad/Shutterstock.com.

Drum Brakes Duo-Servo Incremental Self-Adjuster

Duo-servo brakes use a similar type of self-adjusting mechanism as the leading-trailing brakes. The duo-servo uses a two-piece adjuster assembly with a star wheel screw to increase and decrease its length. An adjuster lever is used to rotate the star wheel and allow the brakes to automatically adjust themselves as the vehicle is operated in reverse.

Drum Brakes Leading-Trailing Incremental Self-Adjuster

An incremental self-adjuster rides between the brake shoes slightly below the wheel cylinder. This self-adjuster consists of an adjuster screw retracting spring, adjuster lever, and a two-piece adjuster assembly with a star wheel. The star wheel is part of a threaded screw, which, when turned, increases or decreases the length of the adjuster assembly. This type of self-adjuster is actuated when the driver moves the vehicle forward for about ten feet and then reverses and applies the brake until the vehicle is completely stopped. This action causes the adjuster screw retracting spring to tug the adjusting lever up over the tip of the tooth of the star wheel. When the linings have worn a particular amount, the adjuster lever engages the star wheel tooth because it tries to maneuver past it. When you take your foot off the brake pedal, the adjuster assembly pushes the adjuster lever out. The lever pivots downward, rotating the star wheel and lengthening the adjuster assembly. Because the adjuster assembly lengthens, it spreads the brake shoes and moves the linings closer to the cylinder.

Dry Manifold

No integral coolant passages are cast into the intake manifold.

Dry Park Test

A test of steering and/or suspension components with the wheels in the straight-ahead position and the vehicle on level ground. The steering wheel is turned while all steering and suspension components are inspected for any looseness.

Dry Sump Lubrication System

An ICE lubrication system that uses two or more oil pumps and a separate oil reservoir, as opposed to a wet-sump system that only has the oil pan sump under the crankcase and a single oil pump. The dry sump system is found on race and off-road applications that make hard turns that could result in loss of oil suction.

Dry Weight

The weight or mass of a weight excluding the weight of all consumables, such as fuel and oils.

DSO (Digital Storage Oscilloscope)

An oscilloscope that samples and stores waveforms in a digital format, but not in real time. It can miss glitches that may occur between samples, so a DSO with a high "sampling rate" is preferred. DSOs have the ability to data log scope traces.

FIGURE D.24 Digital storage oscilloscope.

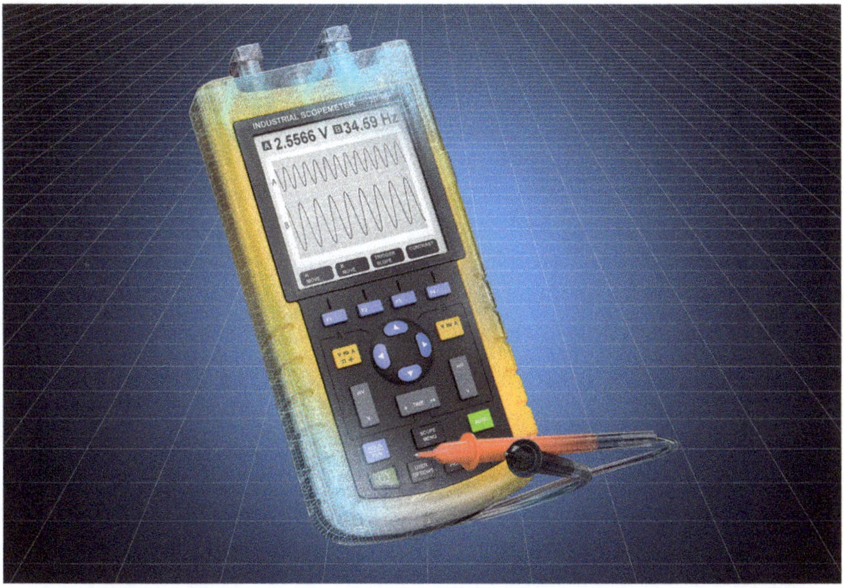

DSRC (Dedicated Short-Range Communications)

One-way or two-way short-range to medium-range wireless communication channels specifically designed for automotive use with a corresponding set of protocols and standards. Under this proposed rule, vehicles would broadcast an outlined data packet, the BSM (basic safety message) up to ten times per second, indicating vehicle location, heading, and speed. In March 2017, General Motors (GM) became the primary United States (US) automaker to produce DSRC as standard equipment on a production automobile, the Cadillac CTS. The USA also has appropriate standards—IEEE (Institute of Electrical and Electronics Engineers) 802.11p and frequency rules in situ. In Europe a frequency harmonized for transport safety and a harmonized standard, called ETSI ITS-G5, is in situ.

DTC (Diagnostic Trouble Code)

Alphanumeric or numeric sequence, indicating a fault in a vehicle operating system. For more information on this, see **J1930DA_202105** [3].

Dual-Mass Flywheel

At low speeds the power strokes are far enough apart to cause speed fluctuations at the flywheel. All, but an excessively heavy, flywheel might allow these fluctuations to reach the driveline as torsion vibrations. These oscillations can produce rattling and other noises in the transmission and driveline due to gear backlash and necessary running clearances. Torsion vibration may even cause the driveline to resonate. A flywheel that is heavy enough to eliminate all low-speed oscillations would add too much mass to the crankshaft, hindering the acceleration ability of the engine. To compensate, a dual mass flywheel may be used to absorb most of the irregular oscillations before they reach the transmission. A dual-mass flywheel may also change the resonant speed of the driveline vibrations. It lowers the resonant speed below any speed at which the vehicle operates, such as below the normal idle speed.

Dual Master Cylinder

A dual master cylinder contains a primary and secondary piston; they sit one behind the other in the master cylinder bore. When the brake pedal is depressed, both pistons are forced down the bore. Each piston draws fluid from a separate reservoir and applies two wheel cylinders. The path the fluid takes is different, depending on whether the brake system is a vertical split or diagonal split. If one system fails, the other system will have partial braking ability to stop the vehicle.

Dual Motor-Generator Hybrid

A hybrid vehicle that uses two separate motor-generators located at different locations within the drivetrain. The two motor generators are usually located inside the transmission or transaxle. For more information on this, see **J1715/2_202108** [10].

Duration

Number of crankshaft degrees that the valve is open during crankshaft rotation. At the top of every stroke, the piston is either at top dead center (TDC) or bottom dead center (BDC). A stroke requires one-half of a crankshaft rotation, or 180°. Duration is measured from zero lift just before valve opening to zero lift just after valve closure. However, since it is important to require up valve train slack and lift together with closing the valve gently lightly, therefore the lobe moves the valve very slowly at the start and end of its cycle. This also makes it very difficult to work out or measure precisely when the valve begins to open or close. To resolve this issue, camshaft manufacturers now measure duration at specific points within the valve lift. The foremost popular specific point is 0.050 in. This measuring point makes it easy to live, and this offers an accurate point at which to start out and stop measuring duration. The upper the lift at which the duration measurement begins and ends, the less rated duration a camshaft will have. The duration of a cam lobe measured between the 0.050-in. lift points is over 50°, but that is measured between the particular opening and shutting points. For more information on this, see **J604_201108** [11].

Durometer

A standardized way to measure the hardness of materials like rubber, elastomers, and plastics. Durometer measurement scales range from 0 to 100. It is a dimensionless measurement.

Duty Cycle

The amount of time (given as a percentage) that current is fed to an ignition coil.

Dwell

The amount of time, recorded on a dwell meter in degrees, which voltage passes through a closed switch also called ignition points. For more information on this, see **J139_202002** [4].

Dwell Section

Dwell is the amount of your time that the current is charging the coil from the transistor-on point to the transistor-off point. The tip of the dwell section marks the start of the following firing line. This time is termed "transistor off" and indicates that the first current of the coil is stopped, leading to a high-voltage spark out of the coil. A part of a secondary scope pattern that starts when the ignition module conducts primary current through the induction coil. For more information on this, see **J139_202002** [4].

Dynamometer (Dyno)

Measures ICE power output and performance. By loading the engine, the dynamometer can check engine acceleration, maximum power output, and on-the-road performance characteristics. This may be used for just the ICE on an engine dynamometer or the entire vehicle on a chassis dynamometer.

FIGURE D.25 Chassis dynamometer.

Pepermpron/Shutterstock.com.

REFERENCES

1. Society of Automotive Engineers, J3152_202005: Taxonomy and Guidelines for Design, Fabrication, and Installation of Brake NVH Fixtures for Inertia Dynamometer Testing, SAE Publishing, Warrendale, PA, 2011.
2. Society of Automotive Engineers, J1939/21_2021: Data Link Layer, SAE Publishing, Warrendale, PA, 2021.
3. Society of Automotive Engineers, J1930DA_202105: Electrical/Electronic Systems Diagnostic Terms, Definitions, Abbreviations, and Acronyms Web Tool Spreadsheet, SAE Publishing, Warrendale, PA, 2021.
4. Society of Automotive Engineers, J139_202002: Ignition System Nomenclature and Terminology, SAE Publishing, Warrendale, PA, 2020.
5. Society of Automotive Engineers, J313_201706: Diesel Fuels, SAE Publishing, Warrendale, PA, 2017.
6. Society of Automotive Engineers, J1145_201109: Emissions Terminology and Nomenclature, SAE Publishing, Warrendale, PA, 2010.
7. Society of Automotive Engineers, J2047_201911: Tire Performance Terminology, SAE Publishing, Warrendale, PA, 2019.
8. Society of Automotive Engineers, J1594_201007: Vehicle Aerodynamics Terminology, SAE Publishing, Warrendale, PA, 2010.

9. Society of Automotive Engineers, J3016_202104: Taxonomy and Definitions for Terms Related to Driving Automation Systems for On-Road Motor Vehicles, SAE Publishing, Warrendale, PA, 2021.
10. Society of Automotive Engineers, J1715/2_202108: Hybrid Electric Vehicle (HEV) and Electric Vehicle (EV) Terminology, SAE Publishing, Warrendale, PA, 2021.
11. Society of Automotive Engineers, J604_201108: Engine Terminology and Nomenclature—General, SAE Publishing, Warrendale, PA, 2011.
12. Society of Automotive Engineers, J923_201207: Nomenclature and Terminology for Truck and Bus Drive Axles, SAE Publishing, Warrendale, PA, 2012.
13. Society of Automotive Engineers, J1390_202202: Cooling Systems Standards Committee: Ground Vehicles, SAE Publishing, Warrendale, PA, 2022.

E

E-85

Gasoline fuel composed of 85% ethanol and 15% gasoline. E85 refers to an ethanol fuel blend of 85% ethanol fuel and 15% gasoline or other hydrocarbons by volume. In the USA the precise ratio of fuel ethanol to hydrocarbon may vary in step with ASTM 5798 which specifies the allowable ethanol content in E85 as starting from 51 to 83%. This may be because the lower calorific value of neat ethanol makes it difficult to crank engines in relatively cold climates without pre-heating air intake, faster cranking, or mixing varying fractions of gasoline in line with the climate. Cold cranking in cold climates is the primary reason ethanol fuel is mixed with any gasoline fraction.

Earthmoving Machines

SAE defines three different types of earthmoving machines:

- Equipment: Set of components mounted to the base machine to fulfill the primary design function
- Attachment: Optional assembly of components that can be mounted to the base machine for a specific use
- Component: Part or assembly of parts of a base machine, equipment, or attachment"

For more information on the *Identification Terminology of Earthmoving Machines*, see **J1057_199905** [12].

EAS (Electrically Assisted Steering)/EPS (Electric Power Steering)

This steering system eliminates all hydraulic components and fluid. An electrical motor replaces the pump. EAS or direct wattage steering completely eliminates hydraulic fluid and also the accompanying hardware from the facility mechanism, creating a complete EPS system. An EPS steering mechanism uses an electrical motor attached either to the steering rack or to the steering column via a mechanism and torque sensor. A microprocessor or electronic control unit and diagnostic software control the steering dynamics and driver effort. Inputs include vehicle speed, steering wheel torque, position, and turning rate. There are four primary styles of electrical power assist steering systems:

1. Column-assist type: During this system, the facility assist unit, controller, and torque sensor are attached to the steering column. The ability assist unit is the motor and the controller is the electronic control unit, and therefore the torque sensor measures the load on the steering wheel.
2. Dual pinion–assist type: During this system, the facility assist unit is attached to the gear pinion shaft. The facility assist unit sits outside the vehicle passenger compartment, allowing assist torque to be increased greatly without raising interior compartment noise.
3. Rack-assist type: During this system, the facility assist unit is attached to the gear mechanism rack using either a dual pinion or a belt drive from a motor to the rack. It is located on the rack to permit greater flexibility within the layout design. Some OEMs use the belt-drive system.
4. Direct-drive type: During this system, the gear mechanism rack and power assist unit form one unit. The mechanism is compact and fits easily into the engine compartment layout. Direct assistance to the rack enables low friction and inertia (resistance to a change in motion), which successively gives a perfect steering feel.

FIGURE E.1 Electric power steering system.

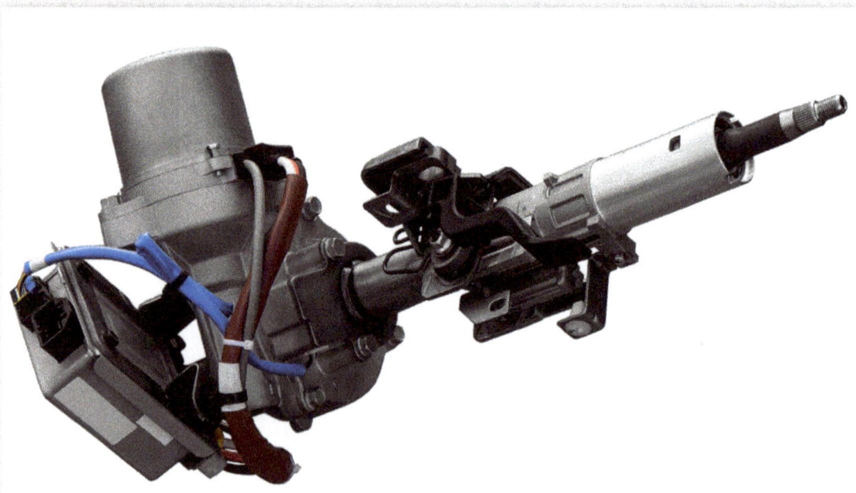

BACHTUB DMITRII/Shutterstock.com.

EBA (Emergency Brake Assist) or BA/BAS (Brake Assist)

Automobile braking technology applies the brakes in an emergency situation through the use of ABS. The system detects the force placed on the pedal in emergency braking situations, and because drivers tend to brake without enough force in these situations, it boosts the brakes to the calculated speed at which the pedal is pressed to achieve full braking power. Brake pressure is then increased to the point just before ABS would kick in, provided the foot pedal remains pressed down. This shortens the stopping distance and reducing the likelihood of accidents. If an emergency develops, a slow reaction and lesser-than-maximum braking input could lead to insufficient time or distance to avoid a collision. EBA or BA can detect panic stops and apply maximum braking effort quickly. It deduces braking behavior by assessing the speed at which the foot pedal is activated. If the system identifies an emergency, it spontaneously initiates full braking faster than any driver can move their foot. Emergency stopping distances will be shortened, reducing the likelihood of collisions. The BA system can enhance braking effort and reduce stopping distances by up to 70 ft at 125 mph. The BA detects circumstances during which emergency braking is required by measuring the speed with which the foot lever is depressed. Some systems additionally take into consideration the speed at which the pedal is released, pre-tensioning the brakes when a panic release of the throttle is noted. When emergency braking is detected, the BA system automatically generates a maximum brake boost to mitigate the driver's tendency to brake without sufficient force. BA can reduce the stopping distance. It may be a lower level of automation than a forward collision alert or avoidance system, which can initiate braking on its own if the on-board computer detects an imminent collision. For more information on this, see **J3063_202103** [8].

EBCM (Electronic Brake Control Module)

A control module that processes all ABS (anti-lock brake system) information and signal functions. It receives and interprets voltage pulses generated by the wheel speed sensors as the exciter teeth pass by the probe. The EBCM uses this data to determine impending wheel lockup and when and how to activate the ABS modulator valves.

E-Coil

Electronic ignition coil in the shape of an "E." For more information on this, see **J139_202002** [2].

Eccentric

The relationship of two round parts having different centers; a part which contains two round surfaces, not on the same center.

ECM (Engine or Electronic Control Module)

The on-board computer of the engine management system that controls fuel and emissions, as well as diagnostics, for the engine management system of the vehicle (also sometimes referred to as the electronic control unit [ECU]). The Society of Automotive Engineers (SAE) standard J-1930 standardizes this term along with PCM (powertrain control module) that provided computer management of both the engine and transmission. The CPU (central processing unit) is mounted on one or more circuit boards and installed in a metal case to provide shielding from EMI (electromagnetic interference). The wiring harnesses that connect the ECM/PCM to the input sensors and output actuators connect to multi-pin connectors. These controllers manage all of the separate and linked electronic systems through a CAN (controller area network) in the vehicle. For more information on this, see **J1930DA_202105** [1].

ECT (Engine Coolant Temperature) Sensor

Threaded in the engine coolant jacket in direct contact with the engine coolant. The coolant sensor contains a thermistor used to measure the engine coolant temperature. The ECT is a thermistor in a circuit which is a variable ground, which uses a voltage divider network where the voltage is divided between the sensor input and a sensor ground inside the computer. The computer provides a 5-V reference signal to the ECT sensor. When cold, the sensor provides high resistance, which the computer reads as high signal voltage. As the engine warms up, the thermistor sensor resistance becomes lower and the signal voltage drops, so that is the difference. For more information on this, see **J1930DA_202105** [1].

FIGURE E.2 Engine coolant temperature.

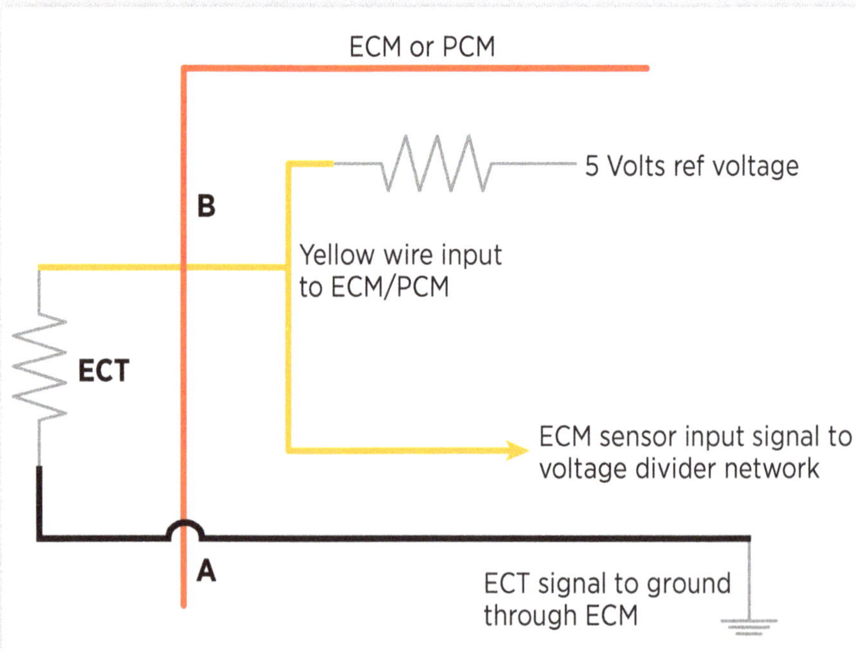

Courtesy of John F. Kershaw Ed.D.

ECU (Electronic Control Unit)

An automotive engine or system management computer generally used by several different OEMs. It is another term that describes an ECM; see **ECM**.

EDR (Event Data Recorder)

A monitoring device that is generally part of a supplemental restraint system. It is used to monitor and record vehicle operating conditions and store them in memory for extraction. It is referred to as the automotive black box. Some OEMs call the EDR an SDM (sensing diagnostic module). For more information on this, see **J1674_201807** [6].

EEPROM or E^{2}PROM (Electronically Erasable Programmable Read Only Memory)

This computer chip can be erased and reloaded with new information using a special software program. EEPROMs can be completely or partially reprogrammed using appropriate software from the engine manufacturer. EEPROMs do not have to be removed from the ECM for reprogramming. The programming information is downloaded from a personal computer to the programmable memory of the ECM through a DLC. All vehicles equipped with OBD II are equipped with EEPROMs. The ability to program customized information and operating parameters means that the same model engine can be tailored to respond to unique operating conditions and commands. GM also calls the EEPROM a Flash prom of E^{2}PROM. For more information on this, see **J1930DA_202105** [1].

EFI (Electronic Fuel Injection)

A fuel injection system that uses an engine management computer to control the opening time of solenoid-controlled fuel injectors that are called actuators. System operation is based on a series of input sensors and the system can be an open loop or closed loop with an oxygen feedback sensor

EFT (Engine Fuel Temperature) Sensor

Some vehicles are equipped with an electronic returnless type of fuel injection. These systems use an engine fuel temperature (EFT) sensor to give the engine management computer data regarding the temperature and, therefore, the density of the fuel. For more information on this, see **J1930DA_202105** [1].

EGR (Exhaust Gas Recirculation)

An emission control device to reduce NOx (oxides of nitrogen) by adding spent gas or exhaust into the combustion chamber to lower combustion temperature. For more information on this, see **J1145_201109** [3].

Elastomer

A natural or synthetic polymer with elastic properties, e.g., latex.

Electric A/C Compressor

A high-voltage 201 V alternating current (AC) motor operates a scroll-type air conditioner (A/C) compressor for use on hybrid electric vehicles (HEV) and electric vehicles (EVs). An AC inverter supplies the AC to operate the compressor and does not require an ICE to operate it. Polyol ester A/C refrigerant oil that is non-conductive is required. Other refrigerant oils will contaminate the A/C system.

Electric Brake Retarder

The electric retarder is used on heavy truck and buses to provide noise-free braking to support the vehicle service brakes. Electromagnets are used to generate drag that is transmitted directly to the vehicle propeller shaft, creating a retarding force. Power for the electric retarder is supplied from the 12 or 24 V electrical system of the vehicle. The Caterpillar BrakeSaver is a hydraulic retarder that bolts directly to the rear of the engine crankshaft flange. When the brake is activated, engine oil is forced under pressure against the rotor vanes in the retarder, thereby slowing the rotor and providing braking power to the rear wheels.

Electric Circuit

An electric circuit consists of a power source, wiring, and load to use electrical energy.

Electric Device

A device that uses electricity to do work, such as motors, switches, relays, solenoids, and light bulbs.

Electric Motor

An electromagnet is created in a loop of wire that is placed between two electromagnetic poles. Motors work on the principle of magnetic repulsion. This magnetic repulsion takes place when a straight-loop wire conductor made up of an armature, commutator, and brushes is located in a magnetic field and current flows through that wire loop. This situation creates two separate magnetic fields. One produced by the magnet (poles of the magnetic field winding) and another produced by the current flowing through the conductor (armature/commutator/brushes)—the magnetic field of the magnet moving from the S-Pole to N-Pole and magnetic field of the conductor flowing around the conductor.

FIGURE E.3 Current flows from the negative (–) battery terminal through the brush and copper ring nearer the S-Pole through the conductor (armature) to the copper ring and brush nearer the N-Pole and back to the positive (+) battery terminal. This electrical flow causes the portion of the loop near the N-Pole to push downward and the S-Pole to push upward. When there is a strong field on one side of the conductor and a weak field on the other side, the conductor will move from the strong side to the weak side. A weaker magnetic field between N and S Poles on one side of the conductor is repulsed by the stronger magnetic field on the other side of the conductor. The armature then rotates.

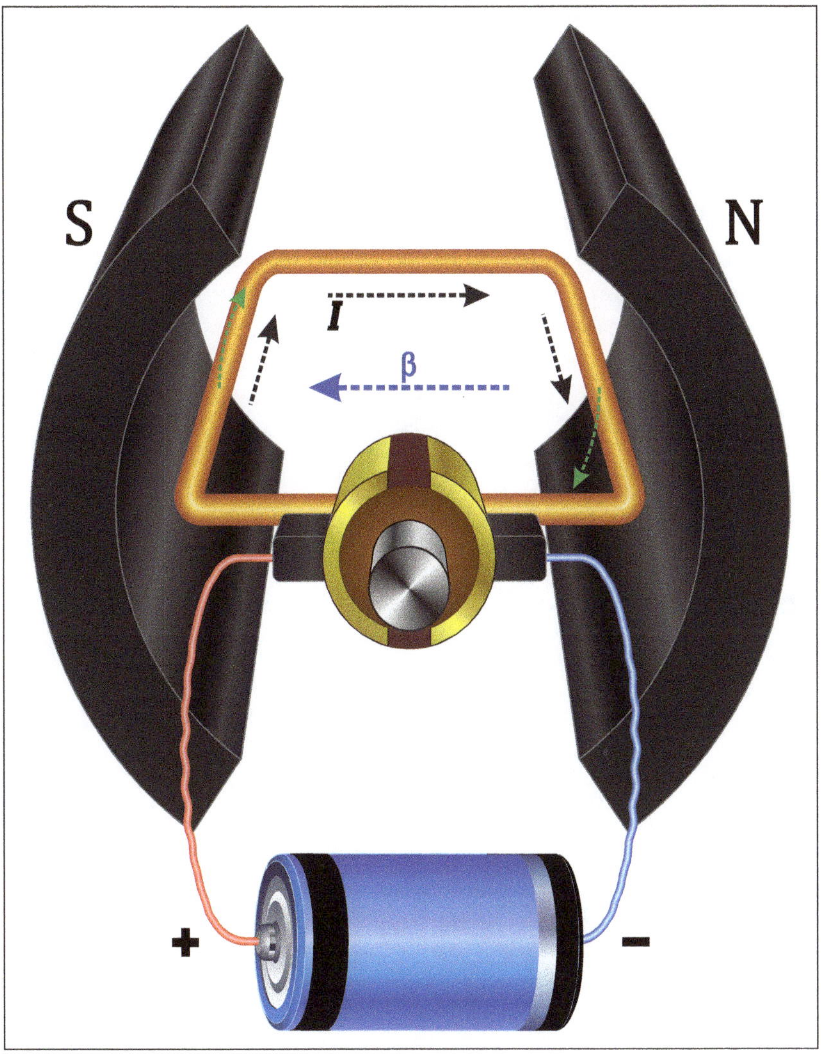

Electric Parking Brake

Electric parking brake, or EPB, uses two different strategies: cable-pulling type that uses an electric motor to pull the parking brake cable instead of a lever-controlled cable and a design that uses an electric motor attached to the brake caliper arm to activate it, that is, controlled by the body control module. Some applications use an EPB that is activated when the vehicle stops and then goes off as soon as the gas pedal is pressed. It is generally controlled by a switch and deactivated either automatically or by pressing the same switch in the opposite direction and stepping on the service brakes. It can be tested using a factory-level bidirectional scan tool.

Electric Shock

The result of current passing through the tissue of the human body.

Electric Water Pump

An electrically operated centrifugal pump that circulates coolant through an ICE. The electric pump cannot compete with the flow of a mechanical pump at high RPM. One major factor is that the electric pump consumes a known quantity of electric power, about 200 W, no matter what the engine is doing. The mechanical pump consumes as much as 15–20 hp at high RPM.

Electrical Burn

A burn caused by the heat from electricity.

Electrical Fire

The result of excess current that heats and burns wire insulation.

Electrical Solvent

A solvent that is used to clean oil and grease from components before soldering. It can also be used to remove soldering flux from a component.

Electricity

Flow of electric current, which is the flow of free electrons from one atom to the next atom.

Electrodes

Metal plates that can donate and receive electrons.

Electro-Hydraulic Brake Assist Systems

The brake assist system that replaces a vacuum booster and its vacuum supply. The brake takeoff rocket takes the driving force applied brake demand and transforms this into an amplified brake pressure and does not depend on the vacuum. The brake booster rocket is intended to control along with an electronic stability control module which is often to blame for ABS, traction control, and stability control. It combines the functions of a brake booster, hydraulic brake cylinder, brake pressure modulator valve (BPMV), and therefore the electronic brake control module (EBCM) into one assembly. The brake fluid reservoir stores brake fluid, are often remote mounted, and uses a brake fluid level sensor, foot lever pushrod, and hydraulic brake cylinder, which is found in line with the pushrod. The EBCM controls all functions of braking including base brake operation, ABS, TCS (traction control system), ESC, automatic emergency braking (AEB), three-phase motor, and plunger to make hydraulic pressure.

Electrolyte

Lead dioxide of the positive plates and the sponge lead of the negative plates are the active materials in the battery. These materials cannot become active until they are absorbed in electrolyte, which is composed of water and sulfuric acid.

Electromagnet

An electrically energized magnet. A metal core surrounded by a coil of wire that has current flowing through it, creating a magnetic field producing a magnet dependent on that magnetic field.

Electromagnetic Clutch

A device used to disengage a drive pulley from a belt to drive an engine accessory.

Electromagnetic Field

The electromagnetic field (EM field, or EMF) is a field produced by accelerating electric charges caused by electromagnetic induction When a magnetic field cuts across an electrical conductor, a voltage is induced in that conductor. See **Electromagnetic Induction** and **Magnetic Field**.

FIGURE E.4 EMF surrounding a wire loop as in a motor.

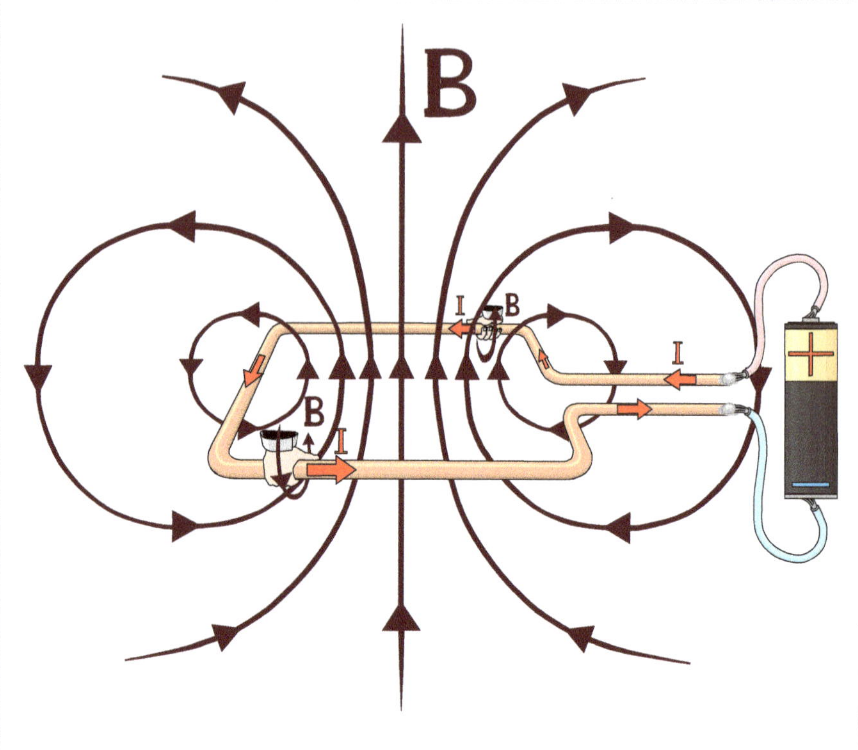

Sergey Merkulov/Shutterstock.com.

Electromagnetic Induction

Principal under which generators and alternators produce current. Michael Faraday's First Law of Electromagnetic Induction describes the induction of EMF (electromagnetic force) in a very conductor and also the Second Law quantifies the EMF produced within the conductor. First Law: A physicist stated that an EMF is induced in a very coil when the magnetic flux across the coil changes with time. Whenever a conductor is placed in a very varying flux, an EMF is induced. If the conductor circuit is closed, a current is induced, which is termed induced current. Second Law: The induced EMF in a very coil is capable of speed change of the magnetic flux. The magnetic flux is the product of the quantity of turns within the coil, and therefore, the magnetic flux is related to the coil. Currents are produced by changes within the flux. These currents are called the induced current because they are produced by a changing field. The EMF that generates this current is called an induced EMF.

Electromagnetic Solenoid

A cylindrical coil of wire acting as a magnet when carrying electric current. The polarity of the coil can be changed to create movement in the other direction.

Electromagnetics

The study of the relationship of magnetic energy to electrical energy.

Electromagnetism

A magnetic field or magnetic flux that is created by current flow through a conductor. Magnetism can also be formed electrically. It was discovered that current-carrying conductors were surrounded by a magnetic field. Current flowing through a conductor such as copper wire creates a magnetic field around the wire. This effect can be observed by passing a compass lengthwise over a copper wire through which current is flowing from positive to negative. The needle will deflect from its North-South orientation when this occurs. Current flow through a wire creates a magnetic field around the wire. The greater the current flow the stronger the magnetic field. This type of magnetism created in the space around a conductor by the current flowing through is called electromagnetism.

FIGURE E.5 The principle of electromagnetic induction shown with a galvanometer reading induced current.

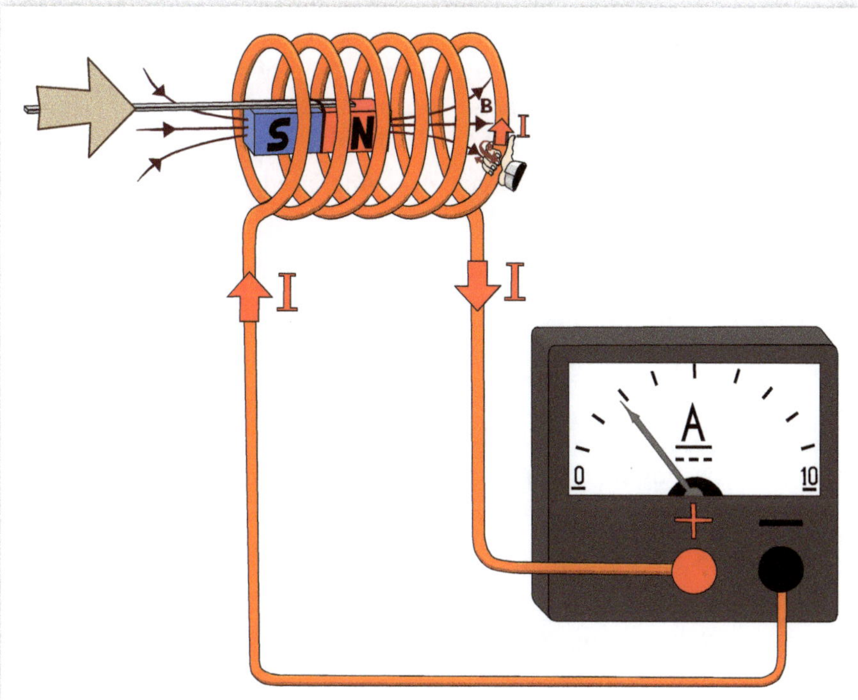

Electron

A negative-charged particle 1/1800 the mass of a proton.

Electron Flow Theory

Electrons leave the negative (–) terminal of the battery, flow through the circuit, and then reenter the positive (+) terminal.

Electronic Device

A solid-state non-mechanical component that is manufactured from semiconductor material, which allows a device to alter the flow of electricity. These devices include diodes, transistors, sensors, and computers. For more terms in electronics, see **J1213_2_198810** [7].

Electronic Fuel Injectors (Diesel)

Diesel engines need from 50 to 1,000 times the pressure utilized in gasoline fuel systems because the injection period is shorter, about 300 ms in duration, and precise metering of fuel is required. The form of the speed of the discharge curve or injection rate must match engine speed and cargo conditions. Robert Bosch, Delphi, and Siemens all manufacture electronically controlled high common rail injectors. There are several generations of injectors to realize the goals of emission compliance, noise reduction, fuel economy improvement, and engine performance. They are designed with the following:

- Injection pressure independent of engine speed
- High injection pressure
- Flexible beginning and ending of injection (timing)
- Multiple injections
- Small, consistent injection quantities for pilot and post-injection sequences
- Rapid nozzle valve opening velocities
- Low injection rate during a period of ignition delay
- Ability to produce maximum injection rates
- Rapid increase in injection rate after the start of combustion
- Rapid decrease to injection rate at end of injection, along with high nozzle valve closing velocity

FIGURE E.6 Bosch piezoelectric high-pressure common rail diesel fuel injector.

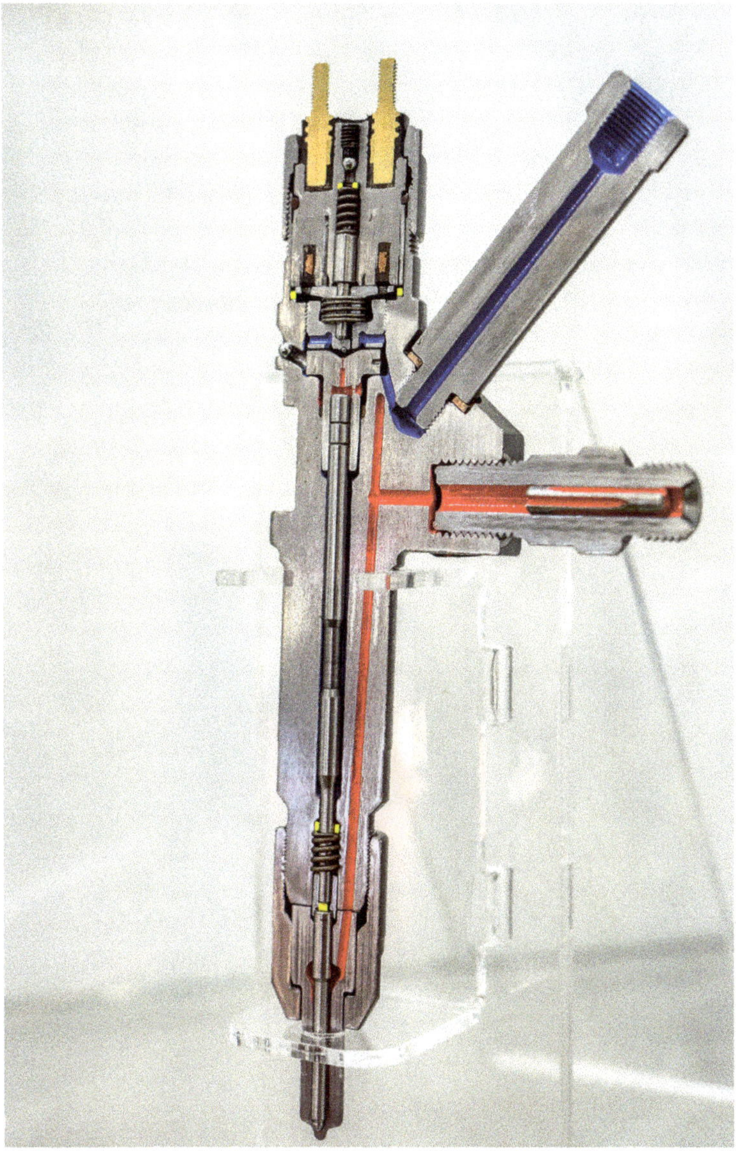

Nordroden/Shutterstock.com.

Electronic Fuel Injectors (Gasoline Fueled)

An electronic fuel injector may be a common actuator found in the gasoline fuel spark-ignited ICE since 1981 and on the current-model diesel system. It was started in 1967 by Robert Bosch. The fuel injectors are electronically controlled by an engine management computer. The control module can vary the quantity of time the injectors are held open, thereby controlling the number of fuel supplied to the engine. The beginning of injection and also the injected fuel quantity are adjusted by electrically triggered injectors. The control module not only injects the fuel but also the proportion of injected fuel. An electronic fuel injector operates at a variable frequency and responds to a modulated pulse width output generated by the control module. The control module changes the injector pulse width with regard to the number of fuel needed by the engine. The longer the heart beat width the longer the injector stays open and therefore the more fuel provided.

FIGURE E.7 An electronically controlled gasoline-fueled engine fuel injector.

Stason4ik/Shutterstock.com.

Electronic Ignition

A general term used to describe any of various types of ignition systems that use electronic instead of mechanical components such as contact points.

Electronic Ignition Module

Computerized module that switches off and on the ignition coil ground to create an ignition spark.

Electronic Listening Device

A tool that allows a technician to hear noises inside of a component accurately. It amplifies sounds so it can be heard through a speaker or headphones. The trade name is Chassis Ears. For more information on this, see **J3152_202005** [5].

Electrostatic Field

The area around an electrically charged body resulting from the difference in voltage between two points or surfaces.

Element

An atom or group of atoms that have exactly the same positive charge in their nuclei. Examples of elements include hydrogen (H) and oxygen (O).

Embittered Ethylene Glycol

Ethylene glycol antifreeze mixtures containing an embittering agent denatonium benzoate which is used to discourage accidental or deliberate consumption.

EMD (Engine Manufacturer Diagnostic)

Systems developed in 2007 as new medium and heavy diesel emission standard for 2007. Prior to 2007, legislated standards for OBD systems were required for medium-duty vehicles and engines with a gross vehicle weight of up to 14,000 lb. Some form of OBD is always used by heavy-duty diesel engine OEMs for electronically controlled engines since 1985 with the introduction of DDEC (Detroit Diesel Electronic Control) I. Introduction of particulate filters, EGR and SCR (selective catalytic reduction) systems added a level of complexity requiring monitoring to make sure they were functioning properly similar to OBD II. EMD standard was less comprehensive than OBD II, so HD-OBD was developed in 2010. Like OBD II, EMD continuously monitors circuit continuity and performs functional monitoring of the injection system, EGR, SCR, and particulate filter. MIL was added along with the useful life or durability requirement.

Emergency Flasher System

A system that consists of a switch, flasher unit, four turn signal lamps, and related wiring. The emergency light switch is normally mounted on the steering column. It is usually a push-pull switch. Also called a hazard warning system.

EMF (Electromotive Force)

Electrical force, or pressure, that pushes free electrons and develops current. It is also called voltage or potential difference.

EMI (Electromagnetic Interference)

An undesirable electronic signal that takes place when an induced voltage enters another system wiring.

Emission Gases

Gases and particles left over from incomplete combustion of an ICE. The primary emissions of concern are hydrocarbons (HC), carbon monoxide (CO), carbon dioxide (CO_2), and oxides of nitrogen (NOx). For more information on this, see **J1145_201109** [3].

Emulsion

A fine dispersion of miniscule droplets of one liquid in another liquid that is not soluble or capable of being mixed in any ratio without separation of the two liquid phases.

Emulsion Tube (Carburetor)

The emulsion tube mixes air and fuel in a carburetor.

Enable Criteria

The OBD II engine management computer requires internal software to keep track of when each diagnostic monitor should run. The different OEMs use different names for this controller, such as the diagnostic executive or the task manager. Each monitor has enabling criteria. Enabling criteria are a set of conditions that must be met before the task manager will allow the monitor to run. For more information on this, see **J1930DA _202105** [1].

Endplay

The total amount of axial play in an automotive component like an engine crankshaft or automatic transmission input or output shaft. Endplay can be measured using a dial indicator or feeler gauges.

Energy

Energy is the capacity or the ability to perform work, which is force times distance.

Energy-Absorbing Steering Column Operation

To reduce the effects of a frontal impact on a driver, most manufacturers incorporate energy-absorbing features into their steering columns. Steering columns can collapse on impact from either the steering gear end or the steering wheel. Most vehicles use a polyacetal resin or other component inside the steering column, steering coupler, and/or column housing that separates or bends when stresses from the steering gear are strong enough. This allows the steering shaft assembly to slide inside the pipe assembly.

Engine Mapping

Operating a vehicle on a chassis dynamometer and manually adjusting the variable factors such as speed, load, and spark timing to determine the optimum output settings for the best drivability, economy, and emission control, which is programmed into the engine management computer.

Engine Speed Sensor

The RPM, or revolutions per minute, signal that comes from the primary signal in the ignition module or crankshaft position (CKP) sensor.

Engine Support Fixture

A support tool used to hold the engine when removing the transaxle in a front-wheel-drive (FWD) vehicle.

Enrichment

The act of adding fuel to the air-fuel mix to create a richer mixture.

EP (Extreme Pressure) Additive

Additive used in gear oil, cutting oil, or grease used to decrease wear of the gears or other components exposed to very high pressures.

EPA (Environmental Protection Agency)

US federal government agency that oversees the enforcement of laws related to the environment such as the Federal Clean Air Act. These laws include exhaust gas emission gas amounts and CAFE (Corporate Average Fuel Economy).

EPHS (Electrically Powered Hydraulic Steering)

This system used on HEVs replaces the customary drive belts and pulleys that drive an influence steering pump during a conventional rack-and-pinion mechanism with a brushless motor. This system still uses a pump, but it is driven by an electrical motor to scale back power drawn from the engine. An electronic controller regulates pump speed to vary pressure and flow. This provides steering efforts tailored for various driving situations. The pump is run at low speed or shut off to supply energy savings during straight-ahead driving. An EPHS system is claimed to use only 20% of the engine power employed by a typical belt-driven pump and improves fuel mileage by approximately 10%. The engine still contributes power to the mechanism through electrical demand on the alternator of the vehicle, but it is greatly reduced from that of hydraulic steering mechanism systems.

EPROM (Erasable Programmable Read-Only Memory)

Compared to the PROM (programmable read-only memory), the data in an erasable PROM can be erased and the memory reprogrammed with new data by a qualified technician. To erase the existing information, the tape covering a window on the EPROM chip is removed and the EPROM is exposed to ultraviolet light. New information is then loaded using an EPROM burner. Like ROM and PROM, an EPROM chip retains the data in its memory even when power is removed. This computer reprogramming is usually called "reflashing." For more information on this, see **J1930DA_202105** [1].

Equivalent Resistance

The total resistance of a series-parallel circuit found by calculating the resistance of each parallel section and then adding the series resistances to these values.

ERFS (Electronic Returnless Fuel System)

Fuel pressure at the rail is sensed by a pressure transducer, which sends a low-level signal to a controller. The controller contains logic to calculate an indication to the pump power driver. The power driver contains a high-current transistor that controls the pump speed using pulse-width modulation (PWM).

Ergonomics

The scientific process of fitting a job to a person.

ESC (Electronic Spark Control)

A computer system equipped with a piezoelectric knock sensor that can retard spark advance, if necessary, to eliminate spark knock. For more information on this, see **J1930DA_202105** [1].

ESD (Electrostatic Discharge)

An electrostatic charge can build up on the surface of a human body. If the body touches something the charge can be discharged to the other surface.

ESI (Electronic Service Information)

An Internet information service that contains vehicle service information and provided by all OEMs and a group of aftermarket companies like ALLDATA, ShopKey Pro, etc.

EST (Electronic Service Tools)

Problems within the engine control system can be identified and diagnosed using a variety of both OBD and EST, also called scan tools. External ESTs can be generic (working with multiple OEM systems) or proprietary, designed to work with specific OEM electronics. OBD II or HD-OBD illuminates a check engine light, MIL (malfunction indicator lamp), or service engine soon light when the vehicle is performing outside required emissions standards. Connecting a scan tool to the DLC (data link connector) allows DTCs (diagnostic trouble codes) to be downloaded. Some systems used dash or ECM-mounted lights that flashed a fault code to identify a problem area. Digital multimeters (DMM) and Digital Storage Oscilloscopes (DSO) are also considered ESTs.

FIGURE E.8 Electronic service scan tool.

guruXOX/Shutterstock.com.

EST (Electronic Spark Timing)

The control of spark timing advance by computer.

ETBE (Ethyl Tertiary Butyl Ether)

An octane enhancer for gasoline, ETBE is added to gasoline up to grades of roughly 13%. It is also a fuel oxygenate that is manufactured by reacting isobutylene with ethanol, which ends in high octane and low volatility. This additive offers equal or greater clean air quality benefits over ethanol, while it is technically and logistically less perplexing with regard to the formulation. It is made by the acidic etherification of isobutylene with ethanol at a temperature of 30–110°C and a pressure of 0-8 to 1.3 MPa.

Ethanol (Grain Alcohol)

An octane enhancer added, at a rate of up to 10%, to gasoline that increases the octane number of the fuel by 2.5 to 3.0 points. Ethanol could be a fuel oxygenate because it contains oxygen.

Ethylene Glycol

A permanent antifreeze as the upper boiling points provided advantages for summertime use as well as during worse weather conditions. The main additives are corrosion inhibitors to safeguard the engine metal, buffers to regulate pH, and colorants to spot the various types. Ethanediol is widely used in automotive applications, but there are also less-toxic alternatives made with humectants available with the newer organic hybrid antifreeze. Antifreeze oxidizes into five organic acids (formic, oxalic, glycolic, glyoxylic acid, and acetic acid). (Glyoxylic acid or oxoacetic acid is a chemical compound that is both an aldehyde and a carboxylic acid.) Inhibited antifreeze mixes are available with additives that can buffer the pH and preserve the alkalinity of the solution to prevent the oxidation of the antifreeze and formation of those acids. Nitrites, silicates, theodin, borates, and azoles may additionally be accustomed to prevent a corrosive attack on metal. It is poisonous to humans and other animals and may be handled carefully and disposed of properly. Its sweet taste can result in accidental ingestion.

EUI (Electronic Unit Injector)

The EUI System was first used by Detroit Diesel Corporation (DDC) in 1985 for its DDEC (Detroit Diesel Electronic Control) system. The EUI system was a full-authority engine management system with a mechanically actuated EUIs. EUIs were computer controlled by an electronic control module (ECM) and operated by a separate engine-mounted injector driver unit called an electronic distributor unit (EDU) injector drivers that activated the electric solenoids on the individual injectors. These were later incorporated into the ECM. When the injector camshaft lobe comes around, it begins to push down the injector tappet and plunger. Control is restricted to the hard limit window created by the cam profile.

FIGURE E.9 EUI that is camshaft activated and electronically controlled.

sima/Shutterstock.com.

EV (Electric Vehicle)

A vehicle that uses one or more electric motors for propulsion. It is powered by a collector system, autonomously using a lithium-ion battery that can be charged with an external charging mechanism. EV batteries can charge from the power grid at home or by using a street recharging point, which is generated from a variety of domestic resources, such as coal, hydroelectricity, nuclear, etc. Home/grid power such as photovoltaic solar panels, microhydro, or wind may also be used and promoted due to sustainability concerns. Using the correct charging equipment it takes about 2–3 h for a full charge, but faster charging can be done. Charging time is limited by the kilowatt capacity of the grid connection. A normal household outlet delivers 1.5 kW in the USA, Canada, and Japan with 120 V supply and 3 kW in countries with 240 V supply. For more information on this, see **J1715_202105** [4].

FIGURE E.10 EV that is powered only by a battery being plugged into a charging station (left) and a Tesla battery EV (right).

buffaloboy/Shutterstock.com.

Courtesy of John F. Kershaw Ed.D.

EVA (Electronic Vibration Analyzer)

An electronic tool used to measure and diagnose vehicle vibrations. For more information on this, see **J3152_202005** [5].

Evaporative (EVAP) Emission Control System

A control system used to prevent fuel vapors in the tank from entering the atmosphere as HC emissions. Evaporative emissions systems capture, control, and eliminate gasoline vapors. The vapors are captured within the charcoal canister and, under certain driving conditions, routed through purge solenoid into the air intake where they are burned within the engine. EVAP purge happens when the engine management computer applies a duty-cycle voltage to the purge control solenoid during closed-loop operation and at cruising speeds. EVAP purge does not happen when the vehicle is in an open-loop operation, at idle, decelerating, or at wide-open throttle. For more information on this, see **J1145_201109** [3].

EV Battery Charging Connectors

The charging power is often connected to the car in two ways. The primary could be a direct electrical connection SAE J1772 conductive coupling connector, which uses the most lead into a weatherproof socket through special high-capacity cables with connectors to safeguard the user from high voltages. This standard for plug-in vehicle charging is the SAE J1772 conductive connector (IEC 62196 Type 1) in the USA. In the European Union (EU) they use the VDE-AR-E 2623-2-2 (IEC 62196 Type 2) for deployment, which, without a latch, means unnecessary extra power requirements for the locking mechanism. For more information on this, see **J1715_202105** and **J1772_201710_EV J Plug** [4, 11].

FIGURE E.11 EV battery charging coupling.

ben bryant/Shutterstock.com.

EV Charging Station

An EV charging station is also called EV (BEV) charging station, electric recharging point, charging point, charge point. The hookup to charge an EV is named the EVSE (electric vehicle service equipment) and must be installed by an electrician. The EVSE supplies electric energy for the recharging of PEV (plug-in electric vehicles), also referred to as BEV (battery electric vehicle) and PHEV (plug-in hybrid electric vehicles). Some EVs have on-board converters which will infix to a typical electric outlet or a high-capacity appliance outlet. Others either require or can use a charging station that has electrical conversion, monitoring, or safety functionality. These stations are used when traveling, and plenty of support for faster charging at higher voltages and currents than are available from residential EVSEs. Public charging stations are on-street facilities provided by electric utility companies or located at retail shopping centers and operated by many private companies. Charging stations provide one or a spread of heavy-duty or special connectors that conform to the range of competing standards. Common rapid charging standards include Combined Charging System and Tesla Supercharger. For more information on this, see **J1715_202105** [4].

FIGURE E.12 EV charging stations.

Sundry Photography/Shutterstock.com.

EV Charging Time

The charging time depends on the battery capacity and also the charging power. It is calculated using the formula: Charging Time [h] = Battery Capacity [kWh]/ Charging Power [kW]. The battery capacity of a completely charged EV from EV OEMs like Nissan is about 20 kWh, providing it with an electrical autonomy of about 100 miles. Tesla released their Model S with battery capacities of 40, 60, and 85 kWh, with the latter having an estimated range of roughly 480 km. As of January 2018, Tesla has two models, 75 and 100 kWh. PHEVs with a capacity of roughly 3 to 5 kWh for an electrical autonomy of 20–40 km but the backup ICE guarantees the complete autonomy of a traditional ICE vehicle. The vehicle should be charged every two or three days. In general, drivers connect their vehicles each night, thus starting day after day with a full charge. For normal charging (up to 7.4 kW), the OEMs have built a battery charger into the car. A charging cable is employed to attach it to the electrical network to produce 120 or 240 V AC. For quicker charging (22 kW, even 43 kW, and more), OEMs have chosen two solutions:

- Use the built-in charger of the vehicle, designed to charge from 3 to 43 kW at 240 V single phase or 400 V three phase (SAE Level 2).
- Use an external charger, which converts AC into DC (SAE Level 3).and charges the vehicle at 50 kW (e.g., Nissan Leaf) or more (e.g., 120–135 kW Tesla Model S).

TABLE E.1 BEV charging times.

Charging time 100 km of BEV range	Power supply	Power (kW)	Voltage (V AC)	Max. current (Amps)
6-8 h	Single phase	3.3	230	16
3-4 h	Single phase	7.4	230	32
2-3 h	Three phase	11	400	16
1-2 h	Three phase	22	400	32
20-30 min	Three phase	43	400	63
20-30 min	Direct current	50	400-500	100-125
10 min	Direct current	120	300-500	300-350

For more information on this, see **J1715_202105** [4].

EV Paddle Connector

A paddle that is inserted into a slot is one winding of a transformer while the winding is part of the vehicle. The paddle completes a magnetic circuit which provides power to the battery pack. The inductive approach removes the chance of electrocution because there are not any exposed conductors. Inductive charging can reduce vehicle weight by moving more charging componentry off the vehicle. For more information on this, see **J1715_202105** and **J1715_202105** [4].

Exhaust ATS (Aftertreatment System)

Exhaust system treatment of emissions from combustion is used by OEMs to comply with federal and state emission regulations. The systems include EGR (exhaust gas recirculation), injection rate shaping from electronically controlled high-pressure systems, and less turbulent combustion chamber designs. Some of these strategies reduce the engine efficiency and were inadequate to achieve the required 90% or higher reduction for particulate emissions needed for 2007 when Tier 2 Bin 5 emission standards were implemented. Later more rigorous reductions in NOx (oxides of nitrogen) emissions were phased in beginning in 2004 and fully implemented by 2010. This action required moving the emission-reduction technologies to the exhaust system to maintain engine efficiency while meeting these very stringent emission values.

In 2010, ATSs used additional components to lower NOx and PM (particulate matter) emissions. The selective catalytic reduction (SCR) system injects a reductant such as ammonia or urea, also known as DEF (diesel exhaust fluid), which increases the nitrogen. SCR systems convert NOx into gaseous nitrogen and water. A DOC (diesel oxidation catalyst) was added to provide additional burning of the HC (hydrocarbon) and CO (carbon monoxide). A DOC converts nitric oxide into nitrogen dioxide through oxidation. Heat is released from this process. SAE Journal Article 02-13-03-0016 states: "With ever tightening emission standards, the automotive industry is continuously seeking novel ways to improve the aftertreatment system (ATS). Exhaust treatment systems using diesel emission fluid (DEF), in conjunction with selective catalytic reduction (SCR) and diesel oxidation converters (DOC), have been gaining popularity in the heavy equipment industry. Spraying DEF (mixture of urea and water) into the exhaust flow can convert harmful NOx gases into N_2 and H_2O. Design of ATSs focuses on high evaporation rate and uniform mixing of ammonia at the entrance to the SCR catalyst.

This study applied support vector regressor (SVR), a machine learning (ML) method to a database of computational fluid dynamics (CFD) simulations to develop a highly efficient mixer with high heat exchange characteristics. Over 500 mixer designs were evaluated using CFD and were then used to train the SVR model. The trained ML model was then used as a surrogate to the CFD and coupled with the genetic algorithm (GA), an optimization technique, to further refine the design parameters. The optimal design obtained from this methodology showed a remarkable performance improvement compared to the baseline." [9]

Exhaust Brake

The engine exhaust brake operates by choking or restricting exhaust flow: the retarding stroke of the piston is therefore the exhaust stroke. Exhaust brakes are managed electrically and actuated pneumatically. Some engines use both an internal engine compression brake and an exhaust brake enabling both the upward strokes of the piston to be retarding strokes. This increases the engine braking capacity by a significant amount.

Exhaust Gas Analyzer

A testing device that uses a gas sniffing probe that is inserted in the engine exhaust that measures the chemical content in an engine exhaust. Most current units measure five types of gases: oxygen, hydrocarbons (HC), carbon monoxide (CO), carbon dioxide (CO_2), and oxides of nitrogen (NOx).

Exhaust Gas Sensor (Oxygen Sensor)

A sensor that measures the oxygen content in the engine exhaust system as a means of checking combustion efficiency. It is located in the exhaust manifold or the exhaust pipe(s). Also called an O_2 sensor, or oxygen sensor.

Exhaust Manifold

A multiple port series of pipes that collects the exhaust gases from multiple cylinders and collects these gases into a collector or single pipe that goes to a series of output pipes to a muffler. Manifold means many. SAE J2515_201712 provides automotive engineers and designers with a basic understanding of the appearance considerations and availability of thermal material for manifold use. For more information on this, see **J2515_201712** [10].

Exhaust Pipe Bending

There are three methods of bending exhaust pipes: program card, pattern, and custom. Program cards are provided by the pipe bender manufacturer and contain all the information necessary to duplicate the exhaust pipes found on most vehicles. The number of the program card is found by looking up the make, model, year, and engine in the card catalog. There are separate cards for each pipe in the exhaust system. The program card lists the length and diameter of the pipe, the distance between the center of the bends, and the rotation and depth of the bends in degrees. There may also be remarks that describe bending instructions for peculiar applications. Using this information, the technician can make a pipe identical to the one which was originally installed by the vehicle manufacture; Pattern bending means measuring and comparing the old exhaust pipe to make a new one. This may be necessary when making a pipe for an older or low-production vehicle for which no program card exists; Custom bending is necessary when there is no program card or existing pipe. Applications for which customizing may be necessary include installing other than the original exhaust pipe.

Exhaust System

A system that channels the spent or exhaust gases coming out of the exhaust valve of an internal combustion engine (ICE) to guide these exhaust gases away from the ICE-controlled combustion. The system conveys spent gases from the ICE and includes one or more exhaust pipes. The role of the automotive exhaust systems is to rid the ICE of spent gases, i.e., what gases are left from the burning of the hydrocarbon fuel, oxygen, and nitrogen. The ICE has the ability to draw in fresh-air charge for combustion and relies on the exhaust system to remove what has been spent. The air that goes in must come out for good engine breathing. Exhaust gases must be removed with as little resistance as possible. The cylinders will not fill with the maximum air (high mass efficiency) if combustion gas pressure builds up in the exhaust system. This means that the cylinders cannot be swept clear of exhaust gases before the next combustion cycle. Exhaust systems that do not remove exhaust gases efficiently may have backpressure.

Exhaust System Hangers

O-ring or fabric devices that attach to brackets mounted on the underside of the vehicle to support the exhaust system. Some are welded to the exhaust system. Hangers can also be made of reinforced fabric, that is, in relation to both brackets on the underside of the vehicle which are bolted or clamped to the exhaust system. Hangers must be flexible enough to allow the exhaust system to expand as it gets hot and move with the engine torque. These hangers will also isolate the vehicle from vibrations in the exhaust system. When installing exhaust system components, make sure you take the necessary steps to properly position them. A hanger should be used to support the exhaust system and not pull components into place. Hangers that are used this way will be under constant stress and will most likely break.

Expansion Tank

A coolant-holding tank that provides cooling system expansion, that is, space for the coolant to expand up to a half gallon more coolant. If an expansion tank is overfilled, it will discharge the coolant at operating temperature. The radiator pressure relieving cap is moved to the expansion tank. The expansion tanks need to be located above the cylinder heads so the additional coolant can flow back into the main cooling system when the engine cools down. When the engine reaches operating temperature, the pressure valve in the cap closes and seals the system. The pressure in the cooling system increases up to the cap pressure to prevent the coolant from boiling. To get the proper expansion tank volume, you need to add together the drawdown capacity requirement and the thermal expansion requirement to get an additional 18% of the total volume required for the expansion tank. Coolant systems are designed to operate with about 13 quarts of coolant in the system. For example, a 13 quart system is calculated as follows: Drawdown = 12% of 13 quarts = 1.56 quarts. Thermal Expansion = 6% of 13 quarts = 0.78 quarts. 1.56 + 0.78 = 2.34 quart expansion tank required.

$$\begin{aligned}\text{Expansion tank size} = {} & 0.12 \times \text{Cooling system capacity} \\ & + 0.06 \times \text{Cooling system capacity}\end{aligned} \tag{E.1}$$

FIGURE E.13 ICE cooling system expansion tank.

J.J. Gouin/Shutterstock.com.

Extended Core Spark Plug

The insulator core and electrodes of this spark plug extend farther into the combustion chamber than they do on other types.

External Combustion Engine

The steam engine is an external combustion engine because the working fluid is separated from the combustion products. The Rankine cycle is used to analyze this combustion process. The term steam engine refers to the railway steam locomotives that is a heat engine performing mechanical work with steam as the working fluid. The steam engine uses the force generated by the steam pressure to push a piston back and forth inside a cylinder and this piston, which is connected to a rod and crank, generating the rotational force to turn the wheels of a steam locomotive.

REFERENCES

1. Society of Automotive Engineers, Standard J1930DA_202105: Electrical/Electronic Systems Diagnostic Terms, Definitions, Abbreviations, and Acronyms Web Tool Spreadsheet, SAE Publishing, Warrendale, PA, 2021.
2. Society of Automotive Engineers, Standard J139_202002: Ignition System Nomenclature and Terminology, SAE Publishing, Warrendale, PA, 2020.
3. Society of Automotive Engineers, Standard J1145_201109: Emissions Terminology and Nomenclature, SAE Publishing, Warrendale, PA, 2010.
4. Society of Automotive Engineers, Standard J1715/2_202108: Hybrid Electric Vehicle (HEV) and Electric Vehicle (EV) Terminology, SAE Publishing, Warrendale, PA, 2021.
5. Society of Automotive Engineers, Standard J3152_202005: Taxonomy and Guidelines for Design, Fabrication, and Installation of Brake NVH Fixtures for Inertia Dynamometer Testing, SAE Publishing, Warrendale, PA, 2011.
6. Society of Automotive Engineers, Standard J1674_201807: Early Acquisition and Preservation of Information in a Motor Vehicle Crash, SAE Publishing, Warrendale, PA, 2018.
7. Society of Automotive Engineers, Standard J1213_2_198810: Glossary of Reliability Terminology Associated with Automotive Electronics, SAE Publishing, Warrendale, PA, 2018.
8. Society of Automotive Engineers, J3063_202103: Active Safety Systems Terms and Definitions, SAE Publishing, Warrendale, PA, 2021.
9. Singh, S., Braginsky, D., Tamamidis, P., and Gennaro, M., "Designing Next Generation Exhaust Aftertreatment Systems Using Machine Learning," *SAE Int. J. Commer. Veh.* 13, no. 3 (2020): 215-220, doi:https://doi.org/10.4271/02-13-03-0016.
10. Society of Automotive Engineers, J2515_201712: High Temperature Materials for Exhaust Manifolds, SAE Publishing, Warrendale, PA, 2017.
11. Society of Automotive Engineers, J1772_201710: SAE Electric Vehicle and Plug in Hybrid Electric Vehicle Conductive Charge Coupler, SAE Publishing, Warrendale, PA, 2017.
12. Society of Automotive Engineers, J1057_199905: Identification Terminology of Earthmoving Machines, SAE Publishing, Warrendale, PA, 1999.

F

FAD (Front Axle Disconnect)

A device that locks and unlocks the front wheels to the differential on a 4WD vehicle. Some 4WD use a vacuum servo and some use a solenoid to move a sleeve, which is splined to the front axle shaft, inside the front axle tube, connecting or disconnecting the front axle to the rest of the drivetrain.

FIGURE F.1 Front axle disconnect servo.

Courtesy of John F. Kershaw Ed.D.

Failure Record

Recorded OBD II failures data reviewed on a scan tool computer. A General Motors (GM) exclusive.

False Lean

If an oxygen sensor reads low as a result of a factor besides a lean mixture, it is often called a false lean indication.

False Rich

When the O_2S reads high as a result of other factors besides a rich mixture, it is often called a false rich indication.

Fanstat

A device that controls the operation of heavy-duty trucks with an on-off fan hub using an air, oil pressure, or electrically actuated clutch. It is generally controlled by the electronic engine management system.

Farad

One farad is the capacity to store 1 C of electrons at 1 V of potential difference. Named after Michael Faraday.

Faraday's Law of Electromagnetic Induction

Faraday's Laws of Electromagnetic Induction have two laws. The first law describes the induction of EMF (electromagnetic force) in a conductor and the second law quantifies the EMF produced in the conductor. In the first law, Michael Faraday stated that an EMF is induced in a coil when the magnetic flux across the coil changes with time. Whenever a conductor is placed in a varying magnetic field, an electromotive force is induced. If the circuit is closed, a current is induced, which is called induced current.

In Faraday's second law, the induced EMF in a coil is equal to the rate of the magnetic flux change. The magnetic flux is the product of the number of turns in the coil and the magnetic flux associated with the coil. Currents are produced by changes in the magnetic field. These currents are called induced current because they are produced by a changing magnetic field. The EMF that generates this current is called an induced EMF.

FIGURE F.2 Current flows when either conductor or magnet is moved; as soon as that motion stops, the current stops flowing. The current and voltage induced in the conductor can be increased in three ways: increasing the speed at which lines of force are cut by the conductor, increasing the number of conductor strands that are cut by the lines of force, or increasing the magnetic field strength.

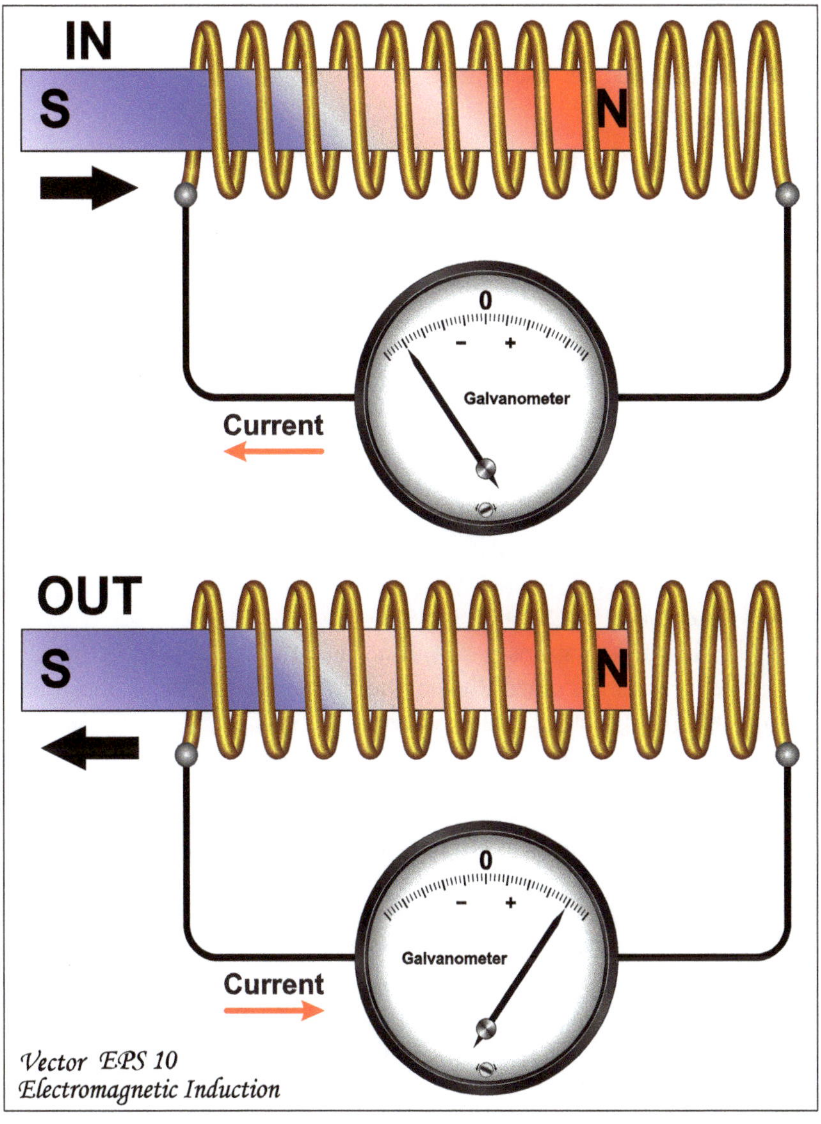

Fouad A. Saad/Shutterstock.com.

$$E = -L\frac{\Delta I}{\Delta t} \tag{F.1}$$

where

E is the EMF generated

$-L$ is the self-inductance measured in henrys

ΔI is the change in current in the coil

Δt is the change in elapsed time

Fast Idle Solenoid

A fast idle solenoid opens the throttle valve during engine operation but allows the throttle plates to close by whatever system operated the solenoid. The throttle is opened to increase the engine speed when the ICE (internal combustion engine) is cold or during certain accessory operation.

Fatigue

State of reduced physical or mental alertness that impairs automotive operation and can be the result of physical or mental issues. SAE J3198_202010 provides definitions and discussions of key terms concerning driver drowsiness and fatigue [1].

F-CAN (Fast Control Area Network)

A computer communications system that operates at 500 kbps (kilobits per second) that is used to control the engine, transmission, SRS (supplemental restraint system), and other vehicle systems.

FCD (Floating Car Data)

Also referred to as floating cellular data, it can be used to determine the traffic speed on a road network. It supports the gathering of localization data, speed, direction of travel, and time information from cell phones in vehicles that are being driven. The data are an important source for traffic information and for many ITS (intelligent transportation systems). This suggests that any vehicle with a fully-charged mobile phone acts as a sensor for the road network. Using this data, traffic holdups are often identified, travel times are calculated, and traffic reports are rapidly generated. In contrast to traffic cameras, number plate recognition systems, and induction loops embedded within the roadway, no additional hardware on the road network is critical. Smartphones having various sensors are often accustomed to track traffic speed and density. The accelerometer data from smartphones used by car drivers is monitored to search out traffic speed and road quality.

Feedback

The reverse flow of electrical current through an electrical circuit or unit that should not normally be operating, which is a reverse-bias current flow that is generally caused by a poor ground connection for the same operating circuit.

Feeler Gauge

Also called a thickness gauge; a series of thin metal blades in a metal holder that are used to measure small clearances, such as the end gap of a piston ring.

Felling Head

Felling heads are attachments that are mounted either to a boom or to the chassis on a base carrier. "There are 3 categories of felling heads: bunching heads, processing heads, and felling heads. Bunching heads are designed to handle multiple stems and don't have processing abilities. Processing heads are capable of felling and processing trees. Felling heads are capable of cutting and felling single stems with no processing capabilities. For more information on the basic types of felling heads, including those with bunching capabilities, that are attachments to a self-propelled machine. Only the major components that are necessary to describe the functions of the felling head, and to apply the principles of the standard are included." For more information on this, see **J1272_202103** [2].

Fender

A term for that part of a car or truck or other vehicle body that frames the front wheel well. It is used to prevent debris and road spray from being thrown into the air by the rotating tire.

FFV (Flexible Fuel Vehicles)

Flexible fuel (flex-fuel) vehicle or dual-fuel vehicle is an alternate fuel vehicle with an ICE designed to run on one fuel usually gasoline blended with either ethanol or methanol fuel, and both fuels are stored within the same tank. Modern flex-fuel engines are capable of burning any proportion of the resulting blend within the combustion chamber as the fuel injection system and spark timing are adjusted automatically per the particular blend detected by a fuel composition sensor. Flex-fuel vehicles are distinguished from bi-fuel vehicles, where two fuels are stored in separate tanks and therefore the engine runs on one fuel at a time, for an example, compressed fossil fuel (CNG), liquefied petroleum gas (LPG), or hydrogen. For more information on this, see **J1297_201710** [3].

Fiber Optics

Field of electronics in which light is used to carry data through small strands of plastic or glass fiber.

Fiberboard

Engineered wood product made out of wood fibers that is used to create free-form shapes such as dashboard panels. For the more information on this, see **J947_201401** [14].

Field Circuit

The charging system circuit that controls current to the alternator field, thereby controlling the alternator output.

Field Circuit Modulation

Modulation where a voltage regulator cycles the rotor field current on and off to control charging system output through an electronic voltage regulator or engine management computer.

Field Coils

Insulated wires wrapped around iron cores that are attached to the inside of the cranking motor housing. When current passes through the sphere coils, a strong field of force or flux is generated. This field uses magnetic repulsion to rotate the motor armature.

Field Relay

A magnetic switch used to open and close the alternator field circuit or, in a charging circuit with a warning lamp, to control the lamp circuit.

Fifth Wheel

Used with semitrailers to allow the truck/tractor to be locked securely to the trailer kingpin under all conditions of movement. The fifth wheel allows the truck/tractor to pivot around the trailer kingpin when turning corners or reversing the combination vehicle. The fifth wheel is mounted on a subframe of its own that is subsequently bolted or welded to a mounting plate on top of the truck chassis and located over the single or tandem axles. Many fifth-wheel assemblies can be slid backward or forward on their subframes by the use of vehicle reservoir air pressure piped to a fifth-wheel air cylinder. This permits variation in the center of gravity of the trailer load applied to the tractor axles and provides adequate trailer swing clearance during turns. Fifth wheels are available in a variety of models for specific applications.

FIGURE F.3 Fifth wheel.

socrates471/Shutterstock.com.

Fillet

A round joint between two parts connected at an angle, which are most often found on an ICE crankshaft.

Film Strength

The ability of a lubricant to maintain a strong covering over the component being lubricated.

Filter (Electrical)

A solid-state device that only passes or blocks certain signal frequencies. An application can be removing noise from a signal.

Final Drive

The rear axle ring and pinion or the last phase of an automotive drivetrain. It is also generally part of a front-wheel-drive transaxle.

FIPG (Formed-In-Place Gasket)

Gasket material that comes from a tube, which is applied to metal surfaces before assembly.

Fire Ring

The flame chamber of a diesel cylinder head has grooves that connect these cylinder head components with each other in one ring. A stainless steel ring is placed inside. By joining the outer edge of the fire ring with the gasket head, more seals can be created.

Firewall (Bulkhead)

In automotive engineering, the firewall or bulkhead is the part of the automobile unibody or body-on-frame that separates the engine compartment from the passenger compartment or cab on a truck.

Firing Line

Part of an automotive spark ignition secondary oscilloscope pattern indicted by a tall spike or line representing the amount of voltage needed to cause an electric arc to jump across the spark plug gap.

Firing Order

The ignition firing sequence of an ICE as determined by the crankshaft and camshaft design.

Firing Section

Part of a secondary ICE ignition oscilloscope pattern that locates problems with the spark plugs, ignition coil(s), plug wires, and distributor.

Firing Voltage

The voltage level to overcome all resistance, ionize, and create a spark in the air gap between the spark plug electrodes.

Flag

System status display of the inspection maintenance (I/M) flags. The system status (I/M) flags are a sign that all diagnostics of critical emissions-related systems have run. This information is also required in some states for I/M test programs, like I/M 240, which is 240 sec. The EPA (Environmental Protection Agency) may check the system status (I/M) flags when auditing vehicles as a part of the FTP (Federal Test Procedure).

Flame Point

The lowest temperature at which a volatile oil will ignite and burn when exposed to a flame.

Flank

Side of a screw thread, rising from the bottom of the groove to the top of the ridge.

Flash Codes

Early OBD (on-board diagnostic) systems use an amber-colored check engine light (CEL) on the instrument panel that tells the driver when a problem occurs in the system. The lamp will then be turned off by the engine management computer at the completion of the static test sequence with no faults found, which may take from 1 to 5 sec. If the warning light remains on or comes on while driving, it indicates a fault in the system that will require further diagnosis. The warning light will flash the diagnostic code digit by digit in numerical order after the last two pins in the DLC (data link connector) are connected using a jumper or the last one is connected to a chassis ground.

Flash Point

The temperature when the fuel evaporates a sufficient amount of flammable vapor that will ignite in the presence of a flame. Flash point does not affect the engine performance. This condition has nothing to do at all with the combustion phase or performance of the fuel in the engine but is rather a measure of the temperature at which the fuel oil vapors will flash when in the presence of an open flame. Safety in handling and storage are the only points warranting consideration for flash point. The flash point of Number 2D diesel fuel is 125°F compared to the flash point of gasoline at −40°F.

Flat Head ICE (Internal Combustion Engine)

The flathead ICE valve-in-block engine as opposed to an OHV, or overhead valve, engine has the poppet valves located within the engine block, instead of in the cylinder head as in the OHV engine.

Flex Fan

Flex fans use a stainless steel frame and blades of either plastic or steel riveted to the frame. These blades will flatten out from their curved shape at a preset RPM to reduce aerodynamic drag on the engine. The more drag produced by the fan, the greater the reduction of available horsepower. Engine-mounted fans are less effective in traffic because they do not pull much air when the engine RPM is low. The biggest advantage of flex fans is their ability to pull more air through the radiator at idle than a viscous clutch fan. It aids in idle or slow-speed operation due to the curvature of the blades scooping air; however, they are noisy at low engine speeds. As the speed of the fan and engine increases, the resistance of the air flattens the blades slightly, so the fan requires less horsepower to turn than does a rigid blade fan. The flex fan does reduce the amount of air the fan can move with each rotation, usually both the fan and the vehicle itself are moving fast enough to maintain a strong air stream through the radiator [4].

Flexplate

A flywheel that is a thin metal plate with a starter ring, which is used to connect the torque converter to the ICE.

Float Bowl

The portion of an ICE carburetor where the liquid fuel is stored. The float mechanism floats in the fuel and its rise or fall opens and closes an inlet valve to the main jets.

Flooding

A condition in an ICE where too much fuel has entered the combustion chamber on a spark-ignition/gasoline-fueled ICE that wets the spark plug and is not allowing the fuel air to ignite. You will need to hold the throttle valve wide open while cranking to dry out the spark plug.

Floorboard

That portion of the car body that lies beneath the feet of the driver and passengers.

Flow Parameters

"Symbols and definitions for air flow parameters were chosen consistent with aerodynamics terminology. The definition of equivalent full scale velocity (VEQ) is included to provide a simple means of relating reduced scale model flow conditions to full scale. Standard day conditions were chosen to correspond to those defined at sea level conditions for the U.S. Standard Atmosphere adopted by NASA, NOAA and USAF in 1976. For high-speed (motorsports) and high-humidity (thermal) applications, references are cited to account for the effects of compressibility on dynamic pressure and relative humidity on air density, if deemed necessary." For more information on this, see **J1594_201007** [5].

Flow Switch (Hydroboost Brakes)

A flow switch used in the hydraulic brake assist booster head outlet port that senses the power-steering fluid flow. The flow switch controls a relay for the electro-hydraulic pump. A pedal switch also activates the pump when the brake pedal is applied and the engine is not running.

Fluid Coupling

An impeller attaches to the engine through a flexplate and turns because the engine rotates. The turbine connects to the transmission input shaft and provides the input to the transmission. The impeller is the driving member, and also the turbine is the driven member. Here goes an easy explanation of a fluid coupling: It is often seen when two fans are placed face to face and just one of them is turned on. The breeze from the primary fan will turn the blades of the other. The air that is a fluid transmits the ability in the same way the ATF (automatic transmission fluid) would transfer motion from the impeller to the turbine. The moving fluid travels along two flow paths that are called the rotary flow and the vortex flow. The rotary flow moves fluid within the same clockwise circular direction as the impeller rotation. This fluid flows around the axis of the engine crankshaft and also the transmission input shaft. The fluid travels along the rotary flow path and the force moves it to the border of the impeller. Because the impeller is curved, the fluid turns slightly as it strikes the outer reaches of the impeller and flows into the turbine. This curvature moves the fluid during a second circular flow path at a right angle to the rotary flow path, and this second flow path is named vortex flow.

FIGURE F.4 Fluid coupling inside a torque converter.

dreamnikon/Shutterstock.com.

Flux

Flux is a lead-free solder joint cleaning agent.

Flux Density

Flux density is the number of flux lines per unit area.

Flux Lines

Flux lines are another term for magnetic flux.

Flywheel

The ICE flywheel smooths out or dampens engine power pulses, absorbs a number of the warmth created by clutch operation, provides the connection point for the cranking motor to engage the engine, and provides the applying surface for the clutch friction disc. It features a large mass which creates inertia. This inertia acts upon the crankshaft rotation to be rid of or dampen engine power pulses. On a running engine, the crankshaft runs as the cylinder fires and then slows down because of internal engine friction until the following cylinder fires. The inertia provided by the flywheel mass tends to stay at a more constant crankshaft speed. The flywheel absorbs some of the heat generated by clutch operation by acting as a conductor for the clutch friction disc. An external ring gear is pressed or welded on to the flywheel along its outer circumference. When the cranking motor is engaged, the starter-drive gear meshes with the flywheel ring gear. Through gear reduction, the flywheel transfers the electric motor rotation to the crankshaft to crank the engine. The face on the transmission side of the flywheel includes a smooth, machined area that provides the engagement surface for the clutch friction disc. This surface must be properly finished to permit adequate slippage as the clutch engages and disengages, and to stop slippage when the clutch is engaged.

FIGURE F.5 Flywheel.

Filippo Carlot/Shutterstock.com.

FM (Frequency Modulation)

The modulation of a sine wave carrier so that its instantaneous frequency is different than the carrier frequency.

FMI (Failure Mode Identifier)

FMI describes the type of failure detected in the subsystem identified by a PID (parameter identifier) or SID (subsystem identifier). The FMI and PID or SID combine to form a given DTC (diagnostic trouble code) as defined in SAE standard J1587. The FMI is used when an active or historic code is read using ProLink 9000 Scan tool or a personal computer (PC) with OEM (original equipment manufacturer) software. For more information on this, see **J1587_200202** [12].

Foam Inhibitors

Reduce the surface of a lubricant by reducing the formation of foam, thereby helping to maintain a good oil barrier between moving parts.

Folding Outside Mirrors

Mirrors that can be electrically folded inward using a control inside the vehicle to fold both mirrors inward when not needed.

Foot-Pound

A measure of work or torque. A one-pound load moved one foot is also called a pound-foot.

FIGURE F.6 The foot-pound.

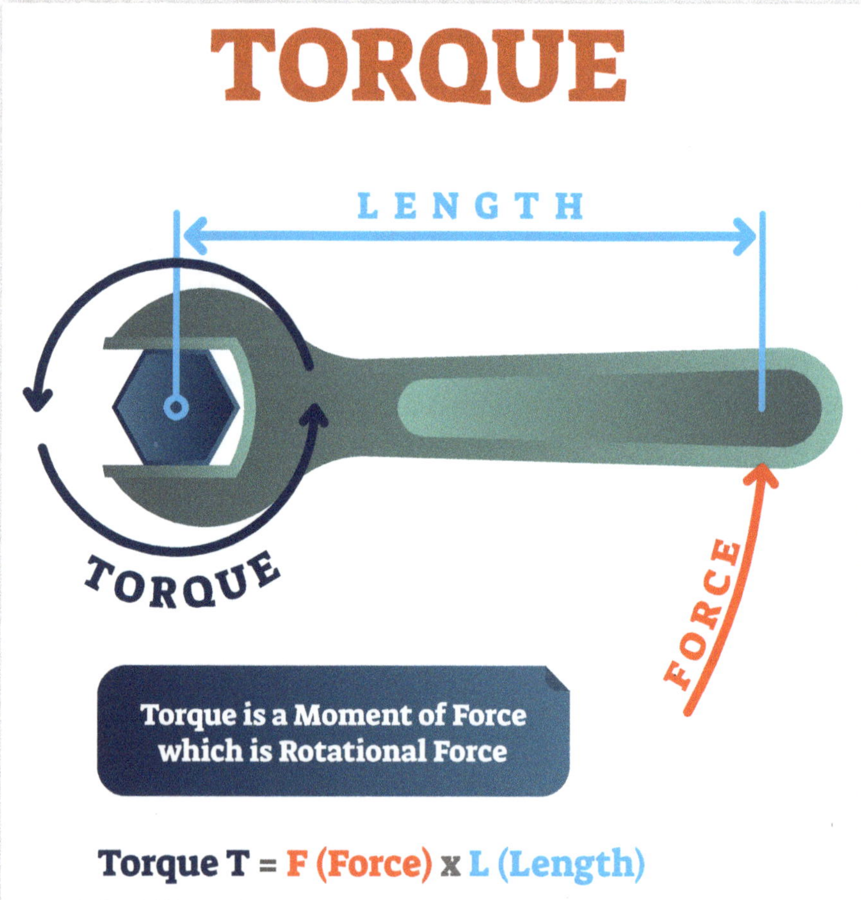

VectorMine/Shutterstock.com.

Force

Newton's second law of motion is F = MA (force equals mass [weight] × acceleration). A force is a push or pull acting on an object, measured in pounds or Newtons.

$$F = MA \quad \textbf{(F.2)}$$

where

F is the force
M is the mass or weight
A is the acceleration or speed increase

Force and Moment Coefficients

"Aerodynamic coefficient definitions were chosen consistent with aeronautical terminology, with one exception. Unlike typical aerodynamics convention, the wheelbase is used to compute moment coefficients. Although it makes more aerodynamic sense to use a body length dimension, this is more likely to change during wind tunnel development than wheelbase. Using wheelbase (WB) provides an additional advantage with the chosen axes system in simplifying the computation of axle loadings. For example, the lift coefficient for the front axle is then equal to CLF = CL/2 + CPM. However, if CPM were based on an overall length (OAL), a ratio of WB and OAL would have to be included in the computation." For more information on this, see **J1594_201007** [5].

Forces and Moments

Primary terminology for aerodynamic force and moment components (drag, lift, side force, pitching moment, yawing moment, and rolling moment) were adopted from aeronautical usage. "The symbols for drag and lift were also taken from aeronautics. To maintain consistency with the symbols for drag and lift, and to provide a mnemonic aid, the other component symbols (S, PM, YM, and RM) were based on terminology." For more information on this, see **J1594_201007** [5].

Forestry Machines

The following machines are forestry machines:

- SAE J1109—Component Nomenclature—Articulated Log Skidder, Rubber Tires
- SAE J1110—Specification Definitions—Articulated, Rubber-Tired Log Skidder
- SAE J1111—Component Nomenclature—Skidder-Grapple
- SAE J1112—Specification Definitions—Skidder-Grapple
- SAE J1116—Categories of Off-Road Self-Propelled Work Machines
- SAE J1255—Specification Definitions—Feller/Buncher
- SAE J1823—Specification Definition—Articulated Rubber-Tired Forwarder
- SAE J1824—Specification Definitions—Clam Bunk Skidder
- SAE J2055—Identification Terminology and Component Nomenclature—Knuckle Boom Log Load

For the *Identification Terminology of Mobile Forestry Machines* see **J1209_200406** [15].

Forged

Forged means that the metal is hammered or forged under pressure into shape.

Forklift Truck

A forklift truck (also called a forklift, towmotor, lift truck, a high/low, a stacker-truck, trailer loader, or a sideloader) may be a powered industrial truck that is accustomed to lift and transport materials. The forklift truck was developed within the 1920s by various companies including the transmission manufacturing company Clark and also the hoist company Yale and Towne Manufacturing. It has since become a really useful piece of apparatus in manufacturing and warehousing operations. Forklift trucks are accustomed to support construction operations whenever there is a necessity to lift, load, or unload materials or supplies. This text only covers the characteristics and basic principles of operations of forklift trucks. By reading the initial equipment manufacturer's operator's manual and repair information, you will be able to obtain detailed information about each make and model. They are classified as materials-handling equipment. The term handling describes an ongoing activity for each project or operation that needs the learning and moving of raw materials, processed parts, finished products, tools, equipment, supplies, or maintenance items. Every operation that requires raising, lowering, or moving an item is assessed as handling.

FIGURE F.7 Forklift truck.

Modvector/Shutterstock.com.

Forward Collision Camera Alert System

The forward collision alert system uses a frontview camera to warn of a possible collision risk exists. The frontview camera is located behind the windshield, looking at the road ahead and detecting vehicles directly ahead, within a distance of roughly 197 ft. The forward collision alert system operates at speeds above 25 mph. When the system detects a vehicle within the path ahead, the green vehicle-ahead indicator in the vehicle direction display is illuminated. When approaching another vehicle too rapidly, the collision alert symbol will flash within the head-up display (if equipped) or a series of red LEDs within the vehicle direction display. Simultaneously an audible and/or haptic alert can sound or the protection Alert Seat pulses. The visual alert cannot be changed, but the motive force can select between audible or haptic alerts within the vehicle personalization menus. The forward collision alert system also can be turned on or off through the vehicle personalization menus. Forward collision alert does nt warn the driving force of any objects that are not detected as a vehicle, e.g., pedestrians, animals, signs, guardrails, bridges, construction barrels, or other stationary or slow-moving objects. The forward collision alert timing sensitivity to manage the space to the vehicle ahead is set using the forward collision alert switch within the left-hand wheel controls. The camera may detect a vehicle that is not in the path ahead, or the system may occasionally provide unrealistic alerts. This might reply to a turning vehicle ahead, guardrails, traffic signs, or other stationary objects. This can be a normal operation, the vehicle does not need service. Forward collision alert may sometimes set unnecessary alerts to turning vehicles, vehicles in other lanes, objects that are not vehicles, or shadows. These alerts are part of normal operation and also the vehicle does not need service. The forward collision alert system is created from the subsequent components:

- Frontview camera module
- Forward collision alert switch
- Instrument cluster/Vehicle direction display

For more information on this, see **J3016_202104** [11].

FIGURE F.8 Forward collision alert.

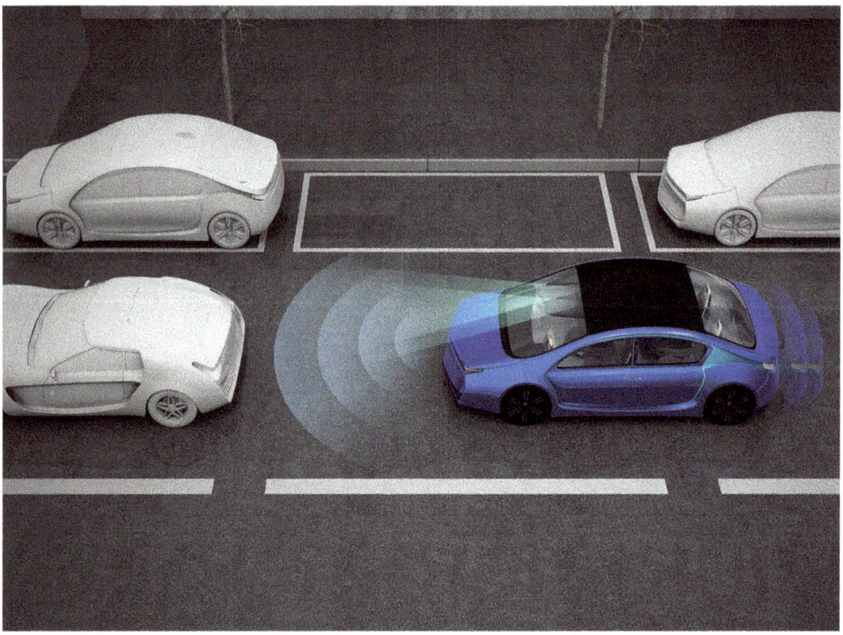

Chesky/Shutterstock.com.

Forward Control Truck

See **COE (Cab-Over-Engine)**.

Fouling Spark Plug

An ICE spark plug condition that causes the spark plug not to fire. It can be caused by dirty air filters, excessive low-speed driving, rich fuel/air mixture, water, or poor fuel injectors.

Foundation Brakes

Braking mechanism located at each end of the axle that stops wheel movement. Foundation brakes use supplied controlled air to force an anchored friction material against a rotating disc or drum to slow or stop a vehicle. When the brake pedal is depressed (brakes applied), air pressure is directed to the brake chambers at each wheel that consist of the air or spring actuator, slack adjuster, or wedge assembly. The mechanical brake mechanism includes shoes and attached friction material, brake pads, and the brake drum or rotor.

Four-Stroke Cycle

See **4-Stroke Cycle**.

Four-Wheel Drive

See **4WD**.

Four-Wheel Steering

Four-wheel steering means the rear wheels are often steered independently of or in conjunction with the front wheels. The flexibility to steer the rear wheels improves high-speed handling and increases maneuverability during driving tasks like U-turns and parallel parking. There are two varieties of four-wheel steering systems—active and passive. Passive systems use compliant rubber bushings that allow a limited amount of rear-wheel steering, typically toward the inside of the turn, under cornering maneuvers. The passive system operates independently of the wheel and driver input, which is why it is called a passive system. In active four-wheel steering systems, the front wheels are controlled normally, and also the rear wheels are typically steered through the use of a computer and electric motors. In previous designs, rear-wheel-steering systems can be controlled mechanically through a direct connection between the front and therefore the rear steering boxes.

In modern computer-controlled four-wheel steering, an actuator, almost like a front rack-and-pinion assembly, attaches to the rear-steering knuckles with tie-rod ends. The actuator turns the wheels when commanded by the steering control module. Originally, some actuators were powered hydraulically by the power steering pump and electronic control valves. Most systems still in use today use a rack-and-pinion assembly driven by an electrical motor.

Some OEMs (original equipment manufacturers) have adopted the quality that at low speeds the rear wheels turn in the wrong direction to the front wheels, providing a substantially reduced turning radius, while at higher speeds, the rear wheels will turn in the same direction because the front wheels, which generate smaller yaw forces during turning maneuvers. This standard provides better control of a vehicle at high speeds and better handling at low speeds, which is beneficial for functions like parallel parking. Other OEMs use systems that disable rear steering at higher speeds and only allow rear steering at slow speeds to permit tighter cornering, like when backing up a trailer.

FIGURE F.9 Four-wheel-steering system.

bsd studio/Shutterstock.com

Four-Wire Oxygen Sensor

A four-wire oxygen sensor is a heated O_2S (HO_2S) that uses an O_2S signal wire and signal ground. The other two wires are for the power and ground for the heater.

Fractured Connecting Rod

This rod is forged from powdered metal. The fracture-split rod could be a technology where the connecting rods and their adjoining end caps are separately forged and machined. Rods may additionally be forged as one piece, with a rather oblong big end. The cap is then cut from the rod and machined for a particular fit.

Frame Design

The material used for constructing a frame will differ based on the operating characteristics of the truck. Materials can range from medium carbon steel, medium carbon heat-treated steel, chrome manganese heat-treated steel, and chrome manganese molybdenum heat-treated steel, or, for lightweight applications, an aluminum alloy.

Frame-Ground Circuit

A circuit that uses the metal structure of the vehicle to return electricity to the ability source. Also called a one-wire circuit.

Free Electrons

Electrons that move from atom to atom.

Free-Floating Piston Pin

The wrist pin in the piston is not a press fit; instead, it moves or floats freely in the piston pin bore and is retained by circlips.

FIGURE F.10 Diesel piston with a free-floating pin.

Courtesy of John F. Kershaw Ed.D.

Free Vibration

The continued vibration in the absence of any outside force. Clamp a thin ruler or yardstick to the edge of a table, leaving about 12 to 20 in. hanging over the edge of the table. You can also use a paint stirrer, pull down on the edge of the stick, and release while observing the movement of the stick. The motion of the stick occurs in repetitive cycles. In the yardstick example, the yardstick continued to vibrate even after the end was released. For more information on this, see **J3152_202005** [7].

FIGURE F.11 The cycle begins at midpoint, continues through the lower extreme of the path, then back past the midpoint, through the upper extreme of the path, and back to the midpoint where the cycle starts over again. The cycle occurs over and once again at the same rate, or frequency, in this case, about 10 cycles in 1 min. If we measure the frequency to reflect the quantity of complete cycles that the yardstick made in one minute, the measure would be 10 cycles × 60 sec = 600 cycles per minute (CPM).

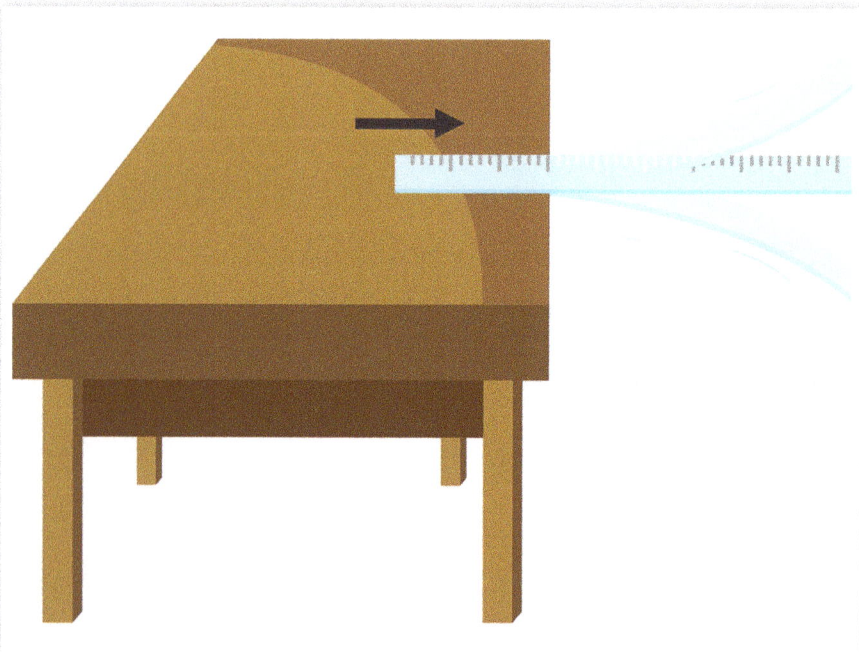

Chesky/Shutterstock.com.

Freeze Frame

On OBD II systems, if the engine management computer recognizes a system failure, the failure is stored in freeze frame. The freeze frame is considered one frame of information taken by the computer.

Frequency

The number of complete waveforms that occur in during a particular time. Also, the quantity of cycles per second, or hertz (Hz). The speed at which an incident occurs during a given period of your time. With vibration, the event may be a cycle, and therefore the period of your time is one second. Thus, frequency is expressed in cycles per second. The correct term for cycles per second is hertz (Hz). This can happen most often thanks to the measured frequency. Multiply the hertz by 60 to induce the cycles or revolutions per minute (RPM). For more information on this, see **J3152_202005** [7].

FIGURE F.12 High and low frequency shown in a waveform.

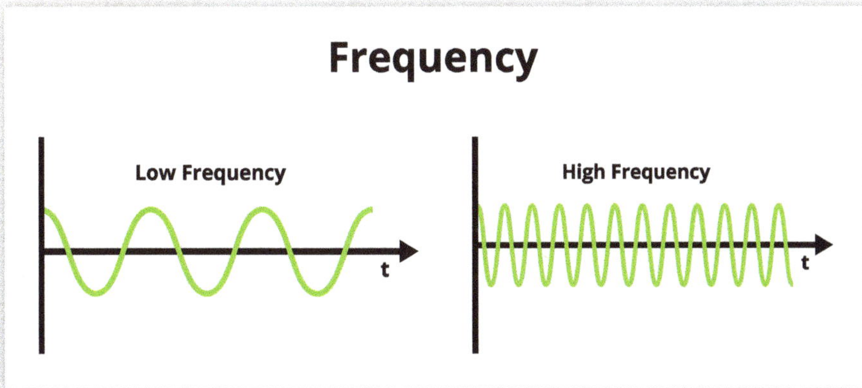

petrroudny43/Shutterstock.com.

Friction

Force that retards the movement of a sliding object. This force acts in the opposite direction to the direction an object wants to move: F = N (normal force) × μ (coefficient of friction).

Friction Clutch

A connection device or clutch where the friction between two moving surfaces is increased until they move together.

Front/Rear Split Brake System

A front/rear split braking system, also called a vertical split system, is most commonly found on RWD (rear-wheel-drive) vehicles. In a front/rear split configuration, the fluid for the front brakes comes from the same hydraulic line. The rear brakes receive their fluid on a separate hydraulic circuit. In this type of system, the rear brakes are controlled independently of the front.

FIGURE F.13 Front/rear braking system typically used on RWD vehicles.

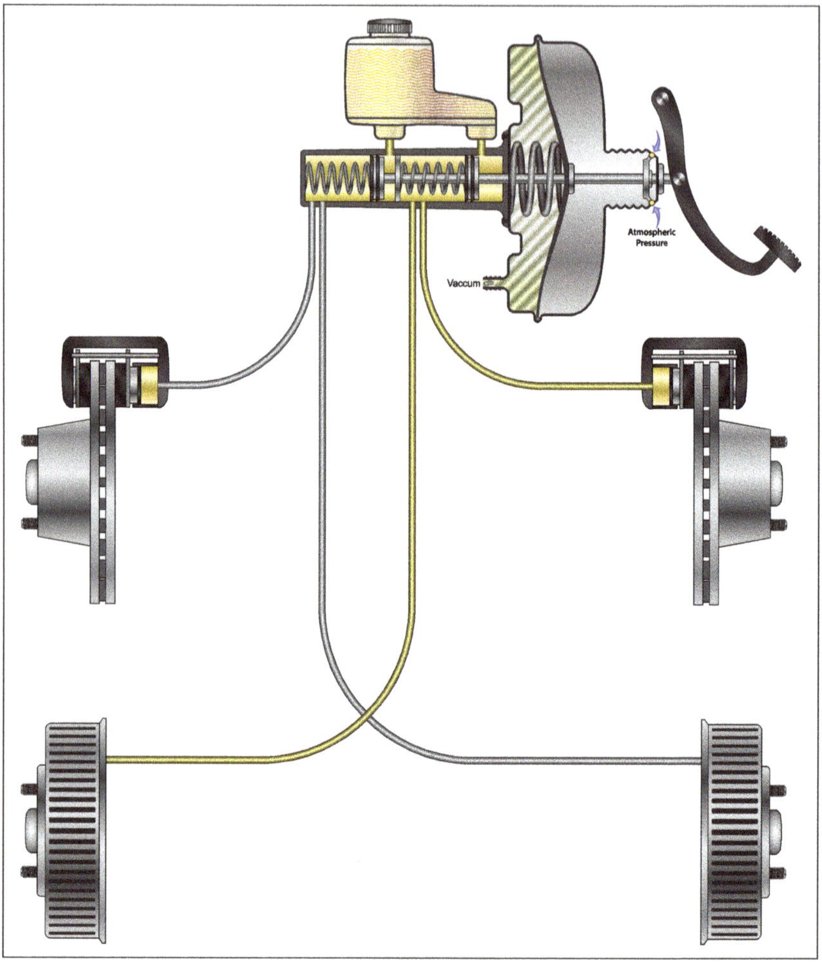

FTP (Federal Test Procedure)

The FTP for exhaust emissions is a test that takes place for 505 sec on chassis rollers (dynamometer) that simulate the urban drive cycle around downtown Los Angeles.

Fuel

A hydrocarbon derived from fossil fuels like gasoline or diesel fuel. Propane or natural gas that is used to fuel an ICE.

Fuel Cell

An electrochemical cell that converts the chemical energy of the fuel and an oxidizing agent into electricity through the use of redox reactions. For more information on *Fuel Cells* and *Pressure Terminology Used in Fuel Cells and Other Hydrogen Vehicle Applications*, see **J2760_201106** [13].

Fuel Compensation Sensor

Sensor or variable fuel sensor used to determine the percentage of alcohol being used in a VFV (variable fuel vehicles).

Fuel Gauge Circuit

The circuit that indicates the amount of gas in a gas tank. A variable resistor is used to operate either a bimetal or magnetic indicator assembly. There is also a 90 ohm resistor on either side of a gauge. The 90 ohm variable resistor is part of the fuel tank sending unit on the ground side of the gauge. Zero sending unit resistance takes place when the tank float is at or near the top, the wiper on the variable resistor rests close to the grounded side, meaning there is little resistance as large amounts of current passes through the sending unit back to the fuel gauge where it registers full. When the float is near the tank bottom, the wiper is at 90 ohms and the needle pulls toward empty. Quick diagnosis is when this ground wire is disconnected, the resistance is infinite and the gauge needle shows empty. Ground the same wire and the resistance is zero and the gauge goes to full.

Fuel Injection

A mechanical or electronic fuel injector, which can be a flow control valve or pressure-creating device that injects fuel into the combustion chamber of an ICE.

Fuel Line

A metal or special elastomer hollow line that carries gasoline or diesel fuel from the tank to the ICE.

Fuel Pressure Regulator

A pressure control valve used in an ICE fuel system that controls the pressure of a fuel to a desired pressure using feedback from an engine management computer or spring force.

Fuel Pump (Mechanical)

The mechanically operated diaphragm pump is used to supply fuel at low pressure usually in the 4 to 7 psi range with a variable output volume. The engine camshaft will actuate this pump through a pushrod or lever operating against an eccentric cam lobe. As the eccentric moves the rod or lever back and forth, the linkage pulls down a diaphragm located between the upper and lower pump halves. While the reciprocating lever or rod movement stretches the diaphragm, it is the spring pressure under the diaphragm that forces it to return to a relaxed position. This reciprocating motion of the diaphragm causes the chamber volume above the diaphragm to alternately rise and fall, creating both a high- and low-pressure area above the diaphragm. Two one-way check valves located at the inlet and outlet of the pump control the direction of fuel flow through the pump. When the pump diaphragm is pulled down, the inlet check valve opens and the outlet valve closes. Relaxing the diaphragm causes the inlet valve to close and the outlet valve to open and deliver fuel at a pressure no greater than the spring force.

FIGURE F.14 Mechanical fuel pump.

ZhdanHenn/Shutterstock.com.

Fuel Rail

Fuel supply lines from the tank to a throttle body or fuel rail are routed to follow the frame along the underbody of the vehicle. Vapor and return lines could also be routed with the fuel flow. All rigid lines are fastened to the frame rail or underbody with screws and clamps, or clips.

Fuel System

The system uses in an ICE to deliver fuel from a tank or memory device to the ICE. The equipment consists of a fuel tank, fuel pump, filter, and injectors or carburetor.

FIGURE F.15 Automotive fuel system.

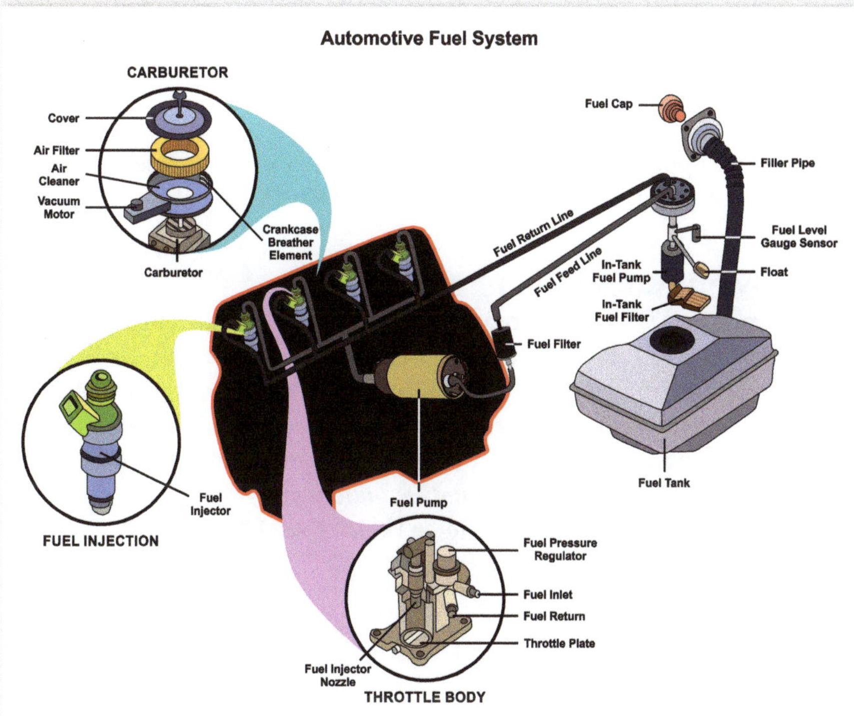

Fuel Tank

A vehicle storage device for the fuel that powers it.

Fuel Trim

A computer function that adjusts fuel delivery during closed-loop operation to bring the air-fuel mixture as close to 14.7:1 as possible.

Fuel Trim Monitor

The fuel control monitor that continuously analyzes the activity of short-term and long-term fuel trim. Fuel trim is an adaptive control strategy employed by the PCM (powertrain control module) for adjusting the pre-programmed injector pulse width. Short-term fuel trim (STFT) allows for temporary mixture adjustments in a closed-loop system and is the idea for long-term fuel trim (LTFT) corrections. For more information on this, see **J1930DA_202105** [8].

Fulcrum

The pivot or supporting point for a first, second, or third class lever. It is the point that the lever rests or is supported and on which it pivots.

Full Driving Automation

"SAE Level 5 Full Driving Automation is the highest classification of driving automation. Level 5 means a vehicle can drive itself everywhere under all conditions with no human interaction. A level 5 vehicle is neither bound by geofencing nor littered with weather and transports people comfortably and efficiently without requiring a driver. The only human involvement are going to be to line a destination." For more *Taxonomy and Definitions for Terms Related to Driving Automation Systems for On-Road Motor Vehicles*, see **J3016_201401** [6].

Full-Floating Axle

On a full-floating axle, the hub and bearing mount on the axle housing are held in place by a bearing retainer or adjustment nuts and safety locks. The inner end of the axle splines into the differential side gears and the wheels mount on the hubs.

Full Hybrid

An HEV (hybrid electric vehicle) that may move at low speeds without using any help from an ICE. HEVs use larger battery packs and more powerful electric motors. This implies they will work solely with an electric motor for brief distances and low speeds. HEVs tend to be more expensive than mild hybrids, but they provide more features and far better fuel economy benefits. For more information on this, see **J1715/2_202108** [9].

Functional Check Procedure

With the information that no DTCs are present, begin symptom diagnosis from information determined from the road test/garage check and line pressure check and by reviewing the transaxle component description and system operation. Reviewing this information enables the technician to understand the operation of the transaxle and helps determine if the condition or malfunction exists. For more information on this, see **J1930DA_202105** [8].

Functional Monitoring

Measures an emission system to make sure it is operating correctly. Monitoring is based on a computerized model of expected or desired system behaviors developed by the OEM. Functional monitoring includes evaluating multiple electrical signals from sensors and doing tests during normal engine operation to evaluate system response. Moving a turbocharger actuator and measuring boost pressure is an example of a functional monitor. For more information on this, see **J1930DA_202105** [8].

Fuse

An electrical safety unit made of a fine tin conductor that will melt and open the electrical circuit if excessive current flows through it beyond its amperage rating.

Fuse Box

A non-conducting plastic container that houses the electrical fuses and circuit breakers in a car or truck.

Fused Jumper Wire

A jumper wire with a fuse in series used for testing live circuits.

Fusible Link

A variety of fuse that will melt and open the protected circuit in the event of a brief circuit, which could cause excessive current flow through the fusible link. Most fusible links are literal wires that are four gauge sizes smaller than the wire of the circuits being protected. SAE standard J156_201901 covers supplemental requirements for low tension primary cable intended for use as fusible links (fuse links) at a nominal system voltage of 60 V DC (25 V AC) or less in surface vehicle electrical systems. These supplemental requirements are intended to qualify cables for an extreme current overload. Update references and procedure. For more information on this, see **J156_201901** [10].

FWD (Front-Wheel Drive)

An FWD vehicle drives the front wheels usually with a transversely mounted engine in the front. The engine transmits its power through the transaxle and half-shafts to drive the front wheels. Unlike RWD (rear-wheel drive), FWD does not have a driveshaft (propeller shaft) or a rear axle assembly.

REFERENCES

1. Society of Automotive Engineers, J3198_202010: Driver Drowsiness and Fatigue in the Safe Operation of Vehicles—Definition of Terms and Concepts, SAE Publishing, Warrendale, PA, 2020.
2. Society of Automotive Engineers, J1272_202103: Felling Head Terminology and Nomenclature, SAE Publishing, Warrendale, PA, 2021.
3. SAE, Society of Automotive Engineers, J1297_201710: Alternative Automotive Fuels, SAE Publishing, Warrendale, PA, 2017.
4. Kershaw, J.F., *High-Performance Automotive Cooling Systems* (Forest Lake, MN: CarTech, Inc., 2019), 78, ISBN:978-1-61325-504-9
5. Society of Automotive Engineers, J1594_201007: Vehicle Aerodynamics Terminology, SAE Publishing, Warrendale, PA, 2010.
6. Society of Automotive Engineers, J3016_202104: Taxonomy and Definitions for Terms Related to Driving Automation Systems for On-Road Motor Vehicles, SAE Publishing, Warrendale, PA, 2021.
7. Society of Automotive Engineers, J3152_202005: Taxonomy and Guidelines for Design, Fabrication, and Installation of Brake NVH Fixtures for Inertia Dynamometer Testing, SAE Publishing, Warrendale, PA, 2011.
8. Society of Automotive Engineers, J1930DA_202105: Electrical/Electronic Systems Diagnostic Terms, Definitions, Abbreviations, and Acronyms Web Tool Spreadsheet, SAE Publishing, Warrendale, PA, 2021.
9. Society of Automotive Engineers, J1715/2_202108: Hybrid Electric Vehicle (HEV) and Electric Vehicle (EV) Terminology, SAE Publishing, Warrendale, PA, 2021.
10. Society of Automotive Engineers, J156_201901: Fusible Links, SAE Publishing, Warrendale, PA, 2016.

11. Society of Automotive Engineers, J3063_202103: Active Safety Systems Terms and Definitions, SAE Publishing, Warrendale, PA, 2021.
12. Society of Automotive Engineers, J1587_200202: Electronic Data Interchange Between Microcomputer Systems in Heavy-Duty Vehicle Applications, SAE Publishing, Warrendale, PA, 2002.
13. Society of Automotive Engineers, J2760_201106: Pressure Terminology Used in Fuel Cells and Other Hydrogen Vehicle Applications, SAE Publishing, Warrendale, PA, 2011.
14. Society of Automotive Engineers, J947_201401: Glossary of Fiberboard Terminology, SAE Publishing, Warrendale, PA, 2014.
15. Society of Automotive Engineers, J1209_200406: Identification Terminology of Mobile Forestry Machines, SAE Publishing, Warrendale, PA, 2004.

G

G-Coolants

G-Coolants come from the trade name Glysantin of BASF in Europe and Valvoline or Zerex in the USA and are listed by their G identifier:

- G05: different from DEX-COOL in certain amounts of additives
- G30 and G34: non-silicate and phosphate free
- G11: blue VW used before 1997
- G12: pink/red VW 1997+ (purple VW 2003+)
- Hybrid organic acid technology (HOAT) formulation phosphate free
- G48: low silicate and phosphate free
- Blue: nitrates, amines, phosphate (NAP) free

Galling

Metal wear that transfers metal from the surface of a rotating metal component to a component in contact with that component. It is caused by metal-to-metal contact without lubrication.

Gap Switch

This switch allows the driving force to work out how closely the control vehicle follows a target vehicle while adaptive control is engaged. When the adaptive controller vehicle speed is being limited because of a slower traveling vehicle, the adaptive controller vehicle speed is automatically controlled by the follow ordinance. The gap switch has three following distance selections that range from 1 to 2 sec. The gap switch following distance between the adaptive control vehicle and therefore the target vehicle is expressed in time as against actual distance. The gap maintained for a specific gap will vary the supported vehicle speed. The faster the vehicle speed the further back you may follow. The gap setting can only be adjusted when the adaptive control system is engaged. The gap switch is hardwired to the BCM (body control module). Supported voltage variations, the BCM is in a position to read the gap switch selection and communicates the switch status on the serial data circuit to the active safety control module. The gap switch may be a passing switch. The primary push of the gap switch recalls this setting and activates the display. Subsequent pushes of the gap switch will change the gap setting. For more information on *Active Safety Systems Terms and Definitions*, see **J3063_202103** [1].

Garter Spring

A circular spring under the lip of a shaft seal that keeps the seal lip in contact with the moving part.

Gas Analyzer

A diagnostic tool that reads exhaust emissions, measuring five different exhaust gases: hydrocarbons, carbon monoxide, carbon dioxide, oxides of nitrogen, and oxygen.

Gas Laws

There are five gas laws:

- Boyle's Law, connection between pressure and volume of a gas
- Charles's Law, connection between the volume occupied by a gas and the absolute temperature
- Gay-Lussac's Law, relationship between the pressure exerted by a gas on the walls of its container and the absolute temperature related to the gas
- Avogadro's Law, which provides a relationship between the volume occupied by a gas and the amount of gas
- Combined Gas Law, which combines the four laws above

Gasket

A compressible material used as a seal between two mating surfaces.

Gasohol

A mixture of gasoline and ethyl alcohol used as fuel in ICEs.

Gasoline

A straight-chain hydrocarbon fuel used to fuel an ICE. It is a combination of hydrocarbon molecules with between 5 and 12 carbon atoms as follows:

- Methane—one carbon atom
- Ethane—two carbon atoms
- Propane—three carbon atoms
- Butane—four carbon atoms
- Pentane—five carbon atoms
- Hexane—six carbon atoms
- Heptane—seven carbon atoms, test octane rating, with an octane rating of 0
- Octane—eight carbon atoms used as a basis for antiknock rating

For more information on this, see **J312_201902** [2].

Gauss

The gauss, symbol G, is a unit of measurement of electromagnetic induction, also known as magnetic flux density. It was named after Carl Friedrich Gauss. One gauss is one maxwell per square centimeter.

Gauss Gauge

Measures a magnetic field in gauss units.

Gay-Lussac's Law

This pressure law was found by Joseph Louis Gay-Lussac in 1809. This law states that for a given mass and constant volume of an ideal gas, the pressure exerted on the sides of its container is directly proportional to its absolute temperature.

GCWR (Gross Combined Weight Rating)

The total mass or weight of a vehicle, including all trailers.

GDI (Gasoline Direct Injection)

A fuel injection system that uses high fuel pressure and varies that fuel pressure to obtain greater fuel delivery using a short pulse width, by 1 ms. GDI uses an almost constant injector pulse width with varying fuel pressure. GDI sprays high-pressure fuel, up to 2,900 psi, into the combustion chamber as the piston approaches the highest compression stroke. With the mix of high-pressure swirl injectors and a modified combustion chamber, almost instantaneous vaporization occurs. This enables a leaner-than-normal air-fuel ratio, which ends up in better fuel economy, higher power output, and reduced exhaust emissions.

Gear Ratio

Speed of rotation of a driven gear to that of a drive gear. You measure this ratio knowing the number of teeth on each gear with this formula:

$$\textbf{Final drive ratio} = \frac{\text{Ring gear}}{\text{Drive pinion}} \text{ or } \frac{\text{Driven gear}}{\text{Drive gear}} \qquad \textbf{(G.1)}$$

Gear Reduction

When the driven shaft turns slower than the driving one. The law of gear reduction says that there is always a decrease in the speed of the output member (driven gear), which is proportional to the increase in the output torque. A 3 to 1 (usually written 3:1) gear reduction, or gear ratio, is the average for the transmission to first move the vehicle. This reduction provides a torque ratio of 1:3. If the drive gear has 24 teeth on it and the driven gear has 8, then the gearset has a 3:1 gear ratio.

Gear Reduction Starter

This type of starter motor uses extra gears to increase the starter motor torque. First, the gear on the armature shaft turns the idler gear. The idler gear then turns a larger drive gear on the clutch, and then the starter drive gear turns the flywheel gear. Even though the starter motor generates a great deal of torque, it is not nearly enough to crank the engine if it were connected directly to the crankshaft. Increasing the starter power requires the use of gear reduction.

Gearbox

A slang or colloquial term for a manual transmission or transaxle.

Gears

Gears are used in many mechanical devices including manual and automotive transmissions, steering gears, and final drives. They provide gear reduction and overdrive. With gear reduction, the output speed can be reduced while the torque is increased.

Gearset

A set of gears, as in a planetary gearset, forming a unit.

Gel Battery

A gel electrolyte battery has silica added to the electrolyte which turns the electrolyte into a gelatin-like substance. This is a valve regulated lead-acid (VRLA) or recombinant design. Recombinant means that the oxygen gas generated at the positive plate moves through the dense electrolyte to the negative plate. When oxygen enters the negative plate, it reacts with the lead which ingests the oxygen gas and prevents hydrogen gas formation. Due to the process of oxygen recombination, VRLA batteries do not use water. For more information on this, see **J1715/2_202108** [3].

Generator

See **Alternator** and **DC Generator**.

Geometric Centerline

Formed between the middle of the front wheels, and therefore the center of the rear wheels. The geometric centerline passes through the midpoints of the front and rear wheel spans or track widths. The geometric centerline could even be drawn through the midpoint of the front and rear axles. The centerline is used to align the toes on all four wheels.

Geometry

The mathematical study of the science of shapes and angles including the Pythagorean theorem.

Glad Hands

Air brake coupling devices are used to connect the air lines from the tractor or truck to the trailer. The couplers use a rubber seal to prevent air from escaping to maintain proper pressure. To connect the glad hands, press the two seals together with the couplers at a 90° angle to one another and then turn the glad hand attached to the hose and lock the couplers together. When making the connections, always be certain to couple the proper glad hands to each other. To keep operators from making a connection error, color-coding of the couplers is sometimes used. Blue is used for supply lines and red for emergency (supply) lines. Metal tags may be used on the brake lines with the words "service" and "emergency" or "supply" stamped on them. For more information on this, see **J2627_200908** [4].

FIGURE G.1 Glad hands are connection devices used to connect the tractor air brake system to the trailer.

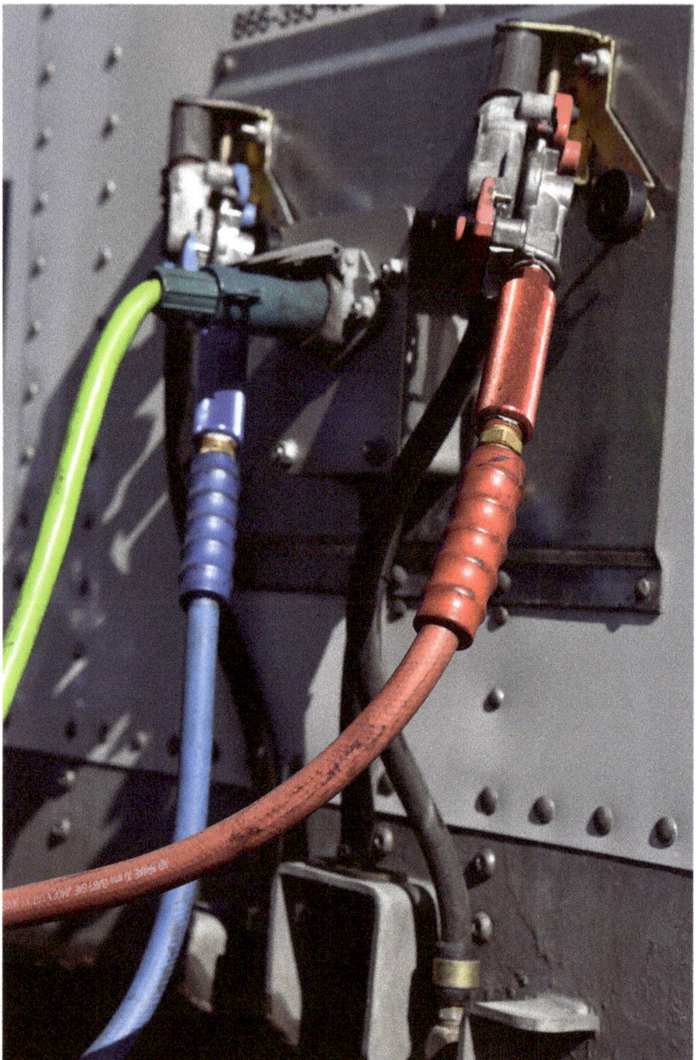

Thomas Trompeter/Shutterstock.com

Glow Plug

A hot plug heating device used to aid in starting diesel engines. High-speed diesel engines can be difficult to start due to the size of the cylinder block and cylinder head, which absorb the heat of compression, preventing ignition. Prechambered engines generally use small glow plugs inside the prechambers. Some direct-injected engines also use glow plugs installed directly in the combustion chamber. There are different types of glow plugs used: Fast glow, which is the more common and that is a 6-V glow plug with a pre-glow of about 8 sec and identified by a 9G on the plug. There is a slow glow system, which used a 12-V glow plug identified by a 7G on the plug. The third type of glow plug is a self-regulating glow plug using a positive temperature coefficient, or PTC glow plugs. The hotter they got the more resistance so they are self-regulating [5].

FIGURE G.2 Glow plug.

Courtesy of John F. Kershaw Ed.D.

Glow Plug Diagnosis

Using an ohmmeter to use glow plug resistance to measure the internal resistance or current to each glow plug. You can detect a dead cylinder because the temperature and resistance of a glow plug in a non-firing cylinder is different than the temperature and resistance in firing cylinders.

Glow Plug Electronic Controller

When you turn the key on, the glow plugs will come on for 4 to 6 sec and then off for about 4-1/2 sec. These will cycle on and off for 1.5 sec and then be on for 4.5 sec and allow you to cycle on and off for a total of 25 sec.

FIGURE G.3 Electronic glow plug controller.

Courtesy of John F. Kershaw Ed.D.

Glow Plug Thermal Controller

The thermal controller located in a coolant passage in the cylinder head which is a thermal controller ground for the glow plug relay. The yellow wire from this thermal controller activates the glow plug relay by providing a ground to operate the glow plugs. Some systems use just a single glow plug light that comes on, indicating the glow plugs are on. Other systems use both a WAIT and START lamp. When you key on the ignition, the WAIT lamp comes on; when it goes off, the START lamp comes on and you can start the engine.

Governor Mechanism

Positions the diesel injection pump metering valve by balancing the opposing forces of throttle shaft position and the speed of the injection pump. The governor mechanism includes the metering valve; linkage connected to the metering valve; governor arm, connected to the linkage and pivoting on a pin in the injection pump housing that contacts the governor arm; governor weight assembly; and a min-max governor assembly. The governor weight assembly has a weight retainer, mounted on the rotor and rotated by the driveshaft and six weights that pivot further outward as the injection pump speed increases. The governor sleeve is moved by the weights moving outward that moves the bottom end of the governor arm to limit metering. The min-max governor assembly mounts in the pump housing by sliding on a guide stud and connects the throttle shaft to the upper end of the governor arm.

FIGURE G.4 Distributor-type diesel fuel injection pump min-max governor that regulated the minimum and maximum position of the metering valve.

Courtesy of John F. Kershaw Ed.D.

GPS (Global Positioning System)

Some vehicles use a GPS receiver. The information is then communicated to the service provider using the regular on-board radio unit or via cellular network data. It is possible that floating car data (FCD) might be used as a surveillance method, although the businesses deploying FCD systems promise that each one data are anonymized in their systems or kept secure to stop abuses, but abuses can still happen.

Gram

Metric unit of weight measurement equal to 1/1000 kg (1 oz × 28 = 1 g).

Graphite

A gray allotropic form of carbon that occurs as a mineral in some rocks and can be made from coke. It is found in some oil additives because it has no affinity with cast iron, from which many cylinder blocks are made.

Graticule

The square sections on an oscilloscope scope face. There are eight graticules vertically (up and down) and ten graticules horizontally (left to right). These reference spaces line on the scope screen and are used as a reference scale.

FIGURE G.5 Graticule square boxes on an oscilloscope screen.

Fouad A. Saad/Shutterstock.com.

Grease

Solid or semisolid lubricant, which is formed as a dispersion of thickening agents in a liquid lubricant that consists of a soap emulsified with mineral or vegetable oil.

Gross Horsepower

An older unit of engine power used for the SAE gross BHP test. A bare engine is employed, i.e., an engine equipped with only those accessories that are necessary for its operation. The pump, alternator, exhaust, etc. are not used; this leads to a better power rating than achievable by the identical engine under real operating conditions.

Ground

In a closed circuit, electrons flow from the positive side of the battery, through the circuit, and then return to the negative side of the battery. We usually call the negative side of the battery the ground. A circuit may be connected to the ground with a wire, or through the case of a component. When a component uses a case ground, the current flows through the metal case to the ground. In an automobile the negative battery terminal is connected to the vehicle chassis with a conductor. As a result, the chassis is at the same electrical potential as the negative battery terminal. It can act as the ground path for circuits throughout the vehicle. If the chassis did not act as a ground, the return sides of vehicle circuits would have to reach all the way back to the ground terminal of the battery.

FIGURE G.6 An electrical ground with the ground symbol.

Ground Circuit Resistance Test

A test performed by placing a voltmeter across the negative battery terminal and alternator housing.

Ground Prong

The round prong on an electrical plug. When a short to the ground develops in the electrical equipment, current passes through this connection to the ground due to the low resistance.

FIGURE G.7 Ground prong on a 120-V electrical cord.

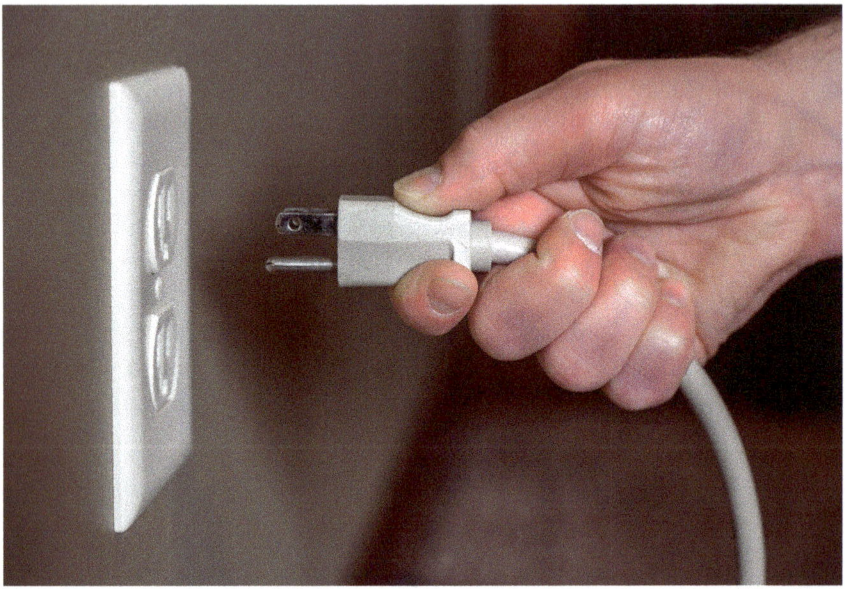

Lost in the Midwest/Shutterstock.com

Grounded Circuit

Circuits can use a negative or positive ground. Most vehicles today use a negative ground system where the negative battery terminal is connected to the vehicle chassis.

GVWR (Gross Vehicle Weight Rating)

Maximum operating weight/mass of a vehicle as specified by the OEM including the vehicle chassis, body, engine, engine fluids, fuel, accessories, driver, passengers, and cargo but excluding the trailers.

REFERENCES

1. Society of Automotive Engineers, J3063_202103: Active Safety Systems Terms and Definitions, SAE Publishing, Warrendale, PA, 2021.
2. Society of Automotive Engineers, J312_201902: Automotive Gasolines, SAE Publishing, Warrendale, PA, 2019.
3. Society of Automotive Engineers, Standard J1715/2_202108: Battery Terminology, SAE Publishing, Warrendale, PA, 2021.
4. Society of Automotive Engineers, Standard J2627_200908: Braking System Definitions—Truck and Bus, SAE Publishing, Warrendale, PA, 2009.
5. Kershaw, J.F., *GM 6.2/6.5 Liter Diesel Engines* (Forest Lake, MN: CarTech, Inc., 2020), Chapter 2, ISBN:978-1-61325-560-5.

H

Half Shaft

External driveshafts (half shafts) have flexible joints on each end. Vehicles with rear- or mid-mounted engines normally use a transaxle in the rear of the vehicle and transfer the drive to the road wheels by external drive half shafts.

Half-Wave Rectification

A circuit where only half of an AC sine wave voltage is rectified to DC.

Hall-Effect Sensor or Switch

Hall-effect sensors (also called a switch) measure the rotational speed of a shaft. They produce a digital-signal-type waveform and a strong signal at low rotational speeds. This is especially useful while cranking an engine when the engine RPM is slow. Current flow through the Hall-effect material is altered in the presence of a magnetic field. The Hall-effect material is a semiconductor that changes current conductivity characteristics in the presence of a magnetic field. When you apply a current to a Hall-effect material, no conduction occurs. However, in the presence of a magnetic field, the material will conduct current. The electrical signal output from the sensor material is analog, but circuits within the sensor will convert and amplify the rising and falling voltage into a square-shaped electrical waveform. Edwin H. Hall discovered the Hall effect.

Hall-Effect Sensor Test

A test performed by checking a Hall-effect sensor output waveform with an oscilloscope.

FIGURE H.1 A Hall-effect switch shown in operation.

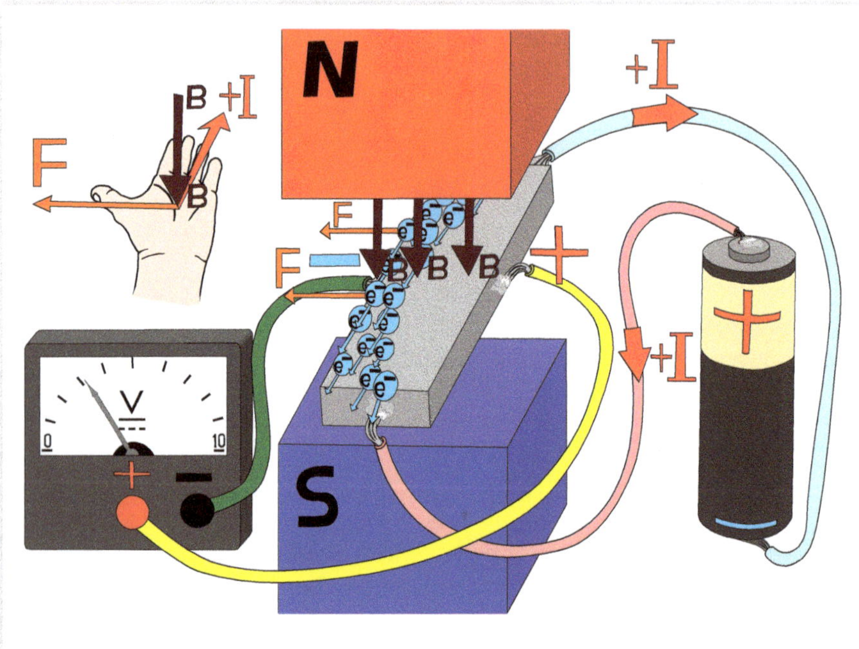

Sergey Merkulov/Shutterstock.com.

Halogen Light Bulb

A bulb that uses tungsten and halogen gas to produce light.

Harmonic Balancer

Most internal combustion engines (ICEs) use a harmonic balancer or torsional damper to dampen or prevent the twisting of the crankshaft caused by power impulses that create dangerous vibrations in the ICE. It is generally an outer heavy ring using a rubber donut in between it and a hub that is fastened to the crankshaft. There are also fluid-type balancers used.

FIGURE H.2 Harmonic balancer.

Courtesy of John F. Kershaw Ed.D.

Harness Connectors

Multi-terminal connectors that join several wires and have a special locking feature. The connector consists of a two-part plastic housing that snaps together to prevent the terminals from working free.

Harness Routing Diagram

A diagram that shows how the major wiring is arranged and where it is located in a vehicle.

HC

Hydrocarbons, which is unburned fuel that when combined with NOx and sunlight creates smog.

HD-OBD (Heavy-Duty On-Board Diagnostics)

On-board diagnostics model for medium and heavy-duty trucks, exceeding 14,000 lb GVWR, which is very similar to car and light-truck OBD II in scope and design. The system uses specific algorithms similar to artificial intelligence for model- or case-based reasoning to identify emissions-related faults, specifically for the truck and bus service industry. It includes a right-to-repair provision for the truck/bus service industry, requiring OEMs to make available any information necessary to make repairs or perform maintenance on OBD systems and other emissions-related engine components. All HD-OBD standards were developed by SAE and adopted by the EPA. HD-OBD must maintain 90–95% reduction to emissions output over the normal engine life cycle. Provisions are:

- All emission control components are monitored for deterioration/malfunctions.
- Other requirements similar to OBD II.
- Phase-in begins in 2010 with full implementation in 2016.
- Threshold monitoring.
- Warning light on when emissions increase.
- Non-threshold monitoring: functional, rational, electrical.
- OBD testing and validation: pre- and post-production; by engine OEM.

For more information on this, see **J3162_201809** [1].

Headliner

The interior roof of any car or truck.

Heat

Increase in temperature above the conventional temperature. A kind of energy related to the kinetic or moving energy of atoms or molecules and capable of being transmitted through solid and fluid media by conduction, through fluid media by convection, and thru empty space by radiation. The transfer of energy from one body to a different one as a result of a variation in temperature or a change in phase.

Heat Crossover

Part of the exhaust gas is directed against the manifold directly under the throttle body.

Heat Engine

Any internal or external system that uses a working fluid to perform work. This system converts heat to mechanical energy, which can then be used to do mechanical work or torque. The working fluid generates work in the working body of the engine while transferring heat to the colder system until it reaches a low-temperature state. During this process some of the thermal energy is converted into work by manipulating the properties of the working fluid. The heat engine follows the first and second laws of thermodynamics.

Heat Range

Measure of the spark plug ability to soak up heat far from the firing end. A spark plug with a good heat range operates between pre-ignition and fouling temperatures to provide the best performance. The spark plug recommended by the engine maker is designed to function in the proper heat range for most operating conditions. The length and configuration of the insulator tip, along with the conductivity of the center electrode, determine the spark plug heat range. Spark plugs can house different heat ranges. The thinking was to accommodate different engine builds with different applications. The reason for a built-in heat range is that the spark plug tip must operate at a high enough temperature to prevent fouling, yet remain cold enough to avoid pre-ignition. Heat range is a measure of spark plug capability to transfer heat received from the engine combustion chamber to the cylinder head. It is the range in which the plug works best thermally and is based on how a spark plug absorbs and dissipates the heat. Many spark plug OEMs like Bosch and NGK use different heat range scales, and that you cannot compare plugs from different manufactures without a conversion table. For more information on this, see **J139_202002** [2].

Heat Riser

Directs exhaust gases under the manifold to heat the air-fuel mixture.

Heat Shield

The heat shield is used to protect the underbody of the vehicle from excessive heat created by the exhaust system. Heat shields are usually riveted or screwed into the bottom of the vehicle floor board above the exhaust system.

Heat Shrink Tubing

Hollow plastic tubing used as an electrical insulator that shrinks when heat is applied.

Heat Sink

A spring-loaded, clamp-on device that is sometimes used to help prevent overheating of electronic components.

Heat Value of Liquid

Measure of the amount of energy to produce work and is measured in British thermal unit (BTU) per pound or per gallon. 1 BTU is the amount of heat necessary to raise the temperature of 1 lb of water by 1°F.

Heat of Compression

Amount of heat generated during the four-stroke cycle compression stroke.

Heat-Treated Truck Frame

Heat-treated frame rails are identified by a stencil marking on the inside middle section of the rail or a stencil mark on one of the crossmembers, although many newer truck frames have a stencil or sticker attached to the outside of the longitudinal frame rail. When a stencil is used, it shows that the rail is heat treated. Therefore, the rail flanges must not be drilled or welded since the heat treatment will be affected at that area. Doing so leads to stresses, which are followed by cracking of the frame. Some manufacturers specify in their truck service manuals that only certain strength frame materials not be drilled. These are generally frames that are manufactured from chrome manganese molybdenum heat-treated steel, which has a maximum safe stress limit of 110,000 psi (758,450 kPa). Some heat-treated frame rails have small patches covering a Brinell hardness test mark along the inside web of the rail. These patches are spaced at 3 to 4 ft (1 to 1.3 m) intervals. The patches can be removed to expose the Brinell hardness markings. Other frame rails may have an "H" stamped on the upper face of the rail flange at about 3 in. (7.62 cm) from the rail end to denote that they are heat-treated frames.

Heated Mirrors

Heats the surface of the outside mirrors, thereby evaporating the moisture on the surface. They may be used on the left-side outside mirror or both the left- and right-side mirrors. The heat helps keep ice and fog off the mirrors and generally operate in the defrost mode.

Heated Rear Window Defroster

Electrically heated car window defroster or defogger system uses an electrical grid baked on the glass that heats the glass to about 85°F (29°C) and clears it of fog or frost. The car window is additionally called a backlight. The car window defogger system is controlled by a driver-operated switch and a timer relay.

Heated Seats

An electric element is found within the seat bottom and seat back for warmth within the cold months. The component is meant to warm the bottom and back of the seat to about 100°F (38°C) or near normal blood heat 98.6°F (37°C). Many heated seats also include a high position or a variable temperature setting so the temperature of the seats can, therefore, be as high as 110°F (44°C). A temperature sensor is found within the cushion to manage the temperature. The sensor may be a resistor, which changes with temperature and is employed as an input to a heated seat control module. The heated seat module uses the seat temperature input, additionally from the high–low (or variable) temperature control, to show this on or off to the constituent within the seat. Some use heated seats in both the front and also the rear seats.

Heater Core

A radiator type device where hot engine coolant flows through that gives heat to the vehicle interior.

Heater Hoses

Hoses that connect the engine cooling system to the heater core.

Heave

Heave occurs when all four tires are in jounce. The condition occurs when the vehicle weight pushes down on the suspension after the vehicle passes a large dip in the road. The condition also occurs when passing over a hill.

HEI

General Motors' name for its high energy ignition. For more information on this, see SAE **J3162_201809** [2].

Height Sensing Brake Proportioning Valve

Special proportioning valve that allows the vehicle to have the correct front/rear braking ratio regardless of vehicle load.

Helical Gear

The teeth on helical gears are cut at an angle to the gear face. When two teeth on a helical gear system engage, contact starts at one end of the tooth and gradually spreads as the gears rotate until the two teeth are in fully engaged.

This slow engagement makes helical gears operate more smoothly and quieter than spur gears. The angle of the teeth on helical gears create a thrust load on the gear when they mesh. They work like a screw pushing or thrusting in. They use thrust bearings to support this thrust load.

Hemispherical Combustion Chamber

This combustion chamber resembles half a sphere and is used with both overhead-camshaft and overhead-valve engines. This design provides improved breathing efficiency because the valves are on opposite sides of the chamber. This placement provides space for very large valves and allows the intake and exhaust to work together effectively during camshaft overlap. When the valves are located on opposite sides of the chamber, the spark plug can be located in the center of the combustion chamber. It breathes well, but lacks combustion efficiency. The classic hemi chamber had no squish or quench areas so there is little turbulence and no area to cool the mixture. Some versions use a squish band machined into the perimeter edge of the piston to improve combustion. They are so large that the spark has a long way to go to ignite the air-fuel mixture. SAE J2612 standard defines the most commonly used terms for pistons. These terms designate either types of pistons or certain characteristics and phenomena of pistons used with different combustion chambers [4].

Hendrickson Equalizing Bar

Rear tandem axle truck suspension with a lower center of gravity since the bar is positioned below the axle centerline. This provides additional leverage that complements the function of the torque rods as they absorb axle torque and road shock. The ends of the equalizer beam attach to the axle housings.

Hertz

A unit measurement of frequency, abbreviated Hz. One hertz is one cycle per second.

Hesselman Helical Loop

The Detroit Diesel two-stroke uses a swirl design called the Hesselman helical loop. Spiral intake ports and masked valves generate a swirl of intake air in a circular motion during the intake stroke. This provides a squish in vertical directions in the concave piston cavity during the compression stroke.

HEUI (Hydraulically Actuated Electronic Unit Injection)

In 1993, Caterpillar and International Truck and Engine Corporation introduced the hydraulically actuated electronic unit injection, or HEUI, system.

These injectors are activated by an engine-driven high-pressure pump using the engine lubricating oil. Most heavy diesel engines used the engine camshaft to operate the injectors. The pressure of the engine oil, that is, 800 to 3,300 psi, controls the injection rate at up to 30,000 psi, while the amount of fuel injected is determined by the engine management computer using pressure up to 30,000 psi. The HEUI injector solenoid is energized by an electronic signal produced in the electronic control module (ECM). Using input from the sensors, the ECM operates these injector. The variable displacement axial high-pressure oil pump features a built-in reservoir to immediately supply oil at cold starts. The injector actuation pressure control valve is an electronically operated valve that controls the oil pump output and injection pressure.

FIGURE H.3 HEUI injector being serviced.

HEV (Hybrid Electric Vehicle)

An HEV uses two power sources to provide energy for propulsion: an ICE and an electric motor drive system. The two power sources are combined into one vehicle powertrain. The ICE and electric drive system work together with computer control to propel the vehicle and operate its electrical accessory systems. Also called a hybrid gas-electric vehicle (HGEV). For more information on this, see **J1715/2_202108** and **J3104_201811** [4, 5].

FIGURE H.4 Types of electric vehicles.

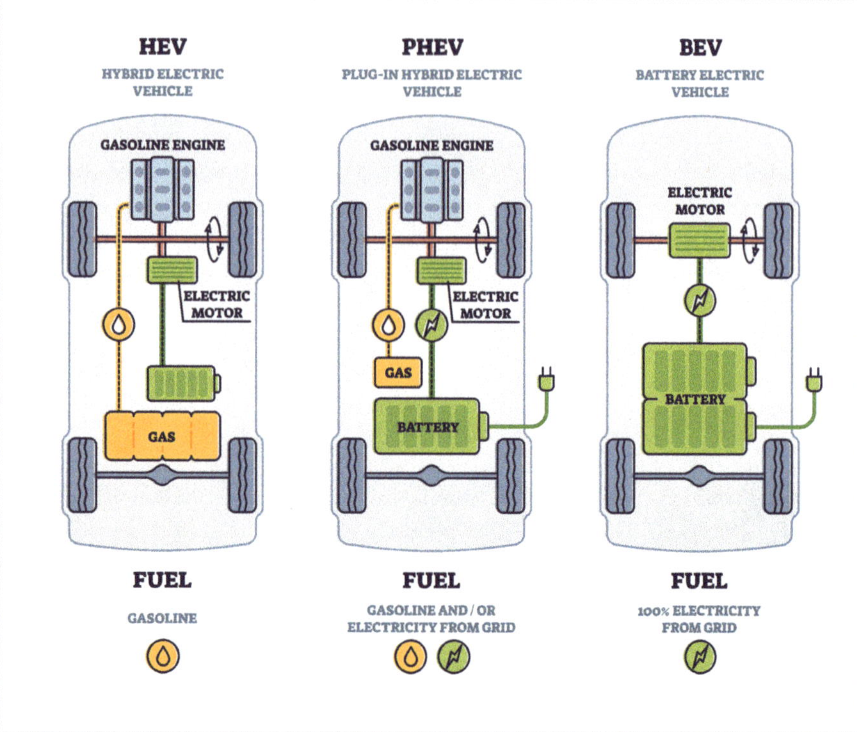

HEV or EV Heater Core

Uses two 165 W positive temperature coefficient (PTC) heaters embedded in the cooling system heater core. When the ICE is not operating, and the coolant temperature is not hot enough to heat the cabin, these PTC heaters are used to increase the temperature [7].

High Driving Automation

SAE Level 4 high driving automation does not require any human interaction within the vehicle operation because it is programmed to help itself within the event of system failure. A Level 4 vehicle might not have a wheel and pedals. At Level 4, you will be able to sleep while riding within the vehicle. Level 4 driving automation technology is to be used in driverless taxis and public transportation services. Such vehicles are programmed to travel between Point A and Point B and restricted to specific geographic boundaries by geofencing technology. Certain conditions may limit or cancel Level 4 autonomous vehicle operation, like severe weather. For more *Taxonomy and Definitions for Terms Related to On-Road Motor Vehicle Automated Driving Systems*, see **J3016_201401** [6].

High-Pressure Common Rail (HPCR) Diesel Fuel Injection

HPCR systems supply a constant high-pressure fuel supply to individual injectors through a common fuel rail. Digital technology is used to switch the fuel on and off using electronic injectors at pressures as high as 30,000 psi with the new speed and metering precision required for clean, efficient combustion. Fuel pressurization is done independent of the engine speed. So no matter what the engine RPM or load is, the fuel is pressurized for injection to the optimal values required for the most efficient combustion, low emissions, and best fuel economy and performance. A high-pressure pump driven by the engine pressurizes fuel to the level required for injection for any given engine load and speed condition. Since the quantity of fuel the high-pressure pump can deliver is more than the engine requirements, pressure and volume output of the pump is electronically controlled. The pump supplies fuel at injection pressure to a common fuel rail, which is connected to the electronically operated fuel injectors located in each cylinder. Since all the injectors are connected to the same fuel rail and supplied the same fuel pressure, the system is termed common rail. Injection timing is controlled by an electrical actuator (solenoid) on each injector while the fuel rate is controlled by the on/off times of the actuator and the fuel pressure in the fuel rail.

FIGURE H.5 High-pressure common rail diesel fuel injection system.

High-Voltage Buffer Zone

A safety zone of about 3 ft (1 m) around the perimeter of the vehicle. For more information on this, see **J1715/2_202108** [9].

High-Voltage (HV) Disconnect

Provided on a Hybrid Electric Vehicle (HEV) to allow the HV battery pack to be physically disconnected from the power control module. For more information on this, see **J1715/2_202108** [3].

HOAT (Hybrid Organic Acid Technology) Antifreeze Coolant

A newer version of OAT technology is termed HOAT, or Hybrid Organic Acid Technology. It is almost like the OAT-type antifreeze because it uses organic acid salts (carboxylates) that are not abrasive to water pumps, yet provide the right pH. If the pH is just too high, the coolant can cause scaling and reduce the warmth transferability of the coolant. If the pH is just too low, the resulting acidic solution could cause corrosion of the engine components exposed to the coolant. HOAT is employed in the newer Ford, Chrysler, and Mercedes vehicles. It uses the most effective aspects of both IAT and OAT [9].

Homogeneous

Substances and mixtures in an exceedingly very single phase, which is commonly in contrast to a substance that is heterogeneous. It refers to the smooth variation of properties with no discontinuities. A homogeneous suspension of cloth implies that when dividing the amount in half, the identical amount of material is suspended in both halves of the substance. However, it might be possible to see the particles under a microscope. An example of a homogeneous mixture is air. Air are often more specifically described as a gaseous solution.

Hood

The front lid or covering over the ICE when mounted in the front of a vehicle. In the UK, it is called a bonnet.

Hooke's Law

Principle of physics regarding spring force. It says that the force represented by F needed to extend or compress a spring by a specific distance represented by x measures linearly with respect to that distance, so the formula is:

$$F_s = kx \tag{H.1}$$

where

k is the constant factor characteristic of the spring-like stiffness

x is the small compared to the total possible deformation of the spring

Horn

An electric device that produces a loud sound and used to let other drivers or persons in the area know of the presence of a moving vehicle. They are made in several different tones, ranging from 1,800 to 3,550 Hz, and must be heard from a minimum distance of 200 ft (60 m). OEMs select from various horn tones for a particular vehicle sound. When two horns are used, each has a different tone, yet the sound combines when both are operated. Most horns on newer vehicle use the BCM to control operation.

Horsepower

Horsepower is the rate of doing work (torque). In the metric system, horsepower is expressed in kilowatts (0.746 kW = 1 hp, or 1.341 hp = 1 kW). The raw horsepower of an engine as measured by a dynamometer is expressed by the formula:

$$hp = \frac{2\pi FRN}{33{,}000} \qquad \textbf{(H.2)}$$

where

F is the dynamometer load, in lb
R is the radius arm of the dynamometer, in ft
33,000 is the conversion factor, ft-lb/min, to produce 1 hp
N is the engine speed, in RPM
π = 3.1416, a constant

The horsepower required to overcome resistance, or drag, is found by the following formula:

$$hp = \frac{DV}{33{,}000} \qquad \textbf{(H.3)}$$

where

D is the velocity, in ft/min
V is drag, or resistance, in lb

Hotchkiss Rear Axle Design

The Hotchkiss drive is a propeller shaft or driveshaft power transmission.

FIGURE H.6 Hotchkiss rear axle design.

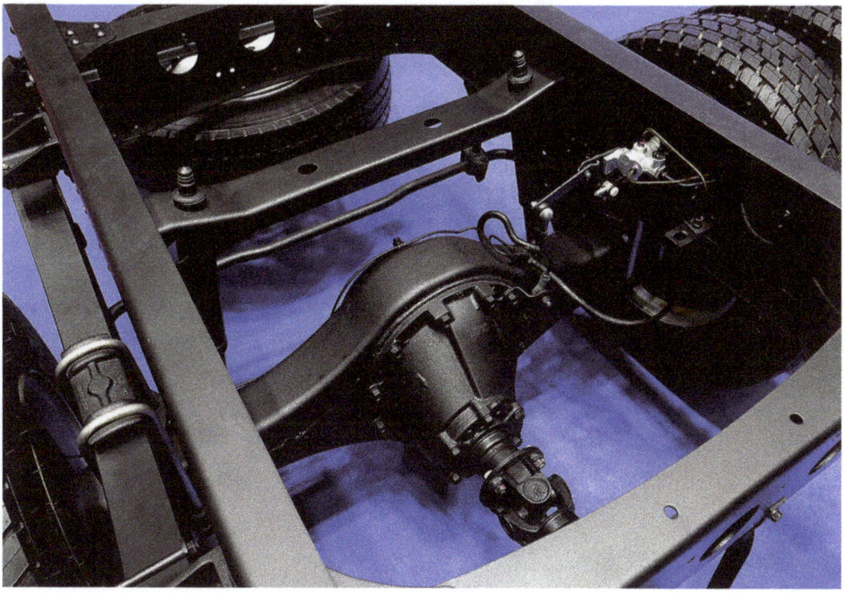

Housing Bore Diameter

Diameter across a bearing bore without the bearing inserts installed.

HSD (High-Side Drivers)

A transistor that controls the power side of the circuit. In these applications when the transistor is switched on, voltage is applied to the device. The device being activated will have a ground.

Hub Units

Use either ball bearings or tapered roller bearings, depending on the vehicle application. Hub units are found on the front wheels of most front-wheel-drive vehicles and four-wheel-drive trucks. Hub units contain bearings that are pre-greased, preloaded, and sealed for life by the bearing manufacturer. They require no additional lubrication or adjustment. However, they do need to be properly installed with the clamp nut properly torqued.

HUD (Heads-Up Display)

A transparent display that shows data projected on the windshield. This allows the driver to look forward, keeping their eyes on the road ahead.

Optical System HUD for Automotive. "The SAE Standard provides measurement methods to determine HUD optical performance in typical automotive ambient lighting conditions. It covers indoor measurements with simulated outdoor lighting for the measurement of HUD virtual images. HUD types addressed by this standard includes w-HUD (windshield HUD) and c-HUD (combiner HUD) with references to Augmented Reality (AR) HUD as needed. It is not the scope of this document to set threshold values for automotive compliance; however, some recommended values are presented for reference. This SAE Standard defines a consistent terminology and metrology for vehicular Head Up Display (HUD). It allows measurement of automotive HUD attributes in accordance to SAE J1757-1 and ISO 15008." For more information on this, see **J1757-2_201811** [8].

Hunting Gears

A set of gears whose tooth counts do not have any common factors. Any tooth on either gear will contact every tooth on the opposite gear before encountering the identical tooth. This spreads the load and wears evenly over all the gear teeth. The hunting tooth frequency of a pair of gears is the gear mesh frequency divided by the smallest amount integer of the numbers of teeth on the two gears. The smallest amount of integer is commonly just the merchandise of the number of teeth.

Hybrid Air Cooling

A cooling system that circulates outside air over an HV battery pack to dissipate heat. For more information on this, see **J1715/2_202105** [9].

Hybrid Battery Pack

Several battery modules wired in series to produce an HV battery source for an HEV. It contains high-efficiency NiMH (sealed nickel metal hydride) or Li-ion (lithium ion) battery cells stacked in a sealed inclusion. For more information on this, see **J1715/2_202108** [3].

Hybrid Driver Ready Light

An indicator light that blinks when you press the brake pedal and turn the ignition key to the start position with the transmission gear selector in park. If the electric drive is engaged, the ready light glows continuously to let the driver know the HEV can be driven. For more information on this, see **J1715/2_202105** [9].

Hybrid Engine-Starting Mode

A hybrid vehicle mode of operation in which the motor-generator spins the engine crankshaft to start the internal combustion engine (ICE). When a hybrid vehicle is started, the ICE will warm up. If necessary, the electric motor acts as a generator, converting energy from the ICE into electricity and storing it in the HV batteries. For more information on this, see **J1715/2_202108** [3].

Hybrid Ground Fault Interrupter

A circuit that constantly monitors a hybrid drive system for HV leakage into the metal chassis of the vehicle. When the HV from the HV battery pack or motor-generator shorts to the frame ground, the management computer will illuminate a warning light and de-energize the main power relays (contactors) to disconnect the battery pack from its electric drive circuit. For more information on this, see **J1715/2_202108** [3].

Hybrid High-Voltage System

A system in a hybrid vehicle that operates on voltage from around 250-V to 650-V AC or DC. For more information on this, see **J1715/2_202105** [9].

Hybrid Low-Voltage System

A conventional 12-V system used to maintain computer memory and provide low voltage to electrical accessories in a hybrid vehicle. For more information on this, see **J1715/2_202105** [9].

Hybrid Master Warning Light

A light in the combination meter that warns if there is a problem in any part of the drivetrain. For more information on this, see **J1715/2_202105** [9].

Hybrid Maximum Voltage

The three-phase AC voltage (up to 650-V AC) fed to the motor-generator from the power control module. For more information on this, see **J1715/2_202108** [3].

Hybrid Motor-Generator

A hybrid vehicle component that functions as both a powerful traction motor and a high-energy alternator in the vehicle powertrain. For more information on this, see **J1715/2_202108** [3].

Hybrid Nominal Operating Voltage

The maximum DC voltage available from the HV battery pack. This voltage typically ranges from about 100 V to more than 300 V. For more information on this, see **J1715/2_202108** [3].

Hybrid Power Display

A dash information display that tells the driver hybrid drive system operation. Some hybrid displays shows when the electric assist is working, how electrical energy is being used, and how much energy remains in the battery pack. The power display often reads in kilowatts (kW) of power output from the HV battery pack. For more information on this, see **J1715/2_202108** [3].

Hybrid Power Splitter

A planetary gearset used to control the engagement of the motor-generators with the transaxle or transmission output gears, chains, belts, and shafts. The hybrid power splitter and transaxle are under computer control using the hybrid electronic control unit (ECU), engine ECU, brake ECU, battery ECU, power control module, and most vehicle sensors. For more information on this, see **J1715/2_202108** [3].

Hybrid Refrigerant Cooling

A cooling system that uses R-134a refrigerant and routed through a separate evaporator located next to the power control module or battery pack. The refrigerant flow cools the evaporator as well as the hybrid controller. For more information on this, see **J1715/2_202108** [3].

Hybrid Water Cooling

The process of circulating the engine coolant through the transaxle and power control module. For more information on this, see **J1715/2_202108** [3].

Hydraulic Force Formula

$$F = p\tilde{A} - A \quad \textbf{(H.4)}$$

where

F is force

p is pressure, in psi

A is the effective area of the cylinder

Hydraulic Valve Lifter

A cam follower that rides on the ICE camshaft and transfers the motion of the camshaft lobe through the action of a hydraulic piston moved by engine lubrication oil pressure. It moves the pushrods and rockers to open and close the poppet valves. The cam lobe under the lifter leverage is multiplied by the ratio of the rocker arms which determines the valve lift and duration. Also see lifter.

Hydraulically Operated Clutch

The drive members and driven members are forced together on engagement by the spring pressure of the pressure plate so that the coupling of power is completed and minimal power is lost through the coupling. Clutch engagement and disengagement are performed by hydraulically operated piston and operating members of the clutch system. The pressure that holds the coupling of the drive and driven members is released through the release or throw-out bearing that moves the pressure plate away from the friction discs when the clutch pedal is depressed.

FIGURE H.7 Hydraulically controlled push-type clutch operation: LEFT SIDE shows clutch disengaged or released with the driver's foot pushing down on clutch pedal. RIGHT SIDE shows the clutch applied or engaged with the driver's foot off the clutch pedal.

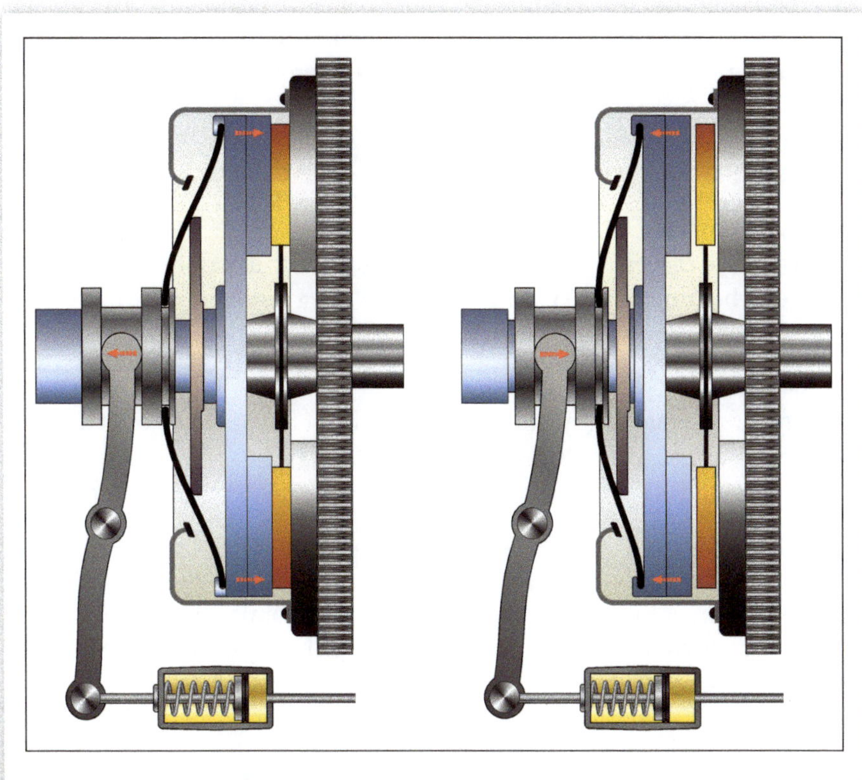

This happens when the release bearing pushes on the fingers of the pressure plate or pulls the finders of the pressure plate. The release bearing is moved either by hydraulic pressure or by direct linkage from the clutch pedal to the clutch fork.

Hydraulics

The study of fluids and their motion to perform mechanical actions. It is a branch of science and technology concerned with the conveyance of liquids through pipes and channels, especially as a source of mechanical force or control.

Hydroboost Brakes (Hydraulic Brake Booster)

Hydroboost brakes apply hydraulic pressure from the facility steering mechanism to help in braking. When the brakes are applied, hydraulic fluid flows from the facility steering pump to extend the braking effort, while fluid also flows from the hydroboost system to the gear to help in the steering system. Mechanical force is converted into hydraulic pressure by the hydraulic brake cylinder, regulated to fulfill the braking system demands by the pressure balance system, and delivered to the brake system calipers. The hydraulic calipers convert the hydraulic pressure into mechanical force, which presses the restraint against the rotor. This hydroboost system works in conjunction with the Hydramax ABS (anti-lock brake system) system. The hydroboost system was first developed by Bendix in 1973 as another vacuum booster brake system to be used on many diesel trucks. This improved braking system took up much less space and was simpler, allowing vehicles to fall in line with new federal regulations requiring decreased stopping distance with less treadle effort.

Hydrocarbons (HC)

Compounds of carbon and hydrogen used as fuel, such as gasoline. High levels of hydrocarbons in tailpipe exhaust emissions are a result of unburned fuel.

Hydrocracking

Cracking or thermal degradation of petroleum in the presence of hydrogen. For more information on this, see **J1715_202105** [9].

Hydrogen

A chemical element that uses the symbol H with an atomic number of 1. It is the lightest element in the atomic table and has diatomic molecules with the valence H_2. It is colorless, odorless, non-toxic, and highly combustible. It is the principal element in water, H_2O.

Hydrogen Hybrid

Because hydrogen is introduced as a gas and not a liquid like the original gasoline design, it was necessary to install new injectors. Fortunately, a company in California has been producing injectors designed specifically for hydrogen. For more information on this, see **J1715/2_202105** [9].

Hydrokinetic

The term hydrokinetic pump means that this design of pump rapidly moves the fuel to create pressure and is a non-positive displacement pump design.

Hydrometer

A device that measures the particular gravity (weight or density) of liquid like engine coolant or battery electrolyte. Some maintenance-free SLA batteries have a special hydrometer built into them. This built-in hydrometer allows a visible check to see the battery state of charge (SOC). Built-in hydrometers have a green ball within a cage that is attached to a transparent rod. The green ball floats at a predetermined relative density that represents a couple of 65% SOC. When the green ball floats, it moves within the cage and positions itself under the rod. This action lets the ball be seen as a green dot within the center of the hydrometer. The built-in hydrometer provides a guide for battery testing and charging. When testing the battery, a visual green dot means the battery is charged enough for testing. If the green dot is not visible and also the hydrometer encompasses a dark appearance, it means the battery must be charged before battery testing will be performed.

With built-in hydrometer-equipped batteries, a yellow or clear spot may also become visible. The appearance of either of these indicates one or more of the following:

- The electrolyte level is low.
- The battery case is broken causing electrolyte leakage.
- Gas bubbles are lodged in the hydrometer.
- The battery is tipped at more than 45°.
- The battery is overcharged.

On some batteries with built-in hydrometers, a red dot indication may also appear. The red dot means the battery is nearing complete discharge and must be charged before being used.

Hypoid Gearsets

Rear axles or final drive gearset with teeth that are curved like a spiral bevel gear. The pinion gear is set below the centerline of the ring gear. This provides maximum gear tooth contact surface area for strength, gradual tooth

FIGURE H.8 Hypoid final drive gearset.

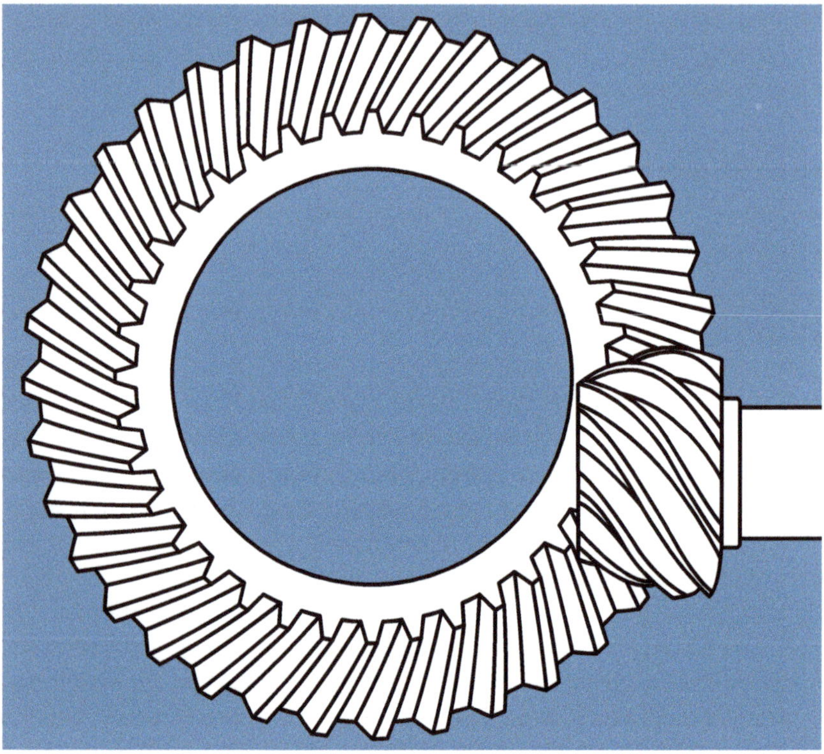

AlexanderZam/Shutterstock.com.

engagement, and quiet operation. It also allows for the drivetrain to be lower that the vehicle floor and only available in matched ground sets of ring gear and pinion.

Hysteresis

The amount of magnetization of a ferrous substance that lags the magnetizing force creating it due to friction. When the value of this force lags behind changes in the effect causing it, such as when magnetic induction, lags behind the magnetic field. A magnet within a magnetic field can have more than one magnetic moment. Movement plots often form a hysteresis curve, where different values of one variable will depend on the direction of another variable. Hysteresis takes place in ferromagnetic and ferroelectric materials along with polymer deformation. It can be related to irreversible thermodynamic change such as phase transitions and with internal friction. Dissipation is a side effect.

REFERENCES

1. Society of Automotive Engineers, J3162_201809: Heavy Duty OBD IUMPR Data Collection Tool Process, SAE Publishing, Warrendale, PA, 2018.
2. Society of Automotive Engineers, J139_202002: Ignition System Nomenclature and Terminology, SAE Publishing, Warrendale, PA, 2020.
3. Society of Automotive Engineers, J1715/2_202108: Battery Terminology, SAE Publishing, Warrendale, PA, 2021.
4. Society of Automotive Engineers, J2612_202002: Internal Combustion Engines—Piston Vocabulary, SAE Publishing, Warrendale, PA, 2020.
5. Society of Automotive Engineers, Standard J3104_201811: Electric Motors for Driveline Actuation Standard Terminology, Test Parameters, and Equipment Requirements, SAE Publishing, Warrendale, PA, 2009.
6. Society of Automotive Engineers, J3016_202104: Taxonomy and Definitions for Terms Related to Driving Automation Systems for On-Road Motor Vehicles, SAE Publishing, Warrendale, PA, 2021.
7. Kershaw, J.F., *High-Performance Automotive Cooling Systems* (Forest Lake, MN: CarTech, Inc., 2019), Chapter 3, ISBN:978-1-61325-504-9.
8. Society of Automotive Engineers, J1757-2_201811: Taxonomy and Definitions for Terms Related to Driving Automation Systems for On-Road Motor Vehicles, SAE Publishing, Warrendale, PA, 2017.
9. Society of Automotive Engineers, J1715/2_202105: Hybrid Electric Vehicle (HEV) and Electric Vehicle (EV) Terminology, SAE Publishing, Warrendale, PA, 2021.

I

I-Beam Steering Axle

The steering axle used for the front axles of most heavy-duty trucks, and referred to as an I-beam since it is forged into the shape of an "I." The steering mechanism is connected to steering levers, steering knuckles, and fixed kingpins. All I-beam non-driving front axles are machined from steel or aluminum forgings. For more information, see **J923_201207**, which outlines the basic nomenclature commonly used for truck and bus drive axle designs [1].

IAC (Idle Air Control)

An ICE IAC system that the engine management computer uses to control engine idle. The IAC is located in the throttle body. The IAC consists of a moveable pintle which is driven by a small electric motor called a stepper motor. The stepper motor is capable of moving in very exact, measured amounts called steps. The engine management computer uses the IAC valve to control idle RPM. The IAC changes the pintle position in the idle air passage of the throttle body. The incoming air is then varied around the throttle plate when the throttle is closed. For more information on this, see **J1930DA_202105** [2].

IAT (Inorganic Additive Technology) ICE Antifreeze

The traditional green coolant used in older vehicles. This solution offers fast-acting corrosion protection, but the additives are quickly consumed, exposing the cooling system to possible corrosion problems if not changed regularly. It can contain silicates, phosphates, and borates. IAT is considered outdated and no longer used because it can cause early failure of ceramic-phenolic seals used in the newer water pumps.

IAT (Intake Air Temperature) Sensor

A thermistor used to measure the air coming into the intake manifold. The IAT is a thermistor in a circuit in a variable ground, which uses a voltage divider network where the voltage is divided between the sensor input and a sensor ground inside the computer. The ECU provides a 5-V reference signal to the

IAT sensor. When cold, the sensor provides high resistance which the computer reads as high signal voltage. As the engine warms up, the thermistor sensor resistance becomes lower and the signal voltage drops. For more information on this, see **J1930DA_202105** [2].

FIGURE I.1 IAT circuit.

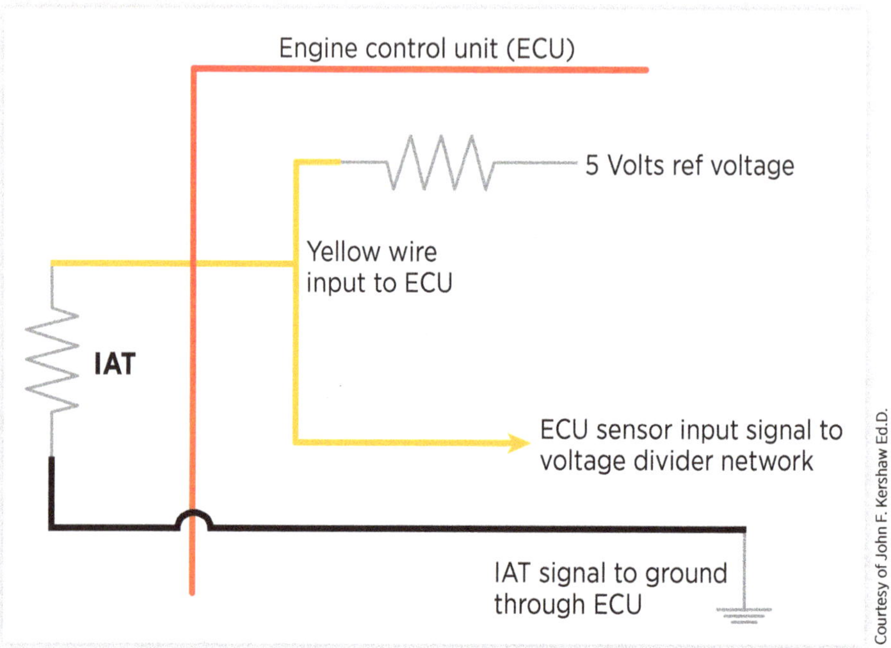

IC (Ignition Control)

OBD II terminology for the output signal from the PCM to the ignition system that controls the engine timing. For more information on this, see **J139_202002** [3].

IC (Integrated Circuit)

An IC or monolithic micro circuit could be a set of electronic circuits on one small flat piece of semiconductor material, usually silicon. Large numbers of little MOSFETs integrate into a little chip. This leads to circuits that are orders of magnitude smaller, faster, and less expensive than those constructed of discrete electronic components. The IC production capability, reliability, and building-block approach to computer circuit design has ensured the rapid adoption of standardized ICs in situ of styles using discrete transistors. ICs are now utilized in virtually all equipment and have revolutionized the world of electronics. Computers, mobile phones, and other digital home appliances are now inextricable parts of the structure of recent societies, made possible by the little size and low cost of ICs like modern computer processors and microcontrollers.

ICE (Internal Combustion Engine)

A heat engine where the combustion of fuel takes place internally in a combustion chamber that is an integral part of the fuel/air or working fluid flow circuit. The expansion of the high-pressure gases produced by combustion applies direct force to the piston or rotor of the ICE.

ICM (Ignition Control Module)

Opens and closes the primary ignition circuit by opening or closing the ground return path of the circuit using a solid-state switching transistor. When the electric switch is on, the voltage should be available at both the positive terminal and the negative terminal of the coil if the first windings of the coil have continuity. For more information on this, see **J139_202002** [3].

IDI (Indirect Diesel Injection)

The IDI diesel uses a prechamber that is joined to the main chamber above the piston by a connecting flow passage. Fuel is injected into the prechamber, which also uses a glow plug to heat the fuel for cold starting. Combustion begins in the prechamber and then spills into the main combustion chamber. As fuel and air are burned, the burning gases emit from the prechamber while the piston descends on the power stroke. Most IDI prechamber engines require a glow plug system to aid in the start-up.

Idle

ICE idle is when the vehicle is stationary and only requires a low rpm air-fuel mixture to maintain engine operation.

Idle Jet

A regulating fuel jet or orifice in a carburetor that provides fuel at idle.

Idle Stop Mode

A mode of operation where the ICE automatically shuts off when the vehicle comes to a full stop and restarts when the vehicle reaches a predetermined speed, or the driver removes his/her foot off the brake.

Idler Arm

This steering linkage component is used in integral recirculating ball-type steering gear linkage. It is attached to the opposite end of the drag link or relay rod from the pitman arm. The idler arm is parallel to the pitman arm and moves in the same plane, which keeps the linkage parallel. It is attached to the vehicle frame with bolts. It supports and guides the outer end of the drag link through the same path (arc) as the pitman arm.

FIGURE I.2 Idler arm.

ZhdanHenn/Shutterstock.com

Igniter

Another name for the electronic ignition control module that uses transistors to turn the ignition coil primary current on and off.

Ignition Circuit

Electrical components and connections that produce and distribute high-voltage electricity to ignite the air-fuel mixture inside the engine. For more information on this, see **J139_202002** [3].

Ignition Coil

An ignition coil is a step-up transformer that uses two windings, a primary and a secondary winding. When a change occurs within the magnetic flux of the primary winding, it will induce a current within the other secondary winding. So if the current is stopped from flowing (circuit open), the collapsing flux cuts across the turns of the secondary winding and creates a high voltage in the secondary winding. When two coils are brought in close proximity to each other, the magnetic field in one coil tends to link with the other coil. This generates a voltage within the second coil. This magnetic property of a coil which affects or changes the voltage in another coil is called mutual induction. The collapsing magnetic flux also creates a high voltage within the coil, capable of jumping the gap in the spark plug. For more information on this, see **J139_202002** [3].

Ignition Coil Output Test

A test that measures the maximum available voltage produced by the ignition coil. For more information on this, see **J139_202002** [3].

Ignition Interval (Firing Interval)

The number of degrees of crankshaft rotation between ignition sparks. For more information on this, see **J139_202002** [3].

Ignition Point/Autoignition Diesel Fuel

The point at which the diesel fuel will spontaneously ignite without an initiating spark under the conditions of density, pressure, and temperature. The ignition point for diesel fuel is about 558°F (292°C).

Ignition Switch

A switch through which electricity flows to energize the starter solenoid that includes contacts that conduct battery current to the magnetic switch.

Ignition System

Components and wiring necessary to create and distribute a voltage up to 40,000 V. The battery voltage is applied to the positive side of the ignition coil and pulse to ground to the negative side. When the negative lead of the coil is grounded, the primary (low-voltage) circuit of the coil is complete and a magnetic field is created in the coil windings. When the circuit is opened, the magnetic field collapses and induces a high voltage in the secondary winding of the ignition coil, which is used to generate a voltage great enough to bridge a gap across the spark plug electrodes to create a spark.

Early systems used a set of contact points to make and break the electrical connection to the ground. The electronic ignition uses a pickup coil or Hall-effect switch to signal an electronic module to make and break the primary ground of the ignition coil. When the ignition switch is turned on, the voltage is both at the positive and negative terminals of the coil. The labeling of positive (+) and negative (–) of the coil indicates that the positive terminal is more positive (closer to the battery positive terminal) than the negative terminal of the coil. This is referred to as coil polarity. The positive or negative coil polarity is established by which direction the coil is wound to the left or to the right. The polarity of an ignition coil is determined by the direction of rotation of the coil windings. The coil output voltage is directly proportional to the ratio of the primary to secondary turns of the wire used in the coil. For more information on this, see **J139_202002** [3].

FIGURE I.3 Automotive point-type ignition system.

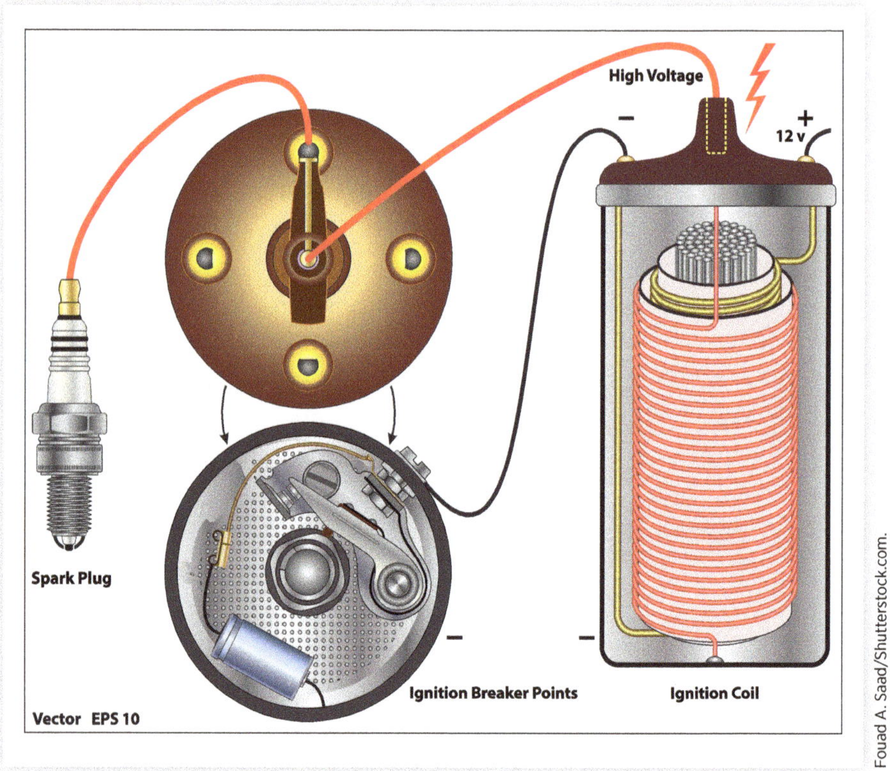

Ignition Timing

The exact point of ignition in relation to piston position.

IHP (Indicated Horsepower)

The power established inside the engine on the face of the piston. The power supplied by an engine to the device it powers is always less than the actual power developed in the cylinders. This takes place because some of the IHP is used to overcome the friction of the moving parts of the engine. The IHP is equal to the sum of the brake horsepower plus the friction horsepower. Using a dynamic pressure transducer and an oscilloscope, it is possible to obtain a pressure-time, or pt, diagram for an engine. From this diagram, with the stroke and connecting rod length known, it is possible to construct a pressure-volume, or pV, diagram. The pV diagram represents the pressures related to the piston position that exist in the cylinder throughout a complete crankshaft revolution.

$$IHP = \frac{PiLANK}{33,000} \quad \textbf{(I.1)}$$

where

Pi is the indicated mean effective pressure, in psi
L is the stroke, in ft
A is the area of the piston, in in.^{2}
N is the power strokes per minute
K is the number of cylinders

ILC (Idle Load Control)

The ILC is a system used by the engine management computer to increase the ICE speed when certain accessories place a load on the ICE, such as air conditioning or power steering.

IMEP (Indicated Mean Effective Pressure)

The pressure developed inside the engine at the face of the piston and used to calculate the indicated horsepower. The symbol Pi is used for IMEP.

Immobilizer

An electronic security device such as a part of a car or light truck that stops the engine operation unless the proper key (transponder or smart key) is present. This prevents the vehicle from being stolen. When the key is inserted, the transceiver sends out an electromagnetic energy pulse. This energy pulse is received by the coil inside the key transponder, which creates a voltage. The data within the magnetic pulses is within the type of a frequency-modulated signal.

A typical immobilizer system consists of a transponder key, coil antenna, key reminder switch, separate immobilizer module, engine management computer (EMC), and security light. Most systems work as follows: The key identification (ID) numbers are stored in an exceedingly nonvolatile memory of the immobilizer module. At each start, the module compares the ID number of the transponder key used with those stored within the memory. If the verification has been successful, the immobilizer module sends a letter of invitation signal to the EMC to match the key ID number with the numbers registered within the EMC. Each immobilizer module has its unique code word that is stored within the PCM. After the verification of the ID number, the immobilizer module requests the code word from the EMC. The immobilizer module controls the starter circuit, and therefore the security light, and signals the PCM to activate the mechanical system and ignition when the ID number and code word verification have been successful. The signals between the immobilizer module and EMC are transmitted via serial data.

Impeller

The impeller is the turning plate of a centrifugal pump.

Included Angle

Combination of the steering axis inclination (SAI) on cars and light trucks or the kingpin inclination (KPI) angle on heavy trucks with the camber angle.

Independent Suspension (also Independent Rear Suspension, or IRS)

A front or rear suspension system where the wheels move independent of each other.

Induced Voltage

Voltage in a conductor when relative motion and proximity exist between it and a magnetic field. When a magnetic field or line of force crosses or cuts across a conductor, a voltage is induced in that conductor.

Inductance

The signal caused by a sudden change in a magnetic field. If you turn off the current through a coil of wire, a voltage spike is generated across the coil.

Induction
Creating electricity in a wire by passing it through a magnetic field.

Inductive Pickup
A current testing lead that is placed around a current-carrying wire to measure amperage by measuring the magnetic field around the wire.

Inductive Reactance
The reactance or resistance to alternating or pulsating current by inductance or capacitance.

Inertia
The tendency of a body at rest to remain at rest or of a body in motion to stay in motion in a straight line unless acted on by another object. This is Newton's First Law of Motion:

$$I = mr^2 \tag{I.2}$$

where

m is the mass or weight
r is the radius

Infinite Resistance
The electrical circuit path is not complete. It is displayed on a digital multimeter as an ∞ (infinity) symbol.

Infrared Digital Handheld Noncontact Thermometer
A thermometer used to measure temperature. You point it at whatever you are going to measure and the temperature reading appears digitally on the tool.

FIGURE I.4 Infrared thermometer used to easily measure engine component temperature.

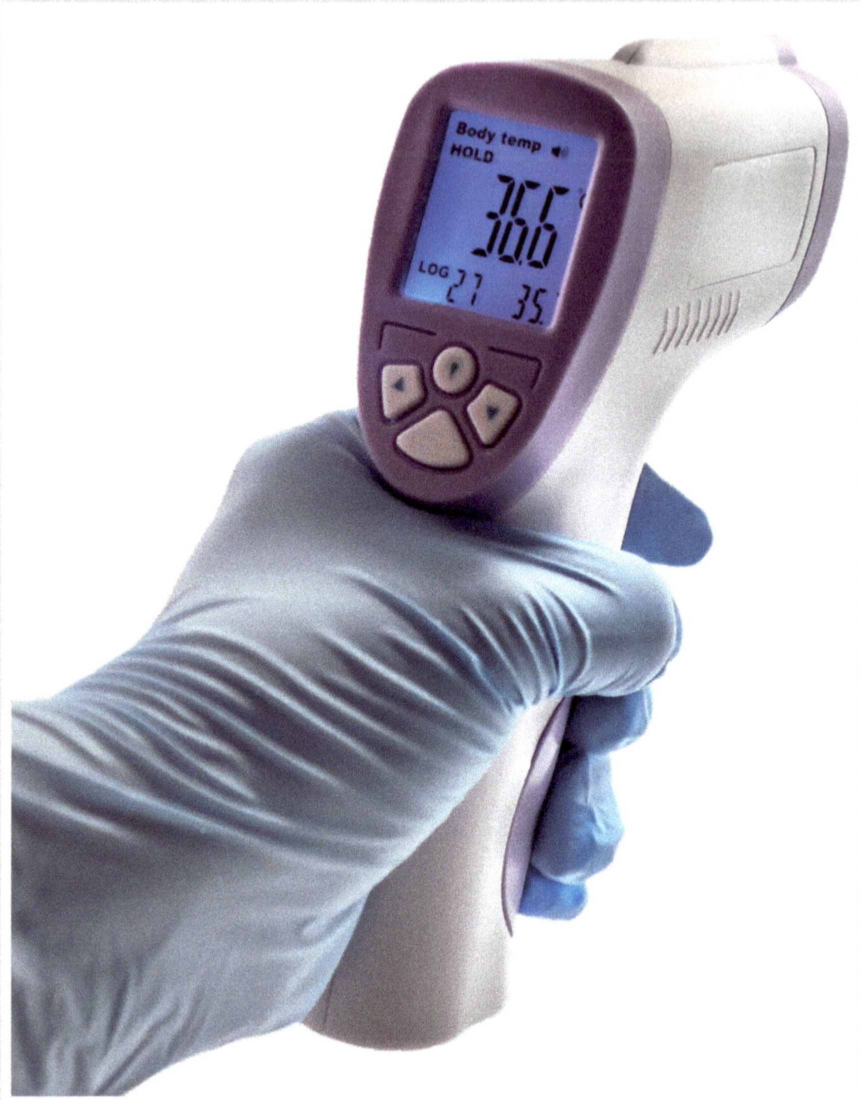

Infrared Radiation

A portion of the electromagnetic spectrum that extends from the longest wavelength, which is RED or the visible end of the light spectrum to the invisible microwave range.

Injection Nozzle

A mechanical valve in a diesel fuel system that is opened mechanically by the camshaft or from pressure from a hydraulic fuel injection pump.

Injector

A mechanical or electrical valve that injects fuel into an ICE.

Injector Pulse Width

The amount of time that voltage is applied to keep an electronic fuel injector open. The engine management computer controls the injector pulse width.

Instant Center

An imaginary point that is considered to be a point of the vehicle lift. On a ladder bar or leaf spring, it is very easy to see because it is the point where the front spring eye or front ladder bar mount attaches to the chassis. On multi-link style suspensions, it would be the point where two imaginary lines extended through the links would converge. The term instant center is related to anti-squat. When a car accelerates, there is weight transfer from the front to the rear. Because of the springs, it can be seen at the rear of the car that the car squats during acceleration. The rear axle, through the suspension links, can be made to counteract this squatting force known as "anti-squat." Anti-squat can keep the rear of the car level and can be made strong enough to raise the rear of the car. Most drag cars have enough anti-squat to raise the rear of the car under acceleration. If anti-squat can increase tire loading without adding to the weight of the car, then there will be more cornering power available. With more cornering power, the driver can get on the throttle sooner without oversteer, and that means faster cornering. Anti-squat is therefore beneficial. For more information on this, see **J670_200801** [4].

Instrument Cluster

Communicates using serial data via an SAE J1939 controller area network display vehicle information such as speed and RPM to the drive. It may also be equipped with a front-view camera controller and will illuminate the amber or green vehicle ahead indicator as requested by the front-view camera module. The instrument cluster also controls the head-up display or the collision alert indicators. For more information on this, see **J1939/21_2021** [5].

Insulator

Some materials hold their electrons very tightly; therefore, electrons do not move through them very well. These materials are called insulators. Insulators are materials with more than four electrons in the outer orbit of their atom. Because they have more than four electrons in their outer orbit, it becomes easier for these materials to acquire (gain) electrons than to release electrons. Examples of insulators include plastics, wood, glass, rubber, ceramics (spark plugs), and varnish for covering (insulating) copper wires in alternators and starters.

Intake Manifold

An ICE component that provides a pathway for air and fuel or just air from the throttle to the individual cylinder using timed passages.

Intelligent Vehicle

A vehicle with autonomous functions and capabilities and uses intelligent vehicle technologies (IVT) or telematics. IVT comprise electronic, electromechanical, and electromagnetic devices working together with computer-controlled devices and radio transceivers to provide precision repeatability functions (robotics or artificial intelligence systems) for emergency warning validation performance. IVT are the precursor to automotive telematics and the autonomous vehicle. IVT apply to car safety systems (ABS brakes, air bags, etc.) and self-contained autonomous electromechanical sensors which generate warnings that can be transmitted within a specified target area of interest, say within 100 m (0.0621371 miles) of the transceiver.

IVT for on-road vehicles are utilized for safety and commercial communications between vehicles or between a vehicle and a sensor along the road. IVT-equipped vehicles can use gathered information from the road to determine lane-specific vehicle usage and scenarios such as lane closures (in-vehicle notification warning), construction zones, and emergency situations. IVT provide on-road instant information to the driver who wishes to map a route to a destination and expects the system to assist in determining the best course of travel, hence the vehicle navigation system. The information provided by the in-vehicle system updates approximately every minute (depending on vehicle speed) all the transmitter beacon information recorded by the vehicle while traveling on the road, that is, all vehicles traveling on the highway update such information to the local mile markers via DSRC (Dedicated Short-Range Communication) telematics.

Interaxle Differential

A power divider or third differential operates much like a center differential in an all-wheel-drive vehicle. The power divider distributes torque from the vehicle driveline equally among two tandem axles. This power divider assembly uses an input shaft, interaxle differential, output shaft, and two constant mesh helical gears, which allow it to compensate for minor variations in speed between the two axles. The power divider also features a driver-controlled, air-operated lockout device. When lockout is engaged, it mechanically prevents interaxle differentiation for better traction.

FIGURE I.5 Interaxle differential.

BACHTUB DMITRII/Shutterstock.com.

Intercooler

A heat exchanger used to reduce the temperature of the air after it has passed through a turbocharger but before entering the intake manifold. Compressing a gas adds heat to it and raises its temperature. As you cool the air going into the intake manifold, the intercooler becomes more compact and dense to achieve more power. It is basically a radiator for hot air and passing air through it cools the air and makes it denser. If your build is for a racing application and you will be using 15–35 lb of boost for work, you will need to install an intercooler to keep the air temperature within a reasonable range. For more information on this, see **J922_201106** [6].

FIGURE I.6 Turbocharger intercooler assembly.

Poungsaed-Studio/Shutterstock.com.

Interference Angle

The difference between the angle of an ICE poppet valve seat and the valve face. The interference angle is used for a tighter seal and to remove carbon.

Interlock

A locking pin used in a manual transmission or transaxle to prevent selecting two different speed gears at the same time, which would lock up the transmission/transaxle.

Intermittent Problem

A fault that only occurs when certain conditions are present. Also called intermittent failure or soft failure.

Inverter Circuit

A circuit in the power control module that changes DC into AC and AC into DC.

Ion

When an atom is unbalanced, it becomes a charged particle called an ion. Ions are atoms that have gained or lost an electron, and they try to regain their balance of equal protons and electrons. An atom with an excess or deficiency of electrons will form either a negatively or a positively charged particle.

Ion-Sensing Ignition System

In this ignition, the spark plug itself becomes a sensor. The ICM applies a voltage of about 100 to 400-V DC across the spark plug gap after the ignition to sense the plasma inside the cylinder. The coil discharge voltage (10 to 15 kV) is electrically isolated from the ion-sensing circuit. The combustion flame is ionized and can conduct some electricity, which might be accurately measured at the plug gap. For more information on this, see **J139_202002** [3].

Ionize

To divide molecules into two or more oppositely charged ions. The air gap between the spark plug electrodes is ionized when the air-fuel mixture is modified from a nonconductor to a conductor.

Iron

Iron is a chemical element with the symbol Fe for ferri (Latin ferrum) with the atomic number 26. It is a metal that is found in group 8 of the periodic table. It is the most common element on Earth. The metal iron is refined from iron ore or ferrous oxide in a furnace.

ISC (Idle Speed Control)

The computer can pulse the ISC or IAC device to maintain the engine idle speed and provide an increased idle speed when needed, such as when the air-conditioning system is on.

ISO (International Organization for Standardization)

An international standard-setting body composed of representatives from various national standards organizations. Founded on February 23, 1947, the organization develops and publishes worldwide technical, industrial, and commercial standards. It is headquartered in Geneva, Switzerland, and works in 165 countries.

Isooctane

Liquid hydrocarbon present in petroleum that serves as a standard in the system of octane numbers.

ITS (Intelligent Transportation Systems)

An advanced system using intelligence to provide services related to different modes of transportation and traffic management and empower users to be well informed and make safer, more coordinated, and smarter use of transportation networks. It is defined as a system in which applied information and communication technologies are used in the field of road transportation, including infrastructure, vehicles, and users, and in traffic management and mobility management, as well as for interfaces with other forms of transportation. ITS may improve the efficiency of transportation in a number of situations, like road transportation, traffic management, and accident prevention.

IVHS (Intelligent Vehicle Highway System)

The IVHS program is applying modern communication and control technology to the needs of highway transportation. IVHS is yet another part of society that is increasingly interwoven with and dependent on modern communications technology. Yet it is vulnerable to cyberthreats, which are difficult to unravel using traditional centralized security approaches. Blockchain is an immutable peer-to-peer distributed database containing cryptographically secured information. Blockchain shows successful use cases in financial applications and smart contact, protecting digital copyright of media contents. It extends to any or all industries including secure IoT devices, embedded systems, etc. The superior feature of blockchain is its decentralized, immutable, auditable database that secures transactions by protecting privacy.

IVT (Intelligent Vehicle Technologies)

Vehicle navigation, traffic signal control systems, variable message signs, and automatic number plate recognition or speed cameras to monitor applications. IVT can also be more advanced applications that integrate live data and feedback from a number of other sources, such as parking guidance and information systems, weather information, and bridge deicing systems.

REFERENCES

1. Society of Automotive Engineers, J923_201207: Nomenclature and Terminology for Truck and Bus Drive Axles, SAE Publishing, Warrendale, PA, 2012.
2. Society of Automotive Engineers, J1930DA_202105: Electrical/Electronic Systems Diagnostic Terms, Definitions, Abbreviations, and Acronyms Web Tool Spreadsheet, SAE Publishing, Warrendale, PA, 2021.
3. Society of Automotive Engineers, J139_202002: Ignition System Nomenclature and Terminology, SAE Publishing, Warrendale, PA, 2020.
4. Society of Automotive Engineers, J670_200801: Vehicle Dynamics Terminology, SAE Publishing, Warrendale, PA, 2008.
5. Society of Automotive Engineers, J1939/21_2021: Data Link Layer, SAE Publishing, Warrendale, PA, 2021.
6. Society of Automotive Engineers, J922_201106: Turbocharger Nomenclature and Terminology, SAE Publishing, Warrendale, PA, 2011.

J

Jack Stands

An adjustable stand used to support an automobile that has been raised by a jack.

Jackknife

A condition in a tractor trailer where the trailer loses traction and swings around to the tractor like a folding pocket knife. "Vehicle jackknifing is generally associated with the loss of yaw stability, and is one of the most common causes of serious traffic accidents involving tractor-semitrailer combinations. . . . [A]n active braking control strategy is proposed for jackknifing prevention of a tractor-semitrailer combination on a low friction road. The proposed control strategy is realized via upper-level and lower-level control structures considering braking of both the units. In the upper-level control, the required corrective yaw moments for tractor and semitrailer are generated using a PID controller aiming to reduce errors between the actual yaw rates of tractor-semitrailer and the target yaw rates deduced from a reference model.

The corrective yaw moments are achieved through brake torque distribution among the tractor and semitrailer axle wheels in the lower-level control. The effectiveness of the proposed jackknifing prevention control is evaluated in a co-simulation environment involving Matlab/Simulink and TruckSim under two different maneuvers on a slippery road surface. Simulation results show that the proposed control approach is effective in jackknifing prevention of the tractor-semitrailer combinations under high speed maneuvers on low friction surfaces." For more information on this, see *Jackknifing Prevention of Tractor-Semitrailer Combination Using Active Braking Control* [1].

Jake Brake

A slang term for a diesel engine compression brake that comes from the manufacturer Jacobs Vehicle Systems.

Jamb Switch

An electrical grounding switch that operates the interior lights of a car or truck when the door is opened. It is located in the door jam.

Jerk Pump

A slang term for the in-line diesel mechanical fuel injection plunger pump.

Jet

An orifice used in automotive carburetors to regulate the metering of gasoline into the ICE.

Jidoka

Part of the Toyota Production System and Lean Manufacturing. It may be described as intelligent automation or automation with a human touch, which builds in quality. This type of automation implements some supervisory functions rather than production functions. At Toyota this usually means that if an abnormal situation arises, the worker will stop the production line. Autonomation (Jidoka) prevents the production of defective products, eliminates overproduction, and focuses attention on understanding the problem. For more information on this, see **J4001_202109** [2].

Jounce

During suspension compression, jounce occurs when the tire travels over a bump in the road and the rod and its piston move into the shock absorber. In extension (rebound), which is when the tire travels over a dip in the road, the rod and piston move out of the shock absorber. For more information on this, see **J1121_201608** [3].

Jump Box

A portable battery pack that eliminates the need for jumper cables. This auxiliary automotive battery is mounted in a box with high-current leads and is used to jump-start a vehicle with a dead battery. For more information on this, see **J1715/2_202108** [4].

FIGURE J.1 Jump box.

Courtesy of John F. Kershaw Ed.D.

Jump Start

A booster battery and jumper cables are used for jump-starting a vehicle with a dead battery. Improper jumper cable connections damage electronic control and charging system components on the stalled vehicle. In addition, connecting jumper cables produces sparks that have the potential of causing a battery explosion. Jump-starting is performed either with a second vehicle that has a fully charged battery and operating charging system or with a separate booster battery that is fully charged. With either method, it is extremely important to properly route the jumper cables and make the cable connections in the correct order. To jump-start a disabled vehicle:

- On the stalled vehicle, switch off the ignition and all electrical accessories, set the parking brake, and place the transmission in neutral or park.
- Attach one end of a jumper cable to the positive terminal of the discharged battery, and then connect the other end of the same cable to the positive terminal of the booster battery.
- Connect one end of the other jumper cable to the negative terminal of the booster battery and attach the remaining jumper cable end to a good ground on the engine block of the stalled vehicle.

- This final cable connection must be 30 cm from the discharged battery. Some vehicles have a remote jump start terminal specifically for this purpose.
- If jump-starting from another vehicle, make sure that the two vehicles are not touching each other. Also make sure that the clamps of one cable cannot accidentally contact the clamps of the other cable.
- Turn on the ignition and attempt to start the stalled vehicle.
- Once the stalled vehicle starts, remove the cables by disconnecting them in reverse order. First, the engine block connection, and then the negative battery terminal connection. Once the negative cable is removed, disconnect and remove the cable linking the two positive battery terminals.

For more information on this, see **J1715/2_202108** [4].

Jumper Cables

A set of high-current cables used to connect two automotive batteries together for the purpose of jump-starting the vehicle with a dead battery. For more information on this, see **J1715/2_202108** [4].

FIGURE J.2 Jumper cables.

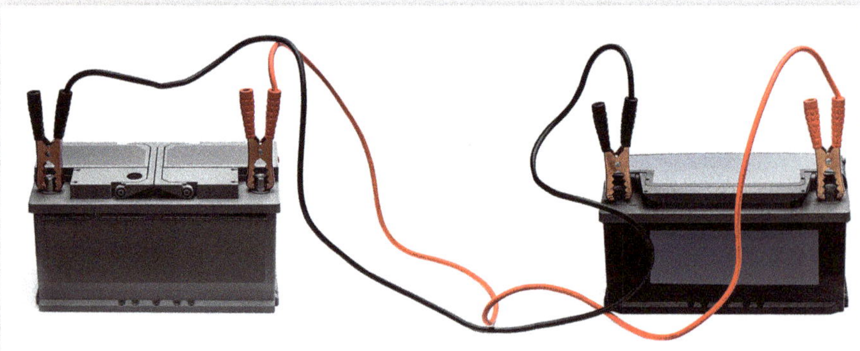

Jumper Wire

A piece of wire with alligator clips on both ends. It is commonly used to bypass components or to apply voltage to a component or section of a circuit.

Junction Block

A component that allows several wires to be connected to a common conductor. It serves as a splice, but the wires can easily be removed from the junction.

REFERENCES

1. Li, B. and Rakheja, S., "Jackknifing Prevention of Tractor-Semitrailer Combination Using Active Braking Control," SAE Technical Paper 2015-01-2746, 2015, https://doi.org/10.4271/2015-01-2746.
2. Society of Automotive Engineers, J4001_202109: Implementation of Lean Operation User Manual, SAE Publishing, Warrendale, PA, 2018.
3. Society of Automotive Engineers, Standard J1121_201608: Helical Compression and Extension Spring Terminology, SAE Publishing, Warrendale, PA, 2016.
4. Society of Automotive Engineers, Standard J1715/2_202108: Battery Terminology, SAE Publishing, Warrendale, PA, 2021.

K

Kaizen

Japanese manufacturing philosophy that stands for continuous improvement. The syllable *kai* means change and *zen* means to become good. This philosophy focuses on continuous improvement in manufacturing activities and business in general. When used in the business sense and applied to the workplace, *kaizen* typically refers to activities that continually improve all functions of a business. By improving standardized activities and processes, *kaizen* aims to eliminate waste. For more information on automatic transmission terms, see reference on Lean Six Sigma [2].

KAM (Keep Alive Memory)

Memory that stores input failures and diagnostic codes and allows the computer to have an adaptive strategy as parts wear and components deteriorate. The information stored in KAM allows the computer to maintain normal vehicle performance with abnormal inputs from sensors. It gives the computer the ability to ignore false inputs to maintain good drivability.

Kanban

Kanban: Kan means visual, and ban means card or board. Kanban is a concept related to LEAN and Just-In-Time (JIT) production. The Japanese word Kanban is a common term meaning "signboard" or "billboard." Taiichi Ohno, the man that developed JIT, has said that Kanban is a means through which JIT is achieved. Kanban is a signaling or scheduling system to trigger action. As its name suggests, Kanban historically uses cards to signal the need for an item. However, other devices such as plastic markers (Kanban squares) or balls (often golf balls) or an empty part-transport trolley or floor location can also be used to trigger the movement, production, or supply of a unit in a factory. If you take one, you make or replace one. For more information on automatic transmission terms, see reference on Lean Six Sigma [2].

Karman Vortex

Vortexes are created when air passed over a pointed surface. A Karman vortex sensor sends a sound wave through the air turbulence created by the incoming air passing through the Karman sensor. Air mass is calculated based on the time required for the sound waves to cross the turbulent air passage.

Kerosene

Kerosene or lamp oil is a combustible hydrocarbon liquid which is derived from petroleum. It is similar to number 1 diesel fuel.

Keyless Entry

A door unlock feature that uses a wireless transmitter built into the key fob or remote. A key fob could be a decorative item on a key chain. The transmitter broadcasts a proof that is received by the electronic control module, which is mostly mounted within the trunk or vehicle. The electronic control unit sends a voltage signal to the door lock actuator(s) located within the doors. Generally, if the transmitter unlock button is depressed once, only the driver's door is unlocked. If the unlock button is depressed twice, then all doors unlock.

Most keyless remote systems use a rolling-code form of transmitter and receiver. In an exceedingly conventional system, the transmitter emits a particular fixed frequency which is received by the vehicle control module. This single frequency may be intercepted and rebroadcast to open the vehicle. A rolling-code style transmitter emits a unique frequency each time the transmitter button is pressed, i.e., rolls over to a different frequency so it cannot be intercepted. Both the transmitter and therefore the receiver must be kept in synchronized order so that the remote functions correctly. If the transmitter is depressed when it is out of range from the vehicle, the correct frequency might not be recognized by the receiver, which did not roll over to the new frequency when the transmitter was depressed. If the transmitter does not work, try and resynchronize the transmitter to the receiver by depressing and holding both the lock and therefore the unlock button for 10 sec when within range of the receiver.

Kickdown (Detent) Lever

A mechanical valve moved by a cable that causes a downshift in an automatic transmission. This lever moves a hydraulic valve called a kickdown or detent valve that applies hydraulic pressure to the opposite side of a shift valve to push back against the governor pressure engaging the lower gear. For more information on this, see **J649_201803** [1].

Kickdown Switch

An electrical switch used on early automatic transmissions that operated a solenoid in the transmission to downshift the unit into a lower gear. For more information on this, see **J649_201803** [1].

Kilo

Metric prefix for 1000; abbreviated k or K.

Kilowatts

Metric prefix kilo represents 1000 W.

Kinematic Formulae

A set of formulae that relate to the five kinematic variables:

1. Displacement: distance moved in a certain direction
2. Time
3. Initial velocity
4. Final velocity
5. Acceleration

$$\text{Average velocity: } V_{avg} = \frac{V_i + V_f}{2} \quad \textbf{(K.1)}$$

$$\text{Average velocity/speed and time: } V_{avg} = \frac{s}{t} \quad \textbf{(K.2)}$$

$$\text{Speed: } s = V = \bar{u}\, t \quad \textbf{(K.3)}$$

$$\text{Acceleration: } a = \frac{V_f - V_i}{t} \; a = \frac{V^2}{2 \times D} \quad \textbf{(K.4)}$$

$$\text{Final velocity: } V_f = V_i + at \tag{K.5}$$

$$\text{Final velocity squared: } V_f 2 = V_{i^2} + 2as \tag{K.6}$$

$$\text{Displacement: } D = V_i t + \frac{1}{2} at^2 \tag{K.7}$$

Kinematic Viscosity

Kinematic viscosity is the ratio of the dynamic or absolute viscosity to density. This is a quantity where no force is involved. Kinematic viscosity can be obtained by dividing the absolute viscosity of a fluid with the fluid mass density.

Kinematics

The analytical and mathematical descriptions of all types of motion at a distance from mass and force. The physics subfield that explains and describes the motion of points, bodies, and systems of bodies or groups of objects without considering the forces that cause them to move. It refers to the study of motion and deals with any type of motion of any particular object and their interrelationships. The science of kinematics relates to Newton's First Law of Motion or inertia which says that an object in motion stays in motion and an object at rest stays at rest unless acted upon by an outside force. It is the alteration to any change in motion.

Kinetic Energy

The energy in any moving object that depends on the weight (mass) of the object and its speed.

Kingpin (Suspension)

A hardened steel pin that attaches the steering knuckle to the front axle, allowing the front wheels to move for steering. This device is used on vehicles with a solid front axle using leaf springs for support.

FIGURE K.1 Front suspension kingpin kit.

Aleksandr178/Shutterstock.com.

Kingpin (Trailer)

Connecting pin used on a trailer that connects to the fifth wheel to lock the trailer to the tractor or tow vehicle and provides a pivot point for turning.

FIGURE K.2 Trailer kingpin.

Horst Bingemer/Shutterstock.com.

Kingpin Inclination (KPI) Angle

An angle that exists between the kingpin, or knuckle pin, and a true vertical plane that runs through the center of the wheel.

FIGURE K.3 KPI angle [3].

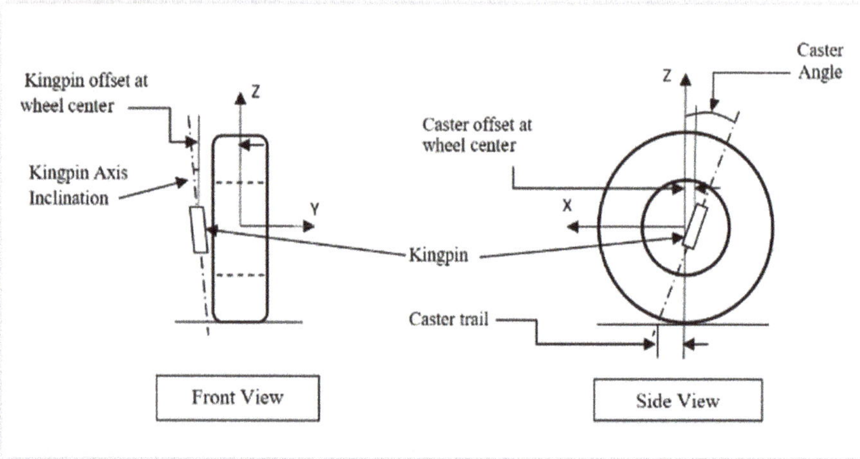

Kirchhoff's First Law of Current

The current entering a junction is the same as the total of the current leaving the junction.

Kirchhoff's Second Law of Voltage Drops

Kirchhoff's second law states that the sum total of all voltage drops in a series circuit equals the applied voltage.

Knock Sensor (KS)

A piezoelectric circuit that detects spark knock, pre-ignition, and detonation.

REFERENCES

1. Society of Automotive Engineers, J649_201803: Automatic Transmission Functions—Terminology, SAE Publishing, Warrendale, PA, 2018.

2. Li, B. and Rakheja, S., "Jackknifing Prevention of Tractor-Semitrailer Combination Using Active Braking Control," SAE Technical Paper 2015-01-2746, 2015, https://doi.org/10.4271/2015-01-2746.

3. Jambukar, S. and Chandramohan, S., "Effects of Kingpin Inclination and Caster Angle on Kinematics and Lateral Dynamics of Long Wheelbase School Bus," SAE Technical Paper 2019-26-0219, 2019, https://doi.org/10.4271/2019-26-0219.

L

Ladder Frame

Most heavy-duty truck frames are of the parallel ladder design, assembled either by bolting or by riveting. The two main parts of the frame are the two longitudinal members, which are often referred to as "rails." These rails actually carry the load, while the transverse or crossmembers tie the two rails together to provide strength and rigidity, as well as weight distribution over both rails equally. The frame must be stiff enough to support the loads placed upon it during operation and must be capable of absorbing the torque transfer from the axles, which attempt to twist the frame. Thus, the frame must be flexible enough to distribute these stresses throughout its length. The parallel-construction-type frame (chassis) is the most common design, yet many trucks may have a frame known as a drop-center frame because the frame side rails drop in height immediately behind or in front of the support member for the fuel tank. This results in a thicker frame that extends throughout the midsection of the construction. This design results in a lower overall center of gravity, plus increased strength.

Lambda Sensor

Oxygen sensor, or O_2 sensor. Lambda is the Greek letter that represents ratio, as in air-fuel ratio.

Landing Gear

A mechanical device used to raise the trailer off of an uncoupled fifth wheel to hold the trailer when the tractor pulls out from under the trailer.

FIGURE L.1 Landing gear.

Kevin Norris/Shutterstock.com.

Lane Departure

A warning system designed to warn the driving force when the vehicle begins to maneuver out of its lane (unless a visual signal is on in this direction) on divided highways and arterial roads. These systems are designed to reduce accidents by addressing the most causes of collisions: driver error, distractions, and drowsiness. There are four styles of systems:

- Warning, if the vehicle is leaving its lane with visual, audible, and/or vibration warnings (lane departure warning, LDW).
- Warn the motive force and, with no response, automatically take steps to confirm the vehicle stays in its lane (lane keep assist, LKA/LKS).
- Assist in oversteering, keep the car centered within the lane, and ask the motive force to take over in challenging situations (lane centering assist, LCA).
- Assist in emergency lane keeping (ELK). The emergency lane keeping applies the correction to a vehicle drifting past a solid lane marking.

The LDW system is a crash-avoidance technology which warns drivers if they are drifting (or have drifted) out of their lane or from the roadway. This warning system is designed to reduce the possibility of a run-off-road crash.

This system will not take control of the vehicle; it will only let the driver know that he/she needs to steer back into the lane. An LDW is not a lane-change monitor, which addresses intentional lane changes, or a blind-spot monitoring system which warns of other vehicles in adjacent lanes. This informational report applies to the OEM and aftermarket LDW systems for light-duty vehicles (gross vehicle weight rating of no more than 8,500 lb) on relatively straight roads with a radius of curvature of 500 m or more, and under good weather conditions. For more information on this, see **J3063_202103** and **J2808_201701** [1, 2].

FIGURE L.2 Lane departure system.

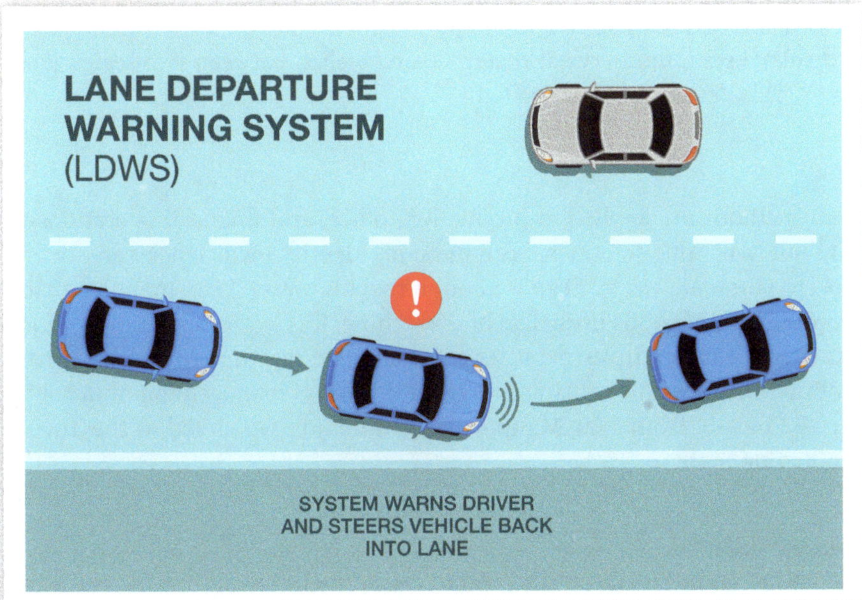

Flat vectors/Shutterstock.com.

Lane Keep Assist Switch

The lane keep assist switch provides an input to the front-view camera to turn the lane keep assist system ON/OFF. The front-view camera provides a signal voltage to the normally open momentary switch. When the switch is pressed, the signal circuit is pulled to ground, indicating to the front-view camera module that the system has been requested to power ON/OFF. The lane keep assist switch also contains the lane keep assist switch indicator, which is controlled by the front-view camera module to indicate the ON/OFF status of the lane keep assist system. When the lane keep assist switch turns on the system, the front-view camera module applies ground to the switch indicator and illuminates the LED. For more information on this, see **J3063_202103** and **J2808_201701** [1, 2].

Lateral Runout

Side-to-side movement of the rotating wheel, tire, or brake rotor as it rotates. Excessive movement will result in vibration and pulsation in the steering wheel or seat of a vehicle.

LCC (Lane Centering Control)

Can continuously steer the vehicle to follow a lane chosen by the driver based on information provided from the vehicle sensor suite under the constraints listed in the requirements below. The LCC feature is engaged and disengaged by the driver using a switch.

LCD (Liquid Crystal Display)

A display uses a fluid crystal material sandwiched between two sealed glass plates. This device can polarize light.

Lean

Lean methods are applied to production, office, and financial operations to trim out non-value activities. Lean methods identify inefficiencies and define tools to eliminate waste. Lean is about removing waste. Lean manufacturing is production of goods using less of everything: less waste, less human effort, less space, less tool investment, and less engineering time to develop a new product. Lean manufacturing is a generic process management philosophy derived mostly from War Manpower Commission which led to the Toyota Production System.

Lean Misfire

A misfire that occurs when the engine is running lean.

Leaf Spring

The leaf spring is made from steel, but some are composed of fiberglass, carbon fiber, graphite, or other materials. The spring curve flattens because the spring is compressed. Some leaf springs are multi-leaf. Each leaf below the main leaf increases in arch. This creates more strength because the spring is compressed. The top leaf is the longest, flattest strip and has eyes (loops) formed at each end for attachment to the vehicle. The U-bolts fasten the spring to the axle. The front eye of the spring is attached to a hanger on the frame. The rear eye is attached to a movable shackle or link that enables most of the leaves to expand long when the vehicle encounters a bump or dip.

FIGURE L.3 Rear leaf spring supporting the rear of a vehicle at the rear axle housing.

Marcel Derweduwen/Shutterstock.com.

Leaf Spring Suspension System Heavy Truck

Leaf spring suspension systems are used on both single and tandem axle straight trucks, as well as tractor trailers due to their weight-carrying ability. Spring suspensions can also be used on both the front and rear axle assemblies; however, more springs are used in the rear to accommodate the extra weight. The main springs, or "leaves," are formed into an eye at each end. A bushing is pressed into the eye, and a shackle pin is inserted through the bushing. There is typically a grease fitting used to lubricate the pin and bushing. The shackles allow movement in the spring suspension. The swinging shackle allows the spring assembly to lengthen or shorten while the truck is in motion, which allows the axle assembly to move up and down as the truck encounters bumps and other road irregularities. The stationary bracket is bolted directly to the frame rail, and the leaf spring is supported by the clamped pin.

LEDs (Light Emitting Diodes)

A semiconductor that emits light when it is energized. LEDs are commonly used as indicator lights in various vehicle systems.

FIGURE L.4 Automotive LED headlight application.

DiPres/Shutterstock.com.

Left-Hand Rule

A method of determining the direction of the magnetic flux lines surrounding a current-carrying conductor, when the electron current flow theory (– to +) is used. If the conductor is grasped with the left hand, the thumb points in the direction of current flow (– to +) and the fingers will point in the direction of magnetic flux or lines of force.

Lenz's Law

The direction of the electric current which is induced in a conductor by a changing magnetic field is such that the magnetic field created by the induced current opposes changes in the initial magnetic field.

$$\varepsilon = -N\frac{\Delta\phi}{\Delta t} \quad \textbf{(L.1)}$$

where

- ε is the induced electromagnetic field
- N is the number of coil turns
- t is the changes in time
- ϕ is the changes in magnetic flux

L-Head

Another term for the flathead engine; see **Flathead ICE**.

Lidar (Light Detection and Ranging)

Lidar measures the distance to a target by illuminating the target with pulsed laser light and measuring the reflected pulses with a sensor. Differences in laser return times and wavelengths can then be used to create digital three-dimensional (3-D) representations of the target. Lidar sometimes is termed laser scanning and 3-D scanning, with terrestrial, airborne, and mobile applications.

The technology is additionally utilized in control and navigation for a few autonomous cars and autopilot control systems just like the Cadillac Super Cruise. Lidar enables an autonomous or self-driving vehicle to watch the planet with some special superpowers: endless 360° of visibility, where human eyes were allowed to work out altogether directions all of the time together with accurate depth information, where you will always know the exact distance of objects relevant to you. Lidar uses ultraviolet (UV), visible, or near-actinic rays to image objects. It can target a good range of materials, including non-metallic objects, rocks, rain, chemical compounds, aerosols, clouds, and even single molecules. A narrow ray can map physical features with very high resolutions, for instance, an aircraft can map terrain at 12 in. resolution or better.

Lidar instruments fitted to aircraft and satellites do surveying and mapping. The National Aeronautics and Space Administration (NASA) used lidar as a key technology for enabling autonomous precision safe landing of robotic lunar-landing vehicles. Wavelengths vary to suit the target: from about 10 μm to the UV. Typically light is reflected via back-scattering as a hostile pure reflection one might find with a mirror. Different types of scattering are used for various lidar applications: the most ordinary are Rayleigh scattering, Mie scattering, Raman scattering, and fluorescence. Suitable combinations of wavelengths can give the mapping of atmospheric contents by identifying wavelength-dependent changes within the intensity of the returned signal. For more information on this, see **J3016_202104** [3].

FIGURE L.5 Lidar self-driving light-object-detection ranging.

shmai/Shutterstock.com.

Lift

The distance the ICE poppet valve is lifted off its seat when fully open.

Lift Axle

An auxiliary truck axle that can be lifted when empty. A lifting or retractable axle is a way to raise the axle so that its tires are not touching the road surface, which improves fuel economy and reduces tire wear. It is lowered when loaded to enhance stability and help meet federal bridge requirements.

Lift Pump

The fuel supply pump used on a diesel engine that creates a low-pressure area that moves the fuel from the tank to the fuel injection pump.

Lifter

A device that converts the cam rotary motion to an up-and-down action in the valve train to open and close the poppet valve. See **Cam Follower**.

Lifter Pump-Up

The time period when the hydraulic lifters fill with oil.

Lightning

Benjamin Franklin proved the electrical nature of thunderstorms in the famous kite experiment. Franklin coined the terms positive and negative and developed the conventional theory of current flow. Originally, he was trying to prove that the positive and negative electrons in the clouds produced the static electricity that causes lightning. Natural negatively charged particles made up of electrons will produce lightning when they find a similar path from negative to positive.

FIGURE L.6 Lightning.

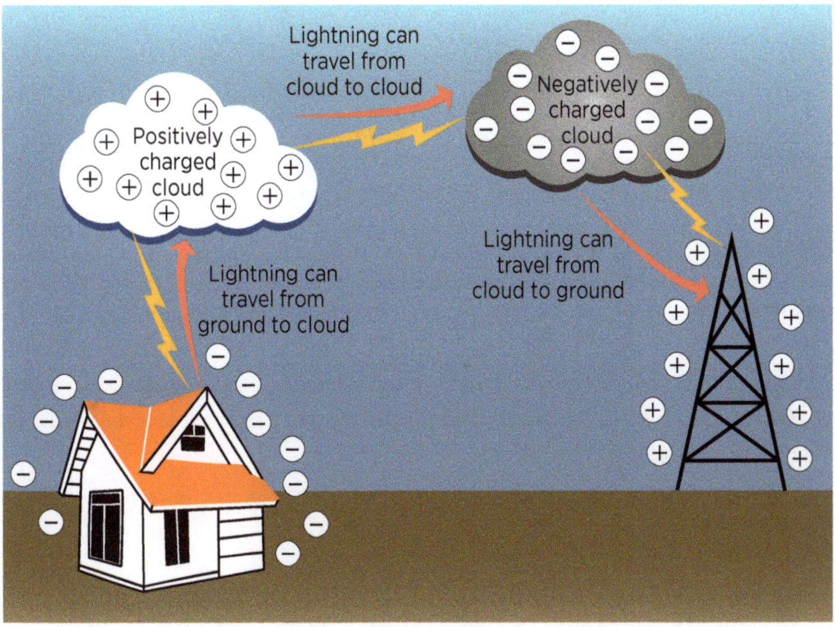

Courtesy of John F. Kershaw Ed.D.

Li-ion (Lithium-Ion) Battery

The Li-ion battery is a rechargeable battery where Li ions move from the negative electrode through an electrolyte to the positive electrode during discharge and back when charging. Li-ion batteries use an intercalated lithium compound because the material at the positive electrode and the negative electrode, typically graphite, need a high energy density, no memory effect (other than LFP cells), and low self-discharge and will be a safety hazard since they contain flammable electrolytes and, if damaged or incorrectly charged, can cause explosions and fires. These batteries are used in most electric vehicles (EVs). For more information on this, see **J1715/2_202108** [6].

FIGURE L.7 Li-ion battery is in the base of this EV.

Gorodenkoff/Shutterstock.com.

Liner

Two types of diesel engine cylinders are used: dry and wet liners. The dry liner is a thin wall sleeve where a wet line is installed in the coolant jacket surrounded by the engine coolant. Dry liners are installed into the block bore usually with a marginally loose fit and retained by the cylinder head. The dry sleeve does not transfer heat as well as a wet liner, but they are easily replaced and do not present coolant-sealing problems.

FIGURE L.8 Cylinder liner.

A.Shvedov/Shutterstock.com.

Lines of Force

A magnetic field that is made up of many invisible lines of force. These lines are called a magnetic field, magnetic flux, or magnetic lines of force. A magnetic flux can be compared to current. Flux lines are directional and EXIT from the north pole and ENTER through the south pole. Flux lines are concentrated at the poles and spread out into the areas between the poles. Magnetic flux is the product of the average magnetic field times the perpendicular area that it penetrates. Flux is used in the operation of solenoids and transformers. In an electric generator where the magnetic field penetrates a rotating coil, the area used in defining the flux is the projection of the coil area onto the plane perpendicular to the magnetic field. Flux density refers to the number of flux lines per unit area. A Gauss gauge can be used to measure a magnetic field. A magnetic field is a vector quantity because it has direction and magnitude. It shows the magnetic influence on moving electric charges, electrical currents, and magnetic materials. The magnetic field of a permanent magnet pulls on magnetic materials such as iron and attracts or repels other magnets.

Lithium Grease

A lubricant with excellent adhesion to metal. This means that once it is applied, it stays there. It is a non-corrosive grease, meaning that it does not corrode the parts/machinery to which it is applied. It is a stable compound and does not chemically react with substances that it comes in contact with and is suitable for use under heavy-load conditions. It demonstrates excellent heat resistance. It is capable of withstanding very high temperatures and has a drip temperature of about 200°C, and it is water resistant.

LNG (Liquefied Natural Gas)

A natural gas mostly methane CH_4 with some mixture of ethane C_2H_6 that has been cooled down to liquid form for the safety of non-pressurized storage or transport. It takes up about 1/600th the volume of natural gas in the gaseous state at standard conditions for temperature and pressure.

Load

A device in a circuit that uses electricity to do its job. For example, the loads in a circuit could include a motor, a solenoid, a relay, and a light bulb. All loads offer some resistance to current flow.

Load Devices

Component wired into a circuit that converts electrical current flow into a useful form.

Loader

A machine used in construction or business to move or load materials such as soil, rock, sand, and demolition debris into or onto a dump truck, conveyor belt, feed hopper, or railroad car. For more information on this, see **J1464_200410** [5].

FIGURE L.9 Loader.

Loadstone

A naturally magnetized piece of magnetite, which is a mineral composed of iron oxide and is ferromagnetic. When a loadstone is suspended, it can turn and will point north. Loadstone became the first magnetic compasses. When suspended it will rotate and always point toward Earth's North Pole. In materials like iron, nickel, and cobalt, the molecules can be aligned so that their north poles all point in one direction and their south poles point in the opposite direction. These materials can become magnets.

Location Grid

Locates components or connections quickly on a wiring diagram. It is an alphanumeric system marked off around the outside of wiring circuits.

Locking Differential

Designed to overcome driving the wheel with the least amount of traction in an open differential by locking both wheels on an axle together. This forces both wheels to turn together, regardless of the traction available to either wheel individually.

FIGURE L.10 Locking differential operated by an electromagnetic clutch.

Pixel B/Shutterstock.com.

Locking Hubs

Manually locking hubs require the driver to get out of the vehicle and manually lock the hubs on the front axle. When the driver turns the locking hub assembly to the "lock" position, a clutch, which is splined to the front axle, is engaged. This mechanically locks the axle to the wheel hub. Switching the hubs to unlock mode disconnects the front wheels from the axle. These hubs were used on all domestic 4WD in the early days, and a version is found on Ford Super Duty trucks.

FIGURE L.11 Manual 4WD truck locking hubs.

kaninw/Shutterstock.com

Lock-Up Clutch

See **TCC (Torque Converter Clutch)**.

Long Block

Engine assembly that includes crankshaft, connecting rods, pistons, cylinder heads, and the valve operating mechanisms.

Low Oil Level Indicator Circuit

A circuit that uses a sensor in the engine oil pan to operate an indicator light. When there is enough oil in the pan, the sensor has high resistance. No current flows through the circuit, and the indicator light remains off. If the oil level drops too low, the sensor resistance drops, and the dash light turns on.

LPG (Liquid Petroleum Gas)

Liquefied petroleum gas (LPG, LP gas, or condensate) is a flammable mixture of propane and butane hydrocarbon gases. LPG is used as fuel gas in heating appliances, cooking equipment, and vehicles.

LSD (Limited Slip Differential)

A mechanical or electronic clutch that locks the rear axles to the differential side gears so that all power goes to both real wheels with limited slip. LSDs counter drive-wheel slippage when different traction conditions exist on either side of an axle and are still widely employed by the automotive industry to improve driving control. In an LSD frictional couplings connect the axle shafts to the differential and provide the means of transmitting power to the wheels. The friction plates in the coupling may contain a variety of friction materials including metal, paper, sintered bronze, and carbon. Each one of these materials has very different frictional and wear characteristics, and each one requires a different response from the gear additive package. Each plate must be durable over the course of the vehicle lifetime irrespective of the material used. As the demands on rear axles increase with the application of greater horsepower and the increasing requirements of aerodynamic engineers, the lubrication of these friction plates remains an ongoing challenge.

Lubricant frictional characteristics are very important in determining the quiet and smooth operation of LSDs. Fully formulated API GL-5 J 2360 gear lubricants are unable to fulfill all the lubrication requirements of LSDs. Special additives have, therefore, been developed to improve the frictional response in the coupling to eliminate noise, vibration, and stick-slip problems. There is a requirement to provide limited slip additives that give excellent friction plate lubrication. The goal of lubricant formulators is to develop new products that retain frictional performance with minimal effect on the thermal stability of the gear lubricant. A test methodology has been developed that correlates friction characteristics with known field performance, and testing has been performed on a variety of friction plate materials. The results of testing with current commercially available limited slip additives and new additives with much improved thermal characteristics in different gear lubricants and with different hardware configurations are discussed. For more information on this, see **J923_201207** [4].

Lubricant

A compound, such as oil or grease, used for minimizing friction, especially in an ICE or gear component.

Lubrication System

All automotive engines use a pressurized lubrication system consisting of the following: oil reservoir and its ventilation, oil pump and pickup, pressure relief valve, filter, galleries and lines, and indicators. The engine design uses both pressure and splash methods of oiling. Oil travels under pressure through galleries to reach the top end of the engine. Gravity flow or splash oiling lubricates many parts. A bypass valve is used in the filter to prevent oil starvation. A pressure regulator is used to maintain the designed oi pressure range.

FIGURE L.12 ICE lubrication system.

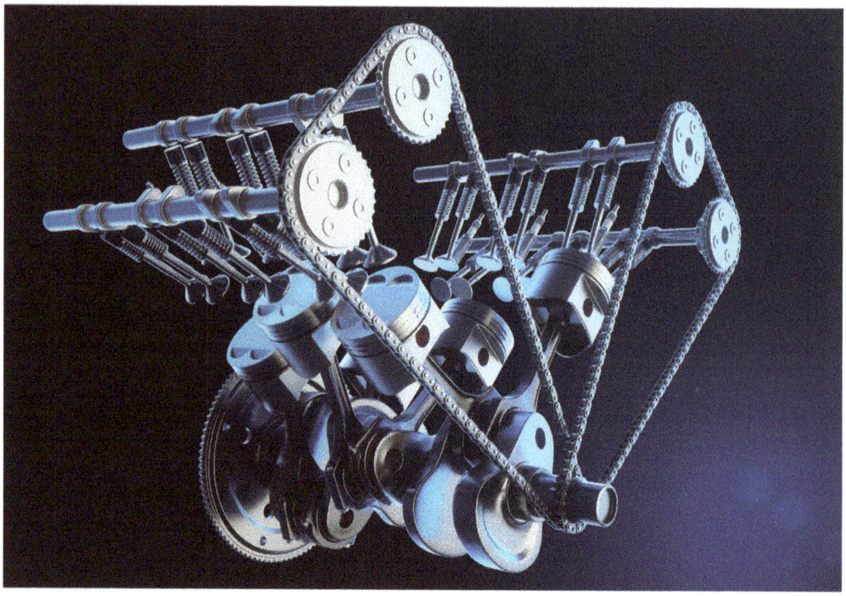

Rost9/Shutterstock.com.

Lubricity

The measure of the reduction in friction and/or wear by a lubricant. Lubricity measures the fuel ability to lubricate the moving parts of the fuel injection pump and injectors.

Lug Nuts

A fastener that holds the wheel and tire to the axle flange or wheel hub. Some lug nuts use a center-tapered cone to center the wheel to the hub. Aluminum wheels may also require lug nuts which are tapered to center the wheel. Some vehicles have left-hand (counterclockwise to tighten) threads. The studs are stamped with an "L" at the end as an indication of the left-hand thread type. Chrome lug nuts are not recommended on chrome wheels because the friction between the two chrome surfaces can cause an incorrect torque reading when tightening. Wheel nuts should be tightened in two steps. In the first step, they are torqued until all lugs are seated. During the second step, they are gradually tightened to specification. This eliminates the possibility of misalignment. Most vehicles use the star pattern for torquing lug nuts. Torque patterns differ in certain vehicles. Five lug wheels are torqued in a star pattern, in multiple steps, to the torque specified in the owner's or service manual. Refer to the owner's or service manual for vehicles with six or eight lugs.

Luminosity Timing Meter

A meter used to check or set the injection timing on a diesel engine. You place a quartz crystal probe in a glow plug hole usually a number 3 cylinder and the meter reads the light of combustion as an indication of cylinder firing. Readings are after top dead center, not before top dead center, as in gas engines. The tester can use this signal to measure ignition (combustion or injection) timing.

FIGURE L.13 Luminosity timing meter.

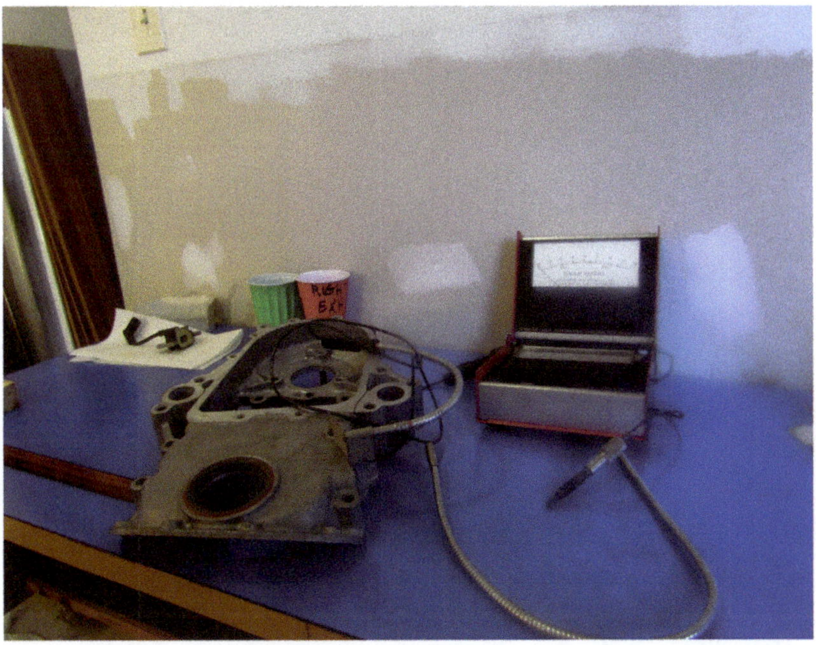

Courtesy of John F. Kershaw Ed.D.

LVD (Low-Voltage Disconnect)

Electrical devices monitor battery voltage and disconnect non-critical electrical loads when the battery voltage level falls below a preset value. These devices reserve battery current to a level enough to start the diesel engine when key-off loads or other parasitic draws are draining the battery. LVD devices then reconnect the electrical loads when the battery level is returned to a high-enough voltage.

FIGURE L.14 Low voltage disconnect.

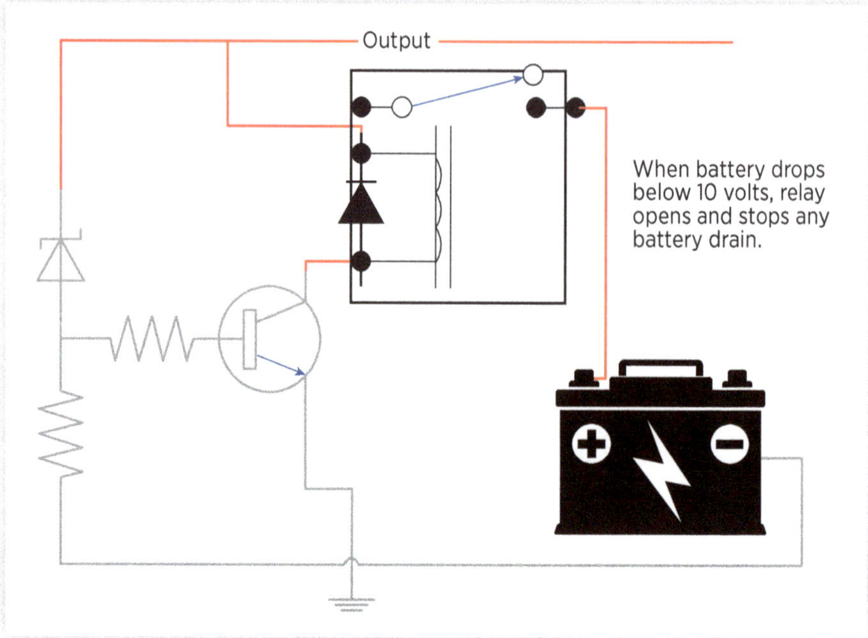

REFERENCES

1. Society of Automotive Engineers, J3063_202103: Active Safety Systems Terms and Definitions, SAE Publishing, Warrendale, PA, 2021.

2. Society of Automotive Engineers, J2808_201701: Lane Departure Warning Systems: Information for the Human Interface, SAE Publishing, Warrendale, PA, 2018.

3. Society of Automotive Engineers, J3016_202104: Taxonomy and Definitions for Terms Related to Driving Automation Systems for On-Road Motor Vehicles, SAE Publishing, Warrendale, PA, 2021.

4. Society of Automotive Engineers, J923_201207: Nomenclature and Terminology for Truck and Bus Drive Axles, SAE Publishing, Warrendale, PA, 2012.

5. Society of Automotive Engineers, J1464_200410: Identification Terminology of Loaders/Tractors with Forks and Rough Terrain Forklifts, SAE Publishing, Warrendale, PA, 2012.

6. Society of Automotive Engineers, J1715/2_202108: Hybrid Electric Vehicle (HEV) and Electric Vehicle (EV) Terminology, SAE Publishing, Warrendale, PA, 2021.

M

M-85

An automotive fuel, M-85 is a mix of 85% methanol and 15% gasoline.

MacPherson Strut Suspension

A front suspension design that eliminates the upper control arm and mounts a coil-type spring over the strut assembly. The MacPherson strut is used on most FWD vehicles with independent front suspensions.

FIGURE M.1 MacPherson strut.

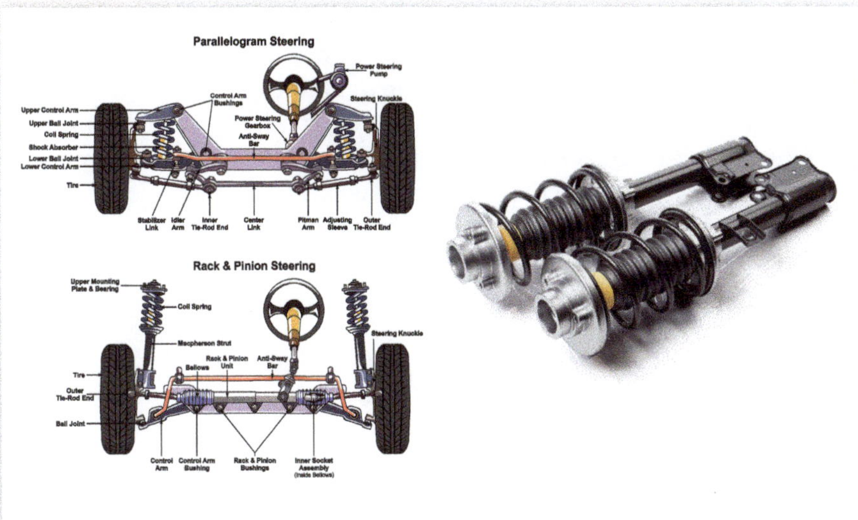

KsanderDN/Shutterstock.com.

udaix/Shutterstock.com.

MAF (Mass Air Flow) Sensor

Measures the volume and density of the incoming air. Because of the way the MAF takes a reading, the sensor takes air temperature, density, and humidity into account. All these variables together determine the mass of the incoming air. The computer reads actual mass airflow to help calculate fuel requirements. For more information on this, see **J1930DA_202105** [1].

Magnetic Coil Gauge

A gauge that uses small coils and magnetic fields to rotate the gauge needle.

Magnetic Crankshaft Position Sensors

Sensors that use the changing strength of the magnetic field surrounding a coil wire to signal the module and computer. This signal is used by the electronics in the module and computer as to piston position and engine speed (RPM).

Magnetic Field

The space surrounding a magnet through which its external lines of force flow. The direction of these lines of force is determined by polarity. Every magnet has a north and a south pole. Poles behave somewhat like electrical charges. Like poles repel; unlike poles attract. Inside the magnet itself, its lines of force travel from south to north. In the space surrounding the magnet, the lines of force move from north to south, returning to the southern pole of the magnet. More lines of force create a stronger magnet. A magnetic field is a vector quantity because it has direction and magnitude. This magnetic field equation is only used to illustrate the relationship between the factors contributing to the magnetic field:

$$B = \mu_0 \frac{N}{1} I \qquad \textbf{(M.1)}$$

where

B is the magnetic field strength, in T
N is the number of turns in the coil
μ is a constant that represents the permeability of the free space
L is the length of the coil wire, in m

FIGURE M.2 Magnetic field surrounding a magnet.

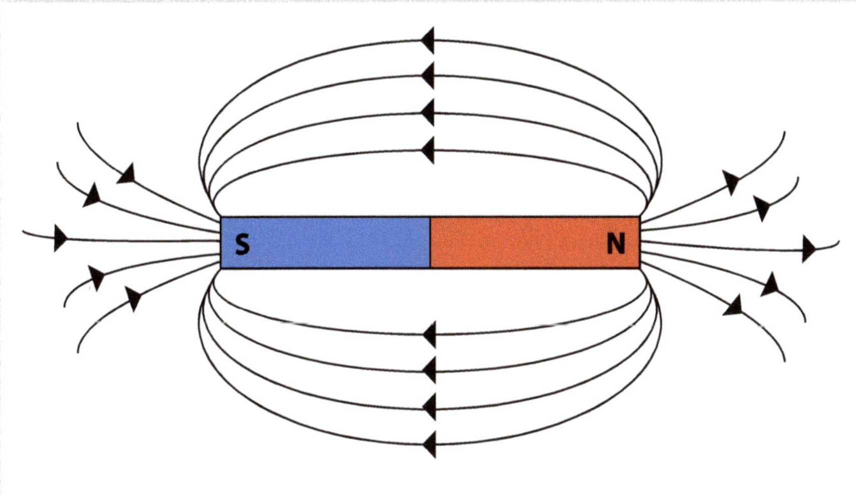

Magnetic Field Intensity

Magnetic field intensity is the strength (force) exerted by the magnetic field and can be compared to voltage in electricity.

Magnetic Flux

Magnetic flux is the same as the magnetic line of force. Flux can be compared to current. Flux lines are directional and EXIT from the north pole and ENTER through the south pole. Flux lines are concentrated at the poles and spread out into the areas between the poles. Magnetic flux is the product of the average magnetic field times the perpendicular area that it penetrates. Flux is used in the operation of solenoids and transformers. In an electric generator where the magnetic field penetrates a rotating coil, the area used in defining the flux is the projection of the coil area onto the plane perpendicular to the magnetic field. Flux density refers to the number of flux lines per unit area.

Magnetic Pulse Generator

A signal-generating device that makes a voltage pulse as the magnetic flux changes around a pickup coil. The device transmits an analog AC signal voltage that varies in frequency and amplitude depending on the speed of rotation. A sort of voltage-generating sensor, the magnetic generator is comprised of many transmission speed sensors, like the vehicle speed sensor and engine speed (RPM) signal.

Magnetic Repulsion

When a straight rod conductor composed of the armature, commutator, and brushes is found during a flux and current is flowing through the rod, this case creates two separate fields: one produced by the magnet that is the pole shoes of the magnetic flux and winding and the other magnetic field produced by this flowing through the conductor, which consists of the armature, commutator, and brushes. This leads to the repulsion of the likes of poles, which moves the conductor, which is the motor armature, and the motor turns.

Magnetic Timing

A method of measuring ignition spark timing that uses a magnetic pickup tool around the spark plug wire or diesel fuel injection line to sense a timing event. For more information on ignition system, see **J139_202002** [2].

Magnetism

Magnetism was first observed in lodestone and also the way iron-based metals reacted to that. Gilbert carved a piece of lodestone into a ball and showed that when a compass is placed anywhere on the ball, it would always point to Earth's North Pole. Magnetism provides a link between energy and electricity. Electricity and magnetism are linked because any electrical current flowing through a conductor creates a magnetic flux. Any conductor moving through a field creates an electrical current. Electricity creates magnetism and magnetism creates electricity. By the utilization of magnetism, an alternator converts some engine mechanical power to electromotive force potential. Magnetism allows an electric motor to convert current from the battery into mechanical power to crank the engine.

Magneto Ignition

A magneto ignition is a self-contained ignition device that was used in alcohol-fueled internal combustion engines (ICEs) to produce a very high current to ignite the alcohol fuel. This type of ignition system is primarily found on lawnmowers, small tractors, chainsaws, alcohol-fueled drag racers, and ICE small aircraft engines. Magnetos have used the term tension in place of the term voltage in the past and the term high tension to mean high voltage. A magneto ignition is like an alternator that generates a spark without the use of outside voltage from a battery. It produces pulses of high current to the spark plugs.

You may find the use of the term tension instead of voltage or high tension, which is high voltage. In a magneto, the engine rotates a magnet inside a stationary primary coil between the poles of a magnet. This coil is like the stator of an alternator. During each revolution, a cam opens the contact breaker points one or more times, which interrupts the primary current and causes the electromagnetic field in the primary coil to collapse. Due to the collapse of the magnetic field, a voltage is induced across the first coil. Due to the opening of the breaker points, an arc goes across the breaker points. A capacitor is placed across the points to suppress the arc. A secondary coil, with more than 100 times as many turns as the primary coil, is located next to the primary coil. So if the primary coil had 200 turns, the secondary coil would have 20,000 turns (100 × 200). When the primary current is opened by the breaker points, a voltage is induced across the secondary coil. The amount of voltage developed by the collapse of the magnetic field across the secondary ignition coils is determined by the turns ratio in a magneto. This voltage will be large enough to arc across the spark plug gap burning any fuel. For more information on this, see **J139_202002** [2].

FIGURE M.3 Magnito line drawing (left) and Magneto ignition unit (right).

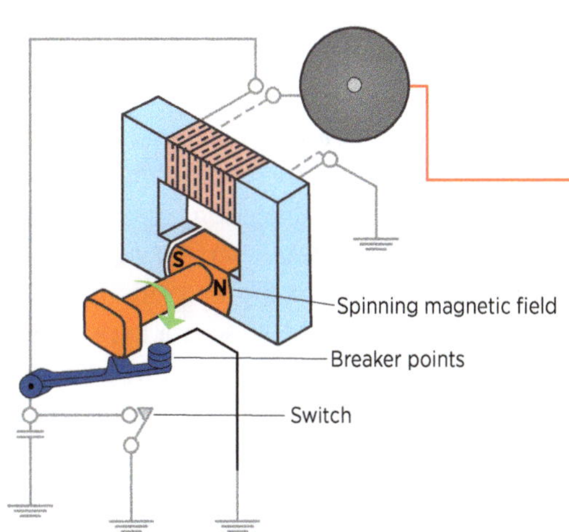

notsuperstar/Shutterstock.com.

Courtesy of John F. Kershaw Ed.D.

Main Bearing Journals

The ICE crankshaft rotates within the cast supported on main bearings. These main bearings support the crankshaft and permit it to rotate without excessive wear. The crankshaft also must be able to absorb loads applied longitudinally (end to end) and thrust loads from the clutch on a manual transmission vehicle or the converter. Thrust loads push and pull the crankshaft forward and rearward within the cylinder block. A bearing supports these loads and maintains the front-to-rear position of the crankshaft within the block. The thrust surface on many engines is typically located at the center or one among the tip main bearings. The bearing insert for most bearings is supplied with bearing flanges that ride against the thrust surface.

MANET (Mobile Ad Hoc Network)

Each device in an exceedingly MANET system moves independently in any direction and may, therefore, change its links to other devices. Each link must forward traffic unrelated to its own use and then be a router. The primary challenge in building a MANET is equipping each device to continuously maintain the information required to properly route traffic. MANETs may operate by themselves or could even be connected online. MANETS are called WANET (wireless accidental network) that sometimes contain a routable networking environment on top of a link layer accidental network. MANETs comprise a peer-to-peer, self-forming, self-healing network. MANETs from 2000 to 2015 communicate at radio frequencies of 30 MHz to 5 GHz. Different protocols are then evaluated, supporting measures such as the packet drop rate, the overhead introduced by the routing protocol, end-to-end packet delays, network throughput, ability to scale, etc. For more information on this, see **J3163_201809** [4].

MAP (Manifold Absolute Pressure) Sensor

The MAP, or manifold absolute pressure, sensor is an engine load sensor. It has a solid-state silicon wafer on one side used and the other side is connected to the intake manifold. As intake manifold pressure changes, the pressure pushing on the silicon wafer changes the output voltage to the engine management computer indicating the engine load. This type of fuel injection control method is known as a speed density system that measures intake air flow by sensing changes in intake manifold pressures resulting from engine load and speed changes. The engine management computer uses the MAP sensor to read these pressures and combines MAP along with temperature, RPM, and throttle position to calculate the injector pulse width or opening time. For more information on this, see **J1930DA_202105** [1].

Master Cylinder (Automotive Brakes)

Foot-operated hydraulic ram that gives the pressure employed in the braking system. It consists of a reservoir, reservoir cover with gasket, housing, piston, rubber piston cup, and return spring. The components may vary slightly with different styles or varieties of vehicles. Its four basic functions are:

- Develops pressure, causing the wheel cylinder pistons to maneuver toward the rotors or drums.
- After the shoes and pads produce sufficient friction, the brake cylinder helps equalize the pressure required for braking.
- Keeps the system filled with fluid because the brake linings wear.
- Maintain a slight pressure to keep contaminants (air and water) from entering the system.

When the foot pedal is applied, the piston and piston cup move forward within the cylinder. Fluid that has entered the cylinder through the compensating or vent port is trapped ahead of the piston cup and is forced along the cylinder bore into the pressure chamber. Hydraulic pressure builds before the first seal because the pushrod moves forward. The rear of the piston is kept filled through the replenishing (inlet) port. This stops any suction (vacuum) from forming behind the piston. The piston moves and the fluid pressure in front of it increases. Fluid within the pressure chamber can only exit into the brake lines, thus increasing pressure within the brake system. Because the piston moves down the bore of the cylinder, the return spring is compressed. When the pedal is released, the force of the spring quickly moves the piston toward its starting position. The piston moves back in the bore and fluid flows past the piston seal against the rear fringe of the piston cup. The piston seal cup deflects, allowing fluid to pass from the secondary area to the pressure chamber. The piston acts as a one-way valve.

This feature is required or the pedal could not be applied again until the fluid that was displaced during pedal application flowed into the pressure chamber. As this happens, the fluid level within the secondary area of the cylinder is replenished from the reservoir by fluid flowing through the fluid inlet port. Once the pedal is released and also the brake cylinder piston returns to its rest position, the shoe return springs (in the drum brakes) and piston seals (in the disc brakes) cause the pistons to retract. The retracting pistons cause the displaced brake fluid to return to the pressure chamber. From there, excess fluid returns to the reservoir through the compensating port.

FIGURE M.4 Automotive braking master cylinder.

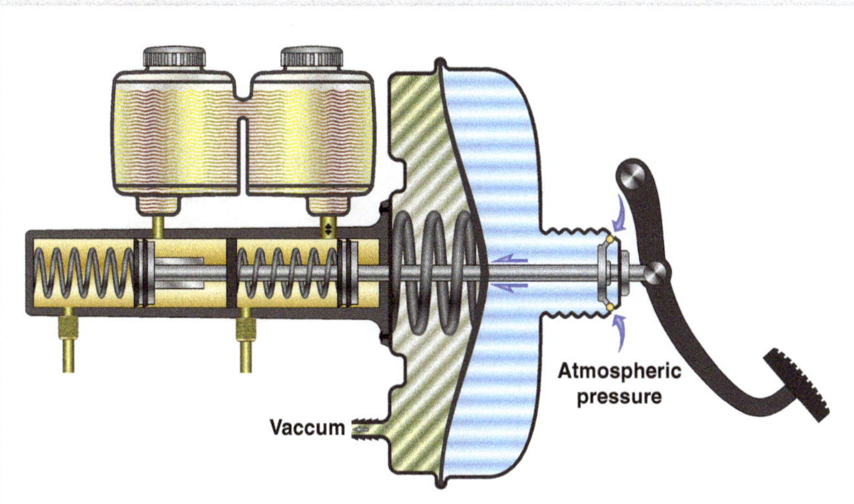

Match Mounting Tires

High and low spots of tires that are matched together to reduce radial forced vibrations. A high spot that becomes excessive transfers its force into the suspension system as a vibration, much the same as a tire and wheel that are out of balance. When runout is excessive, it is referred to as "radial runout" Match mounting relocates the high spot of the tire to the low spot of the wheel. Tires may often be in balance and still create some vibration, which could stem from tire/wheel runout or radial-force variations. Tire/wheel radial runout is a condition in which the tire does not have the same radius at all points around its circumference, meaning high or low spots may be present. If the runout is high at one spot and the tire is rotating, it would have to raise the spindle that distance for each revolution or the vehicle weight would try to force the tire into the correct position (decrease the high spot).

Matter

Anything that occupies space and has mass, or weight. It is the generic name given to everything in the physical universe. Matter is made up of elements, which are comprised of atoms.

Measurement Prefixes

Each metric system unit has a prefix that is used to describe the unit of measurement. The most common prefixes—micro, milli, kilo, and mega. These are found in the are shown in the lower right corner of a digital multimeter display. Another way to remember this is that the decimal point moves three places with each prefix. For example, to convert from milliamps to amps, you move the decimal point three places to the left.

Mechanical Efficiency

The mechanical efficiency of an ICE is the fraction of the indicated horsepower delivered at the crankshaft. Since the brake horsepower is always less than the indicated horsepower by the amount of the friction horsepower, the mechanical efficiency must be less than one, or less than 100%. You can calculate the mechanical efficiency by dividing the brake horsepower by the indicated horsepower and multiplying the result by 100 to find the mechanical efficiency. The formula is:

$$n_m = \frac{BHP}{IHP} \times 100 \tag{M.2}$$

where

n_m is the mechanical efficiency, in %

BHP is the brake horsepower of the engine

IHP is the indicated horsepower of the engine

Mechanical Fuel Injection Nozzles

These are poppet valve mechanical injection nozzles that open at a high nominal opening pressure from delivered to them by a mechanical fuel injection pump. Some of these nozzles are fitted with a low-pressure return line that carries excess fuel and any trapped air back to the fuel tank. The injector body requires a metal gasket in order to seal the cylinder head.

FIGURE M.5 Diesel mechanical fuel injection nozzle.

Konstantin Shadrin/Shutterstock.com.

Medium-Duty Truck Hydraulic ABS

The medium-duty truck hydraulic ABS uses a booster system that makes sure that the brake assist is out there if the engine stalls or if there is a malfunction within the power-steering/brake assist system. Mechanical force is converted into hydraulic pressure by the piston chamber, regulated to satisfy braking system demands by the pressure balance system, and delivered to the brakes wheel circuits by the pipes and versatile hoses. For more information on this, see **J2627_200908** [3].

The hydraulic boost assist system consists of the subsequent components:

- Electronic brake control module
- Pump
- Pump motor
- Motor control relay
- Fluid flow switch

Memory Seat

A seat position memory that uses a potentiometer to sense the position of the seat. The seat position can be programmed into the body control module or memory seat module and stored by position number 1, 2, or 3. The driver pushes the desired button and the seat moves to the stored position. On some vehicles, the memory seat position can also be programmed into the remote keyless entry key fob.

MEP (Mean Effective Pressure)

A quantity relating to the operation of a reciprocating internal combustion engine (ICE) and the measure of the ICE capacity to do work that is independent of engine displacement.

$$pme = W / V_d \qquad \textbf{(M.3)}$$

where

pme is the mean effective pressure
W is the work performed in a complete engine cycle
V_d is the engine displacement

Metering Valve

Equalizes the braking action on vehicles that combine the front disc and rear drum brakes. The metering valve prevents the front disc brakes from being applied until the rear brakes are applied. This is necessary because disc brakes are fast acting, while drum brakes must overcome the spring tension and spend time in the linkages before they work. It is generally part of the combination valve.

Methanol

Wood alcohol or methyl alcohol is a simple alcohol with the formula CH_3OH. It is a light, volatile, colorless, flammable liquid with a distinctive alcohol odor similar to that of ethanol.

MIC (Multiplex Integrated Control)

A computer system that allows several vehicle electronic control modules to communicate with each other by exchanging digital signals over a single wire or conductor.

Microorganisms

Bacteria and fungi found in the vehicle or bulk fuel storage tanks. When water is at the bottom of the fuel tank, bacteria will live in it and feed off the hydrocarbons in the fuel. Waste from these microorganisms is acidic and can cause fuel injection component corrosion. Just cleaning out the tank is not enough, the technician must treat the fuel tank with a biocide such as Biobor JF.

MID (Message Identifier)

MID stands for the message Identifier and is used to describe a major vehicle electronic system, usually with a separate processing capability. The MID is the first character of every message, which is used to identify the source of the data transmission. The MID is specific to a vehicle system. The MID identifies the microcomputer where the data originated from on a serial communication link. For example, MID 128 identifies an engine computer, whereas MID 130 identifies a transmission computer. For more information on this, see **J1587_200202** [5].

Mild Hybrid

Hybrid electric vehicles (HEVs) using an ICE equipped with an electric motor that allows the engine to shut off whenever the car is coasting, braking, or stopped, yet can be restarted quickly. Mild hybrids may employ regenerative braking and some level of power assist for the ICE. They do not have an electric-only mode of propulsion. For more information on this, see **J1715/2_202108** [6].

Mild Steel

A low-carbon steel is a common form of steel due to its low cost. It contains approximately 0.05–0.25% carbon, making it malleable and ductile. Mild steel has a relatively low tensile strength, but surface hardness can be increased through carburizing.

Miscibility

The property that allows one fluid to blend with other fluids.

Miscible

A term that means "capable of being mixed."

Misfire

A condition that occurs when complete combustion does not happen in one or more cylinders due to fuel, ignition, or cylinder compression.

Misfire Monitor

An OBD II monitor to detect engine misfires. The engine management computer receives this crankshaft position sensor to calculate the time between the edges of the reluctor, as well as the rotational speed and acceleration. By comparing the change in position of each firing event, the computer can tell if a cylinder is not firing correctly. For more information on this, see **J1930DA_202105** [1].

MMT

Methyl cyclopentadienyl manganese tricarbonyl (MMT) is an anti-knock additive.

Mo (Molybdenum)

Chemical element with the symbol Mo and atomic number 42. It is a brittle silver-gray metal of the transition series that is found in some alloy steels.

Mobile Data

Wireless data communications using radio waves to send and receive real-time computer data to and from devices utilized by the driving force of the vehicle. These devices may be fitted solely to be used while within the vehicle (fixed data terminal) or to be used in and out of the vehicle (mobile data terminal). The common methods for mobile electronic communication for telematics were supported by private vendors RF communication infrastructure. Early manufacturers of mobile data terminals tried to supply cheaper ways to transmit telematics information and a wider range to support the country's full coverage of cellular providers. However, with cellular providers offering low 2.5G and later 3G rates, mobile data is delivered almost exclusively to telematics customers via cellular communication.

Mobility (Shared)

"Shared mobility is defined as the shared use of a vehicle, motorcycle, scooter, bicycle, or other travel mode; it provides users with short-term access to a travel mode on an as-needed basis." "This standard provides a taxonomy and definitions for terms related to shared mobility and enabling technologies. Included are functional definitions for shared modes (e.g., carsharing, bikesharing, ridesourcing, etc.). Public transit services and other incumbent services—such as car rentals, shuttles, taxis, paratransit, ridesharing (carpooling/vanpooling), and pedicabs—are also included in the ecosystem of shared mobility services." This recommended practice also provides a taxonomy of related terms and definitions (e.g., station-based roundtrip and free-floating one-way) but does not provide specifications or otherwise impose requirements on shared mobility. For more information on this, see **J3163_201809** [4].

Modified Strut Suspension

Unlike the MacPherson strut suspension, the modified strut suspension does not include a spring as a part of the assembly and is used on the front of some vehicles and the rear of others. The modified strut system uses one control arm with the strut mounted in the same manner as a MacPherson strut. The biggest difference is that the spring is mounted on the lower control arm, which is typically an A-frame. This enables the placement of the spring to control fore and aft movements. The strut still is the upper steering pivot, but the burden of the vehicle is supported by the spring and carried by the lower control arm and ball joint.

FIGURE M.6 Modified strut suspension.

Modular Fuel Sender

A combination fuel tank unit containing a replaceable fuel level sensor, a turbine pump, and a jet pump. The reservoir housing is attached to the cover containing fuel pipes and the electrical connector. Fuel is transferred from the pump to the fuel pipe through a convoluted (flexible) fuel pipe.

Module

A group of electronic components functioning as a component of a larger system.

Module Programming

The body control module (BCM) and other electronic modules are used to operate the various vehicle systems. The number of these modules and their purpose varies among the original equipment manufacturers (OEMs). These will also vary among the different models offered by these OEMs. All these systems use different controller area networks (CANs) to connect all these modules together. If there are communication errors on the various networks, some or all of these modules may not function. You will find one or more "U" diagnostic trouble codes (DTCs) will be set, which will indicate a fault in the communication network. If this network is active or partially active, you will need to use a bidirectional factory-level scan tool to diagnose the DTC(s). You may need to reprogram the current or new BCM with the latest operating software and calibrations. You will need a factory-level bidirectional diagnostic scan tool for accurate diagnosis. You will have to use the on-screen reprogramming directions and may need to perform the programming/setup procedure using a specific reprogramming tool or some OEM scan tool systems can perform the reprogramming.

MON (Motor Octane Number)

This is a similar octane number determining method to RON, or research octane number; see **Research Octane Number**. The difference is that it is determined at 900 RPM engine speed instead of the 600 RPM for RON. A cooperative fuel research test engine is also used, but with a preheated fuel mixture and variable ignition timing to increase the stress on the fuel knock resistance. MON will be about 8 to 12 octane numbers lower than the RON, but there is no direct link between RON and MON. Gasoline pumps typically require both a minimum RON and a minimum MON.

Monitor Controller

A controller used as an interface between the main body ECM and the LCD that processes the data from the body computer module, other modules, and control switches to drive the LCD screen.

Monitors

A diagnostic test run by the engine management computer to determine if the engine is operating in accordance with FTP standards in an OBD II system. For more information on this, see **J1930DA_202105** [1].

Monograde

Single-grade engine oil that meets the requirements for only one SAE viscosity grade or classification.

Motor Assist Mode

A mode of hybrid electric vehicle operation in which the motor generator and the internal ICE apply torque to the drivetrain for propulsion. For more information on this, see **J1715/2_202108** [6].

Motor Generator

Some varieties of HEV use an ICE to turn an electrical generator to either recharge the vehicle batteries or directly power its electric drive motors; this is called a motor generator. For more information on this, see **J1715/2_202108** [6].

Motor-Generator Armature

A set of permanent magnets mounted inside a steel segmented disc in the motor-generator of an HEV. The armature spins inside a set of electrical coil windings. For more HEV and EV information see J1715/2_202108 [6].

Motorcycle

A two- or three-wheeled motor vehicle powered by one-, two-, or four-cylinder engines and a manual foot-shifted transmission. The V2, also known as V-twin, engine is the most common along with single-cylinder models and some boxer-type four-cylinder engines. "Due to the uniqueness of motorcycles and the more active role the motorcycle rider has while riding, and the various potential operating environments, SAE Terminology was expanded to define terms and phrases of usage, rider safety and behaviors, and other areas that provide clarity for researchers and other not familiar with motorcycling. The definition provided within this document are directed and intended to be applied to certain motorized cycles." For more information on this, see **J3133_201909** [7].

MR (Magnetorheological)

MR fluid is employed in monotube-type shock absorber. It is a type of smart fluid in a carrier fluid inside a mineral oil. When subjected to a magnetic field, the fluid greatly increases its apparent viscosity or thickness, to the point of becoming a viscoelastic solid. Magnetic saturation is the condition when a magnetic field or flux reaches full strength and maximum flux. For more information on this, see **2022-01-5018** [8].

MRFS (Mechanical Returnless Fuel System)

MRFS uses a bypass regulator to control rail pressure that is located in close proximity to the fuel tank. Fuel is sent by the in-tank pump to a chassis-mounted in-line filter with excess fuel returning to the tank through a short return line.

MTBE (Methyl Tertiary Butyl Ether)

A fuel oxygenate that is permitted in unleaded gasoline up to a level of 15%. MTBE is an organic compound using the structural formula (CH_3) $3COCH_3$. It is a volatile, flammable, and colorless liquid that is somewhat soluble in water. It is used as a gasoline additive to increase knock resistance due to the removal of tetraethyl lead. It is made using the chemical reaction of methanol and isobutylene. Methanol is primarily derived from natural gas, where steam reforming converts the various light hydrocarbons in natural gas into carbon monoxide and hydrogen.

Muda

A Japanese term for waste and does not add value. A process consumes resources and waste occurs when more resources are consumed than are necessary to produce the goods or provide the service that the customer actually wants.

Muffler

A device used in the exhaust stream to help deaden the sound of engine combustion. The muffler is simply a can or container with an inlet and an outlet pipe attached. Inside the muffler is a series of baffles designed to deflect the exhaust inside. The baffling or deflecting of the exhaust gases inside the muffler deadens the sound of combustion before the gases leave out of the tailpipe. Some mufflers are equipped with a small hole at the rear part lower end to drain accumulated water. Water vapor will condense on the cooler portions of the exhaust system unless the engine is run long enough to fully warm the muffler above the boiling point and vaporize it. A universal muffler may require cutting existing pipes, pipe adapters, and new clamps and hangers. When you use an OEM exact fit replacement muffler, this eliminates the need for pipe adapters.

Muffler Clamp

These U-bolts come in different sizes and correspond to the pipe diameter to secure the slip joint connections that attach some exhaust system components and may also be used to attach hangers in some applications. Always use the correct size clamp for the pipe in question or the pipe may be distorted when the clamp is tightened, which will cause a leak. Also make sure the inside pipe is fully inserted in the slip joint connection and the clamp is positioned around both pipes, but not too close to the edge of the outer pipe. Position the clamp so that the ends of the U-bolt will not contact other components or snag objects that may pass under the vehicle. Tighten the clamp enough to secure the connection. Overtightening may distort the pipes and also cause a leak.

Multigrade (Multiviscosity) Oil

An engine oil that meets the requirements for more than one SAE viscosity grade or classification.

Multimeter

See DMM. Also called a volt-ohm-milliammeter, or VOM.

Multiplexing

Carrying several current control signals in a single wire.

Mura

A traditional general Japanese term for unevenness in physical matter or in the human spiritual condition. *Mura* is avoided through kanban systems with little supported or no inventory. You supply the method with the right part, at the right time, in the correct amount. JIT (Just In Time) systems create a "pull system" within which each sub-process pulls its needs from the preceding sub-processes, and ultimately from an outdoor supplier. When a preceding process does not receive a request or withdrawal, it does not make more parts. This kind of system is meant to maximize productivity by minimizing storage overhead.

Muri

A Japanese term for overdoing something. *Muri* may be avoided through standardization. *Muri*, a regular condition or output, must be defined to assure quality. Then every process and performance must be reduced to its simplest elements for examination. The method must then be standardized. This can be done by taking simple work elements and mixing them one by one into standardized work sequences.

Mutual Induction

When two coils are brought in close proximity to each other, the magnetic field in one coil tends to link with the other coil. This generates a voltage within the second coil. This magnetic property of a coil affects or changes the voltage in the other coil. The changing current in the primary coil creates a changing magnetic flux through the secondary coil that leads to an induced electromotive force (EMF) in the secondary coil. The equation for mutual inductance is shown below:

$$E = -M\frac{\Delta I_p}{\Delta t} \tag{M.4}$$

where

E is the EMF induced in the secondary coil B2
–M is the mutual inductance measured in Henry
ΔI_P is the change in current in the primary coil
Δt is the change in elapsed time

FIGURE M.7 Mutual induction coils.

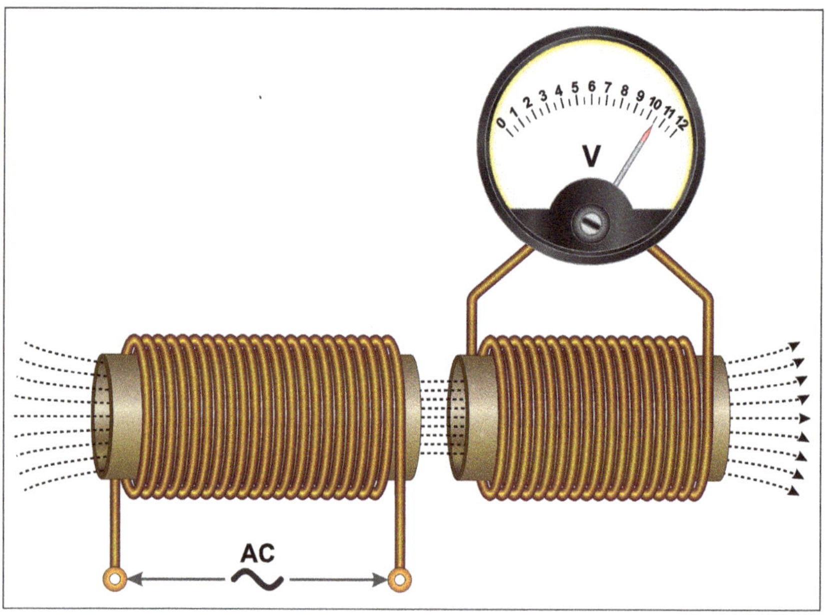

REFERENCES

1. Society of Automotive Engineers, J1930DA_202105: Electrical/Electronic Systems Diagnostic Terms, Definitions, Abbreviations, and Acronyms Web Tool Spreadsheet, SAE Publishing, Warrendale, PA, 2021.
2. Society of Automotive Engineers, J139_202002: Ignition System Nomenclature and Terminology, SAE Publishing, Warrendale, PA, 2020.
3. Society of Automotive Engineers, Standard J2627_200908: Braking System Definitions—Truck and Bus, SAE Publishing, Warrendale, PA, 2009.
4. Society of Automotive Engineers, Standard J3163_201809: Taxonomy and Definitions for Terms Related to Shared Mobility and Enabling Technologies, SAE Publishing, Warrendale, PA, 2018.
5. Society of Automotive Engineers, J1587_200202: Electronic Data Interchange between Microcomputer Systems in Heavy-Duty Vehicle Applications, SAE Publishing, Warrendale, PA, 2002.
6. Society of Automotive Engineers, J1715/2_202108: Hybrid Electric Vehicle (HEV) and Electric Vehicle (EV) Terminology, SAE Publishing, Warrendale, PA, 2021.
7. Society of Automotive Engineers, J3133_201909: Motorcycle Terminology, SAE Publishing, Warrendale, PA, 2018.
8. Metered, H., "Enhancement of Semi-active Vehicle Suspension System Performance Using Magnetorheological Damper," SAE Technical Paper 2022-01-5018, 2022, https://doi.org/10.4271/2022-01-5018.

N

Natural Frequency

The frequency at which an object tends to vibrate. Bells, guitar strings, and tuning forks are all samples of objects that tend to vibrate at specific frequencies when excited by an external force.

Naturally Aspirated

Non-turbocharged engines that depend on air pressure to push an air-fuel mixture into the combustion chamber vacuum created by the downstroke of a piston.

Navigation System

A system that uses video display and satellite information to indicate a road map of the car location and direction. Employing a global positioning system (GPS), the navigation system can plot the vehicle location in real time to assist the driver in finding his or her location on a map of the geographical area.

Needle Bearing

A roller-type bearing that uses rollers which are longer and thinner than traditional needle bearing elements. This design is found in applications where it reduced the overall bearing height and size.

Needle Valve

A valve with a little port and a threaded, needle-shaped plunger. It allows precise regulation of flow at low flow rates.

Neoprene (Polychloroprene)

A synthetic polymer produced by polymerization, which is the process to form a polymer (plaster form) of chloroprene. This material has good chemical stability that maintains flexibility over a large temperature range, so it is an excellent material for automotive drive belts.

Net Brake Horsepower

Brake power (power available at the flywheel or output shaft—usually the crankshaft) of a completely equipped engine fitted with all the accessories necessary to perform its intended functions unaided.

Net Valve Lift

The base ICE poppet valve lift.

Network

System of the intercommunication of computers or processors.

Networked

When several computers are connected wirelessly or by cable to work together toward a common goal.

Neutral Safety Switch

A switch that prevents the engine cranking motor from operating by opening the circuit from ignition/cranking control to the cranking motor solenoid unless the shift selector is in neutral or park. On a vehicle with a manual transmission, the safety switch nullifies the starting system when the clutch pedal is not depressed. Safety switches eliminate the possibility of the vehicle unexpectedly lurching forward or backward after ignition.

Neutral Steer

The distinctive feature of the slip angles where both front and rear slip angles are the same. During slip conditions, both the front and rear tires will break loose at the same time so that the vehicle slides sideways. This happens rather in oversteer when the front end steers more or sharper than desired (fishtailing) and loses traction in the rear. The rear end of the vehicle wants to swing out or rather understeer when the front end moves less than desired in relation to the actual turning position of the front tires. This occurs when the front wheels lose traction in a turn (plowing).

Neutron

A neutral-charged particle, which is one of the basic particles of the nucleus of an atom. The atomic mass or weight minus the number of protons.

Night Vision

An infrared system to locate images in the dark. The night vision control module sends the processed image output directly to the cluster via a dedicated cable. The active safety control module communicates via serial data with the night vision camera module from the BCM, which uses the night vision camera module to turn on when conditions are correct and perform detections. The night vision camera module will send a message to the active safety control module if a warning or alert is required. Warnings and alerts are posted on the cluster or head-up display, depending on vehicle equipment. For more information on this, see **J3063_202103** [2].

NiMh or Ni-MH (Nickel Metal Hydride Battery)

A rechargeable battery employed in hybrid electric vehicles (HEVs). The natural action at the positive electrode is analogous to the nickel-cadmium cell (NiCd), with both using nickel oxide hydroxide (NiOOH). Yet the negative electrodes use a hydrogen-absorbing alloy instead of cadmium. NiMH batteries can sometimes have the capacity of NiCd batteries of the same size, with significantly higher energy density. They are typically used as a substitute for similarly shaped non-rechargeable alkaline batteries as they feature a rather lower, but generally compatible, cell voltage, and are less prone to leaking and explosion. This was the dominant battery utilized in parallel-type HEVs. For more information on this, see **J1715/2_202108** [1].

FIGURE N.1 Nickel metal hydride battery (NiMh/Ni-MH).

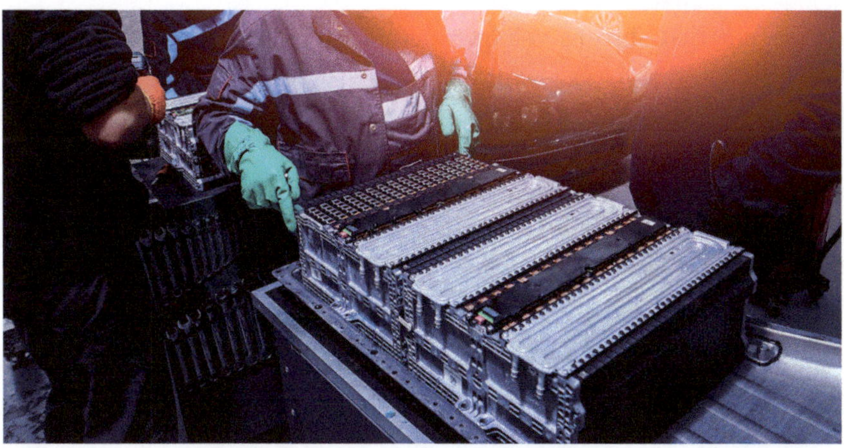

Nitriding

A metal hardening process used on ICE crankshafts, where the crank is heated to about 1,000°F (540°C) in a furnace that is filled with ammonia gas and then allowed to cool. This process adds nitrogen to the surface of the metal, which forms hard nitrides on the surface of the crankshaft to a depth of approximately 0.007 in. (0.8 mm).

No Driving Automation

"SAE Level 0 (zero) refers to a vehicle that has no driving automation technology. The driving force is entirely to blame of operating the vehicle's movement, including steering, accelerating, braking, parking, and the other necessary maneuver to maneuver the car in any direction. At Level 0, driver support systems which will temporarily intervene during driving could also be present. Examples include stability control, forward-collision warning, automatic emergency braking, blind-spot warning, and lane-keeping assistance. These technologies are considered Level 0 because they do not drive the vehicle but offer alerts or momentary action in specific situations. The bulk of vehicles on American roadways are Level 0." For more information on this, see **J3016_202104** [3].

No-Load Voltage

Voltage at the battery with the engine running and the electrical accessories off.

Noid Test Light

A test light that is installed in place of an electronic fuel injector and used to check electronic fuel injector feed circuits.

Noise Suppressor

A capacitor that absorbs voltage fluctuations in an electrical system to reduce noise.

Non-Continuous Monitor

An OBD II or HD-OBD feature that requires certain conditions to be met before evaluation of an emissions system. For example, the monitor may not run if power take-off is operating. It can operate once per vehicle drive cycle. Once this monitor has run to completion, the readiness status on a scan tool computer shows "complete" or "done." Monitors that have not run to completion show on the scanner as "incomplete." For more information on this, see **J1930DA_202105** [4].

Non-Hunting Gearset

When one tooth on the gear will mesh with the same tooth on the gear again and again, rather than hunting for all of the rest of the teeth. It is a form of timing.

Non-Interference ICE

A non-interference ICE valve timing is specified if the timing chain failed, the pistons would not hit the valve. In a very non-interference engine, the valve lift or angle prevents the valve heads from entering the cylinder volume swept by the piston. In an interference fit engine, the valve lift or angle would allow the valve heads to enter the cylinder volume swept by the piston if the timing chain failed.

Nonvolatile Memory

The computer memory capability that is not lost when power is removed. This memory can retain its information even when the battery is disconnected. One use for this type of memory is the storage of odometer information in an electronic speedometer. The memory chip retains the mileage accumulated by the vehicle. When speedometer replacement is necessary, the odometer chip is removed and installed in the new speedometer unit. Also, see **Read-Only Memory (ROM)**.

NOP (Nominal Opening Pressure)

Pressure required to unseat the nozzle valve on a fuel injector. This term can also be referred to as the popping pressure. The injector spring tension defines the actual NOP value. The injector spring tension presses the nozzle needle valve to its seat and determines the pressure needed to open the needle valve. Many injector designs have a way to adjust the injector spring tension so that the NOP value can be set to OEM specifications. These adjustment methods use either shims or an adjustment screw. The NOP measurement is made using a bench-mounted pop tester and is almost always the first thing you check when testing an injector.

Normally Aspirated

An ICE that only relies on the pressure differential between the atmospheric pressure of 14.7 psi at sea level and the negative pressure of the intake stroke in the four-stroke cycle to induce the air-fuel mixture into an ICE.

Nose

The peak of the cam lobe.

Nose Circle

The nose circle intersects the very top of the lobe; the highest point on the cam lobe in relation to the valve train.

NOx

Oxides of nitrogen that create smog when combined with hydrocarbon and sunlight.

NTC (Negative Temperature Coefficient)

A thermistor-type temperature input sensor. Usually used in reference to a temperature sensor (coolant or air temperature). As the temperature increases, the resistance of the sensor decreases.

Nucleus

The central part of an atom which has a positive charge and contains almost all the mass of the atom.

REFERENCES

1. Society of Automotive Engineers, J1715/2_202108: Hybrid Electric Vehicle (HEV) and Electric Vehicle (EV) Terminology, SAE Publishing, Warrendale, PA, 2021.
2. Society of Automotive Engineers, J3063_202103: Active Safety Systems Terms and Definitions, SAE Publishing, Warrendale, PA, 2021.
3. SAE, Society of Automotive Engineers, J3016_202104: Taxonomy and Definitions for Terms Related to Driving Automation Systems for On-Road Motor Vehicles, SAE Publishing, Warrendale, PA, 2021.
4. SAE, Society of Automotive Engineers, J1930DA_202105: Electrical/Electronic Systems Diagnostic Terms, Definitions, Abbreviations, and Acronyms Web Tool Spreadsheet, SAE Publishing, Warrendale, PA, 2021.

O

O_2 Sensor

A sensor that measures the oxygen content in the engine exhaust system as a means of checking combustion efficiency. It often fits into the exhaust manifold or exhaust pipe(s). Also called an oxygen sensor or exhaust gas sensor.

OAT (Organic Acid Technology) Ethylene Glycol-Based Antifreeze

Found in DEX-COOL and is the antifreeze/coolant of choice for General Motors (GM), Volkswagen (VW), and many Japanese/Asian vehicles. This formula is engineered to offer long-life corrosion protection. Originally made by Texaco and used exclusively by GM. DEX-COOL antifreeze uses two inhibitors, sebacate and ethylhexanoic acid (2-EH), as corrosive inhibitors. 2-EH tends to damage plastics like Nylon 6.6 used in intake manifold gaskets and radiators. G30 OAT (G-Coolants come from the trade name Glysantin of BASF in Europe and Valvoline (Zerex) in the USA) and Peak Global OAT do not use 2-EH [1].

OBD (On-Board Diagnostic)

A feature of the powertrain control module to check itself, sensors, and actuators as well as to store information about malfunctions. To define an inventory of anomalies associated with OBD communications. Misinterpretations of varied OBD communications standards and recommended practices have resulted in OBD "no-communications" situations within the field. This information report identifies the foremost prevalent of those. The SAE technical report covers technology, products, or processes which are mature and unlikely to alter within the foreseeable future. Therefore, no further changes are expected because of the SAE J1979 updated protocol requirements. For more information on this, see **J1930DA_202105** [2].

OBD II (On-Board Diagnostics Generation II)

OBD II includes many changes such as misfire detection and advanced diagnostics. For more information on this, see **J1930DA_202105** [2].

OBD II Communications Anomaly List

A list that defines anomalies related to OBD communications. Misinterpretations of various OBD communications standards and recommended practices have resulted in OBD "no-communications" situations in the field. This information report identifies the most prevalent of these. The technical report covers technology, products, or processes which are mature and not likely to change in the foreseeable future, therefore, no further changes are expected due to the SAE J1979 updated protocol requirements. For more information on this, see **J1930DA_202105** [2].

OBD II Drive Cycle

This test is normally performed whenever the battery or engine management computer has been disconnected, or after diagnostic trouble codes have been erased. Additionally, some states require the drive cycle to be performed before an emissions test. For more information on this, see **J1930DA_202105** [2].

Octane Rating

The measurement of the ability of gasoline to resist engine knock and a measure of fuel stability. These ratings are based on the pressure at which a fuel will spontaneously combust (autoignite) in a cooperative fuel research engine. This is a research internal combustion engine with a variable compression ratio testing engine. The octane number is the average of two different octane rating methods: motor octane rating and research octane rating, which are different primarily in the specifics of operating conditions. A higher octane number means a more stable fuel. There are three main grades of gasoline based on the octane level:

- Regular (low octane fuel 87)
- Midgrade (middle range octane fuel 89–90)
- Premium (high octane fuel 91–94)

The higher the octane rating, the less prone the gasoline is to cause engine knock (detonation).

Odometer

A mechanical or digital meter that is connected to the vehicle speed measuring device in the instrument cluster or actual speedometer either using a vehicle speed sensor or an output gear on the transmission output shaft and a mechanical cable that goes to the speedometer assembly. Can also be a software algorithm based on vehicle speed or GPS or telemetry systems.

Odometer Memory Chip

Integrated circuit chip that electronically stores the mileage on a car.

OEM

Original equipment manufacturers like General Motors, Ford, Fiat-Chrysler, Honda, Hyundai, Toyota, and Mercedes.

Offset

An amount or distance by which something is out of line. For example, this main shaft is offset shaft by 2 mm.

Ohm

Electrical unit for resistance and represented by the omega (Ω) symbol. It was named after Georg Ohm of Ohm's Law.

Ohmmeter

An analog or digital meter that measures the resistance to current in ohms (Ω). It can be a stand-alone meter or part of a digital multimeter.

Ohm's Law

Ohm's Law states that voltage equals current times resistance and is expressed as $E = I \times R$. Ohm's Law is based on the fact that it takes 1 V of electrical potential to push 1 A of current through 1 Ω of resistance.

$$I = ER \tag{O.1}$$

FIGURE O.1 Ohm's Law formula.

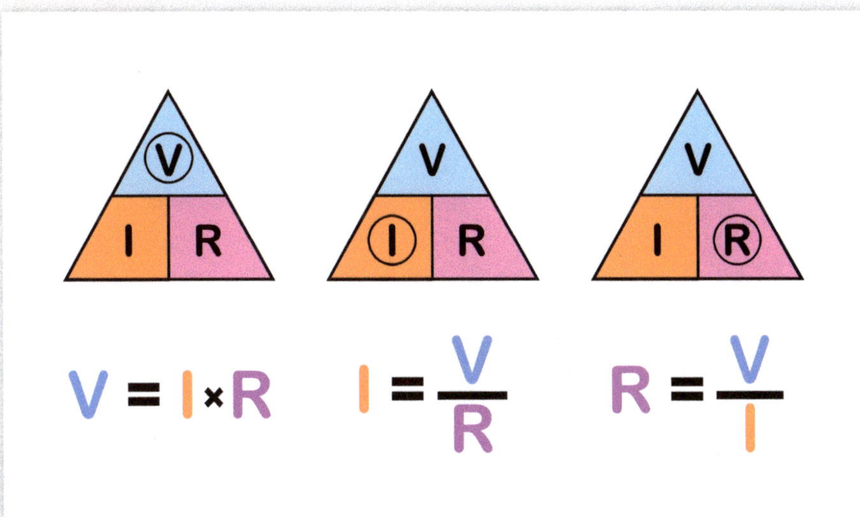

Ohm's Law General Rules

Ohm's Law shows that both voltage and resistance affect current. Current never change on its own—it changes only if the voltage or resistance changes. Current cannot change on its own because voltage pushes the current through a conductor, and all conductors have resistance. The amount of current can only change if the voltage or the conductor changes. Ohm's Law states if the voltage in a conductor increases or decreases, the current will increase or decrease in proportion. When the resistance in a conductor increases or decreases, the current will decrease or increase in inverse proportion. The general Ohm's Law Rule assuming the resistance does not change is as follows:

- As voltage increases, current increases.
- As voltage decreases, current decreases.

FIGURE O.2 Ohm's Law chart.

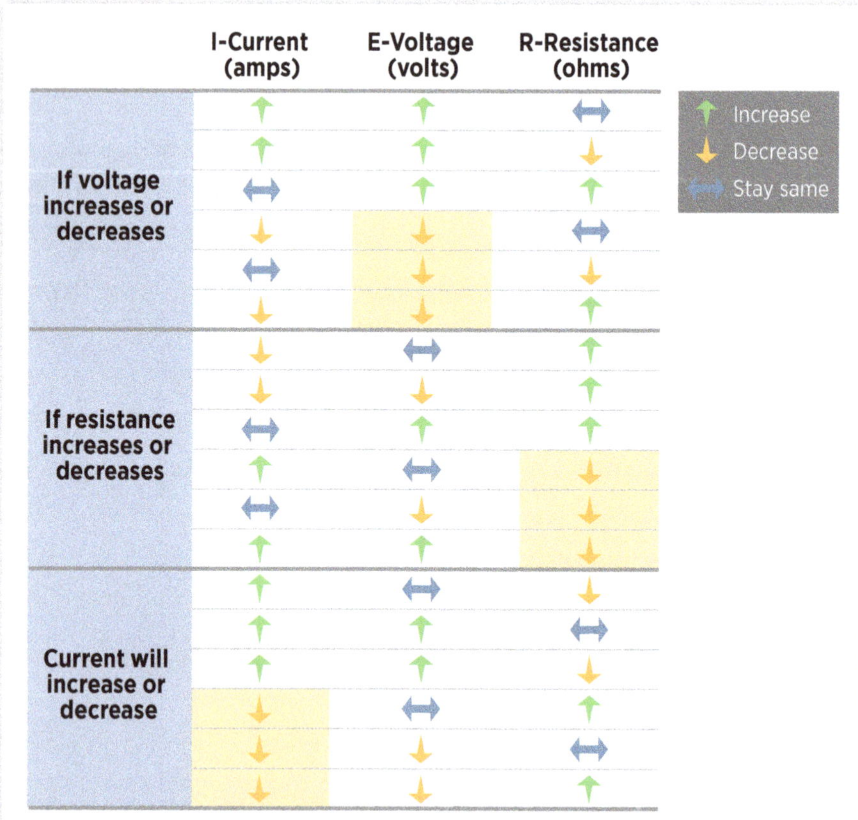

Courtesy of John F. Kershaw Ed.D.

Ohm's Law Solving Circle

A way to remember how to solve any part of the equation. To use this "solving circle," in Figure O.3 cover the letter you do not know. The remaining letters give the equation for determining the unknown quantity.

FIGURE O.3 Ohm's Law solving circle.

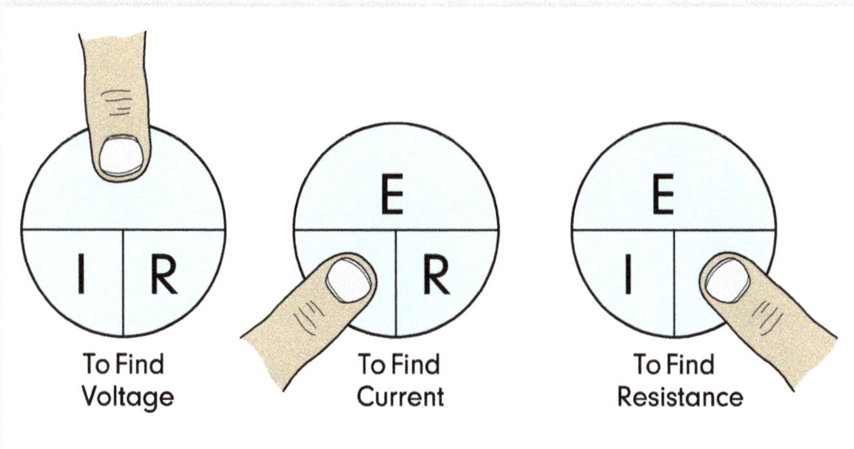

Oil Control Ring

The oil control ring is designed to control the flow of oil onto the cylinder wall on the upstroke of the piston for lubrication and to scrape the oil back off on the downstroke.

Oil Cooler

The engine lubricating oil must be kept within the proper temperature range. If the oil is too cold, it does not flow freely. If it is too hot, it cannot support the bearing loads and cannot carry away enough heat, possibly resulting in too great an oil flow. The lubricating oil absorbs a considerable amount of heat that is dissipated by an oil cooler. The oil cooler consists of a housing containing a series of plates or fins and is connected to the engine cooling system to use the engine coolant to carry the heat away by the convection process.

Oil-Filled Truck Hub

Oil-lubricated wheel hubs are used on both front and rear axles. Lubricant within the wheel hub cavity is visible through an observation port. Rear wheel hubs are fed from the oil contained in the front or rear axle.

Oil Pump

An engine lubrication system component that pumps oil under pressure to the moving parts of the engine. It pumps oil from the sump through the galleries to the main bearings, then the crankpin to the connecting rod bearings, and to the camshaft bearings and oil jets. The primary purpose of this pump is to circulate pressurized lubricating oil within the engine moving parts.

FIGURE O.4 Crankshaft drive oil pump used in an ICE.

Konstantin Shadrin/Shutterstock.com

Oil Slinger

A circular metal ring or washer that is mounted onto the inside of a shaft that is oil lubricated and used to prevent possible oil leakage from the oil wiper packing to the outside area.

OL DMM (Digital Multimeter) Symbol

This DMM display symbol does not mean the DMM is not reading any signal. Many DMMs indicate OL on the display to indicate out of limits. This may be called *over limits* or *overload*. Some DMMs show this condition with a "1" on the left side of the screen. This means that the reading is over the maximum value that can be displayed for the selected range. For example, the DMM will display OL if 12 V is measured and the meter has been set to read a maximum of 4 V. Autoranging meters adjust the range to match whatever is being measured. Here OL means a value higher than the meter can read or infinity when measuring resistance. So OL means infinity when measuring resistance, or an open circuit is being indicated. The meter will read 00.0 if the resistance is zero, so no reading indicates continuity (zero resistance), whereas OL indicates infinity resistance.

FIGURE O.5 DMM OL out-of-limits symbol (left) and the "1" out-of-limits symbol (right).

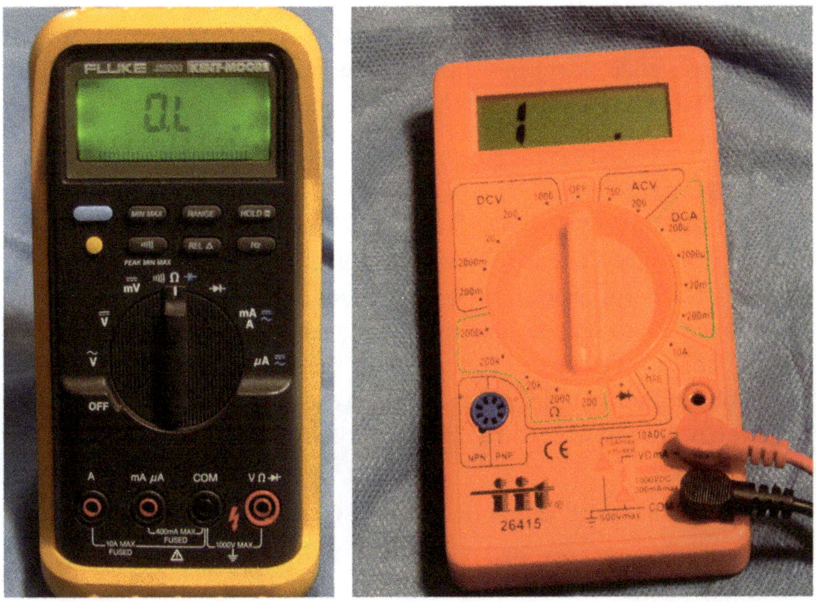

Courtesy of John F. Kershaw Ed.D.

Omega (Ω)

The last letter of the Greek alphabet; the symbol for ohm, the unit for electrical resistance.

One-Wire Circuit

A circuit that uses the metal structure of the vehicle to return electricity to the power source. Also called a frame-ground circuit.

One-Wire Oxygen Sensor

The one wire of a one-wire oxygen sensor is the O_2S signal wire. The ground for the O_2S is through the shell and threads of the sensor and through the exhaust manifold.

OnStar

General Motors (GM) was the first US automaker to bring connected vehicle features to market with the OnStar System in 1996 in the following vehicles: DeVille, Seville, and Eldorado. OnStar was formed in 1996 as a collaboration between GM, Electronic Data Systems (EDS), and Hughes Electronics Corporation, which were all part of GM at the time. The OnStar service relied on cellular phone voice and data communication, via Verizon Wireless in the USA and Bell Mobility in Canada, as well as location information using GPS technology. You can use its audio interface to contact OnStar representatives for emergency services, vehicle diagnostics, and directions. The original purpose was safety and to get emergency medical and law enforcement help to a vehicle when there was an accident. A cellular phone call is made from the vehicle to a call center where the agent sends help.

Open Circuit

A break in the current flow of what should be a closed circuit. The current cannot flow because there is no complete path. On the two sides of the open circuit, a difference in electrical potential does exist, but there is no path for electron flow. Current can flow across a gap when the voltage potential is great enough. A corroded electrical connection is the most common cause of open circuits. An automotive spark plug is a perfect example of an open circuit. Open circuits can occur anywhere in a circuit and can be the result of the following possible causes:

- Open fuse or fusible link or a tripped circuit breaker
- Loose, dirty, or corroded connection
- Broken wire
- Failed component

Testing for continuity is the best way to find an open. Continuity is the presence of a continuous path for the current. As observed with a test light, continuity causes the test light to glow. As observed with a DMM, continuity causes a low resistance measurement (close to 0 Ω), if a load is not located in the portion of the circuit being measured. When you check continuity, you are checking the resistance across two points in a circuit. The resistance across a good connection or conductor should measure close to 0 Ω, but you would measure infinite or extremely high ohms across an open circuit. The service information sometimes specifies the resistance values of some circuit loads.

FIGURE O.6 Open circuit when the switch in the illustration is off, the circuit is OPEN.

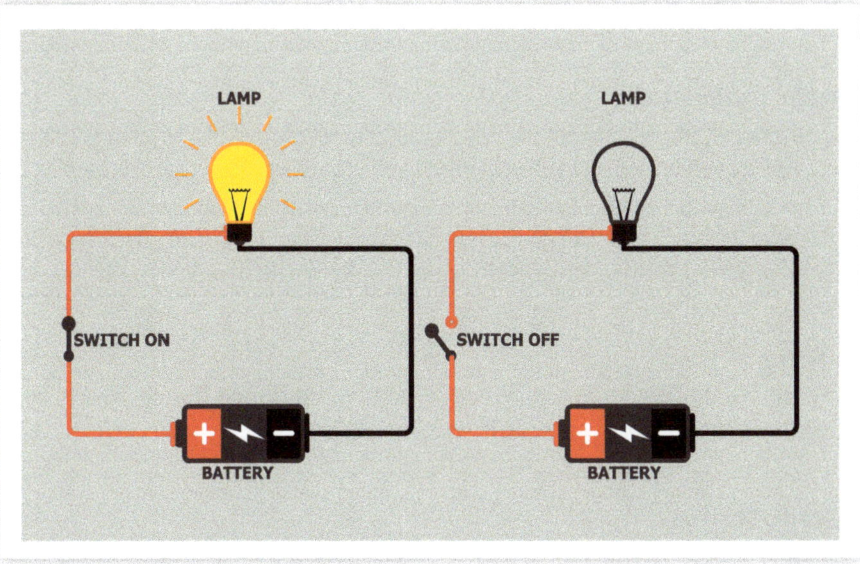

Open-Circuit Voltage Test

Measuring battery voltage from positive to negative with a digital multimeter with no load. On a 12-V automotive battery, it is 12.6 V, which is 2.1 V per cell on a 6-cell battery.

Open Diode

A diode that has a high (infinite) resistance in both directions.

Open-Loop Mode

A mode of operation in which the electronic injection system operates on information stored in the engine management computer, called fixed commands, to manage the air-fuel mixture.

Opening Ramp

As the cam continues to rotate, the hydraulic lifter or cam follower moves up on the camshaft opening ramp, which performs the actual opening of the poppet valve. The rate at which the valve is lifted is accelerated significantly on the opening ramp, up to a fairly constant opening speed.

Optical Sensor

An input sensor that generates a voltage change using light change over a phototransistor.

Optical Signal Generator

A switching device that uses the principle of light beam interruption to generate voltage signals from a photosensitive transistor using a light emitting diode.

Order (Vibration)

The number of times an event occurs during one revolution of a rotating component. For example, a tire with one high spot would create a disturbance once for every revolution or a first-order vibration. An oval-shaped tire with two high spots would create a disturbance twice for every revolution. This is called second-order vibration. Three high spots would be third-order vibration.

O-Ring

A circular seal made from a polymer or rubber compound that is used to seal a circular opening between a shaft and a cylinder.

Oscilloscope

Displays voltages in a waveform with the voltage related to time. An oscilloscope may be part of a diagnostic scan tool or multimeter or it may be a stand-alone unit. See **DSO**.

FIGURE O.7 Digital storage oscilloscope (DSO).

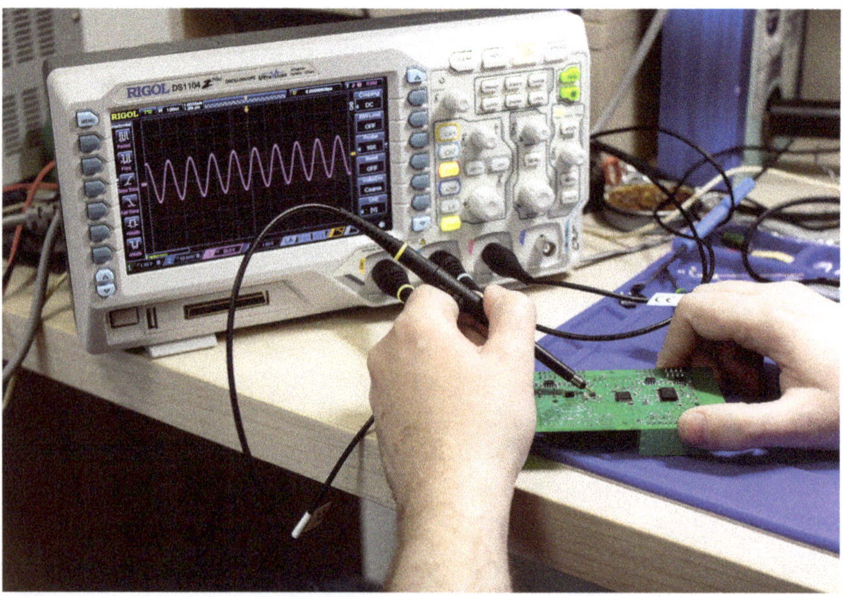

Otto Cycle

The Otto-cycle engine was named after its inventor, Dr. Nikolaus Otto, who built the first successful four-stroke engine in 1876. Most engines are four-stroke, spark-ignition engines. Other types of engines include two-stroke and compression-ignition (diesel) engines.

Out of Limits

A DMM (digital multimeter) will display OL to point out of limits. This might be called over limits or overload by some companies. On some DMMs this can is shown with a "1" on the left side of the screen. "Out of Limits" or "1" simply implies that the reading is over the maximum value that can displayed for the chosen range.

Output Drivers

Power transistors in a computer that control the current through an actuator. When energized by the computer microprocessor, the drivers ground the actuator circuits. This allows the current to operate the actuator.

Outputs

Signals sent by and controlled by the ICE management computer, such as fuel injectors and shift solenoids.

Overdrive

When the output member of a transmission turns at a greater speed than the input member. There is an increase in the speed of the output member with a proportional decrease in the torque of the output member.

Overhang

Part of a component that sticks out or hangs over another thing.

Overhead ICE Camshaft

When the ICE camshaft is mounted over the engine cylinders. A single overhead camshaft (SOHC) uses just one camshaft to operate all the poppet valves. A dual overhead camshaft (DOHC) uses two camshafts, one for the intake poppet valves and one to operate the exhaust poppet valves.

Overlap

Overlap is when the intake valve opens before the exhaust valve has closed. Many four-stroke engines have at least a short overlap period. A long-duration camshaft usually extends the overlap period, which helps make large amounts of power available at high RPM. The overlap period packs more air and fuel mixture into the cylinder by opening the intake valve near the end of the exhaust stroke while the exhaust valve is still open. The burst of exhaust gas rushing out of the cylinder helps pull in the intake mixture, even though the piston has not yet reached top dead center. The exhaust valve will is still open when the piston starts down on the intake stroke. This does not allow exhaust gases to be sucked back into the cylinder at high RPM because air velocity is high and the outward movement rush of exhaust gases is still pulling the air-fuel mixture into the cylinder.

Overrunning Clutch

A freewheeling or overrunning clutch is a component in a transmission that disengages the driveshaft from the driven shaft when the driven shaft rotates faster than the driveshaft. An overrunning clutch is part of the cranking motor starter drive or Bendix drive. See **Starter Motor**.

Oversquare Engine

The term used to describe an engine in which the cylinder bore diameter is larger than the stroke dimension.

Oversteer

Oversteer occurs when the front end steers more or sharper than desired (fish-tailing), there is loose traction in the rear, and the rear end of the vehicle wants to swing out. If oversteer is detected, the ABS (anti-lock brake system) will apply braking force to the outside front wheel. This braking action will correct the oversteer condition and the slip angle is small.

Oxidation

A process where one element of a compound is combined with oxygen in a chemical process to produce another compound.

Oxidation Catalysts

Platinum and palladium are utilized in the convertor to mix oxygen (O_2) with hydrocarbons (HC) and carbon monoxide gas (CO) to make non-harmful tailpipe emissions of water (H_2O) and carbonic acid gas (H_2CO_3).

Oxidation Reaction

The combustion process is an oxidation reaction. This means oxygen is combined with gasoline to create new chemical compounds.

Oxides of Nitrogen (NOx)

A primary emission produced within the combustion chamber under high temperatures when nitrogen combines with oxygen. NOx is a generic term for the oxides of nitrogen that are most relevant for air pollution, namely, nitric oxide (NO), nitrous oxide (N_2O), and nitrogen dioxide (NO_2). It is a colorless, tasteless, odorless gas that reaches the atmosphere and mixes with O_2 (oxygen) to form NO_2 (nitrogen dioxide) and NO (nitric oxide) grouped together and referred to as NOx. The "x" represents any number of oxygen (O_2) atoms. NOx gases are produced from the reaction among nitrogen and oxygen during the combustion of fuels, such as hydrocarbons. These gases contribute to the formation of smog and acid rain, affecting the tropospheric ozone. NOx gases are also produced naturally by lightning.

Oxygenated Fuels

An octane component containing hydrogen, carbon, and oxygen in its molecular structure. This includes all kinds of oxygenates such as ethers, like MTBE, and alcohols, like ethanol and methanol.

Ozone

Ozone, or tri-oxygen, is an inorganic molecule with the formula O_3. Oxygen-rich gas created by sunlight reacting with unburned hydrocarbons (HC) and oxides of nitrogen (NOx) is termed smog. It is a pale blue gas having a pungent odor. It is an allotrope of oxygen that is much less stable than the diatomic allotrope O_2 and breaks down in the lower atmosphere to O_2 (oxygen).

REFERENCES

1. SAE, Society of Automotive Engineers, J1930DA_202105: Electrical/Electronic Systems Diagnostic Terms, Definitions, Abbreviations, and Acronyms Web Tool Spreadsheet, SAE Publishing, Warrendale, PA, 2021.
2. Kershaw, J.F., *High-Performance Automotive Cooling Systems* (Forest Lake, MN: CarTech, Inc., 2019), Chapter 3, ISBN:978-1-61325-504-9.

P

Packing

A gasket material used for sealing, that is, packed into an opening to prevent leakage.

Panhard Arm or Bar

An automotive or truck suspension component used on the rear suspension that connects the frame to the axle. The Panhard arm or bar mounts on the frame on one end and on the axle housing on the other end. This prevents the frame and axle from moving side to side independently. The Panhard arm swings the axle from side to side. The shorter the Panhard arm the worse the movement. Also, the more level the Panhard bar is at ride height the better the axle stays in place. It can be mounted above the axle housing or behind the axle. The ideal mounting position, from an engineering position for the Panhard bar, is to have it at the same height as the axle centerline and as long and level as possible. This provides excellent control and keeps the side swing of the axle housing to a minimum. The Panhard arm or bar is generally found on road racing vehicles and some circular track applications. It is not found on regular production vehicles.

Paraffin

A flammable and translucent wax compound consisting of a mixture of saturated hydrocarbons. It is derived from the distillation of petroleum or shale and used in fuels, candles, cosmetics, polishes, and sealing and waterproofing of compounds.

Parallel Circuit

A circuit that has more than one path for the current. A parallel circuit has more than one electrical path between the power source and the ground or return path. These are characteristics of all parallel circuits:

- More than one path for current flow. Each current path is called a branch.
- All of the branches connect to the same positive terminal and the same negative terminal. This means the same voltage is available to all of the branches.
- Each branch drops the same amount of voltage, regardless of resistance.
- The current flow in each branch can be different, depending on the resistance. The total current in the circuit equals the sum of the branch currents as in Kirchhoff's law.
- Total resistance is always less than the smallest resistance in any branch.

FIGURE P.1 Parallel circuit.

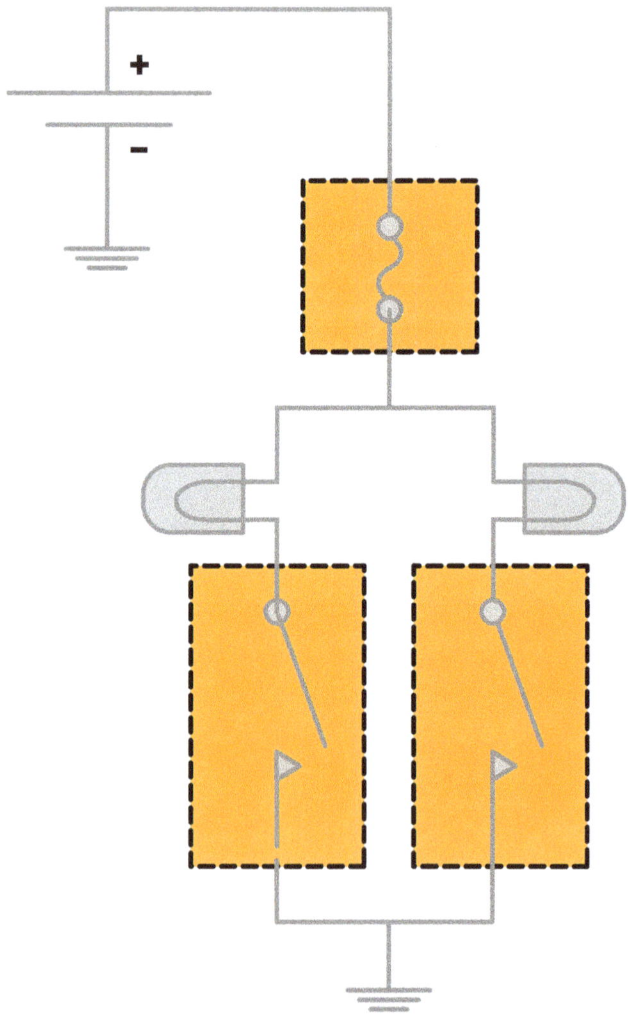

Courtesy of John F. Kershaw Ed.D.

Parallel Hybrid

Uses the internal combustion engine (ICE) or the motor generator to directly drive the vehicle. The electric motor is located between the ICE and transmission. The electric motor may assist the ICE in vehicle acceleration or it may act as a generator to charge the high voltage (HV) batteries. When the motor is used to assist in vehicle acceleration, the electrical energy to power it comes only from the HV batteries. The electric motor may charge the batteries by regenerative braking or the ICE may run to turn the motor as a generator while the vehicle is stationary if the battery state of charge is low. For more information on *Hybrid Electric Vehicle (HEV) and Electric Vehicle (EV) Terminology*, see **J1715_202209** [1].

FIGURE P.2 Parallel HEV configuration.

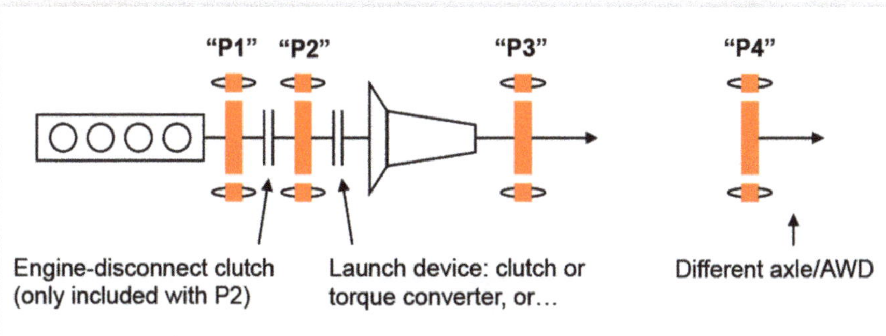

Parallelism

A disc brake service term coined to indicate that the brake rotor is parallel, that is, the measurement around the circumference of the rotor is even and within OEM specifications.

Parallelism (Steering Angle Alignment)

Refers to the tread centerlines of the wheels being parallel to the geometric centerline. When the rear toe is adjusted, the thrust line should be parallel with the geometric centerline. The steering wheel is straight and also the front toe is adjusted to the thrust line (now the centerline). When the vehicle moves, the front tires assume a parallel direction with the rear. If the toe on the rear is aligned, the front tires will follow a parallel path with the rear, creating centerline steering.

Parallel Wire

A wire that has two or more individually insulated conductors or wires molded into one flat or round body of multicolor insulation.

Parasitic Load or Draw

Current draw that is present when all electrical-electronic devices are shut off. This load includes anything that requires a small current when the engine and ignition key are turned off, such as on-board computers and the dash clock. Also known as key off load.

Park

An automotive term used primarily with automatic transmissions. A mechanical arm, called a parking paw, locks the drivetrain to prevent the vehicle from moving. The transmission gear selector will have a "P" on it to indicate Park. Sometimes this shifter display is referred to as PRNDL (pronounced prindle).

FIGURE P.3 Automatic transmission park indicator.

FLUKY FLUKY/Shutterstock.com

Park Control Valves (Air Brakes)

The spring-brake park control valve may be a dash-mounted control valve that permits the operator to use or release the truck or tractor spring parking brakes. The valve model used can vary among different air system circuits. They are generally a push-pull dual-circuit model. The valve button must be pushed in to release the parking brakes by exhausting air from the spring brake chambers. Different models of spring brake chambers are available. If atmospheric pressure is lost within the brake, the buttons will come out. When the vehicle is parked and therefore the dash control button is pulled out, air is exhausted from the hand brake chamber, and therefore the heavy volute spring forces the brake slack adjuster forward to use the shoes against the drum or restraint against a disc. No air is left within the emergency brake chamber. The hand brake system of a tractor-trailer combination manages hold-off pressure delivered to the spring brake chambers. FMVSS 121 requires that the shape of every valve is as follows—the color of every valve is an industry standard:

- Red octagonal knob: The trailer supply valve. The system spring-brake control valve must have already been actuated before this valve may be. It supplies all system air to the trailer(s) and must vent to release the trailer spring brakes.
- Yellow diamond/square: The system spring-brake (park) control valve. This valve masters the parking brakes on a combination. It is pushed inward to release the parking brakes. When pulled dead set its park position, it cuts the air supply to the trailer (thus applying its park/emergency brakes) and puts the tractor into park/emergency mode.
- Blue round: Isolates the tractor park brake function from that of the trailer. This valve is mastered by the spring-brake control valve.

FIGURE P.4 Air brake parking valves. Yellow diamond/square: the system spring-brake (park) control valve. This valve masters the parking brakes on a tractor-trailer combination. Red octagonal knob: the trailer supply valve. This valve is responsible for supplying all system air to the trailer(s) and must be ON to release the trailer spring brakes.

Fazakas Mihaly/Shutterstock.com.

Parking Assist

Two types of parking assist systems are used in cars and light trucks: The first is a very sophisticated system called intelligent/automatic parking assist system, which will actually park the vehicle. The second system is called parking assist, which uses ultrasonic sensors to warn the diver of the proximity of obstacles in the path of the vehicle. For more information on this, see **J3063_202103** [2].

Parking Brake

The parking brake, also known as a handbrake or e-brake, is used to keep the vehicle securely motionless when parked. Parking brakes often consist of a cable connected to two wheel brakes, which is then connected to a pulling mechanism. In most vehicles, the parking brake operates only on the rear wheels, which have reduced traction while braking. The mechanism may be a hand-operated lever, a straight pull handle located near the steering column, or a foot-operated pedal located with the other pedals. There are also four wheel disc brake systems that use a lever-operated rear wheel caliper piston to engage the rear brake pads or small brake shoes within a caliper drum to apply the parking brake.

FIGURE P.5 Parking brake caliper (left) and parking brake lever (right).

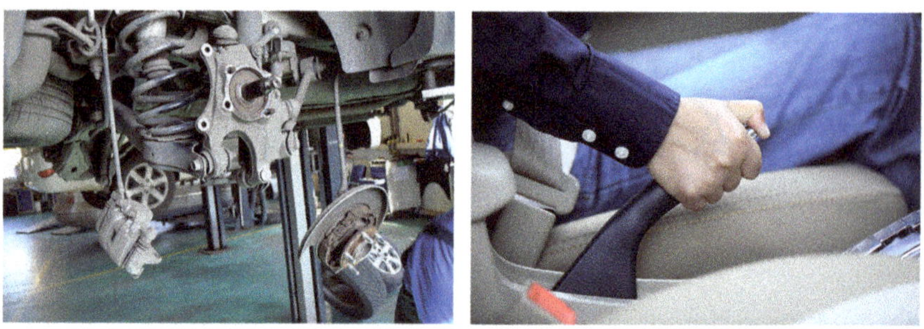

Dima Aletskyi/Shutterstock.com.

Partial Driving Automation

"SAE Level 2 driving automation applies to vehicles with advanced driving assistance systems (ADAS) which will take over steering, acceleration, and braking in specific scenarios. But, although Level 2 driver support can control these primary driving tasks, the driving force must remain alert and is required to actively supervise the technology in the slightest degree times. An example of Level 2 driving automation is Highway Driving Assist, installed in Genesis, Hyundai, and Kia vehicles. It requires the driving force to own her hands on the wheel but actively steers, accelerates, and brakes the vehicle when traveling on highways. BlueCruise from Ford SuperCruise from GM are a brand new hands-free partial driving automation technology from Ford. They are more sophisticated than Highway Driving Assist, allowing the motive force to require her hands off of the wheel on specific, approved highways within the U.S. and Canada. Both of those samples of Level 2 driving automation require the motive force to stay alert, engaged, and prepared to require control at any time. For the record, and in line with what the automaker told the state of California, Tesla's new Full Self Driving Capability technology may be a Level 2 system, and it'll remain so when Autosteer for city streets arrives as an over-the-air software update." For more information on this, see **J3016_202104** [3].

Particulates (or Particulate Matter)

The quantity of all solid and liquid particles suspended in air, many of which are hazardous. It is basically soot composed of very small particles of solid/semisolid material. This complex mixture includes both organic and inorganic particles, such as dust, pollen, soot, smoke, and liquid droplets. This is the black soot you sometimes see coming out of an older diesel engine. For more information on this, see **J1145_201109** [4].

Pascal's Law

Pressure on a confined fluid is transmitted equally in all directions and acts with equal force on equal areas. Following this law, hydraulic systems are able to transmit force and motion via liquids, simply due to the fact that a liquid cannot be compressed. No matter how much pressure is placed on a liquid, its volume remains the same. This allows a liquid to transmit force much like a mechanical lever. Pascal's Law also states that a liquid can also decrease or increase the force it transmits depending on the area of the output surface to which the force is applied. The larger the area the greater the force.

$$\Delta p = \rho g \times \Delta h \quad \textbf{(P.1)}$$

where

- Δp is the hydrostatic pressure in pascals in the SI system or the difference in pressure at two points within a fluid column due to the weight of the fluid
- ρ is the fluid density, in kg/m^3 in the SI system
- g is acceleration due to gravity 9.8 m/s
- Δh is the height of the fluid above the point of measurement or the difference in elevation between the two points within the fluid column (m)

Passive Keyless Entry

A keyless entry system that does not require direct action by the driver. It uses the key fob as a transmitter which communicates with the vehicle as it comes close. The key is identified using one in every of several antennas around the vehicle and a radio generator within the key housing. The vehicle is automatically unlocked when a button or sensor on the door handle or trunk release is depressed. The passive system also can be accustomed to unlock and open a rear lift gate on an SUV if a foot is moved under the sting of the vehicle allowing it to open the slider or the rear hatch. Vehicles with a passive (smart) key system can even have a mechanical backup, usually within the type of a key blade built into the key fob. Vehicles with a sensible key system are often started without inserting a key within the ignition, provided the motive force has the key fob inside the vehicle. On most vehicles, this can be done most frequently by pressing a start button. When leaving a vehicle equipped with a wise key system the vehicle is locked, depending on the make, model, and year of manufacture of the car, by doing one of the following:

- Pressing a button on one of the door handles
- Touching a capacitive area on a door handle
- Walking far from the vehicle as the door locks when the key fob is further away than 15 ft (5 m)

Passive Sensor

A sensor that depends on an external source of voltage to return a signal to the computer. The internal resistance of the sensor changes with a change in a condition, but it does not generate its own voltage signal. Often the sensed signal comes back to the source via a voltage divider.

Passive Systems

Passive infrared systems do not use an infrared light source; instead, they capture thermal radiation already emitted by the objects using a thermographic camera. They have a range of about 300 m or 1,000 ft.

PCD (Pitch Circle Diameter)

The PCD is the diameter of a circle drawn through the middle of the bolt holes of the wheel. It is a hard and fast measure set during manufacture that cannot be altered. PCD is measured in both inches and millimeters, and it also indicates the quantity of studs or bolts the wheel will have. One in every of the most common configurations has 4 studs and a PCD of 100 mm, so the scale is 4 × 100 designation. To live the diameter of the bolt circle with a good number of studs, like four or six, measure from the middle of one stud to the middle of an opposite stud. Technicians generally use a bolt pattern gauge or a template to urge an accurate measurement on a five-stud bolt circle or on other odd-numbered patterns. To get an approximate measurement manually, you will be able to measure from the middle of one hole to the rear of the third-hole diameter of the imaginary circle created by passing a line through the middle of the lug holes

PCM (Powertrain Control Module)

An engine and transmission management computer that receives, processes, and transmits input signals, both voltage and current, for both engine and transmission operation. It uses these inputs to operate a series of output actuators to operate the engine. In an HEV, it is a set of electronic circuits for altering and feeding current between the HV battery pack and motor generator. It controls how much and what kind of current flows through the large power cables in the electric drive system. The hybrid power control module can also control the speed and torque of the motor generator by varying the AC frequency and phase shift. For more information on this, see **J1930DA_202105** and **J1715_202209** [1, 5].

PCV (Positive Crankcase Ventilation)

Clean, filtered air is drawn into the crankcase, and crankcase vapors are recycled to the intake manifold. PCV systems also provide better crankcase ventilation by using a valve from the intake manifold to the crankcase. At idle, when the intake manifold pressure is low (vacuum), when there is time for more gasses to leak past the rings, the valve is open allowing blowby gases to flow into the engine and be burned in the combustion chamber. At larger throttle openings, the intake manifold pressure is high and the valve is closed and blowby gases are not burned. This system replaced the road draft tube that just vented the blowby gases to the atmosphere. Since the crankcase is vented to the intake manifold, the engine vacuum works to evacuate the crankcase of vapors.

PDOF (Principle Direction of Force)

The PDOF or vertical overlap "which is the measure of damage overlap, expressed as a percentage, determined by the contact damage length plus one-half of the difference between the undeformed end width and the overall width relative to the overall width." For more information on *Terminology and Measurement Methodology for Determining Vehicle Overlap in Offset-Frontal Crashes*, see **J3227_202105** [16].

Peak-and-Hold Fuel Injector Waveform

The waveform of a peak-and-hold fuel injector has two induced voltage spikes: one when the engine management computer lowers the current by switching to the limiting resistor circuit and another when the computer shuts off the current flow to the injector.

Pedestrian Collision Mitigation Braking

A subset of front impact mitigation that aims to scale back the likelihood of collisions or reduce the impact speed with pedestrians within the forward direction by the following:

- Determining the forward path of the car
- Monitoring the path with reference to pedestrians in or near the forward path
- Provides pedestrian detection and alerts the driver and, if certain conditions are met, provides autonomous braking to assist in avoiding or reducing the impact speed of a collision with the pedestrian

This system is not intended to exchange the responsibility of the driving force for paying careful attention to the forward scene for pedestrians, the vehicle, and other potential hazards. Its function is restricted to supplemental use only to help instead of replacing the motive force in responding to pedestrians within the forward scene. It can detect and respond to up to 10 pedestrians in forward gear at speeds between 5 and 50 mph. During daytime driving, the system detects pedestrians up to a distance of roughly 131 ft. Pedestrians must be a minimum of 31.5 in. tall to be detected. No added components are required for front automatic braking or pedestrian collision mitigation braking. For more information on this, see **J3063_202103** [2].

Pendulum Belt Retractor

A seat belt retractor that uses the inertia of a small-hinged metal weight to engage the seat-belt locking mechanism during vehicle deceleration. If the vehicle stops quickly, a small pendulum in the retractor is thrown forward, causing the belt-locking mechanism to engage. For more information on restraint systems, see *Glossary of Automotive Inflatable Restraint Systems* [7].

Pentroof Combustion Chambers

The pentroof combustion chamber design resembles a wide, inverted "V." It is flat for a small volume, allowing high compression ratios with flat-topped pistons. The small volume also decreases the chance of detonation, permitting high compression ratios with today's relatively low-octane fuel. The shape provides a favorable surface-to-volume ratio for low emissions of unburned hydrocarbons and has room for large valves to help breathing. The intake and exhaust valves are opposite one another, like the hemispherical chamber, to further enhance breathing. The spark plug is centrally located and provides squish areas for turbulence and combustion efficiency. For more information on the most commonly used terms for pistons, see **J2612_202002** [8]. These terms designate either types of pistons or certain characteristics and phenomena of pistons used with different combustion chambers.

Petrochemical

A substance or chemical that is derived from the refining of petroleum or natural gas.

PHOAT (Phosphate Hybrid Organic Acid Technology) Antifreeze Coolant

PHOAT was developed for Mazda-based Fords (2008+), same as Mazda FL-22, and is ethylene glycol based. This coolant is available in a 55% coolant and 45% water premix. PHOAT properties are as follows:

- Concentration 55%
- Boiling point (with 15 psi pressure cap) 270°F (132°C)
- Freezing point −47°F (−44°C)
- Color: Dark green
- Embittered (made to taste bitter so animals will not drink it)

The use of PHOAT coolant in Ford Mazda engines is required to be assured of proper protection of the internal coolant passages in the engine. It is also only available in premix containers to make sure that the water used meets specifications [6].

Photoelectricity

When certain metals are exposed to light, some of the light energy is transferred to the free electrons of the metal. This excess energy breaks the electrons loose from the surface of the metal. They can then be collected and made to flow in a conductor. Photoelectricity is used in light-measuring devices such as photographic exposure meters and automatic headlamp dimmers.

FIGURE P.6 Photoelectricity action.

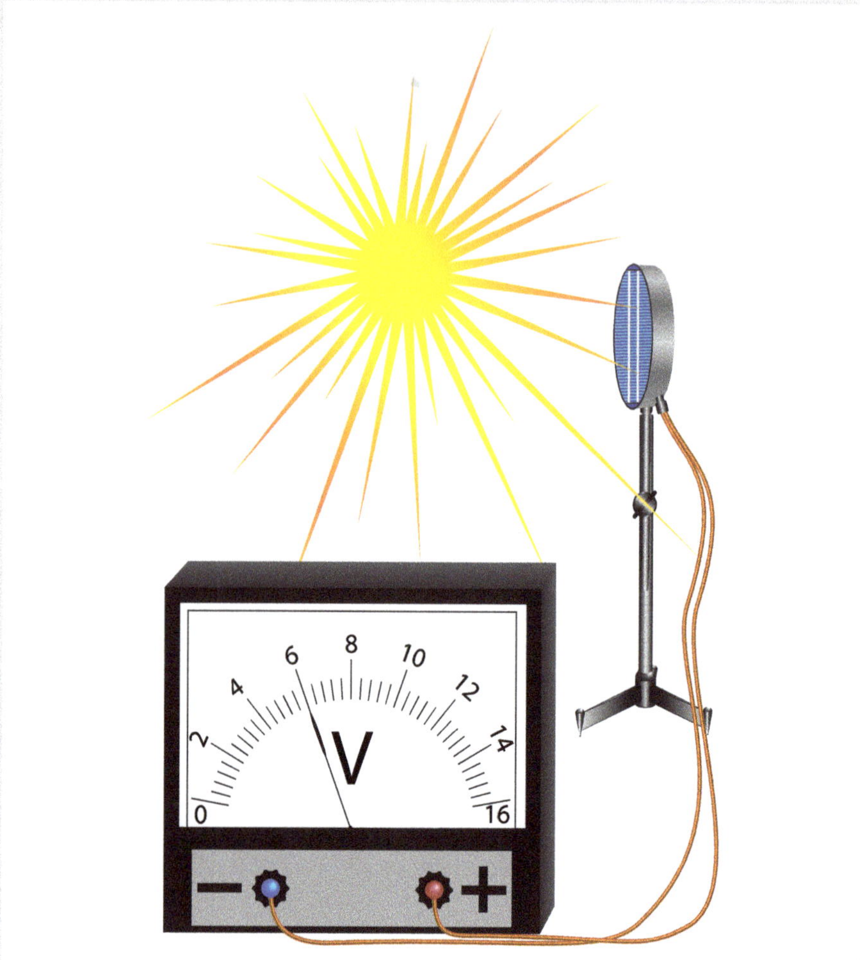

Pickup Coil

An ignition system triggering device, that is, a transducer. A coil of wire wrapped around a metal pole piece with a number of arrow-shaped points that equal the number of ICE cylinders. There is a rotor or timer core that revolves in the center of the pole piece with the same number of arrow-shaped points that will alter the magnetic field to indicate RPM and cylinder position. It is an ICE camshaft position sensor.

FIGURE P.7 Pickup coil.

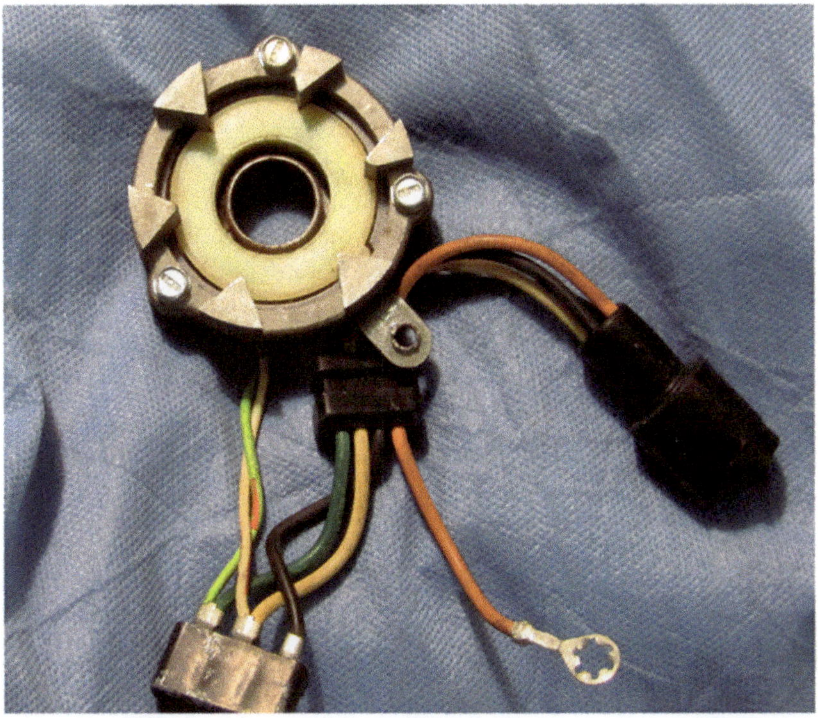

Courtesy of John F. Kershaw Ed.D.

PID (Parameter Identifier)

PID stands for parameter identifier, and it is used to identify components within the electronic subsystem. It is a two- or three-digit code assigned to each component that identifies data using a data link to the ECM/PCM. SAE standard J1587 says that a single byte character identifies the data bytes following the MID. The PID is the main number in an SAE diagnostic trouble code (DTC) (i.e., P111-3 is a coolant level sensor with high input voltage). The "P" in an SAE

DTC identifies it as a PID, 111 is the identifier, and the 3 is a failure mode indicator. For more information on this, see **J1587_200202** [9].

Piezoelectric Principle

When subjected to pressure, certain crystals, like quartz, develop a possible difference, or voltage, on the crystal faces. This current is employed in phonograph pickups, crystal microphones, underwater hydrophones, and certain stethoscopes. The voltage created is termed piezoelectricity. Many engine control sensors use piezoelectricity to form the voltage or vary the resistance and control a computer signaling. Piezoelectric sensors are used as detonation sensors and fuel injectors. The knock sensor produces about 300 mV of electricity and vibrates at a 6,000-Hz (cycles per second) frequency, which is the frequency that the cylinder walls vibrate at during detonation.

FIGURE P.8 Piezoelectricity action.

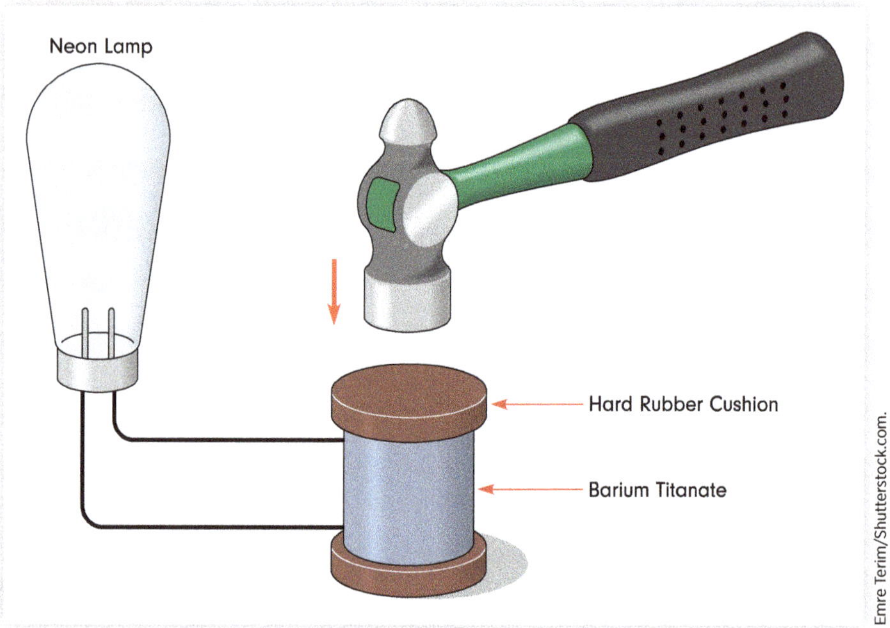

Piezo-Resistive Sensors

A sensor that relies on the ability of certain mineral crystals to produce voltage or change resistance when compressed. These sensors will have a piezo-resistive crystal arranged with a Wheatstone Bridge to measure the change in resistance of the piezo crystal. These sensors can measure very high pressures. Because of the sturdiness of the crystal, piezo-type sensors are better adapted to measuring vibration and dynamic or continuous pressure changes. For example, knock sensors measuring abnormal combustion signals are a common application of piezo-resistive sensors.

Piezo-Resistor Chip

A chip that changes internal resistance in response to pressure.

Pigment

Pigment is the color in automotive paint.

Pilot Jet

This is an ICE carburetor jet or orifice that meters the amount of fuel when idling.

Ping

Secondary rapid burning of the last 3–5% of the air-fuel mixture in the combustion chamber of an ICE causing a second flame front that collides with the first flame front which causes the cylinder walls to vibrate from the localized pressure, producing a knock noise. Also called detonation, spark knock, and pre-ignition.

Pinion Gear

The small gear that slides over the armature shaft of the starting motor and engages a large gear on the engine flywheel. The overrunning clutch allows the pinion gear to turn the flywheel ring gear for starting.

Pintle Nozzle

A type of nozzle used in a fuel injector where the stem of the nozzle valve is extended to form a pin or pintle which protrudes through the opening of the nozzle. The size and shape of the pintle can be varied according to the specifications and application. It provides a spray operating at low injection pressures of 8–10 MPa. The spray cone angle is generally 60°. The main advantage of this nozzle is that it avoids weak injection and dribbling. It prevents the carbon deposition on the nozzle hole.

FIGURE P.9 Pintle nozzle in a fuel injector.

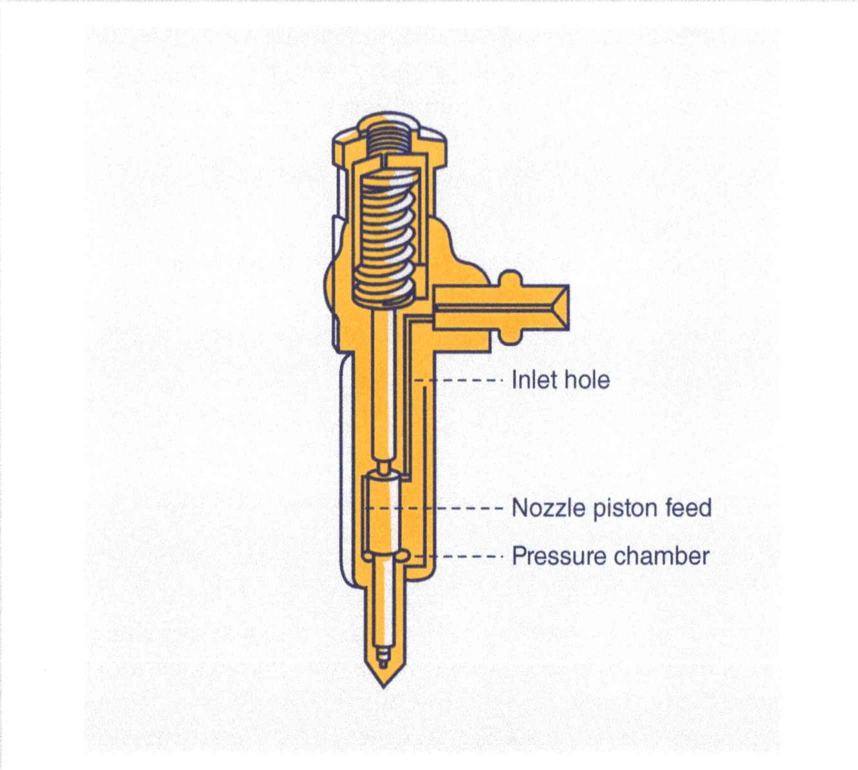

Piston

In an ICE piston engine, the piston is a circular plug moving in a cylinder that provides compression and power. The action of combustion pushes the piston down in a reciprocal action to cause the rotation of the crankshaft. For more information on this, see **J2612_202002** [8].

FIGURE P.10 Piston ICE four-stroke cycle.

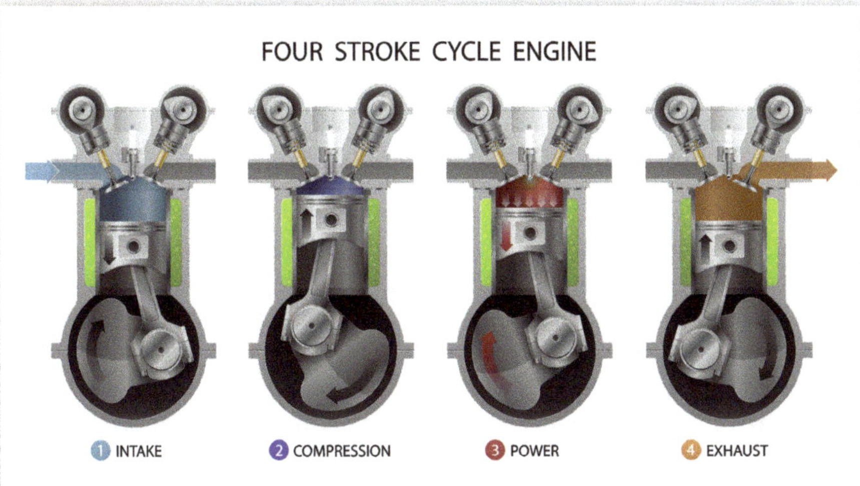

Farber/Shutterstock.com.

Pitch

This angle occurs when the vehicle weight is transferred from the front tires to the rear tires and is reciprocal. Pitch normally occurs under heavy acceleration. The condition improves traction, in a rear-wheel-drive vehicle, during straight-ahead acceleration. For more information on this, see **J1594_201007** [11].

Planar Zirconia Oxygen Sensor

An oxygen sensor with a zirconia element, electrodes, and heater combined in a flat, laminated strip that can reach its operating temperature in 10 sec and is more resistant to contamination and vibration than conventional zirconia sensors.

Planetary Gearset

A simple planetary gearset consists of three primary components:

- Sun gear
- Planet carrier assembly
- Internal ring gear

The sun gear gets its name from its position at the center of the gearset. The planet carrier contains the pinion gears, also known as planet gears. Pinion gears revolve around the sun gear. The outermost member of the gearset is an internal ring gear, having gear teeth along its inside circumference. The pinion gears are in simultaneous mesh with both the sun and ring gear. The pinion gears are free to rotate on pins that are part of the carrier while the entire assembly rotates to direct power flow or the path of drive torque through the transmission or transaxle. The pinions mesh fully with both the sun gear and internal ring gear at all times. The planetary gears never disengage to change gear ratios; power is simply redirected.

The power flow through a planetary gearset, both input and output, occurs along a single axis. In any planetary gearset, each gear always meshes with several other gears. Therefore, driving one gear will drive all of the other gears as well. To transmit torque through a planetary gearset, you drive one member while holding a second, which causes the third to be driven. Each member of a planetary gearset can play any one of these three roles to transmit torque. The various combinations of drive, held, and driven members that are possible result in the number of gear ratios available. Certain combinations of drive, hold, and driven can change the direction of rotation as well. Power flows through a planetary gearset in several steps to get from the drive action of the first member to the driven action of the last member. When three or more gears are involved, the second gear is a driven gear in relation to the first, but a drive gear in relation to the third. For this reason, the drive member of a planetary gearset is called the input member, the held member, the reaction member, and the driven member, the output member. For more information on this, see **J646_201807** [10].

FIGURE P.11 Three members of a simple planetary gearset are the sun gear, internal ring gear, and planet carrier assembly.

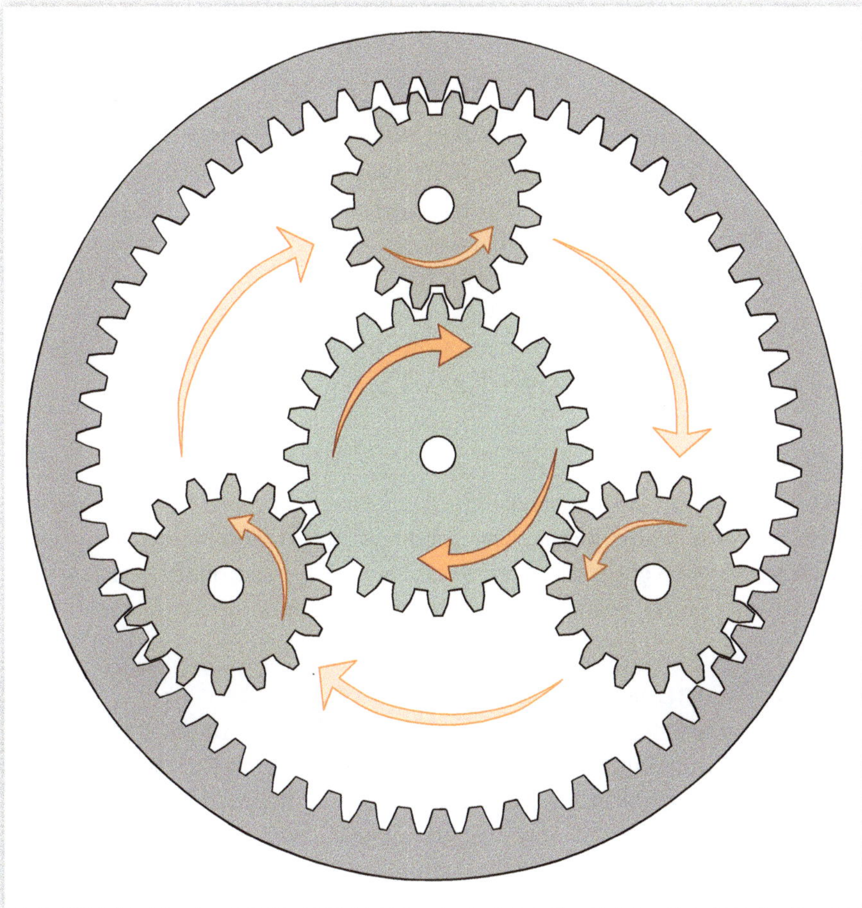

Sergey Merkulov/Shutterstock.com.

Plasma Screen

A dash display that uses a layer of phosphor cells between two panels of glass. The phosphor cells emit visible light (red, green, and blue light) when excited.

Plastigage™

A soft plastic that flattens out to predetermined widths when squeezed and torqued to specifications between the shaft and the bearing. These widths equal a specific clearance. This material is used to check the main and rod-bearing clearance. It is sold in a paper sleeve that is also the scale on which it is measured (in thousandths of an inch on one side and millimeters on the other side).

Plate Strap

Used in a battery to join several negative plates to form a negative plate group. For more information on this, see **J1715/2_202108** [13].

Plug-In Hybrid

A hybrid vehicle that can be connected to a 120-V AC home wall outlet and extension cord to fully recharge the HV battery pack. With a charged battery, the vehicle can operate in an all-electric mode without starting the gas engine. You get all the benefits of an electric vehicle, without the biggest drawback—a limited range. You will be able to go all-electric for 90% of the drive. When the electric charge runs out, an ICE kicks in and your vehicle drives like a non-plug-in hybrid. For more information on *Hybrid Electric Vehicle (HEV) and Electric Vehicle (EV) Terminology*, see **J1715_202209** [1].

PM Motor

Permanent-magnet electric motor is a type of electric motor that uses permanent magnets to create the magnetic field that causes motor rotation through magnetic repulsion of the magnetic fields between the stationary fields and the rotating fields. These permanent magnets replace the wound field coils.

PMAC (Permanent Magnet AC) Motor

The PMAC uses functionalities that partially overlap with those of both an AC induction motor or larger, higher-end applications requiring precisely metered torque, speed, or positioning. Permanent magnets mounted on or embedded within the rotor are united with the current-induced internal magnetic fields of the motor generated by electrical input to the stator. The rotor contains permanent magnets which are either surface-mounted to the rotor lamination stack or embedded within the rotor laminations. As in common AC induction motors, power is supplied through the stator windings. Permanent-magnet fields are, by definition, constant and not subject to failure, except in extreme cases of magnet abuse and demagnetization by overheating. PMAC, PM synchronous, and brushless AC are synonymous terms. The Nissan Leaf and

Chevrolet Bolt use an inside magnet motor. Tesla uses magnet motors within the Model 3 and Semi-tractor. For more information on *Hybrid Electric Vehicle (HEV) and Electric Vehicle (EV) Terminology*, see **J1715_202209** [1].

PNP (Plug and Play) Fuel Injector Waveform

The waveform of a plug and play (PNP) fuel injector is an inverted form of a saturated switch waveform.

Polarity

The condition of being positive or negative in relation to a magnetic pole.

Pole

Area of a magnetized body where the lines of magnetic force are concentrated. One end of a magnet.

Poppet Valves

A valve used to control the timing and quantity of gas or vapor flow into an ICE. It consists of a port or open-ended chamber, usually round or oval in cross- section, and a plug, usually a disc shape on the end of a shaft known as a valve stem. They work by linear motion.

Potential

The ability or capacity to become or develop into something in the future. This term is used in electricity to define voltage as potential, but not yet in use. Potential voltage is the voltage between two points before a circuit is closed (not operated).

Potential Difference

The electrical force, or pressure, that pushes free electrons and causes current. It is also referred to as voltage or electromotive force (EMF).

FIGURE P.12 Electrical potential or EMF.

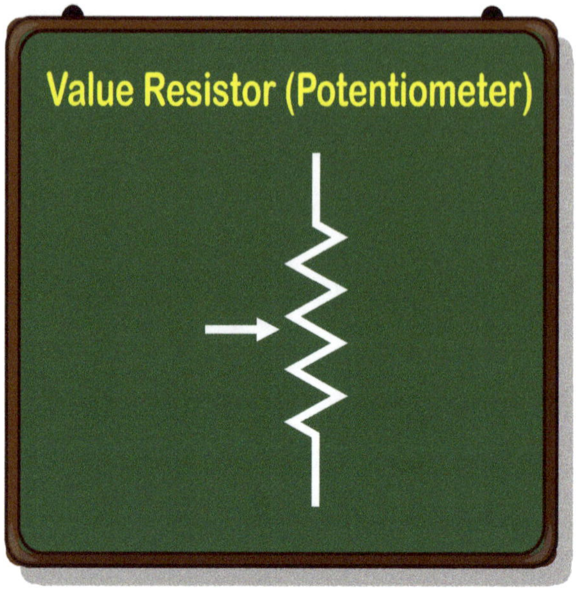

Potential Energy

Capacity to perform work, or energy at rest. When the gasoline is injected into the engine and ignited, the potential energy of the liquid gasoline changes states and becomes heat. The engine converts this energy into motion and then, using various devices (crankshaft, transmission, etc.), transfers this motion to the wheels. The wheels rotate, using this motion to maneuver the vehicle. Finally, the longer you drive, the hotter the tires become. This can be the energy released in the form of heat.

Potentiometer

Variable resistors that sense motion or position. The potentiometer uses three terminals that are used for reference: voltage, signal voltage, and ground return.

Power

Power is the term we give to the speed of labor done by any variety of machine. The output of automotive engines is sometimes expressed in horsepower. So it is the output of electrical motors. Many electrical devices are rated by the quantity of electric power they consume, instead of by the quantity power they produce. Power consumption is expressed in watts: 746 W = 1 hp.

Power Balance Test

A test that measures the power output from each of the engine cylinders. Also called a cylinder balance test.

Power Distribution Diagram

A diagram that shows how battery current feeds to the main components or sections of the electrical system of the vehicle.

Power Divider

Performs the initial task of splitting the power from the driveshaft between each of the rear drive axles. It also works like an inter-axle differential, providing a variable power split between each of the rear drive axles, like a conventional rear axle differential splits power between the left and right drive wheels. The power divider is mounted on the forward rear axle, which is the front axle in a tandem assembly, and features a series of gears which allow it to split power between each rear drive axle. The power divider assembly is a two-gear design consisting of an input shaft, inter-axle differential, output shaft, and two constant-mesh helical gears. Each axle features a spiral bevel gear design, with the drive pinion located at the centerline of the ring gear. The differential and drive pinion are mounted on tapered roller bearings.

Power Door Locks

Electrically operated door lock system that uses a static magnet, or permanent magnet (PM), reversible motor to lock or unlock all vehicle door locks from a bearing switch or switches. The motor uses a built-in circuit breaker and operates the lock-activating rod. PM reversible motors do not require grounding because, like power windows, the control is decided by the polarity through the two motor wires. Some two-door vehicles do not use a door lock relay because this flow for less than two PM motors will be handled through the door lock switches. Yet most four-door vehicles and vans with power locks on rear and side doors use a relay to manage this flow necessary to work four or more power door lock motors. The door lock relay is controlled by the door lock switch and is often the placement of the one and only ground connection for the complete door lock circuit.

Power Seats

The power-operated seat uses a reversible motor and a transmission assembly that has three solenoids or motors, and six drive cables that turn six seat adjusters. A six-way power seat offers forward and backward seat movement, plus up and down cushion movement at the front and also the rear. The drive cables rotate inside a cable housing and connect the ability output of the seat transmission to a gear or jackscrew assembly that moves the seat. The jack assembly is termed a gear nut and accustomed to move the front or back of the cushion up and down. A rubber coupling, usually located between the electrical motor and therefore the transmission, prevents motor damage in the event of a jammed seat. This coupling is intended to forestall motor damage. Most power seats use a magnet motor which will be reversed by reversing the polarity of the present sent to the motor by the seat switch. Some seats now include a back massage feature.

Power Steering Drive Belt System

The drive belt system may use one serpentine belt or two belts. The drive belt is thin so that it can bend backward, and has several ribs to match the grooves within the pulleys. There is also a V-belt style accustomed to drive certain accessory drive components. The drive belts are made from different types of rubbers. Each side of the drive belt could also be accustomed to drive the various accessory drive components. When the rear side of the drive belt is employed to drive a pulley, the pulley is smooth. The crankshaft balancer pulley pulls the drive belt across the accessory drive component pulleys. The spring-loaded drive belt tensioner keeps constant tension on the drive belt to stop the drive belt from slipping. The drive belt tensioner arm will move when loads are applied to the drive belt by the accessory drive components and also the crankshaft. The drive belt system may have an idle pulley, which is employed to feature wrap to the adjacent pulleys. Some systems use a pulley block in place of an adjunct drive component when the vehicle is not equipped with the accessory.

Power Steering Pump

A vane-type positive displacement pump that develops a continuous stream of fluid, usually in the 1,500 psi range. The fluid reservoir may be either integral to or built into the pump, and it could also be remotely mounted and connected to the pump by a hose. The power steering fluid reservoir is usually made of plastic or stamped metal. Some pumps use an integral reservoir that is part of the pump.

FIGURE P.13 Power steering pump.

Art Konovalov/Shutterstock.com.

Power Steering Rack-and-Pinion Gear

Rack-and-pinion gear features a pinion connected to the underside of the steering column. This pinion runs in mesh with a rack that is connected to the steering tie rods. The pinion is connected to the underside of the steering column. This pinion runs in mesh with a rack that is connected to the steering tie rods. This connection gives more direct operation, which enables the driving force to feel the road better. Turning the steering wheel rotates the pinion and moves the rack from side to side. On end takeoff racks, ball sockets at the top of the rack locate the tie rods and permit movement within the steering and suspension. On center takeoff racks, the tie rods connect with the middle of the rack and pinion. The middle of the rack is what moves on this sort of rack and pinion.

FIGURE P.14 Rack-and-pinion steering gear.

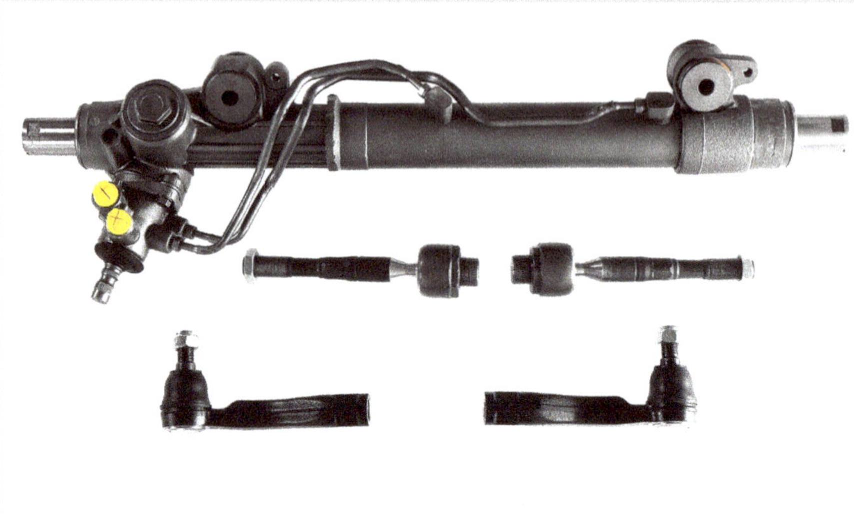

Stason4ik/Shutterstock.com.

Power Steering System

Power steering systems employ hydraulics, which is the study of liquids and their use to transmit force and motion. Hydraulic systems transmit force and motion through the use of fluid pressure. Force is a push or pull acting on an object and is usually measured in pounds or newtons. Pressure is a force applied to a specific area. Pressure is usually measured in force per unit of area (psi) or kilopascals (kPa). One psi is equal to 6.895 kPa. A pump is used to supply pressure to either a rack piston or hydraulic cylinder to assist in the movement of the steering gear.

FIGURE P.15 Rack-and-pinion power steering system.

temp-64GTX/Shutterstock.com.

Power Windows

Electric motors are accustomed to lifting and lowering door glass. They are operated by both a master control switch located beside the driving force and additional independent switches located at each electric window. Some systems use a lockout/child-safety switch located on the driver's controls to prevent operation of the power windows from the independent switches. Power windows are designed to regulate only with the controller in the on (run) position, although some manufacturers use a time delay for accessory power after the switch is turned off. This feature permits the driver and passengers an opportunity to shut all windows or operate other accessories for about 10 min or until a vehicle door is opened after the ignition has been turned off. This feature is often called retained accessory power.

PPM (Parts Per Million)

Parts per million is used to describe the quality of performance or concentration of a substance in a larger mixture.

Pre-Adjusted Hubs

A hub assembly has a long, precision-manufactured bearing spacer that is used to preset the special tolerance dimension of the inner and outer bearings. The pre-adjusted wheel-end assembly is identified by a spacer between the inner and outer single-row tapered roller bearings within the hub. Always refer to the OEM instructions to set up the correct bearing pre-load and end-play for this system.

Prechamber (Antechamber)

A small combustion chamber in an indirect injected (IDI) diesel engine where the fuel is first injected and combustion begins. The IDI diesel prechamber is joined to the main chamber above the piston by a connecting flow passage. Fuel is injected into the prechamber, which also uses a glow plug to heat the fuel for cold starting. Combustion begins in the prechamber and then spills into the main combustion chamber. As fuel and air are burned, the burning gases emit from the prechamber while the piston descends on the power stroke. Most IDI prechamber engines require a glow plug system to aid in start-up. Most use swirl-type (high turbulence) design prechambers. These swirl prechambers have a spherical shape that mixes the air and fuel by air swirl and assist in promoting high turbulence by creating a swirling mass of air in the prechamber.

FIGURE P.16 Diesel IDI engine showing a prechamber.

Predictive Systems

Predictive systems modify speed-supported predictions of the behavior of other vehicles. Such systems can make earlier, more moderate adjustments to the anticipated behavior, improving safety and passenger comfort. One example is predicting the likelihood of a vehicle in a very adjacent lane to get to the front of the controlled vehicle. One system predicts a lane change up to five seconds before it occurs. For more information on this, see **J3016_202104** [3].

Pre-Ignition

This is an ICE combustion event that is similar to detonation (see **Detonation**). There is confusion because pre-ignition and detonation often take place at the same time. Pre-ignition can raise pressure and heat in the cylinder to the point of causing detonation. If the point of pre-ignition occurs at a place in the combustion chamber far from the spark plug, two flame fronts consume the intake charge. The unburned portion of the air-fuel mixture that is trapped between the two advancing flame fronts is compressed and superheated which causes it to detonate resulting in a noise called ping, resulting from the vibration of the cylinder walls. Common causes are:

- Hot spots inside the combustion chamber like spark plug threads can protrude into the combustion chamber, or sharp edges on the piston top can become hot enough to ignite the charge.
- Carbon deposits in combustion chamber can ignite the charge.
- Spark plugs with too hot a heat range, which means the electrode does not dissipate heat fast enough and may get hot enough to ignite the charge without a spark.
- Cross-firing—electrical induction between spark plug wires that takes place because, when the plug wires are cracked, the spark can jump from wire to wire.
- Prolonged pre-ignition can create enough heat to melt spark plug electrodes, valve heads, and combustion chamber surfaces—or burn holes in the piston crowns.

Pressure Differential Valve

The pressure differential valve is located in the combination valve. The combination valve may also house the proportioning and metering valves (this valve will be discussed later in this lesson). During normal operation, pressure to the primary and secondary brake circuits balances the valve. As long as this pressure exists, the plunger on the brake warning lamp switch stays extended in a groove in the center of the piston. If a leak develops in either the primary or secondary brake system, unequal pressure acts on each side of the pressure differential valve piston. This causes the piston to move to one side. Upon moving a certain amount, the piston forces the plunger on the brake warning lamp switch out of the groove, depressing it. The switch closes, completing an electrical circuit and illuminating the brake warning lamp.

FIGURE P.17 Automotive brake system.

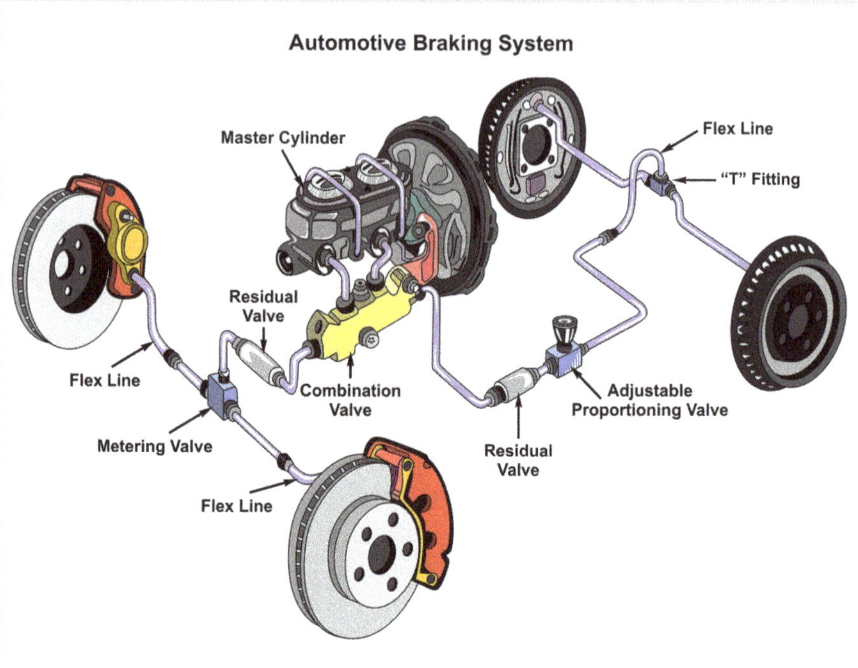

udaix/Shutterstock.com.

Pressure-Limiting (Ratio) Valve

A valve to limit the output pressure of an air brake system to a preset value. The pressure-limiting valve is set so that its output pressure on the low-pressure side is limited. A spring acts on the piston and holds it in the upper end position where it is in contact with the housing to regulate the pressure. All trucks today are required to have front-wheel brakes. Some much older trucks had a manual front axle limiting valve that enabled the driver to toggle between dry roads for full front-wheel braking or slippery roads to reduce by 50% application pressure to the front-wheel brakes. Some trucks are fitted with automatic front-wheel proportioning valves also known as ratio valves. A ratio valve manages air pressure to the front brakes by proportioning the application pressure delivered to it. For more information on *Braking System Definitions—Truck and Bus*, see **J2627_200908** [14].

Pressure Plate

The clutch component that holds pressure against the clutch disc to connect the ICE to the transmission. There are diaphragm and finger spring-type pressure plates. The diaphragm pressure plate with a single Belleville spring provides the force necessary to clamp the friction disc to the flywheel. The Belleville spring is a round piece of spring steel that has slots cut in the disc from the center toward, but not completely reaching, the edges of the disc. It works like the bottom of an oil can. These slots form fingers at the center of the Belleville spring. The outer edges of the Belleville spring are held against the pressure plate by the clutch cover housing.

As the diaphragm spring fingers are pushed toward the flywheel by the release bearing, the fingers pivot on their fulcrum points and the spring pressure is moved away from the pressure plate, thereby releasing the pressure from the driven disc. The spring fingers of the pressure plate may be flat or bent at an angle away from the flywheel. A diaphragm spring with bent fingers exerts more clamping force between the pressure plate, friction disc, and flywheel when the clutch is engaged. As the ICE speed increases, the centrifugal effect forces the fingers away from the flywheel and thus increases the spring pressure and clamping force. The single diaphragm Belleville spring clutch cover cannot generate the required clamping force to hold the friction discs of a multiple-disc clutch system used on heavy-duty trucks. For heavy-duty applications, coil spring design pressure plates are used. The coil spring design can have more clamping force than a diaphragm-type pressure plate.

Most heavy-duty coil spring clutches use multiple clutch discs and have an adjustable pressure plate assembly. Turning the eyebolts located at the release lever pivots increases or decreases the clearance, or air gap, between the flywheel, the friction disc, and the pressure plate when the clutch is disengaged. Adjustment keeps the clutch operating efficiently even as the friction disc material wear. For more information on *Automotive Pull-Type Clutch Terminology*, see **J1479_20121** [15].

FIGURE P.18 Clutch pressure plate types: diaphragm type (left) and finger spring type (right).

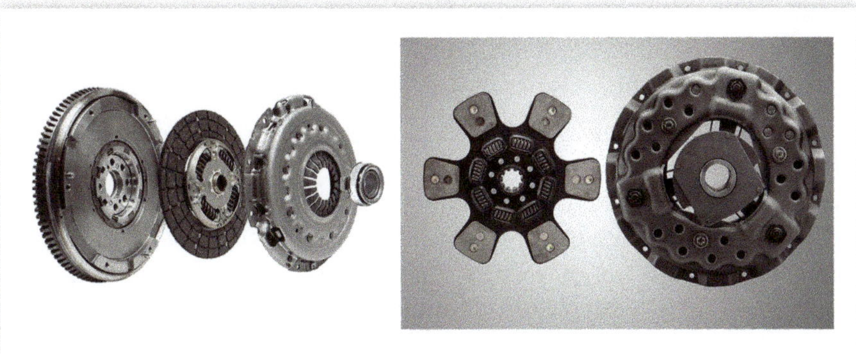

Stason4ik/Shutterstock.com.

Prabhjit S. Kalsi/Shutterstock.com.

Pressure Sensor

Active-type sensors that produce analog output signals. Strain gauge measurements are made by measuring small changes in the resistance of tiny wires referred caused by stretching or contraction of the wires. It is a pressure-sensing device that uses strain gauge wires embedded in a flexible glass block. Behind the block is a vacuum chamber providing a reference point for measurement of absolute pressure. If the device measures in gauge pressure, the chamber will have atmospheric pressure as the reference value. When the glass plate flexes under pressure, the small resistive wires in it will slightly change dimensions. As the plate dimensions change, the resistance of the wire changes slightly, too. By measuring this small change in the wire resistance, the pressure applied to the plate is determined. For more information on this, see **J1930DA_202105** [5].

Primary Coil Windings

Windings that surround the secondary windings of an ignition coil that are approximately 150 turns of heavy wire. For more information on this, see **J139_202002** [17].

Primary Ignition Circuit

The ignition components that turn on and off the current ground in the coil primary winding. For more information on this, see **J139_202002** [17].

Primary Ignition Scope Pattern

A scope pattern that shows the low-voltage or primary-voltage waveform in an automotive ignition system. For more information on this, see **J139_202002** [17].

Printed Circuit

Copper foil bonded onto a plastic or fiber base to form a conductor.

Process

A series of logically sequenced tasks to complete the work. You need to understand how inputs to a process are integrated and how value is added to a product, information, or services. Understanding the combination of what and how inputs and other key variables come together to produce the deliverables enables a more accurate forecast of whether requirements (targets) will be met. Prior results alone are poor predictors of future outcomes, and without knowledge of the process, any forecast is blinded, and any successful forecast would be by chance. Engineering prefers accurate forecasts of performance, so a process-centric view becomes an important component.

Process Map

Workflow diagram to develop a clearer understanding of a process. It is a flowchart that shows schematically a set of activities or steps to transform inputs into outputs. It identifies the sequential steps in order and its associated inputs and outputs. It may vary in degree of detail from a high level, depicting only the critical few process categories or milestones, to the minutest detail, describing the lowest level tasks. The flowchart centers on the sequenced activities but also may include information about its inputs, outputs, metrics, and the process players.

Processor

The central processing unit, or CPU, which is an integrated circuit that interprets inputs from input sensors and produces logical outputs to actuators to perform actions.

Program

The instructions a computer uses to perform its job. The program consists of mathematical instructions called algorithms and may include fixed data and require data from vehicle input sensors.

PROM (Programmable Read-Only Memory)

A computer chip that contains information that is specific to a vehicle or automotive engine application. The use of PROM allows the OEM to use the same basic engine management computer for different vehicles. Information programmed into a PROM is generally used to fine-tune the ROM (read-only memory) data to a specific vehicle or system application. The PROM might contain information on, for example, engine calibration, transmission specifications, vehicle weight, rear axle ratio, and so on. Like the ROM, the information is permanent. It can be read, but it cannot be erased or altered. PROM is nonvolatile, which means it is not lost when the system is shut down or when the battery is disconnected. For more information on this, see **J1930DA_202105** [5].

Propane

See **LPG (Liquid Petroleum Gas)**.

Propeller Shaft (Driveshaft)

Front propeller shafts were fitted with a slip joint incorporated into the shaft. Yokes are welded to the ends of each shaft and are usually configured as a single cardan universal joint at the front axle and a double cardan universal joint at the transfer case. The front and rear drive axles on four-wheel-drive (4WD) vehicles transfer engine torque to the front and rear wheels while allowing the front wheels to turn. The front axle on a typical 4WD is either designed for use with an independent front suspension or configured for a solid axle setup.

FIGURE P.19 Propeller shaft.

Proportioning Valve

Functions to balance brake pressure between the front and rear brakes. Proportioning valves are calibrated to a split point and slope and located in the rear braking system. Some proportioning valves have these values stamped on them. The split point is the point at which the valve restricts fluid flow to the rear brakes.

Propylene Glycol

A less toxic antifreeze ICE coolant formulated to provide benefits beyond those provided by conventional antifreeze and coolant products. Unlike ethylene-glycol-based products, which are toxic, propylene glycol coolant is biodegradable, and its low toxicity limits the threat to children and animals. It provides maximum cooling system protection in extreme temperatures and operating conditions. The Food and Drug Administration (FDA) has classified propylene glycol as an additive that is generally recognized as safe for use in food [6].

Proton

Positively charged particle; one of the basic particles of the nucleus of an atom.

Pro-Torq® Spindle Nut

The Pro-Torq® spindle nut used on heavy-duty trucks is manufactured by Stemco and used to hold bearing end-play near zero while maintaining exact cup and cone alignment on the spindle. The Pro-Torq® nut keeps the bearing properly aligned so it runs cooler and performs reliably for a longer time. The tight control of the nut on the wheel bearing adjustment helps prevent premature seal failure and extends seal and brake lining service life.

PTC (Positive Temperature Coefficient)

Usually used in reference to a conductor or electronic circuit breaker. As the temperature increases, the electrical resistance also increases.

PTO (Power Takeoff)

A gear-driven device that is attached to a driven gear inside a transmission, final dive, or transfer case to supply external power for an accessory device like a winch. This device is used on trucks, farm equipment, and industrial machines. For more information on *Auxiliary Power Take-off Drives for Agricultural Tractors*, see **J717_201912** [12].

P-Type Material

Silicon or germanium doped with boron or indium.

Pulse

A voltage signal that increases or decreases from a constant value and then returns to the original value.

Pulse Generators

An electromagnetic device that generates a voltage signal that is used to trigger the ignition control module that controls (turns on and off) the ignition current of an ignition coil using a switch transistor.

Pulse Train

DC voltage that activates on and off during a series of pulses.

Pulse Width

The amount of "on" time of an electronic fuel injector or other electrical component. Also called PWM, or Pulse Width Modulation.

FIGURE P.20 Pulse width.

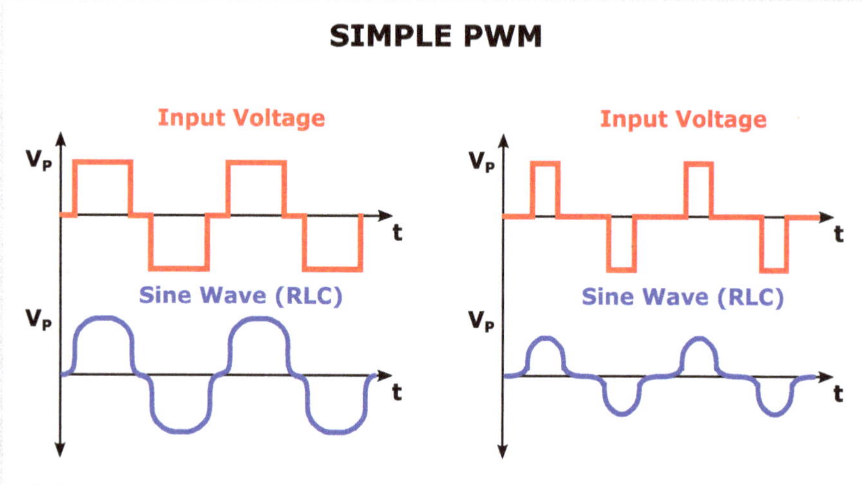

Maikon Lucian Lenz/Shutterstock.com.

Pump

A device that creates or increases fluid flow. Most oil pumps are mounted directly behind the torque converter and usually driven by the converter drive hub.

Pumping Losses

Power is lost when pumping air into and out of the engine. The throttle restricts air flow into the engine; thus, the engine creates a vacuum within the manifold. It takes power to compress the air-fuel mixture and force the exhaust out of the cylinder. These are cited as pumping losses.

Pushrod

A long hollow tube rod that connects the rocker arm on an overhead valve ICE to the lifter or cam follower.

PWM (Pulse Width Modulation)

PWM is an electrical signal that varies on/off. It shares similar characteristics with both digital and analog signals. PWM. It is digital in one aspect because it represents data in two states only: either ON or OFF high or low. However, information is represented by the amount of time that the signal stays on or off. Time on or off is variable, which gives it an analog characteristic. The units for measuring pulse width are always expressed in units of time. Time is the measure of how long the signal is high or on. PWM signal is a digital signal, usually 0 and 12 V, i.e., cycling at a fixed frequency. Varying the length of time that the signal is on produces a signal that can vary the on and off time of an output.

The ratio of on-time relative to the period of the cycle is referred to as duty cycle. Examples include solenoids, injectors, and light circuits. The units of measurement are milliseconds. Output drivers of microprocessors are types of switches that conveniently use PWM signals to operate devices in an ON/OFF state. The computer can easily vary the duration of driver opening and closing. Some OEMs use sensors which use PWM signals to transmit data. A throttle position sensor will transmit pedal position data using PWM signals. This type of data is unaffected by voltage drops encountered through long runs of wiring harnesses and multiple connectors between the sensor and controller.

REFERENCES

1. Society of Automotive Engineers, J1715_202209: Hybrid Electric Vehicle (HEV) and Electric Vehicle (EV) Terminology, SAE Publishing, Warrendale, PA, 2021.
2. Society of Automotive Engineers, J3063_202103: Active Safety Systems Terms and Definitions, SAE Publishing, Warrendale, PA, 2021.
3. Society of Automotive Engineers, J3016_202104: Taxonomy and Definitions for Terms Related to Driving Automation Systems for On-Road Motor Vehicles, SAE Publishing, Warrendale, PA, 2021.
4. Society of Automotive Engineers, J1145_201109: Emissions Terminology and Nomenclature, SAE Publishing, Warrendale, PA, 2010.
5. Society of Automotive Engineers, J1930DA_202105: Electrical/Electronic Systems Diagnostic Terms, Definitions, Abbreviations, and Acronyms Web Tool Spreadsheet, SAE Publishing, Warrendale, PA, 2021.
6. Kershaw, J.F., *High-Performance Automotive Cooling Systems* (Forest Lake, MN: CarTech, Inc., 2019), Chapter 3, ISBN:978-1-61325-504-9.
7. Society of Automotive Engineers, J1538_201504: Glossary of Automotive Inflatable Restraint Systems, SAE Publishing, Warrendale, PA, 2019.
8. Society of Automotive Engineers, J2612_202002: Internal Combustion Engines—Piston Vocabulary, SAE Publishing, Warrendale, PA, 2020.

9. Society of Automotive Engineers, J1587_200202: Electronic Data Interchange between Microcomputer Systems in Heavy-Duty Vehicle Applications, SAE Publishing, Warrendale, PA, 2002.
10. Society of Automotive Engineers, J646_201807: Planetary Gears—Terminology, SAE Publishing, Warrendale, PA, 2002.
11. Society of Automotive Engineers, J1594_201007: Vehicle Aerodynamics Terminology, SAE Publishing, Warrendale, PA, 2010.
12. Society of Automotive Engineers, J717_201912: Auxiliary Power Take-off Drives for Agricultural Tractors, SAE Publishing, Warrendale, PA, 2019.
13. Society of Automotive Engineers, Standard J1715/2_202108: Battery Terminology, SAE Publishing, Warrendale, PA, 2021.
14. Society of Automotive Engineers, Standard J2627_200908: Braking System Definitions—Truck and Bus, SAE Publishing, Warrendale, PA, 2009.
15. Society of Automotive Engineers, J1479_20121: Automotive Pull-Type Clutch Terminology, SAE Publishing, Warrendale, PA, 2012.
16. Society of Automotive Engineers, J3227_202105: Terminology and Measurement Methodology for Determining Vehicle Overlap in Offset-Frontal Crashes, SAE Publishing, Warrendale, PA, 2021.
17. Society of Automotive Engineers, J139_202002: Ignition System Nomenclature and Terminology, SAE Publishing, Warrendale, PA, 2020.

Q

Quarter Panel

The rear body panel of a car between the rear door and the trunk wraps around the wheel well. Quarter panels are typically made of sheet metal, but are sometimes made of fiberglass, carbon fiber, or sheet molding compound. It is a welded-on component of the unibody structure. Replacement requires it to be cut off the vehicle and a replacement part to be welded or bonded to the vehicle.

FIGURE Q.1 Quarter panel.

Darren Brode/Shutterstock.com.

Quench Areas

The squish area in the combustion chamber of an ICE is created by the shallow portion of the combustion chamber. The piston goes up to compress it. The squish area might also be called quench area. A knocking or detonation can occur if the top gases reach a high-enough temperature to self-ignite before the flame front reaches it. To lower detonation or spark knock, the quench area is intended to keep the air-fuel mixture below the ignition point. For more information on *Internal Combustion Engines—Piston Vocabulary*, see **J2612_202002** [1].

Quick-Release Valves (Air Brakes)

Quick-release valves are mounted near the brake chambers or components they serve. They are used throughout the air brake and other vehicle air systems to hurry release times. The standard quick-release valve features a single inlet port and two outlet ports. When air is charged to the inlet port, the valve acts as a sort of a T fitting and easily divides the air into the two outlets. However, when the gas pressure supplied to the inlet is exhausted at the applying or control valve, the air being outputted from the short release to the brake chambers is exhausted at the fast-release valve. A fast-release valve can also be used on the service gladhand supplying the trailer(s). This feature greatly reduces the discharge lag that may occur in commission braking by exhausting the service signal at a way higher speed. For more information on *Braking System Definitions—Truck and Bus*, see **J2627_200908** [2].

Quill

The front nose of a manual transmission input shaft retainer that the release bearing and clutch brake are mounted on.

FIGURE Q.2 Clutch release bearing mounted on the quill of a manual transaxle.

BACHTUB DMITRII/Shutterstock.com.

REFERENCES

1. Society of Automotive Engineers, J2612_202002: Internal Combustion Engines—Piston Vocabulary, SAE Publishing, Warrendale, PA, 2020.
2. Society of Automotive Engineers, Standard J2627_200908: Braking System Definitions—Truck and Bus, SAE Publishing, Warrendale, PA, 2009.

R

R1234yf Refrigerant

R1234yf is a colorless HFO (hydrofluoroolefin) refrigerant gas also called tetrafluoropropylene. It has an extremely low greenhouse gas damaging effect. "J2843_201907standard is required to facilitate service of contaminated mobile air-conditioning (MAC) R-134a and/or R-1234yf refrigerant systems. If the MAC system refrigerant has been contaminated such that the refrigerant in the MAC system cannot be on-site recycled and meet J2099 with equipment that meets SAE J2788, J2843, or J3030, then it should be recovered and sent for proper disposal. If a refrigerant identifier that meets J2912 or J2927 indicates that the refrigerant has been contaminated, it should be removed only with equipment that meets this standard and then disposed of by a qualified facility." For more information on this, see **J2843_201907**, **J2888_201902**, and **J2911_202002** [1, 2, 3].

R134A Refrigerant

R134a is the commonly used refrigerant for automotive applications. Since 1993 R134a has been the refrigerant for automotive; before 1993, R-12 or Freon was used for mobile. Tetrafluoroethane is a hydrofluorocarbon and haloalkane refrigerant with thermodynamic properties similar to R-12 but with insignificant ozone depletion potential and a lower 100-year global warming. It uses the formula CF_3CH_2F and a boiling point of –26.3°C at atmospheric pressure. R134a cylinders are colored light blue.

R134a will develop a minimum to maximum refrigerant charge range in which the heating, ventilation, and air conditioning (HVAC) system can maintain proper operation. Operating conditions and characteristics of the equipment will influence the optimum charge. Since these conditions and characteristics vary greatly from one application to another, careful consideration should be taken to determine the optimum R134a refrigerant charge for the HVAC system. "Specific minimum equipment performance requirements for recovery and recycling of HFC-134a that has been directly removed from, and is intended for reuse in, mobile air-conditioning (A/C) systems. It also is intended to establish requirements for equipment used to recharge HFC-134a to an accuracy level that meets section 9.0 of this document and SAE J2099." For more information on this, see **J2788_200612** [4].

Rack

A toothed gear that is straight.

Rack-and-Pinion Steering

See **Power Steering Rack-and-Pinion Gear**.

Radar

Acronym for radio detection and ranging. It can determine a vehicle speed by measuring the Doppler (speed induced) shift of the return of its transmitted frequency (think of the sound you sometimes hear of an approaching or receding train or emergency vehicle). Doppler radar could also be a selected radar that uses the Doppler effect to produce velocity data about objects at a distance. It does this by bouncing a microwave signal off a desired target and analyzing how the motion of the object has altered the frequency of the returned signal. This variation gives direct and highly accurate measurements of the radial component of a target velocity relative to the radar. Doppler radars are employed in aviation, sounding satellites, Major League Baseball Statcast system, meteorology, radar guns, radiology, and healthcare (fall detection and risk assessment, nursing or clinic purpose), and bistatic radar (surface-to-air missiles). Because of its common use by television meteorologists in on-air weather reporting, the precise term Doppler radar has mistakenly become fashionable for the sort of radar utilized in meteorology. The hottest weather radars use the pulse-Doppler technique to showthe motion of precipitation, but it is only a part of the processing of their data. So, while these radars use a highly specialized kind of Doppler radar, the term is way broader in its meaning and its applications.

FIGURE R.1 Doppler radar.

Pepermpron/Shutterstock.com.

Radial

This term explains a component or device that is arranged like rays or radii (plural of radius) in a circle. This device could have spokes, bars, lines, etc., arranged like radii.

Radial Engine

A multicylinder ICE where the pistons are arranged in a circle and drive a single crankshaft. To balance the engine, the number of cylinders is odd like three or five. It was primarily an aviation application ICE.

FIGURE R.2 Radial aircraft engine.

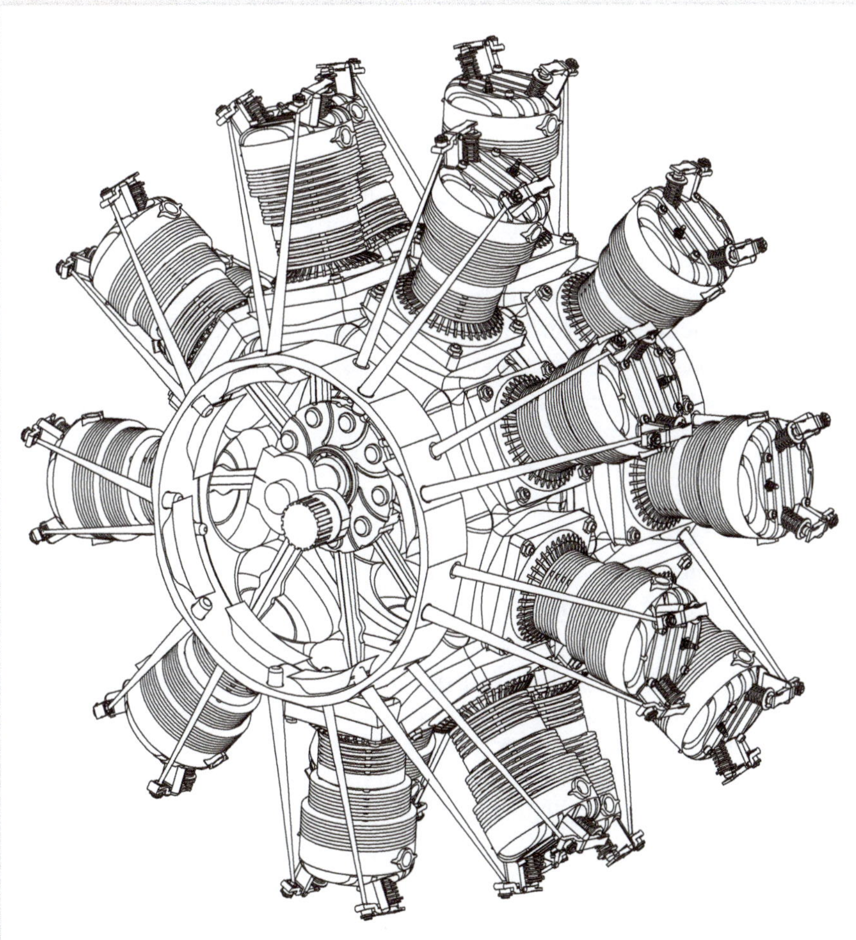

Radial Tire

Radial tire construction uses the identical body ply cords, or casing plies, as bias-ply tires, but these are laid across the tread extending from bead to bead radially; therefore, the cords find themselves parallel to every other at approximately 90° to the centerline of the tread. The cords give the tire shape and strength with the cord loops, which are loops of high-strength material inlaid with rubber, running radially from bead to bead. The cords that form up the plies have high lastingness, which could be a measure of the innate strength of a fabric. The cords resist stretching, but are still flexible under load. Plies have high strength in one direction and are flexible in other directions.

Newer, and stronger, modern cord materials, as described previously, require the utilization of fewer plies than tires made with cotton cords. So now the older term "ply rating" is predicated on the same strength as compared to the older bias-ply tires. A steel-belted radial-ply tire with a six-ply rating may have six plies within the tread but just two plies in its sidewall. Another pneumatic tire with a ply rating of 10 may only have one body ply and 4 steel belts. Having fewer plies makes the tire more flexible. The sidewalls are more flexible because the casing plies do not cross over one another.

Higher numbers of plies make a tire response to bumps harsher, but the tire can withstand punctures far better. Radial tires require a bracing layer of two or more steel belts that run horizontally beneath the tread to strengthen and stabilize the tire. The cords of the bracing layers could also be of material (rayon, polyester, fiberglass, nylon, or aramid) or steel and are placed at 12 to 15° to the circumference line. This design forms triangles where the belt cords cross over the radial cords. The stiff bracing layer connects the cord loops together to allow stability. The cord plies flex and deform only within the area above the road contact patch. There are not any heavy plies to distort, and flexing of the skinny casing generates little heat, which is definitely dispersed.

Most tires have a minimum of two belts, and a few have a further layer of Kevlar to supply puncture protection. High-speed tires also include a nylon belt cover to shield the warmth generated by the steel belt when it is subjected to higher speeds (greater centrifugal force). Without the belt cover, the tire might break apart. A properly inflated pneumatic tire runs cooler than a comparable bias-ply tire, increasing tread life. Also, a pneumatic tire has less rolling resistance because it moves over the paved surface, increasing fuel economy. The sidewalls are the lightly reinforced sides of the tire between the tread and also the bead that provide lateral strength to the tire and stop it from ballooning. SAE J2047_201911 establishes the terminology that aims to encompass all terms and definitions pertaining to the road performance of pneumatic tires designed for over-the-highway use, such as passenger car, light truck, truck and bus, and motorcycle tires. Not included are terms specific to the performance of agricultural, aircraft, industrial, and other off-highway tires. However, many terms contained in this document also apply to non-highway tires. For more information on this, see **J2047_201911** [5].

FIGURE R.3 Comparison between a radial tire and a bias-belted tire.

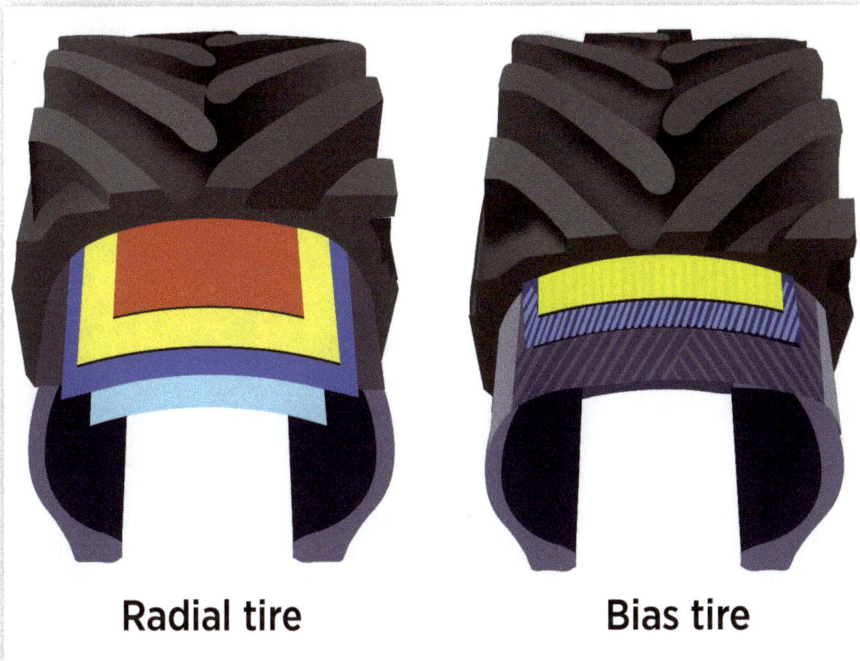

Radiation

Transfer of heat through space. This takes place only in a vacuum (absence of pressure or very low pressure) but can also take place through a solid and fluids that are transparent to wavelengths in the visible and infrared range. A small fraction of the heat transferred to the engine cylinder walls from the hot combustion gases transfers via radiation.

FIGURE R.4 Radiation is the process of emitting radiant energy in the form of waves or particles. It is the combined processes of emission, transmission, and absorption of radiant energy. The heat source has no physical contact with the object receiving the heat and the heat is transferred to this object.

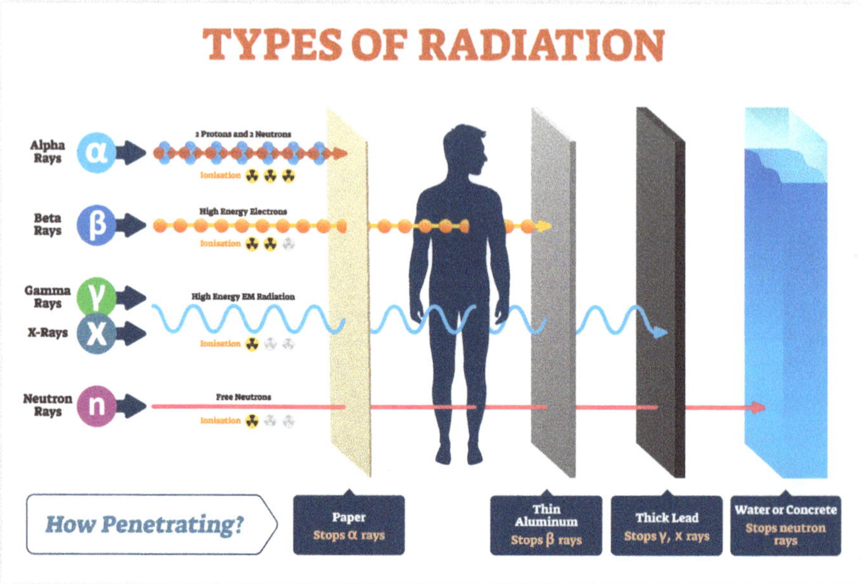

VectorMine/Shutterstock.com.

Radiator

A device that lowers the temperature of the liquid coolant after it leaves the ICE. The radiator cools the coolant which absorbs the warm energy from the engine. Radiators are constructed from header and collector tanks attached to a core with narrow passages that provide a high area. This radiator core is sometimes manufactured from stacked layers of metal sheet, pressed to make channels, and soldered or brazed together. The recent coolant is then fed into the header tank of the radiator that is located either on the highest of the radiator or along one side. From the header tank, it is distributed through the radiator core through tubes to the collector tank on the other end of the radiator. Due to the fact that the coolant passes through the radiator tubes, thanks to the other tank, it transfers much of its heat to the tubes which transfer the warmth to the fins that are between the row of tubes. The cooling fins release the warmth to the skin in the surrounding air. Cooling fins are accustomed to increasing the contact surface of the tubes, therefore increasing heat transfer efficiency. The cooled coolant is returned to the engine, and the cycle repeats. A radiator does not return the temperature of the coolant back to air temperature, but it is enough of a transfer to stay the engine from overheating.

Radiators can be either a downflow or crossflow direction. The downflow was the common variety of radiator utilized in older vehicles. The core is attached to an upper header tank and a lower collection tank. The header tank resolves the expanding coolant and provides a reservoir to manage the coolant lost. The collector tank collects coolant that is returned to the engine after the heat has been removed and which is then pumped back to the engine. The crossflow radiator was developed because the automotive industry needed a radiator design that will provide enough air flow through the newer lower radiator grills with a core that was low and wide. General Motors invented the crossflow radiator in the mid-1960s to complete these new grill designs. It absolutely was first utilized in the Chevrolet Corvette. This radiator had an inlet or header tank on one side and an outlet or collector tank on the opposite side to supply a coolant reserve and to stop aeration. In some applications a separate header tank was connected to the upper end of the outlet tank. In other applications the separate header tank was not used. To stop aeration the filler cap and therefore the overflow parts are located at the upper end of the outlet tank because this is often where air tends to gather [6].

FIGURE R.5 Automotive cooling system radiator heat exchanger.

Radiator Fan Shroud

The primary purpose of a fan shroud is to direct the air flow over the radiator, you can experience heat issues when you do not use one on a high horsepower engine. The shroud directs the incoming cool air directly at the radiator as well as help reduce previously heated air from circulating back through the radiator. Air flow through the radiator and condenser is vital for engine cooling. The shroud and air dams are responsible for between 10 and 30% of the system performance. The air shroud and dam create a low pressure behind the radiator and high pressure in front, drawing cool air through the condenser and radiator fins. The lack of a fan shroud could reduce cooling by up to 30%.

Radiator Hose

Hoses used to connect the ICE coolant inlet at the water pump to the outlet at the top of the engine. The inlet is attached to the bottom of the radiator where the cooled coolant is sucked out of the radiator via the lower radiator hose and goes out the outlet into the engine via the upper radiator hose.

FIGURE R.6 Upper radiator hose on an ICE.

J.J. Gouin/Shutterstock.com.

Radiator Hose Clamp

Hose service will depend on the type of hose being replaced and the connecting device being used. Many OEMs like GM use spring-type clamps, which can be difficult to remove and install due to tight locations. Aftermarket screw-type clamps are commonly used on robust cooling systems "AN" fitting and braided steel or Cool-flex™ hoses. Screw-type hose clamps should tighten it with an inch-pound torque wrench. The Mubea cooling hose clamp is a constant tension hose clamp designed to reduce the amount of coolant leakage from radiator and heater hoses. Screw-type clamps do not maintain a constant load on the hose connection. Some of the load is lost under certain temperature changes. The Mubea clamp exerts a relatively even and constant pressure that is maintained under varying temperature conditions [8].

FIGURE R.7 Mubea-type cooling system hose clamp.

Courtesy of John F. Kershaw Ed.D.

Radius

Straight line spreading from the center of a circle to the outside edge or from the center of a sphere to the surface.

Radius Arm or Rod

A suspension component or link used to control wheel motion in the longitudinal direction. The link is connected with a bushing on one end to the axle to the chassis of the vehicle. Radius rods, torque rods, or torsion bars help maintain the proper angles for the axle assembly and also help absorb and manage driving and brake force. The torque rod is also referred to as a radius rod by some truck manufacturers. Hendrickson refers to it as an Ultra Rod.

FIGURE R.8 Radius or torque rod.

kckate16/Shutterstock.com

Rain-Sense Wiper Systems

A wiper system that uses a sensor located on, within, and at the highest level of the windshield to detect rain droplets. This sensor is named the rain-sense module (RSM), and it controls and adjusts the time delay of the wiper supported by the quantity of moisture detected on the windshield. The wiper switch is left on the sense position all the time, and if no rain is sensed, the wipers do not swipe. The microprocessor within the RSM sends a command to the body control module (BCM). The RSM may be a triangle- or rectangle-shaped black plastic housing. The unit contains four infrared (IR) diodes, two photocells, and a microprocessor. The IR diodes generate IR beams that are aimed by four of the convex optical lenses near the bottom of the module through the windshield glass. Four additional convex lenses near the highest of the RSM are focused on the IR ray on the skin of the windshield glass and permit the two photocells to sense changes within the intensity of the IR light. When sufficient moisture accumulates, the RSM detects a change within the monitored IR beam intensity. The RSM processes the rain intensity signal and sends it over the info BUS to the BCM, which then commands a swipe of the wiper.

RAM (Random Access Memory)

A non-permanent type of memory. RAM stores temporary values and data, such as a coolant temperature reading or a manifold pressure signal that has changed. The computer then writes these values to the RAM.

Ram Air

Ram air is air that is forced to enter a moving ICE. The ram air at 0 mph would not exist and, like aerodynamic drag, would increase exponentially with speed. The faster the vehicle moves the more air is forced into the ICE intake.

Ram Duct

A portion of an ICE air intake when ram air can engine the air cleaner.

Ramps

As the camshaft rotates, the cam lobe ramps lift and lower the valve rapidly but smoothly. The lash ramp is the initial portion of the lobe that meets the lifter as the cam rotates off its base circle. The lash ramp compensates for small deflections and takes up slack in the valve train without actually lifting the valve. However, cams ground for hydraulic lifters do not need lash ramps because they operate at zero clearance without lash. As the cam continues to rotate, the lifter moves up to the opening ramp, which does the actual opening of the valve. The rate at which the valve is lifted is accelerated significantly on the opening ramp, up to a fairly constant opening speed.

Rationality

While input signals to the engine management computer are constantly being monitored for electrical open and short circuits, they are also tested for rationality. This means that the input signal is compared against other inputs and information to see if it makes sense under the current conditions.

Rationality Monitoring

Another feature of OBD II or HD-OBD that evaluates the accuracy of an electrical input signal compared with other available data. The ECT (engine coolant temperature) sensor should sense a temperature similar to the air temperature sensor at initial start-up if an engine has not operated for several hours. If there is a large difference between two different temperature sensor values, then a DTC (diagnostic trouble code) for either sensor performance or rationality, depending upon the engine manufacturer. For more information on this, see **J1930DA_202105** [9].

RBM (Resistance to Bending Moment)

A term used to express a measure of the frame strength that accounts for both the section modulus and the strength of the material used. It is, therefore, the product of both the section modulus and the yield strength.

Reactance

Resistance in a magnetic field for flux. The opposition to the flow of alternating current.

Read-Only Memory (ROM)

Computer memory that cannot be changed, and the memory data cannot be not lost because it is nonvolatile memory.

Rear Axle

The second or back axle of a car or truck.

Rearview Mirror

An interior mirror in a car or truck that provides the driver with a view out the back window or backlight.

Rear-Wheel Disc Brakes

Rear-wheel disc brake assemblies are essentially the same as front disc brakes; however, rear-wheel disc brakes incorporate a parking brake mechanism. The parking brake mechanism mechanically actuates through the parking brake lever or pedal inside the vehicle. When applied, the lever, or pedal, pulls on the parking brake cable, which is attached to the actuator lever on the parking brake mechanism. The actuator lever rotates the operating shaft, which has three indentations that contain small balls. As the operating shaft rotates, these balls move out of their indentations, forcing the thrust screw toward the piston. This moves the piston out of its bore, forcing the brake pads against the rotor. When the parking brake releases, the balls move back into their indentations and the piston retracts, allowing the pads to release the rotor.

Rebound

Downward motion of an automotive wheel, when the tire enters a hole in the road, the stored energy of the suspension spring device is released from compression due to the hole.

Recapped Tire (Retread)

A recap tire is one where the original tread is worn out but the tire carcass is still good, so it has had the remnants of the old tread ground off and a new tread (cap) applied through different bonding processes. See **Retreading a Tire**.

Reciprocating Engine

This is a reciprocated piston engine where the pistons move up and down or back and forth as a result of combustion on the top of the cylinder to turn a crankshaft in rotary motion to perform work.

Recirculating-Ball Steering Gear

The worm-and-nut gear, recirculating ball gear, or integral non-rack-and-pinion gear. The recirculating ball gear box contains a gear wheel inside a block (nut) with a threaded hole in it and kit teeth drawn outside that engage the arena shaft to maneuver the pitman arm. Within the recirculating ball gearbox, both ends of the worm shaft are supported within the housing by angular bearings, which are preloaded to scale back lateral movements called end-play and side thrust movements of the gear wheel when it is under load. A ball nut rides on the cogwheel, supported on the spiral grooves of the worm and therefore in the nut by many balls. The balls form a friction internal thread, which causes the nut to string up or down on the worm because it rotates. With rotation, the balls are rolled along the grooves, partly within the worm and partly within the nut. After they get to the top of the groove, they circulate by passing through ball-return guides to the opposite end of the nut. Ball-return guides are simply special passages through which the balls move.

FIGURE R.9 Worm-and-nut steering gear or recirculating ball steering gear or integral non-rack-and-pinion steering gear.

Nitish Rajendra gupta/Shutterstock.com.

Recombinant Battery

Recombination electrolyte technology or VRSLA (Valve Regulated Sealed Lead-Acid) batteries provide improved grid materials to allow some sealed, maintenance-free batteries to develop fully charged, open-circuit voltage of 2.1 V per cell, or 12.6 V total for a 6-cell 12-V battery. Separators made from a porous material reduce the internal resistance and generate higher voltage and current. The electrolyte is housed inside plastic envelope-type separators around the plates. The entire case is not flooded with electrolyte, which eliminates sloshing of the electrolyte or leaks from a cracked battery. There are two types AGM (absorbent glass mMat) or gel. For more information on this, see **J1715/2_202108** [7]. Also see **VRLA Battery**.

Rectifier

An electrical component like a diode that changes the alternating current into a direct current by only allowing the current to flow in one direction.

Reductant

See **DEF (Diesel Exhaust Fluid)**.

Reed Tachometer

Two rows of reeds arranged side by side, where each reed is tuned to vibrate or resonate when it is excited by a specific frequency. The reeds are arranged by their specific resonant frequency, increasing from left to right, ranging from 10 to 80 Hz. This arrangement allows for a visual display of the most dominant frequencies which fall within this range. The reed tachometer can be a helpful diagnostic tool; however, it is extremely sensitive to external inputs that are not related to the vibration concern, such as rough road surfaces, etc., and it is difficult to master its use. Due to these conditions, the reed tachometer has limited diagnostic capability. For more information on this, see **J3152_202005** [10].

FIGURE R.10 Reed tachometer.

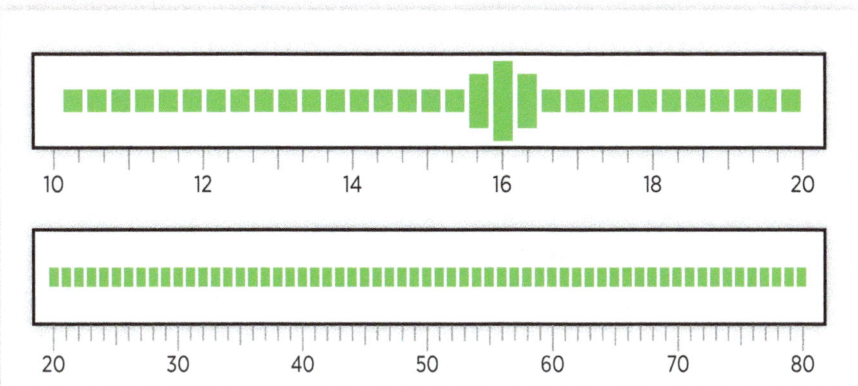

Courtesy of John F. Kershaw Ed.D.

Reed Valve

A one-way blade-like valve where the blade is spring closed and pressure opens the valve. These valves are found in air-conditioning compressors and small two-stroke ICEs.

Refer

The industry term for a trailer refrigeration unit. The technical name is TRU, or Transportation Refrigeration Unit. The TRU is a self-contained, diesel-powered cooling unit installed on a specially constructed trailer. The TRU operates like an air-conditioning system, but it is more like a freezer with a compressor powered by a diesel engine. The refrigeration principles are the same an automotive air conditioning. In this course we will cover the unique differences. The TRU is used for the transportation of loads that must be kept cool. These units include:

- Complete diesel engine with automatic starting
- Battery and charging system
- Complex refrigeration/heating unit
- Microprocessor control head

FIGURE R.11 TRU, or transportation refrigeration unit.

Reference Voltage (Vref)

A known voltage (5 or 7 V) provided by the engine management computer to the switching and variable resistance input sensors. Changes in input sensor resistance will be read by the computer. The sensor effect on the reference voltage becomes a signal used by the engine management computer as input data to judge the condition(s) at the sensor.

Refining Oil

The process of converting crude oil into gasoline, diesel fuel, and other petrochemicals.

Reflector

A non-electric panel that is mounted on a vehicle that reflects incandescent light.

Reformulated Gasoline (RFG)

The US EPA states that RFG is gasoline blended to burn more cleanly than conventional gasoline and to reduce smog-forming and toxic pollutants in the air we breathe. The RFG program was mandated by Congress in the 1990 Clean Air Act amendments. The first phase of the RFG program began in 1995 and the second (current) phase began in 2000. For more information on this, see **J312_201902** [11].

Regenerative Braking

Regenerative braking is a brake-system-based charging system employed in HEV (hybrid electric vehicle), PHEV (plug-in hybrid electric vehicle), or PEV (plug-in electric vehicle)/BEV (battery electric vehicles) from the utilization of the electrical drive motor. When an electrical motor runs in one direction, it converts current into energy accustomed to performing work as a motor. The motor runs within the other way of motoring rotation (motor runs backward) when the facility inverter turns off the facility to the motor and connects the motor output to the high-voltage batteries. At this time, the motor has become an electrical generator and converts energy into voltage. The power developed when the motor is generating electricity slows down the powertrain and becomes a braking mechanism. Once you put the motor into reverse, this slows the vehicle.

It is also an energy recovery system that slows the vehicle by converting its kinetic energy into a form that might be either used immediately or stored within the batteries. The electrical motor is using vehicle momentum to recover energy that may be lost to the brake discs as heat. This can be different than a hydraulic braking system, where the surplus mechanical energy is converted to unwanted and wasted heat by friction within the brakes. Regenerative braking can extend the lifetime of the brake friction elements. In emergency braking situations, regenerative systems can neither provide the mandatory stopping power nor handle the amount of electricity generated from a maximum-deceleration stop. Regenerative brakes also lose their stopping power and efficiency at lower speeds. For more information on *Hybrid Electric Vehicle (HEV) and Electric Vehicle (EV) Terminology*, see **J1715_202209** [12].

FIGURE R.12 Regenerative braking.

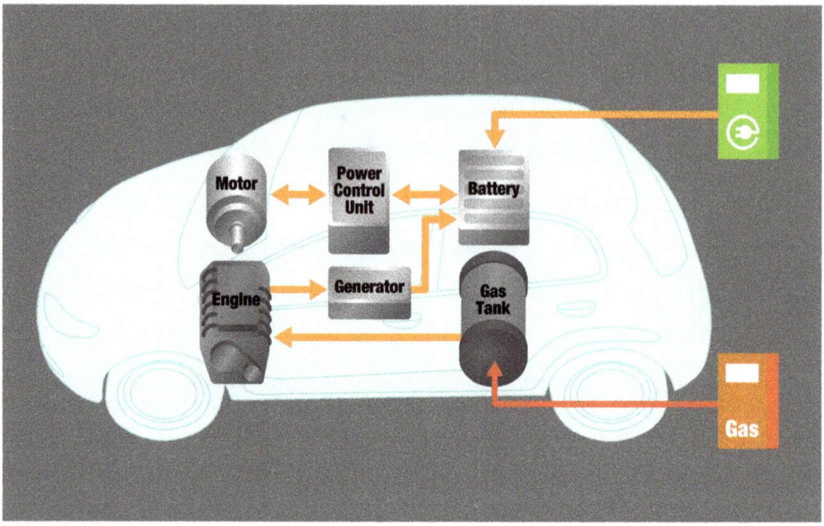

metamorworks/Shutterstock.com.

Relative Humidity

Percentage of water vapor in the air relative to the amount that could be in the air.

Relay

A device that uses current from one circuit to switch current on and off in a second circuit. It is used to allow a low-current circuit to operate a high-current circuit. Certain circuits and components are designed for high current. Circuits and components such as engine management computers that are not designed for high current can be damaged if high current is allowed to pass through them. Therefore, the use of a relay provides a level of protection. Relays are often used to control actuators that have battery voltage applied to them. The computer controls a relay, which, in turn, controls the ground side of the actuator circuit. To activate the actuator, the computer sends a signal to the relay, which closes the ground, completing the circuit and allowing battery voltage to the actuator.

FIGURE R.13 Simple electrical relay.

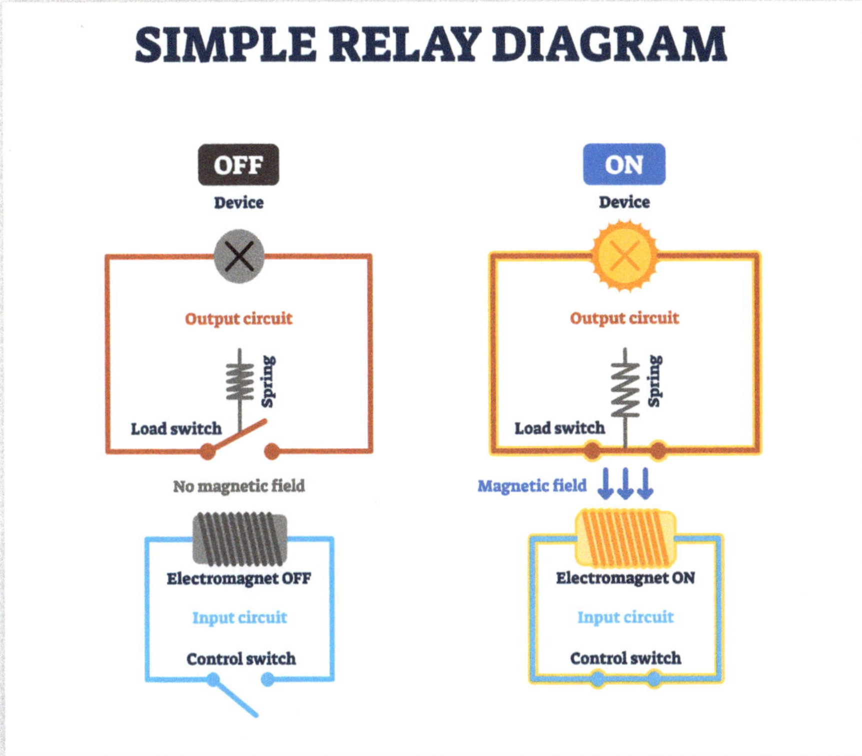

Relay Rod

A parallel automotive steering rod in a parallelogram-type automotive steering system that connects the tie rods to the steering gear pitman arm. See **Drag Link**.

FIGURE R.14 Relay rod.

Warut Wetsanarut/Shutterstock.com.

Relay Valves (Air Brakes)

Relay valves used in a truck air brake system permit a remote air signal (from application or trailer valve) to allow service braking with an air supply close to the brake chambers. This relay valve works like an electrical relay to quickly connect a bigger source to a different larger source employing a smaller signal. The valve is controlled by a sign that is connected to its service port. At the provision port, system gas pressure is offered to the valve from an air reservoir. When the relay valve receives an air signal, this pressure acts on the relay piston. When the relay piston is displaced, it modulates the air from the reservoir at its supply port to actuate the service brake chambers connected to its delivery ports. One relay valve can typically manage two to four service brake chambers.

Relay valves are employed in conjunction with ABS (anti-lock brake systems) controllers to manage brake application. The valve is typically designed in order that the signal value is just like the output value modulated to the delivery ports. When the signal pressure is dropped or relieved, the relay valve exhausts the service supply air getting back from the service chambers and a retraction spring returns the relay valve to its neutral position. The relay valve is additionally accustomed to operate parking brakes. Some relay valves use a bobtail proportioning valve at the highest to be used when no trailer is attached and operates to cut back bobtailing. There is also a pilot-operated relay valve that is used with another trailer to hasten baseline control signal application and release on long, non-towing trailers, towing trailers, dollies, and long-wheelbase tractors. For more information on this, see **J257_201907** and **J2627_200908** [13, 14].

FIGURE R.15 Truck air brake system relay valve.

Release Bearing

The clutch release bearing disengages the clutch disc from the ICE. It rides on the input shaft collar of transmission and is moved by clutch fork through either direct linkage or hydraulic cylinder. This collar can also be the front bearing retainer or quill shaft, which is a long, hollow tube extending toward the engine. The release-bearing sleeve slides on the outer surface of this bearing retainer or quill shaft. The release bearing is moved by shift linkage to release the pressure of the pressure plate that is acting on the clutch disc to break the engine torque so the transmission can be shifted. The clutch operating system moves the clutch release bearing when the driver presses down on the clutch pedal to release the pressure of the pressure plate on the friction discs. For single disc systems, the bearing is pushed against the finger. For more information on this, see **J1479_20121** [15].

FIGURE R.16 Clutch release bearing (left) and release bearing on transmission quill (right).

Palitsyn Evgenii/Shutterstock.com. snapper8S8/Shutterstock.com.

Reluctance

Reluctance is the resistance to the movement of magnetic lines of force. Using permeable (susceptible to penetration) materials within magnetic fields reduces reluctance. The permeability of matter is rated by ascribing a rating of 1 for air, generally considered to be a poor conductor of magnetic lines of force: in contrast, iron would be ascribed a permeability.

Reluctor

A circular metal ring with a number of segments around the circumference and with one missing or a gap. It is used with a Hall-effect or electromagnetic sensor. Current flow through Hall-effect material is altered in the presence of a magnetic field. Hall-effect material is a semiconductor that changes current conductivity characteristics in the presence of a magnetic field. When you apply a current to Hall-effect material, no conduction occurs. However, in the presence of a magnetic field, the material will conduct current. The electrical signal output from the sensor material is analog, but circuits within the sensor will convert and amplify the rising and falling voltage into a square-shaped electrical waveform. The opening in the reluctor allows a change in the magnetic field as there is no path and the current will also be reduced.

FIGURE R.17 Reluctor pulley.

Fouad A. Saad/Shutterstock.com.

Remolding Tires

See **Retreading a Tire (Recap)**.

Remote Start System

A system that allows starting the engine from a distance by using a remote control. Most factory-installed remote start systems use the same key fob that it used to unlock the doors. Once the engine is started, you have a limited amount of time before it will shut off before you enter the vehicle and engage the start switch.

Replenishing Port (Master Cylinder)

The replenishing port behind each piston primary cup seal is open when the brakes are not applied. The replenishing port is larger than the compensating or vent port and provides a passage between the fluid reservoirs and the low-pressure chambers of the master cylinder. The replenishing ports keep the low-pressure chambers filled with brake fluid that will be required when the brakes are released.

Resistance

Resistance opposes current flow or the movement of electrons through a circuit. All electrical devices and wires have some resistance. Materials with low resistance are called conductors; materials with high resistance are called insulators. As resistance works to oppose current flow, it changes electrical energy into some other form of energy.

$$\text{Formula } R = \frac{E}{I} \tag{R.1}$$

where

R is resistance
E is the voltage
I is the current

Resistance Factors

Determine the resistance of a conductor by a combination of the following:

- Atomic structure: The number of free electrons, the more free electrons the less resistance it offers to current flow.
- Length: The longer a conductor the higher the resistance.
- Width (cross-sectional area): The larger the cross-sectional area of a conductor the lower the resistance.
- Temperature: The higher the temperature the higher the resistance. There are a few materials whose resistance go down as temperature goes up.

The conductor condition can also have an effect on its resistance. Broken strands of wire, corrosion, and loose connections can cause the resistance of a conductor to increase.

Resistors

Devices used to provide specific values of resistance in electrical circuits. A common type is the carbon-composition resistor. These are available in many specific resistance values. They are generally marked with colored bands. These bands make up a code that expresses each resistor value. The size of the resistor determines how much heat the device can dissipate and, therefore, how much power it can handle.

Resistor-Type Spark Plug

A spark plug that has a resistor in the center electrode to reduce the inductive portion of the spark discharge, which reduces radio frequency interference. For more information on this, see **J139_202002** [16].

Resolver

Resolvers as position and speed sensors are widely used in many areas including the automotive industry, especially in the application of an electric drive for a hybrid electric vehicle. One of the disadvantages of using resolvers is that their complexity usually leads to quality issues during development that requires engineering resources and time to solve, which could be better served in solving the complex system issues of improving the hybrid system functionality.

"This paper describes the creation of a full resolver system simulation that can be used to prevent development issues, improve the hybrid system DFMEA, and act as a catalyst for solving development issues when they do happen. The system simulation model utilizes an integrated systematic approach using co-simulation between several commercial simulation tools to simulate the resolver's electromagnetic characteristics using a transient solver of finite element method, the excitation and termination circuits using an electric/electronic circuit simulator, and the behavior model of a Type II tracking closed loop algorithm to simulate the dynamics of the R/D IC using MATLAB SIMULINK™ The simulation model is verified with test results and then used to predict the behavior during failure modes." [18]

Resolving Center

Center of gravity (CG) and body geometry-defined resolving centers used in vehicle dynamics and aeronautics, respectively, are not satisfactory for road vehicle aerodynamics applications. A large portion of automotive aerodynamics development testing is performed before the vehicle CG is known. The CG location can also vary significantly with vehicle option content and loading. Relating the axis center to the body geometry is also problematic when major body geometry changes are explored during wind tunnel tests. These situations are avoided by placing the resolving center at ground level, positioned at mid-wheelbase and mid-track. An added advantage of this location is the direct translation of aerodynamic loading to tire contact patch ground reactions. For more information on this, see **J1594_201007** [17].

Resonance

The intersecting point where the frequency of an unbalanced rotating object intersects with the natural frequency of the support mechanism like an automotive suspension. This causes the suspension to vibrate. For more information on this, see **J3152_202005** [10].

FIGURE R.18 When two unbalanced forces intersect, it is call resonance.

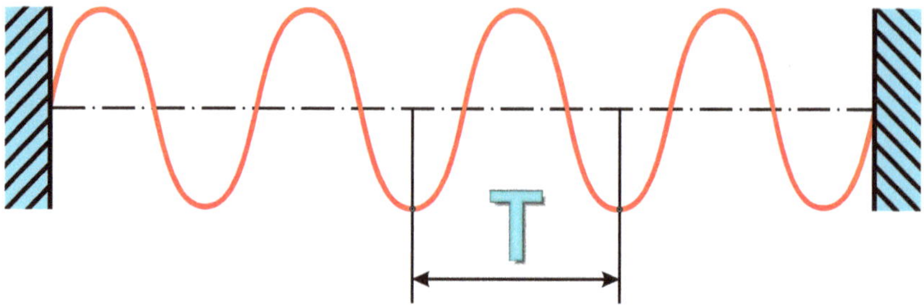

Photobond/Shutterstock.com.

Resonator

A silencer that is similar to a muffler and is usually connected in the exhaust stream before the muffler. The resonator helps to quiet the sound of combustion before the exhaust gases reach the muffler.

Retarder

See **Electric Brake Retarder**.

Retread Identification

In January 1971, the US Department of Transportation established a requirement in 49 CFR part 574 for a tire identification number (TIN) that must be labeled on one sidewall of each tire that is newly manufactured or retreaded. The purpose of the TIN is to facilitate notification of purchasers of defective or noncompliant tires. Federal law requires all tire casings that are retreaded to be branded with a retread manufacturer plant code every time that particular casing is retreaded. When a tire casing is retreaded, each tire is branded with an "R," which stands for retread; the manufacturer's plant code; and then the two-digit week and two-digit year it was retreaded. For example,

R_ABC_0117

- R indicates that it is a retreaded tire.
- ABC is the retread manufacturer plant code.
- 0117 represents the first week of 2017.

For more information on *Tire Performance Terminology*, see **J2047_201911** [5].

Retreading a Tire (Recap)

Retreading is a tire remanufacturing process that replaces the tread on worn tires. Retreading is not remolding, which is a higher quality process. There are two types: cold retreading and hot retreading. In cold retreading, molds are not used; instead, they are replaced by tire liners and use heat >212°F (100°C). Hot retreading uses a mold where the tire is above 150°C. This process is used when you need to get the right shape of the tire. When you vulcanize tires at >302°F (150°C), it is possible to change the threads and tire shape, which is the purpose of hot retreading. The retreading process is as follows:

- Tires go through an inspection process to find out if the tire is suitable for a retread. Tires go through a laser digital shearography, where a laser scans through the whole tire to see if the intricate portions have been compromised before retreading.
- Tire buffing is used to take out or cutout portions of the tires to remove needless elements and improve tread flexibility. This process can remove portions of the tread between the casing and the grooves of the treads. A computerized buffer is used by a skilled technician.
- Repairing is used to find out the cause of the damages and treating this damage in a way to avoid any further failure in the future. For example, small holes in the tires are identified by passing high-voltage electricity through the tire. This also gives the tire the ability to regain its strength.
- Tire material is added to the casing through the aid of the computerized tire building machine to improve the natural properties.
- The tire is heated before checking the level of the tire. This is done to find out the quality of the rubber and to verify whether it is still vulnerable or not.
- The final inspection determines if the tire is prepared for road service. A series of inspections are done to look for any imperfections. The tire goes through two tests: visual and tactile inspection, and inflation test. After all the tests are done they are then sent to the first test of shearography where, if there are any minute errors, they may be identified.

For more information on *Tire Performance Terminology*, see **J2047_201911** [5].

Reverse

The gear position in a manual or automatic transmission that reverses the gear motion to allow a vehicle to move in rearward direction.

Reversible Kingpin

A trailer kingpin that can accommodate different fifth-wheel assemblies.

Reynolds Number

A quantity with no dimensions utilized in mechanics accustomed to help predict airflow patterns. At low Reynolds numbers, flows tend to be dominated by laminar (sheet-like) flow; while at high Reynolds numbers, turbulence results from differences within the air speed and direction, which can sometimes intersect or perhaps move counter to the direction of the airflow. The Reynolds number applications range from liquid flow in a very pipe to the passage of air through an engine cooling radiator. It is accustomed to predict the transition from laminar to flow. The predictions of the start of turbulence and also the ability to calculate scaling effects may be accustomed to help predict airflow behavior. The Reynolds number may be a ratio indicating the link between viscous and inertial forces within the air which will be accustomed to predict the transition from laminar to flow. This number and related concepts are applied to studying flow in systems. The speed at which the air flows past an object varies with the distance from the object surface, the position of the radiator, fan, and shroud is important to attain the most cooling. The region surrounding an object where the air speed approaches zero is understood because the physical phenomenon. It is here that surface friction most affects flow; irregularities in surfaces may affect the physical phenomenon and disrupt airflow.

RFG

See **Reformulated Gasoline**.

RFI (Radio Frequency Interference)

Form of electromagnetic interference created in the ignition secondary circuit which disrupts radio and television transmission.

Rheology

Physics that deal with the deformation, viscosity, and flow of matter, especially the non-Newtonian flow of liquids and the plastic flow of solids.

Ride Height Adjustment

Ride height is the height of clearance the car has between the underside of the car and the road. The ride height has a control on the vehicle center of gravity and on its performance when cornering, braking, or any shift in weight.

Ridge Reamer

A tool used to remove the ridge at the top of each cylinder bore.

Right-Hand Rule

States that if you wrapped your hands around a current-carrying conductor together with your thumb pointing within the direction of current flow (positive to negative), your curved fingers would point within the direction of the lines of force. It is important to notice at this time that we use the traditional theory of current positive to negative or ground (+ to –), we use the right-hand rule to see the direction of the magnetic flux lines.

FIGURE R.19 The right-hand rule.

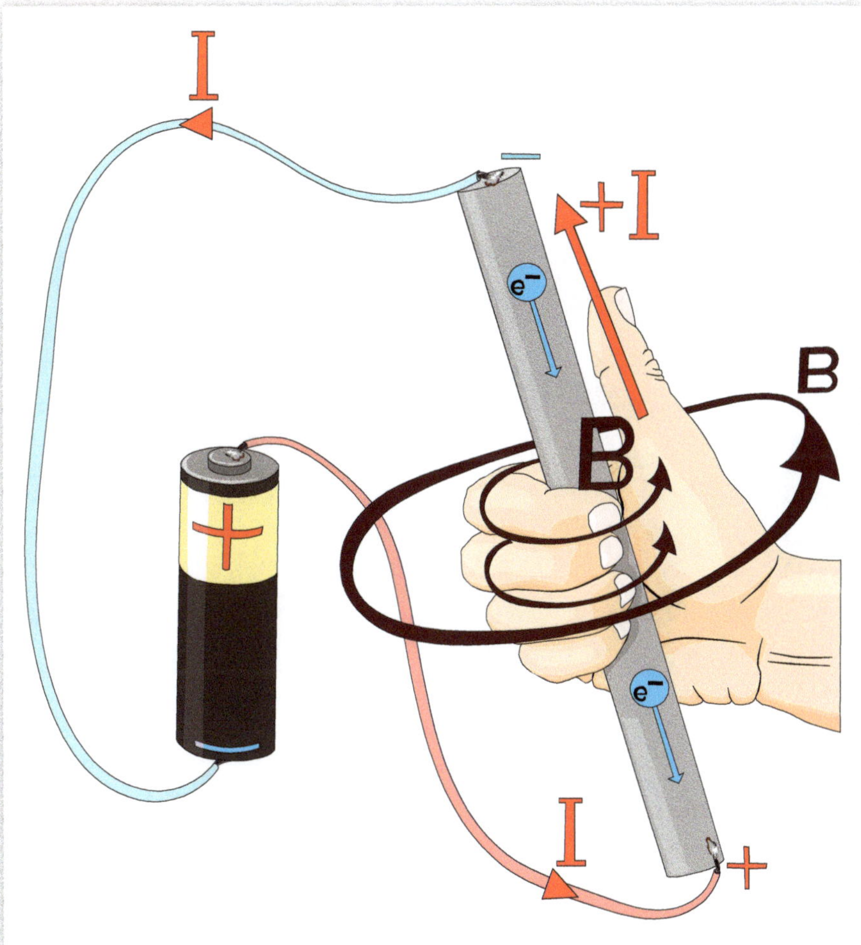

Ring Gap

The opening or gap in the circular or annular shape of an ICE piston ring that allows it to be installed on the piston. To check the piston ring end gap, begin by fitting the rings into the cylinder bores and measuring the space between the ends. Fit a ring into the cylinder, invert a piston and use it to squarely push the ring down into the cylinder within the area of ring travel, and then use a feeler gauge to measure the gap between the ends of the ring. The ring gap is usually listed in the OEM specifications as a tolerance value that corresponds to piston size. You can expect to see about three thousandths of gap for each inch of piston diameter.

Ring and Pinion

The final drive gears in a car or truck. The pinion is the smaller of the two gears driving the ring gear, which is larger in reduction.

FIGURE R.20 Ring-and-pinion final drive type bevel gears.

Pixel B/Shutterstock.com.

Rise Time

The time for the output of an ignition coil to rise from 10 to 90% of its maximum output.

Road Draft Tube

A vent tube that vents blow-by gases from the ICE crankcase into the atmosphere.

Rocker Arm Geometry

Mechanics of the movement of the rocker arm to open the poppet valve.

Rocker Arm Ratio

In overhead valves and overhead camshaft engines with rocker arms or followers, gross and net valve lift are greater than lobe lift because of the rocker arm ratio. One side of the rocker or follower is longer than the other so that the valve lift is multiplied.

Roll Bar (Cage)

A tubular metal bar running up the sides and across the top of a vehicle, like those used for racing or off-road applications. It strengthens the vehicle frame and protects occupants should the vehicle overturn.

Roll Center

It can be defined as the pivot around which a body roll will occur and the point at which the lateral forces transfer from the axle to the spring mass. It is the point where the cornering forces within the suspension react to the vehicle body. There are two definitions of roll center. The geometric (or kinematic) roll center that is dictated by the suspension geometry, and might be found using principles of the moment center of rotation. A force-based roll center, in line with the SAE definition, is the point within the transverse vertical plane through any pair of wheel centers at which lateral forces are also applied to the sprung mass without producing suspension roll.

The lateral location of the roll center is usually at the centerline of the vehicle when the suspension on the left and right sides of the car are mirror images of every other. The roll center can only be valued when the vehicle center of mass is additionally a part of the calculation. When there is a difference between the positions of the middle of mass and the roll center, an instant arm is formed. When the vehicle experiences angular velocity because of cornering, the length of the instant arm, combined with the stiffness of the springs and possibly the sway or stabilizer, defines what proportion the vehicle will roll. For more information on this, see **J1594_201007** [17].

Roller Chain

Drives a single or double overhead camshaft (OHV). The roller chain is made of links, pins, and rollers. Each link has two plates held together by pins and separated by a roller around the pin. The length of the rollers determines the chain width. The roller chain is stronger, but noisier. Many OHC and heavy-duty overhead valve engines use the roller chain. Double roller chains are used in high-performance or diesel applications.

Roller (Tappet) Lifter

A valve lifter with a roller on the bottom that contacts the cam lobe to provide a lower resistance in rotation force.

FIGURE R.21 Roller lifter.

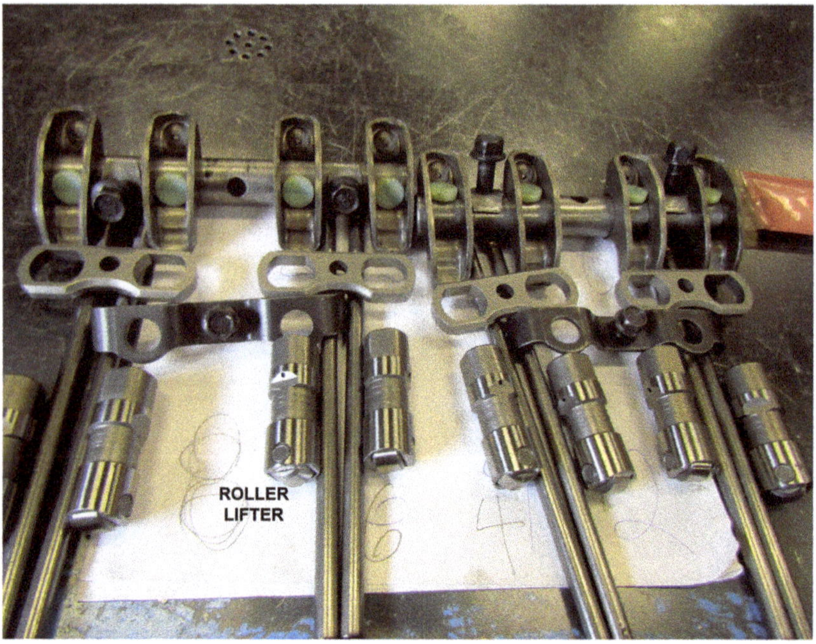

Courtesy of John F. Kershaw Ed.D.

Rollover

Overturning of a moving vehicle can take place under the following conditions: Traveling at high speed on a curved road; severe cornering maneuver; traveling on collapsing road and suddenly providing steering input for a vehicle with a low level of roll stability; losing control due to a rapid decrease in friction, such as driving on an icy road; laterally sliding off the road as in a tripped rollover; or sliding from a cliff. Factors that affect rollovers are tires, vehicle characteristics, environmental conditions, and drivers. Rollovers can take place on a flat road, cross-slope road, or off road. The lateral acceleration at which a rollover begins is called the rollover threshold. It would be 0.94g for a top-heavy vehicle. A Corvette, which has a very low center of gravity (CG), the rollover threshold would be 1.2 to 1.7g. The rollover threshold is measured in g-force (9.8 m/sec or 33 ft/sec). Rollover can be divided into two categories: Tripped rollover and untripped rollover.

- Rollover threshold is the lateral acceleration at which a rollover begins
- t is the width of the vehicle tread
- h is the CG
- y is the rollover threshold
- Tread width 65 in. = t
- CG, 34.65 in. = h = 2h = 69.3
- Ground clearance = 7 in.

$$\frac{a_y}{g}=\frac{t}{2}+\frac{t}{2}+\varnothing\frac{h}{h}$$

Rollover threshold equation derived from previous equation:

$$y=\frac{t}{2h}=\frac{65}{69.3}=0.94\,g \qquad \textbf{(R.2)}$$

For more information on this, see **J1594_201007** [17].

Rollover Threshold

The lateral acceleration at which a vehicle rollover begins. For more information on this, see **J1594_201007** [17].

ROM

See **Read-Only Memory**.

RON (Research Octane Number)

This is one of two different methods of determining the gasoline octane number for anti-knock properties: the research measure (RON) and the motor octane number, or MON. A single-cylinder CFR (cooperative fuel research) engine is used with a test fuel where the beginning of pinging is observed. It is commonly known that branched-chain paraffins have a better measure than their straight-chain analogs in order that one in every foremost operations of a refinery is to convert straight-chain hydrocarbons to branched-chain alternatives. Arbitrarily, pure isooctane (2,2,4-trimethylpentane) is accorded a number of 100, while the straight-chain paraffin n-heptane is given an octane rating of zero. So the gasoline fuel sample with the identical anti-knock quality as that of a combination containing 90% isooctane and 10% *n*-heptane is alleged to own a RON of 90. Gasoline is a mix of mostly branched-chain paraffins, with suitable additives, to relinquish a RON within the range 90–100. Antiknock was added to delay the onset of self-ignition. Due to the toxicity of organic lead, it was replaced by less harmful additives like methyl tertiary butyl ether (MTBE). For more information on this, see **J312_201902** [11].

Roots Type (Blower)

The roots-type supercharger uses twisted blade rotors that turn in the opposite direction to each other. The gap between these rotors helps to determine the boost pressure. The pressure ratio (i.e., the pressure before compression to the pressure after compression of a mechanical supercharger) does not depend on its rotation speed. However, the volumetric flow (volume of air) is dependent on the blower rotational speed. If the engine and the blower have the same speed ratio, a bigger blower that has more volume will make more boost than a smaller one on the same size engine. If the engine displacement increases, the boost will decrease if the blower speed and blower size remain the same. With a specific blower size and a given size engine, the boost may be increased by running the blower faster by using a gearset to overdrive the blower. It can be decreased by running it slower.

FIGURE R.22 Roots-type blower.

Rotor
A hub with a number of radiating airfoils that is rotated in an approximately horizontal plane to provide the lift for a helicopter. Rotating assembly in a turbine. The rotating part of an ICE ignition distributor that opens and closes a path to ground for a step-up transformer or ignition coil used to generate a large voltage to jump the gap in a spark plug. The power driving member of a Wankel engine.

Rotor Gap Voltage
The rotor-gap voltage test measures the voltage required to jump the gap (0.030 to 0.050 in. or 0.8 to 1.3 mm) between the rotor and the inserts (segments) of the distributor cap. Select the display (parade) scope pattern and remove a spark plug wire using a jumper wire to provide a good ground connection. For more information on this, see **J139_202002** [16].

RPM (Revolutions per Minute)
A measure of how fast an object is rotating around an axis.

Run-Flat Tire
Allows a driver to keep up vehicle control if a tire suffers a rapid pressure loss when in motion. They are also called zero pressure tires or extended mobility tires that enable the driving force to continue the journey within the specified speed and distance limits (about 50 mph [80 kph] and a minimum of 50 miles 80 [km]). This alleviates the necessity to exchange the wheel on the side of the road or in an unsafe area. But like any tire, the run-flat tire cannot still be driven if the sidewall has been compromised or blown out. In these instances, the tire must be changed. A tire pressure monitoring system is generally mandatory for all run-flat technology applications to watch the drivability of the run-flat tires of the vehicle. For more information on *Tire Performance Terminology*, see **J2047_201911** [5].

Rust Inhibitor
A type of corrosion inhibitor designed to slow down the oxidization process in metals which have been exposed to oxygen and water and can begin to rust.

RVP (Reid Vapor Pressure)
A method used to determine the vapor pressure of gasoline and other petroleum products. It is used as an indicator of the volatility of gasoline.

RWD (Rear-Wheel Drive)

When the engine is mounted longitudinally and drives the rear wheels through a driveshaft or propeller shaft to a hypoid bevel gear final drive assembly.

FIGURE R.23 An RWD arrangement.

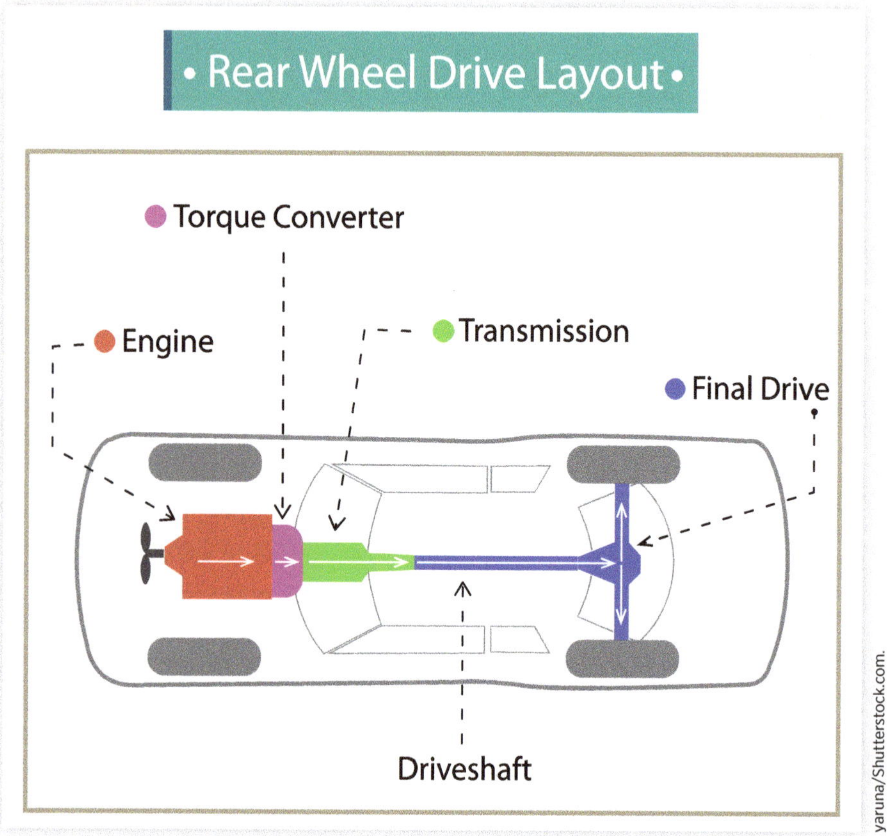

REFERENCES

1. Society of Automotive Engineers, Standard J2843_201907: R-1234yf (HFO-1234yf) Recovery/Recycling/Recharging Equipment for Flammable Refrigerants for Mobile Air-Conditioning Systems, SAE Publishing, Warrendale, PA, 2019.
2. Society of Automotive Engineers, Standard J2888_201902: R-1234yf Service Hose, Fittings and Couplers for Mobile Refrigerant Systems Service Equipment, SAE Publishing, Warrendale, PA, 2006.
3. Society of Automotive Engineers, Standard J2911_202002: Procedure for Certification that Requirements for Mobile Air Conditioning System Components, Service Equipment, and Service Technician Training Meet SAE J Standards, SAE Publishing, Warrendale, PA, 2020.
4. Society of Automotive Engineers, Standard J2788_200612: HFC-134a (R-134a) Recovery/Recycling Equipment and Recovery/Recycling/ Recharging for Mobile Air-Conditioning Systems, SAE Publishing, Warrendale, PA, 2006.
5. Society of Automotive Engineers, Standard J2047_201911: Tire Performance Terminology, SAE Publishing, Warrendale, PA, 2019.
6. Kershaw, J.F., *High-Performance Automotive Cooling Systems* (Forest Lake, MN: CarTech, Inc., 2019), ISBN:978-1-61325-504-9
7. Society of Automotive Engineers, Standard J1715/2_202108: Battery Terminology, SAE Publishing, Warrendale, PA, 2021.
8. Kershaw, J.F., *GM 6.2/6.5 Liter Diesel Engines* (Forest Lake, MN: CarTech, Inc., 2020), 11-13, ISBN:978-1-61325-560-5
9. Society of Automotive Engineers, Standard J1930DA_202105: Electrical/ Electronic Systems Diagnostic Terms, Definitions, Abbreviations, and Acronyms Web Tool Spreadsheet, SAE Publishing, Warrendale, PA, 2021.
10. Society of Automotive Engineers, Standard J3152_202005: Taxonomy and Guidelines for Design, Fabrication, and Installation of Brake NVH Fixtures for Inertia Dynamometer Testing, SAE Publishing, Warrendale, PA, 2011.
11. Society of Automotive Engineers, Standard Automotive Gasolines, SAE Publishing, Warrendale, PA, 2019.
12. Society of Automotive Engineers, Standard J1715_202209: Hybrid Electric Vehicle (HEV) and Electric Vehicle (EV) Terminology, SAE Publishing, Warrendale, PA, 2021.

13. SAE, Society of Automotive Engineers, Standard J2627_200908: Braking System Definitions—Truck and Bus, SAE Publishing, Warrendale, PA, 2009.

14. Society of Automotive Engineers, Standard J257_201907: Brake Rating Power Requirements—Truck and Bus, SAE Publishing, Warrendale, PA, 2019.

15. Society of Automotive Engineers, Standard J1479_20121: Automotive Pull-Type Clutch Terminology, SAE Publishing, Warrendale, PA, 2012.

16. Society of Automotive Engineers, Standard J139_202002: Ignition System Nomenclature and Terminology, SAE Publishing, Warrendale, PA, 2020.

17. Society of Automotive Engineers, Standard J1594_201007: Vehicle Aerodynamics Terminology, SAE Publishing, Warrendale, PA, 2010.

18. Shao, L., Maki, K., Saikalis, G., Moore, J. et al., "Simulation and Analysis of a Resolver for HEV Electric Drive Application," *SAE Int. J. Passeng. Cars – Electron. Electr. Syst.* 2, no. 1 (2009): 383-388, doi:https://doi.org/10.4271/2009-01-131.

S

Sac Cavity (Diesel Fuel Injector)

This type of diesel fuel injector has a cavity directly below the discharge orifice in the nozzle. Sac volume is the amount of diesel fuel that is retained in the sac and is located between the valve seat (fuel shutoff point) and the entrance to the final injector metering orifice. A SAC-style nozzle has a teardrop shape. The sac feeds into the injector and a small needle at the end of the valve closes the cavity. The sac cavity keeps a small amount of diesel fuel in the tip that cools the nozzle. SAC-style nozzles are found in marine applications that have lower emission standards. When the needle valve moves, the fuel cavity immediately fills with fuel and provides a crisp, clean injection. These nozzles tend to have more idle noise and more of a knock sound.

SAE (Society of Automotive Engineers)

SAE, or the Society of Automotive Engineers, is a nonprofit educational and scientific organization dedicated to advancing mobility technology to serve humanity. Engineers and scientists, who are SAE members, develop technical information on all varieties of self-propelled vehicles including automobiles, trucks and buses, off-highway equipment, aircraft, aerospace vehicles, and marine, rail, and transit systems. SAE disseminates this information through its meetings, books, technical papers, magazines, standards, reports, professional development programs, and electronic databases.

SAE/ATA J1587/J1939/J1708 Codes and Protocols

SAE/ATA diagnostic codes have been adopted by all the diesel engine/electronics engine original equipment manufacturers (OEMs). SAE standard J1587 covered common software protocols in electronic systems. SAE standard J1708 covers common hardware protocols, and the more recent SAE standard J1939 covers both hardware and software protocols.

The acceptance and widespread usage of these protocols enables the interfacing of electronic systems manufactured by different OEMs on truck and bus chassis, as well as provides any manufacturer's software to at least read the electronic systems of other OEMs. The controller area network (CAN) was not originally stated in J1939, which covered cars and tractor-trailer units and with some dual- and triple-use 8-bit addresses assigned by the SAE J1939 board. Since 2000, the CAN has been included, the chipset for J1939 has been clocked faster, and 16-bit addresses parameter group number (PGN) have replaced 8-bit addresses. J1939 attempts to define standard PGNs to encompass a wide range of automotive, agricultural, marine, and off-road vehicle purposes. A range of PGNs (00FF0016 through 00FFFF16, inclusive) is reserved for proprietary use. PGNs define the data which are made up of a variable number of suspect parameter number (SPN) elements defined for unique data. For example, there exists a predefined SPN for engine RPM [2, 3, 4].

SAE EV (Electric Vehicle) Charging Standards

"There are three levels of recharging for a BEV (Battery Electric Vehicle) or EV (Electric Vehicle). The requirements to charge a mid-sized EV is comparable thereto of an electrical stove connected to a 40 A, 240 VAC circuit developing up to 9.6 kW. Most mid-sized EVs carry a 6.6 kW on-board charger designed for a 4- to 5-h charge (6.6 kW comes by multiplying 220 V by 30 A). On-board chargers are limited by cost, size, and thermal issues. With the supply of three-phase AC power in most European residences, on-board chargers may be made smaller than with a two-phase system." For more information on *Hybrid Electric Vehicle (HEV) and Electric Vehicle (EV) Terminology*, see **J1715/2_202108** [1].

FIGURE S1 EV charging stations.

Chesky/Shutterstock.com.

SAE Horsepower

Net horsepower available from the engine with specified induction specifications and engine speed:

- Rating guaranteed within 57%
- Corrected to SAE standard ambient conditions
- Air temperature 85°F (29.4°C)
- Elevation 500 ft (159.4 m)
- Dry air density 0.0705 lb/ft^3 (11.29 g/m^3)

May be without accessories required for the application (compressor, fan, generator, etc.)

SAE Mobilus

SAE Mobilus is a database of all vehicle standards. SAE International standards are internationally recognized as the best scientific data to globally optimize the processes, practices, and products that advance technology in the mobility industry.

SAE Viscosity Grade

Expressed as a number and refers to oil resistance to flow. Typical oil viscosity grade numbers are 0 W, 5 W, 10 W, 15 W, 20 W, 20, 30, 40, and 50. Lower viscosity grade numbers indicate thinner oils that flow more easily. Higher numbers indicate thicker oils with greater resistance to flow.

Sag Frame Damage

This damage takes place when the frame rails are bent downward in relation to the rail ends. The following are some causes:

- Excessive loads
- Uneven weight distribution
- Holes drilled in frame flanges
- Too many holes drilled in the frame web
- Holes drilled too close together in the frame web
- Welding on the frame
- Cutting holes in the frame with a torch
- Cutting notches in the frame rails
- Fire
- Collision damage

SAI (Steering Axis Inclination) Angle

The axis around which the wheel assembly swivels as it turns to the right or left is called the steering axis and is formed by drawing a line through the center of the upper and lower pivot points of the suspension assembly. When seen from the front of the vehicle, it is tilted inward at the top. The angle formed between this line and the vertical provides the SAI angle. SAI is a diagnostic alignment angle and, when not within specifications, indicates a fault. It is not adjustable. If the camber angle is correct, then the SAI should also be correct, that is, it should match specifications. If it does not meet OEM specifications, this almost always is a sign of a bent component. If the SAI angle, camber angle, and included angle are below factory specifications, the possible causes are a bent control arm, strut tower out at the top, bent spindle, and/or strut.

FIGURE S.2 The SAI angle is a diagnostic angle and when not within specifications indicates a fault.

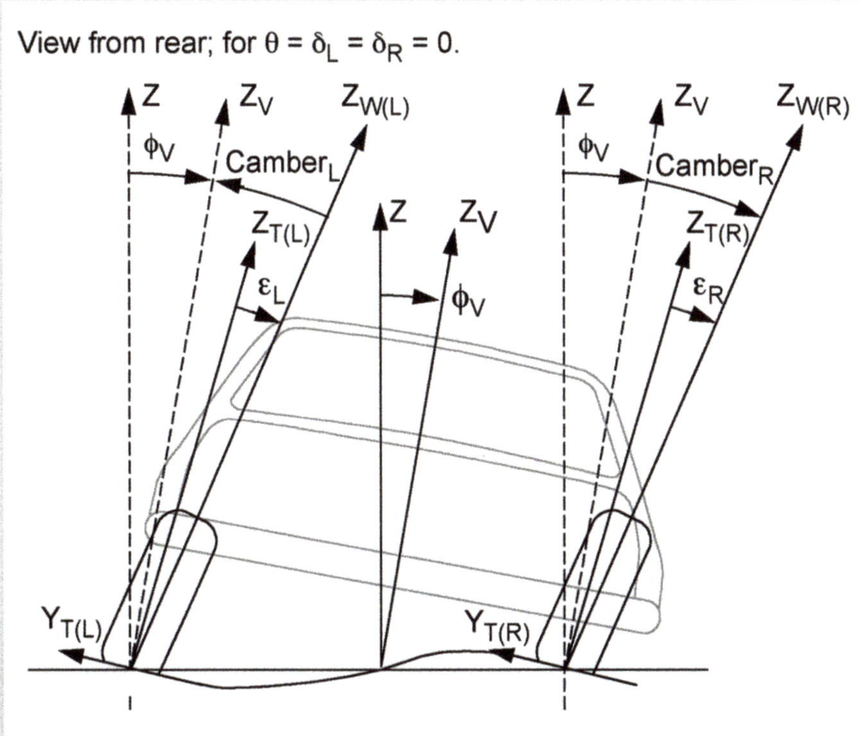

Satellite Radio

A type of radio system that operates off a radio transmitter orbiting in outer space at an altitude of about 22,000 miles (35,000 km) above Earth's surface. The satellite broadcasts use an FCC (Federal Communications Commission)-allocated broadcast frequency in the S-band (2.3 GHz). Also called digital audio radio service (DARS).

Saturated Switch Fuel Injector Waveform

The waveform of a saturated switch fuel injector that shows a voltage trace that almost drops to zero as the injector is turned on and a 30-V to 50-V inductive voltage spike as the injector is turned off.

Saw Cut Loop Detectors

Detectors used for vehicle detection are buried within the pavement at this intersection as seen by the oblong shapes of loop detector sealant on the underside. Inductive loops are placed in an roadbed to detect vehicles as they withstand the force field of the loop. The best detectors simply count the amount of vehicles during a unit of your time (60 sec) that leave out the loop, while more refined sensors estimate the speed, length, and sophistication of vehicles and also the distance between them. Loops will be placed in a very single lane or across multiple lanes and that they work with very slow or stopped vehicles additionally as vehicles moving at high speed.

Scaling a Race Car

Method of weighing a race car. All the fluids, the driver's weight, and all ride heights should be set in the car before the actual scaling begins. The car should also be scaled on a flat surface, and the scale should also be calibrated, if necessary, to the manufacturer's specifications. Scaling the car keeps track of the weight of the car and keeps track of modifications made.

S-Cam

S-cam brakes are named because the shape of the brake camshaft forms the letter S. This was the most widely used air brake model. When the brake chamber pushrod is stroked by a brake application, the S-shaped camshaft rotates and forces two rollers outward at one end of the shoes to push the brake shoes against the rotating drum. The opposite side of the shoes pivots around two roller pins.

FIGURE S.3 S-cam brakes.

Siwakorn1933/Shutterstock.com.

S-Cam Brake System

When the driver steps on the treadle, the application valves send air signals to the quick-release valve and relay valve/anti-lock braking system modulators. These devices then connect the air supply to the flexible diaphragms of the service brake chambers that forces a pushrod assembly forward against the force of a coil return spring to move an S-cam to apply the disc or drum brakes. When the air is released from the chamber, the coil return spring pulls the pushrod back to the released position. The brake chamber is mounted on a bracket welded to the axle housing assembly. The chamber pushrod is connected by a yoke and a clevis pin to a slack adjuster. The slack adjuster acts as a lever to multiply the force from the diaphragm and pushrod, which in turn rotates the S-cam brake shaft. The S-cam is rotated to permit two cam rollers to be raised. This action, in turn, causes the brake shoes to be forced outward against the spinning brake drum bolted to the wheel. On air disc brakes, as the actuating beam moves the inner brake pad into contact with the rotor, the caliper slides away from the rotor pulling the outer brake pad into the rotor.

Scan Data (Parameters Identification)

The diagnostic information available from the control modules or computers of vehicles that is delivered in a serial form. It contains the parametric readout of the input and output sensors used for vehicle operation.

Scan Tool

A tool to retrieve diagnostic trouble codes from an automotive computer memory and to display these codes as a number. Also checks the operation of vehicle sensors, actuators, wiring, and the computer itself. They connect to the DLC (data links connector) under the dash. They measure voltages or computer signals, providing the technician with information about the condition of numerous components and circuits. Also called an electronic service tool, it can read and transmit serial data carried on diagnostic data links. Reading diagnostic codes and monitoring real-time data are common functions of the scan tool.

FIGURE S.4 Factory-level scan tools are bidirectional communicating tools designed to read and reprogram all proprietary systems.

Africa Studio/Shutterstock.com

Scan Tool Snapshot

An instantaneous scan tool reading of operating parameters that are present when the customer concern takes place.

Schrader Valve

A one-way pneumatic check valve placed in the wheel rim that is used to inflate a tire. It can be found in all pneumatic tires and tubes.

FIGURE S.5 Schrader valve core.

Aleksandr N/Shutterstock.com

SCR (Select Catalyst Reduction)

The SCR system sprays a metered amount of liquid of the diesel exhaust fluid (DEF) into the exhaust gas stream to destroy NOx molecules. The DEF mixes with the exhaust gases to reduce NOx. This exhaust catalyst helps speed up the conversion of NOx into harmless nitrogen and water. DEF is a urea ammonia-type substance that is injected into the exhaust stream to eliminate NOx from the exhaust stream. The reaction between ammonia and NOx, producing water and nitrogen, is enhanced by passing the mixture over a catalyst substrate that causes the reduction reaction to happen faster at lower temperatures. SCR specifically targets the oxygen making up NOx while ignoring atmospheric oxygen in the exhaust, the method is said to selectively remove NOx, hence the name selective catalyst reduction (SCR).

FIGURE S.6 SCR catalyst on an engine.

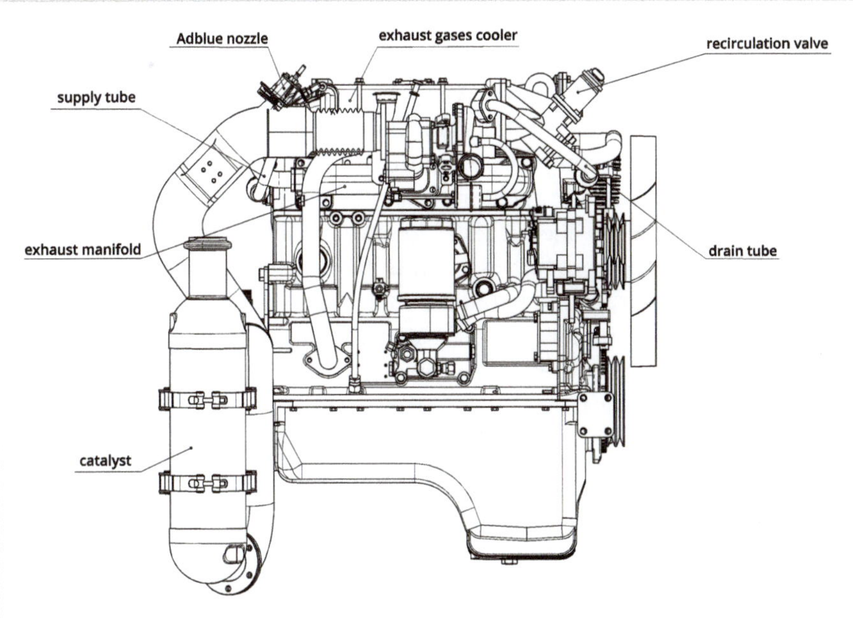

Scrub Radius

The distance between a perpendicular line drawn through the middle of the tire and therefore the steering axis inclination (SAI) or kingpin angle line, where both lines touch the road. If you alter the alignment angles by changing the wheel width and diameter, it can affect the offset and alter the purpose where the flange centerline and steering axis touch the bottom. Scrub radius is additionally called steering offset or scrub geometry. If these two lines intersect at the middle of the tire, at the paved surface, then the vehicle is claimed to own zero offset, or zero scrub radius. If the camber line is outside of the SAI line, then it is a positive offset or positive scrub radius. If the camber line is within the SAI line, then it is a negative offset or negative scrub radius. For more information on *Tire Performance Terminology*, see **J2047_201911** [5].

Scuffing

A form of abrasive damage to the cylinder wall surface of an ICE.

SDS (Safety Data Sheets), Formerly MSDS (Material Safety Data Sheet)

SDS or MSDS is provided for all chemicals used in the industry. The Occupational Safety and Health Administration requires that MSDSs "shall be maintained and kept in a readily accessible area." SDSs for workplace hazards must be readily available. Any SDSs received with shipments or received separately should be placed in the local MSDS binder or file.

In addition to the compliance requirements, MSDSs must be kept and maintained because they provide necessary, helpful, and useful information on the properties of the chemical or chemical product. Technicians should be familiar with those properties, such as flammability, corrosiveness, and toxicity, as well as storage and handling information, before working with any chemical. Also, it is vital for personal safety that technicians can refer to that MSDS immediately in the event of an emergency such as a spill, fire, or physical contact with the chemical.

Sealed Axle Bearing

Sealed front-wheel drive bearings are often part of a hub unit that supports a stub axle, tire, and wheel. This hub unit is classified as a Type 3 hub and is sealed for life. The use of Type 3 hub units allows for lighter rotors because the hub and studs are part of the hub unit, not the rotor. Many 4WD and AWD vehicles also use a Type 3 hub unit. The entire unit is replaced if a problem is found related to the bearing. This type of unit has a non-adjustable bearing preload. The axle nut must be torqued to a high value, depending on the vehicle and specifications.

FIGURE S.7 Sealed wheel bearing.

Kekyalyaynen/Shutterstock.com.

Sealed Beam Headlamp

A headlight lens that has its lens, bulb, and reflector fused in one unit.

Seat Memory Module

A computer module that may "remember" previous seat positions.

Seat Position Sensor

A sensor that feeds data to the seat memory module to reposition the seat.

Secondary Battery

A battery during which chemical processes may be reversed. A secondary battery may be recharged so it will still supply voltage.

Secondary Coil Winding

Windings surrounding the laminated core that are approximately 20,000 turns of fine wire (approximately 42 gauge). For more information on *Ignition System Nomenclature and Terminology*, see **J139_202002** [6].

Secondary Ignition Circuit
Components necessary to make and distribute the high voltage produced within the secondary windings of the coil. For more information on *Ignition System Nomenclature and Terminology*, see **J139_202002** [6].

Secondary Ignition Scope Pattern
An automotive ignition oscilloscope pattern that shows the high voltages needed to fireplace the spark plugs. For more information on *Ignition System Nomenclature and Terminology*, see **J139_202002** [6].

Secondary Nonresistance Wire
A wire that incorporates a metal conductor with no extra resistance. This kind of wire is employed only in high-performance applications, like off-road racing, because it causes static in nearby radios.

Secondary Pickup
An accessory that may be clamped on the high-voltage coil wire accustomed to measure secondary ignition patterns. For more information on *Ignition System Nomenclature and Terminology*, see **J139_202002** [6].

Secondary Wire
Wire accustomed to carrying high voltage within the mechanism.

Section Modulus
Measure of the strength of a frame, which is based on its height, width, thickness, and shape of the side rails. Section modulus does not have a bearing on the strength of the material used in the construction of the side rails. Section modulus is also called the "modulus of elasticity."

Sediment
The solid material that settles to the bottom of a liquid like gasoline or diesel fuel.

Seize
The sudden frictional locking of components like a piston in a boundary lubricated sliding contact due to surface welding or clamping.

Selector Mechanism

Gear or range selecting mechanism composed of a gear lever, selector fork, and rails.

Self-Induction or Self-Inductance

Induction of an electromotive force (EMF) in a circuit when the current in that circuit is varied, not constant. It resists or opposes the change of current flowing through it. This occurs due to the self-induced EMF produced in the coil. Self-inductance is the induction of a voltage in a current-carrying wire. Lenz's Law states that this self-induced voltage tends to oppose the current that produces it. If the current continues to increase, the second voltage opposes the increase. When the current stabilizes, the counter voltage is no longer induced because there are no more expanding flux lines (no relative motion). When current to the coil is shut off, the collapsing magnetic flux lines self-induce a voltage in the coil that tries to maintain the original current. The self-induced voltage opposes and slows down the decrease in the original current. The self-induced voltage that opposes the source voltage is called counterelectromotive force (CEMF) or a back EMF. Just for reference, the equation for self-inductance is shown below.

$$E = -L\frac{\Delta I}{\Delta t} \tag{S.1}$$

where

E is the EMF generated

–L is self-inductance measured, in Henrys

ΔI is the change in current in coil

Δt is the change in elapsed time

Semiconductor

A material with exactly four electrons in their outer shell so it cannot be classified as insulators or conductors. The resistance of some elements is such that they are neither good conductors nor good insulators. If an element falls into this group but can be changed into a useful conductor, it is a semiconductor. Selenium, copper oxide, and gallium arsenide are all semiconductors, but silicon and germanium are the most common. Pure semiconductors have tight electron bonding; there is no place for electrons to move. In this state, the elements are not useful for conducting electricity.

Semiconductor Doping

Silicon has four electrons in its outer shell. Silicon is grown into large crystals by applying heat to melt the silicon followed by cooling. Pure silicon is not very useful in electronics components. For silicon to be useful, it must be doped, that is, to add small quantities of impurities. The impurities affect the number of free electrons. Depending on which impurity is added, the resultant material will have either to many free electrons or a shortage of free electrons. The doping agents are usually phosphorus and boron. The doping intensity will define the electrical behavior of the crystal. After doping, you will be able to slice the silicon crystals into thin sections called wafers.

The kind of doping agent to produce silicon crystals will define the electrical properties of the crystals produced. A boron atom has three electrons in its outer shell. The outer shell is thought of because the valence and an atom with three electrons in its outer shell is understood as trivalent. A boron atom in an exceedingly crystallized cluster of silicon atoms will produce an outer shell with seven electrons rather than eight. This vacant electron opening is understood as a hole. The outlet makes it possible for an electron from a closeby atom to comprise the opening. In other words, the holes can move, permitting a flow of electrons. Silicon crystals doped with boron (or other elemental atoms with three electrons within the outer shell trivalent forms a P-type silicon. The phosphorus atom for example has five electrons in its outer shell. It is pentavalent. Within the bonding between the semiconductors and therefore the doping material, there is room for under eight electrons within the center shell. Even when the fabric is in an electrically neutral state, the additional electron can move through the crystal. When a silicon crystal is manufactured employing a doping material with five electrons within the outer shell (pentavalent), it forms an N-type silicon or semiconductor material. Germanium also has four electrons in its outer shell.

Semi-trailer

A tractor-trailer rear half or trailer where the front end is supported by and attached to the tractor using a kingpin or pintle and a fifth wheel. This creates an articulated vehicle. For more information on this, see **J3085_201701** [7].

Sensor

A device that gives input to an automotive computer. They will detect part movement, temperature changes, pressure variance, chemical content, part location, rotation speed, sounds, and other conditions. Sensing functions collect data about operational conditions or the state of a tool by measuring some value like temperature, position, speed, pressure, flow, angle, and then forth. Sensors are a kind of transducers because they convert physical conditions or states into electrical data. A small distinction may be made between sending units and sensors. Sensors provide information to electronic control units, whereas sending units provide information to instrument gauges.

Serial Communication or Serial Data

Electronic communication that uses a single or differential ended wire to exchange bits of data. The various system control modules of the vehicle communicate with each other as well as with the off-board diagnostic equipment. The communication between control modules and communication with the off-board diagnostic equipment is achieved on the serial communication lines using serial data. The serial data transfers information one bit at a time, in a linear fashion, over a single line, hence the term "serial." The serial data line is typically referred to as the data bus.

FIGURE S.8 Serial data patterns.

90miles/Shutterstock.com.

Series Circuit

A series circuit is a complete circuit that has more than one electrical load, where all of the current has to flow through all of the load. The following are characteristics of all series circuits:

- Voltage drops add up to the source voltage.
- There is only one path for current flow.
- The same current flows through every component. In other words, you would get the same current measurement at any point along the circuit.
- Since there is only one path, an open anywhere in the circuit stops current flow.
- Individual resistances add up to the total resistance.

FIGURE S.9 Series electrical circuit.

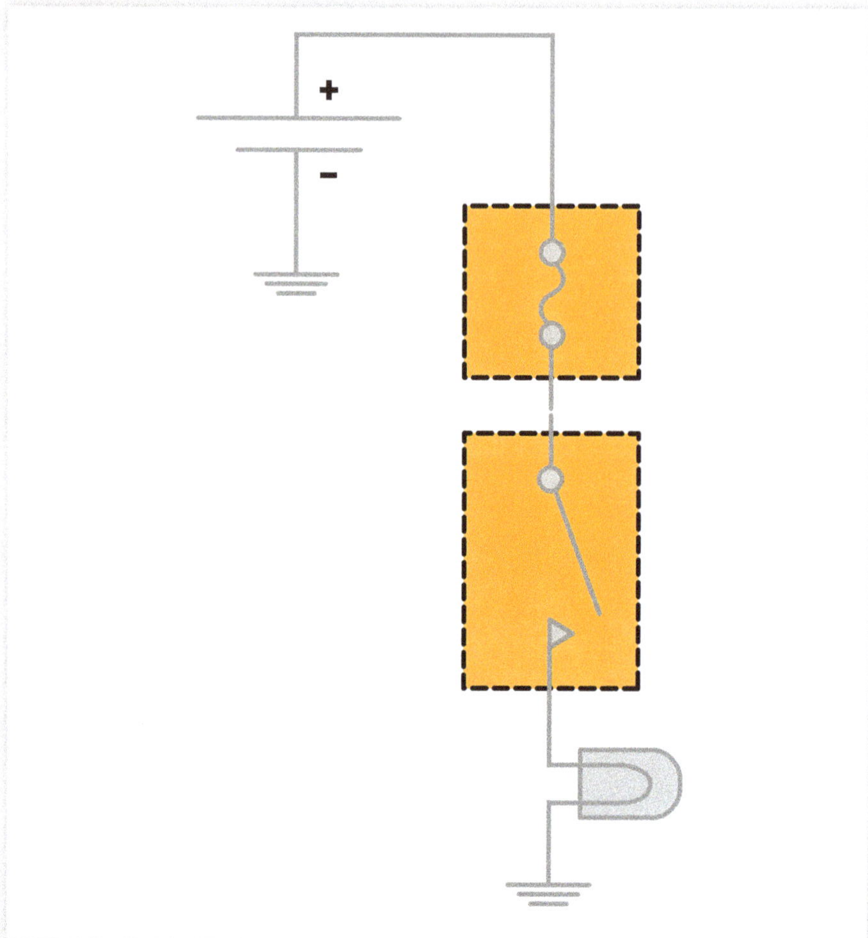

Courtesy of John F. Kershaw Ed.D.

Series Hybrid

A hybrid vehicle that has a separate generator and traction motor. It does not use a motor generator. The traction motor is the only method used to apply torque to the vehicle drivetrain. The engine has no mechanical connection to the drivetrain. The ICE never directly powers the vehicle because the electric motor is capable of powering the vehicle by itself. The series configuration requires higher voltage electric drive systems similar to those found in full-use electric vehicles (EVs). Series hybrids use technology similar to that of diesel locomotives that use diesel engines to power a generator that supplies electrical energy to an electric motor that drives the ship propellers or the locomotive wheels. For more information on *Hybrid Electric Vehicle (HEV) and Electric Vehicle (EV) Terminology*, see **J1715/2_202108** [1].

FIGURE S.10 Series hybrid vehicle.

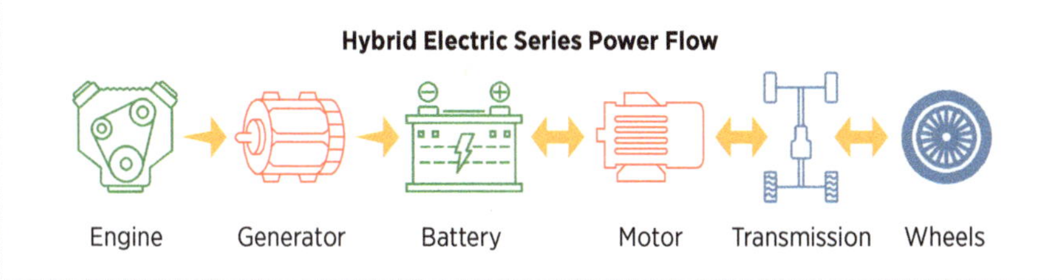

Courtesy of John F. Kershaw Ed.D.

Series Motor

A motor with only one path for current flow through the field and armature windings.

Series-Parallel Circuit

A series/parallel circuit is a circuit that contains both series circuits and parallel circuits. The circuit in Figure S.11 has a 2-ohm resistor in series from the battery. Then it splits into two parallel branches of a 6-ohm resistor and a 3-ohm resistor before recombining and returning to the battery. There is no specific law or formula that pertains to the whole series/parallel circuit for voltage, amperage, and resistance. Instead, it is a matter of determining which branch loads of the circuit are in series and which are in parallel, simplifying the circuit where possible, and using the Ohm's laws that apply to each of these branches to find the value totals.

FIGURE S.11 Series-parallel circuit.

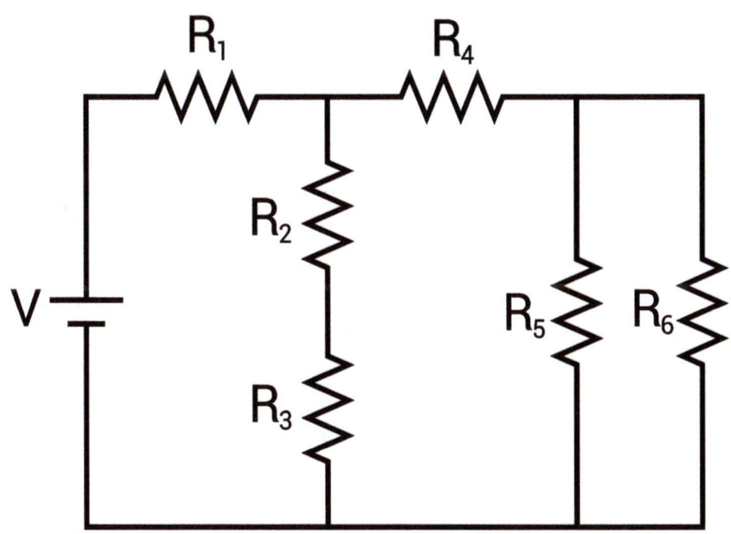

zizou7/Shutterstock.com.

Series-Parallel Hybrid

A series-parallel hybrid, also called a combination hybrid, is a vehicle that merges the advantages of the parallel hybrid with those of the series hybrid. The internal combustion engine can drive the wheels mechanically but can be disconnected from the drivetrain so that only the motor generator drives the vehicle. For more information on *Hybrid Electric Vehicle (HEV) and Electric Vehicle (EV) Terminology*, see **J1715/2_202108** [1].

Series Parallel Switch

A switch used on truck starting systems that allows the use of 24 V to start and 12 V to run.

Series-Wound Motor

A motor that develops the most torque during start-up. The torque decreases as motor speed increases. The sector coils and the armature are wired serial. All this flows through the sector coils, the new brushes, the armature, and then to the bottom through the bottom brushes.

Serpentine Belt

A drive belt system that uses a single belt that is wound around the accessory drive pulleys like a snake to connect them to the crankshaft.

Service Brake Chamber (Air Brakes)

Compressed air through lines or hoses goes to the brake chamber to move a slack adjuster or caliper piston to apply the brakes. There are a variety of brake chamber sizes used for different sizes of trucks. The size needed depends on the force required to activate the brake shoes successfully against the drums or the calipers on a disc brake system. The overall brake chamber diameter is classified by size, based on the area on which the compressed air can work. Common sizes for the diaphragm areas are 9, 12, 16, 20, 24, 30, and 36 in.^{2}. Compressed air from the various control valves within the brake system is directed through lines or hoses to two or more brake chambers. Pressurized air sent through the valve system acts on a flexible diaphragm, which forces a pushrod assembly forward against the force of a coil return spring. When the air is released from the chamber, the coil return spring pulls pushrod back to the released position. For more information on *Braking System Definitions—Truck and Bus*, see **J2627_200908** [8].

Service Brakes
The primary vehicle brake system that is activated when the drive of a vehicle steps on the brake pedal to stop the vehicle. For more information on *Braking System Definitions—Truck and Bus*, see **J2627_200908** [8].

Servomechanism
A powered device producing motion or forces at a higher level of energy than the input level, for example, a pneumatic or hydraulic brake assist servo that assists the driver in applying the service brakes.

Setback Angle
The distance one wheel is set back from the wheel on the opposite side of the axle, relative to imaginary lines running through the center of each wheel and perpendicular to the vehicle center.

SFI (Sequential Fuel Injection)
Sequential operation of the fuel injectors is the activation of the injections according to engine firing order of the ICE. The fuel injectors are timed and pulsed individually in the same order as the spark plugs.

Shackle
The swinging shackle in a truck suspension allows the leaf spring assembly on a solid axle to lengthen or shorten while the truck is in motion, which allows the axle assembly to move up and down as the truck encounter bumps and other road irregularities. The stationary bracket bolts directly to the frame rail, and the leaf spring is supported by the clamped pin.

Shift
The action in a transmission or transaxle to change gears or the speed range.

Shift Collar
A splined sliding clutch that connects a spinning gear to a fixed gear on a countershaft in a manual transmission or transaxle.

Shift Fork
The component in a manual transmission transaxle, or transfer case that moves the shift collar on a synchronizer clutch hub to connect a spinning gear to a fixed gear on a countershaft.

Shift Selector

A device used on the automated transmission in an Class 7, 8, or 9 heavy truck, which also uses an Internal mode switch (IMS) mounted internally on the gear range selector shaft. The IMS transmits selector position information to the transmission control module (TCM). The IMS detects the spatial relation of the shift selector shaft. This position is communicated to the TCM that certain vehicle control functions are often coordinated with the position of the shift controls. The IMS has redundant circuitry to alert the TCM in the event of one wire or switch failure. The neutral signal output of the IMS is usually used as confirmation that the transmission is in neutral before the engine starter is engaged. The IMS is interfaced with the starter circuit. The operator chooses the transmission range by moving the selector lever to the acceptable gate position. When properly adjusted, the shifter gates prevent inadvertent shifting between ranges and correspond to the interior transmission detent positions. A positive detent is provided within the transmission to keep up the selector shaft within the selected position. The TCM shift calibration determines the available forward ranges for every selector position. Although specific installations vary, typical selector positions for Allison transmission include the following:

- P—PARK: parking pawl is engaged. The transmission is in NEUTRAL.
- R—REVERSE is chosen so as to maneuver the vehicle backward.
- N—NEUTRAL is also used when starting the engine and for stationary operations. The TCM disables the starter switch if a variety apart from NEUTRAL or PARK is chosen before starting the vehicle.
- D—DRIVE is the highest forward range used for normal driving. The transmission shifts to first range for starting, then automatically upshifts through the ranges, as operating conditions permit, until the very best range is attained.
- M—Manual selection of ranges is provided by moving the selector to the M—MANUAL position.

A driver shift request switch has been added to the column shift lever. The motive force information center (DIC) then shows the six available ranges with the present range bracketed. The bracketed range is the highest attainable range with all gears below accessible, as an example, when the fourth range is chosen, ranges 1 through 4 are available. By using the +/– buttons on the gear select lever, the motive force can select the range of gears desired for his or her current driving conditions. 1—Manual LOW provides low operation only.

Shock Absorbers

Provide damping of automotive suspension spring oscillations. When a wheel hits a bump, the spring is compressed; because the wheel passes over the bump, the spring expands. Without a damping device, the spring would still "oscillate," or contract and expand repeatedly, until all the energy of the bump has been dissipated. A cushion dampens spring oscillation by pushing and pulling a piston through a column of oil. The piston contains orifices that allow the oil to flow through at a hard and fast rate.

$$\text{Reaction force } Q\left[N\right]Q=\left(1.5\cdot W\,3\right)/s \tag{S.2}$$

where

W is the weight or mass
N is the speed
s is time, in sec

FIGURE S.12 Shock absorber.

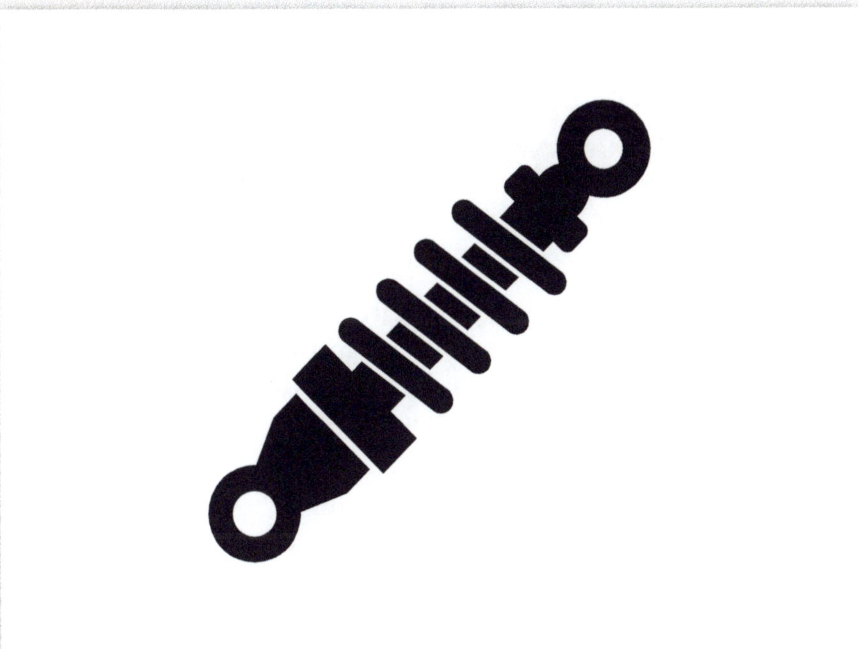

Short Block

A term that defines the cylinder block or case along with the pistons, connecting rods, and crankshaft in an ICE. This component is a new replacement unit.

FIGURE S.13 Subaru Boxer ICE short block.

Short Circuit

When the intended direction of the electrical current takes a shorter path, often the path of least resistance. This cause a higher current flow than design, resulting in a blown fuse or circuit breaker opening. It can be a short to ground or short to voltage. See **Short-to-Ground** and **Short-to-Voltage**.

Shorted Diode

A diode that has a low (almost zero) resistance in both directions.

Short-Term/Long-Term Fuel Trim

Short-term and long-term fuel trim data is read with a scan tool and can be useful information when diagnosing engine performance conditions. Short-term/long-term fuel trim data is a reflection of the engine management computer changing the pulse width of the fuel injectors to keep the air/fuel ratio of the engine as close to the optimal 14:7:1 stoichiometric. The important difference between them is that the short-term fuel trim indicates current changes of short-term duration, while the long-term fuel trim indicates learned changes over a long-term period. If the short-term value climbs above 128, the computer is adjusting for a lean condition so more fuel is being added to the engine. If the short-term values fall below 128, the computer is adjusting for a rich condition, so less fuel is being added. The short-term fuel trim number is an indication that the computer is compensating for changing fuel demands. If the computer readings indicate the short-term fuel trim is becoming overloaded (meaning the fuel requirements of the engine are changing too much toward lean or rich), the long-term fuel trim becomes involved.

The long-term fuel trim shows the computer "learned" fuel correction. The long-term fuel trim reading shows how much the computer has compensated. Even though the short-term fuel trim can make a wide range of fuel delivery corrections frequently, a long-term fuel trim can indicate a trend in the lean or rich direction the short term is taking. A long-term fuel trim can make a more significant fuel delivery change after a longer period of time. Long-term numbers are also based on 256 counts (0 to 255), with 128 as the baseline.

A block of cells contains information arranged in combinations of engine RPM engine load for the full range of vehicle operating conditions. The computer checks the appropriate block for data to use in calculating the correct injector pulse width. Ideally, each block value would be 128. If the short-term fuel trim is far enough from 128, the long-term fuel trim changes its value and resets the short term to 128.

Short-to-Ground

A short-to-ground occurs when a circuit finds a path to ground before going through the load. Since the resistance of the load does not control the current, the large amount of current usually blows a fuse or trips a circuit breaker. If a fuse blows or a circuit breaker trips, it will be easy to find. The short-to-ground is located either in the load itself or in the circuit between the fuse/circuit breaker and the load. The problem cannot be on the ground side of the load. To diagnose a short-to-ground, connect a load (such as a test light or headlight) in place of the fuse while you isolate the problem. Technicians usually use a 12-V test light. But remember: not just any test light will work! If the fuse circuit you are testing has a lot of parallel branches that are on whenever the ignition is on, a typical test light will stay on all the time, even after you fix the short-to-ground.

Short-to-Voltage

When the power side of one circuit is electrically connected to the power side of another circuit. Short-to-voltage occurs when the power sides of two electrical circuits are connected.

Siamese ICE Cylinders

The Siamese bore ICE cylinder block has no coolant passages between the cylinders. These blocks have thicker walls, so a capacity for a larger overbore. They do have more engine cooling problems.

FIGURE S.14 Siamese ICE cylinder block.

Sergey Ryzhov/Shutterstock.com

SID (Subsystem Identifier)

SID stands for subsystem identifier, and it is a one, two, or three-digit code that identifies a section of the computer control system that does not have a related parameter identifier (PID). It identifies the major subsystems of an electronic circuit. The SID is a single byte character used to identify repairable or replaceable subsystems for which you cannot detect or isolate failures. The SID is used in conjunction with SAE standard diagnostic codes identified in SAE standard J1587. The PID 194 contains a transmitter system diagnostic code and an occurrence count. For example, S250-12 identifies a J1587 data link fault and S22-4 indicates a timing sensor problem with voltage below normal or a shorted sensor. The number "3" is a failure mode indicator. For more information on this, see **J1587_200202** [2].

Side Blind Zone Alert System

When the vehicle is started, both side blind zone indicators located on the left and right outside rearview mirrors will briefly illuminate to point out that the system is working. The system will detect objects in a neighborhood about 1.5 to 6 ft off the bottom. The side blind zone alert monitors the blind-spot area alongside the vehicle, at about 11 ft from the side of the vehicle, extending 16 ft behind the vehicle midpoint. The system may remove darkness from an indicator thanks to guardrails, signs, trees, shrubs, and other non-moving objects. This is often a normal system operation; the vehicle does not need service. When the system detects a vehicle within the side blind zone while driving forward, an amber warning symbol will illuminate the suitable car mirror. This means that it is going to be unsafe to vary lanes. If the driving force activates the amber warning symbol starts flashing as an additional warning to not change lanes. For more information on *Active Safety Systems Terms and Definitions*, see **J3063_202103** [9].

Sidesway Frame Damage

Takes place when one or both frame rails are bent inward or outward. The following are some causes: collision damage, fire damage, and using the truck for purposes other than those for which it was originally designed.

Silent Chain

A sprocket and link-type chain that pivots around pins. Each link is a series of plates held together with pins. The number of plates determines the chain width. The word silent is from a comparison to the roller chain on a bicycle.

Single Grade

An oil with just one viscosity rating, such as 10 and 20.

Single Motor-Generator Hybrid

A hybrid vehicle that uses only one large motor-generator to assist the gas engine. For more information on *Hybrid Electric Vehicle (HEV) and Electric Vehicle (EV) Terminology*, see **J1715/2_202108** [1].

Single-Phase Current

The alternating current created by a single-phase winding.

Single-Phase Voltage

The sine wave voltage induced within one phase of winding by one revolution of an alternator rotor.

Single-Reduction Axle

A rear final drive axle with only one gear reduction, which occurs through the differential ring gear and drive pinion. The differential gear ratio refers to the relationship between the ring gear and the pinion gear. This ratio is calculated by simply dividing the number of ring gear teeth by the number of pinion gear teeth. For example, if you divide a 41-tooth ring gear by a 10-tooth pinion gear, the gear ratio is 4.10:1 (41/10 = 4.10).

Skid

When a vehicle slides sideways, or obliquely, on slippery ground or as a result of stopping or turning too quickly.

Slack Adjuster

Slack adjusters connect the air brake chambers with the brake shoes on drum brakes or the ability shaft or air screw disc brakes. They are connected to the pushrod of the brake chamber by means of a clevis yoke (threaded to the pushrod) and pin. Slack adjusters are spline mounted to the S-cams on drum brakes and positioned by shims and an external snap ring. The slack adjuster converts the linear force of the brake chamber rod into rotary force or torque and multiplies it.

The gap between the slack adjuster S-cam axis and the clevis pin axis will define the leverage factor: the greater this distance the greater the leverage. Drum brakes using slack adjusters use the term AL factor, which is the relationship between the service brake chamber size and the slack adjuster arm length. The AL factor is a way of comparing the brake torque output of the friction materials. A 165 AL factor may be a type 30 service chamber employing a 5.5-in. slack adjuster and is calculated below as follows: 30 × 5.5 = 165.

$$\text{AL factor} = \frac{\text{Tire rolling radius} \times 6 \times \text{Weightonwheel} \times \text{Cam radius}}{\text{Drum radius} \times \text{Lining friction} \times 2 \times \text{Air pressure}} \quad \textbf{(S.3)}$$

For more information on *Braking System Definitions—Truck and Bus*, see **J2627_200908** [8].

FIGURE S.15 Truck air brake slack adjuster.

Sleeve

See **Liner**.

Slip Angle

Tire slip angle describes the principle that automotive tires use to grip the surface of the road, which allows cornering around a curve. It is a force that makes a vehicle able to turn a corner. The more cornering force your vehicle has, the sharper the vehicle can turn. It is the angle between the wheel direction the wheel is pointing and the vehicle traveling direction. It is the angle between the forward velocity vector v_x and the vector sum of wheel forward velocity v_x and lateral velocity noted as "vy." A slip angle results in a cornering force that is in the plane of the tire contact path and perpendicular to the joining of the contact path and the midplane of the wheel. For high-performance tires, the ideal slip angle is around 6 to 10°. This number is lower for street tires. It is also called the distortion angle. For more information on this, see **J670_200801** [10].

FIGURE S.16 Slip angle.

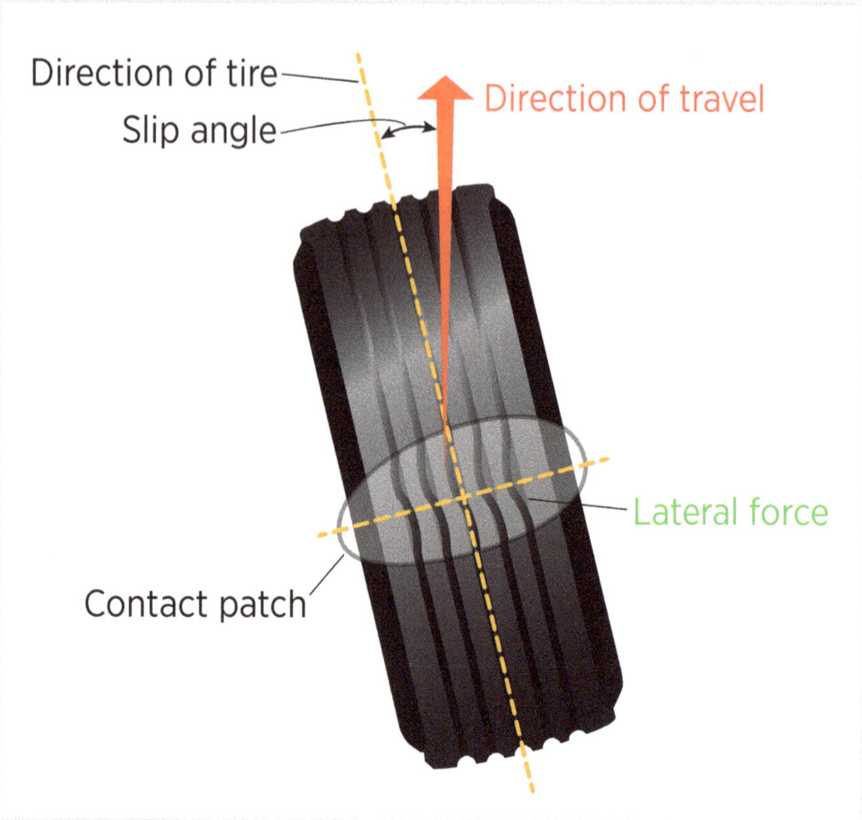

Slip Yoke

Drive axles also tend to move forward and backward as the vehicle travels over bumps; the driveline must also be able to small changes in the actual length of the driveline. The amount of horizontal movement in a drive axle is relatively small compared to the vertical travel. The main reason why truck driveshafts have slip yokes is to accommodate vertical movement and suspension travel.

FIGURE S.17 Slip yoke.

Slippage

The sliding movement between two moving or coupled components, such as a clutch disc and flywheel or tire and road.

Sludge

The deposit formed by water vapor and the fuel in blowby gases will mix with cold oil and form sludge.

Sludge Trap

These are usually threaded plugs that have been staked in place to seal the ends of the oil passages. They are removable if you drill out the punch mark before you can unscrew the plug.

Smog

The term used to describe a combination of smoke and fog. It is created by oxides of nitrogen and hydrocarbons with sunlight.

Smoke

A visible emission from the ICE exhaust in the form of a vapor or particulate: See **Black Smoke**, **Blue Smoke**, or **White Smoke**.

Snubber

A device that prevents excessive movement in a component. It can also be a pressure reduction orifice used in fuel injection systems.

Sodium-Cooled Exhaust Valve

An ICE poppet exhaust valve that has an internal core of sodium within the valve stem. The movement of the sodium removes heat from the valve head to the stem where it is conducted to the valve guide and cylinder head.

Solder

An alloy (mixture of metals) that has a relatively low melting point.

Soldering

Involves using heat and solder to join electric components or wires.

Soldering Gun

A trigger-operated tool that has a transformer that sends a high current through a high-resistance tip to generate heat to melt the solder.

Soldering Iron

A tool that must be plugged into a wall outlet and heated up for several seconds before using it to solder. The tip on a soldering iron is generally smaller and more suited to delicate work on a circuit board than the tip on a soldering gun.

Solenoid

A device that uses electromagnetism to provide motion. A solenoid produces small, individually controlled back-and-forth movements. However, the standard solenoid produces plenty more holding force than a relay. Meaning it can do over close electrical contacts, which is what a relay does. When current flows through a solenoid coil, electromagnetism pulls an iron core into the coil; when current stops flowing through the coil, the core returns to its base position. These back-and-forth movements of the core is attached to a rod that is attached to a tool sort of a door lock to maneuver it in two different directions to lock and unlock a vehicle door.

Spark Gap

The clearance or space between the center electrode and side or ground electrode of an ICE ignition spark plug. This gap can be adjusted in most spark plugs. For more information on *Ignition System Nomenclature and Terminology*, see **J139_202002** [6].

Spark Knock

Secondary rapid burning of the last 3–5% of the air-fuel mixture in the combustion chamber causing a second flame front that collides with the first flame front and produces a knock noise. See **Detonation**.

Spark Line

A short horizontal line connected to the firing line as shown in an ignition oscilloscope. The peak of the spark line waveform represents the voltage required to take care of the spark across the electrical device after the spark has started. The spark line peak should be one-quarter of the peak of the firing line (between 1.5 and 2.5 kV). For more information on this, see **SAE J139_202002** [6].

Spark Plug

An ICE spark ignition device that delivers electric current from an ignition system to the combustion chamber of a spark-ignition ICE. This spark ignites the compressed air-fuel mixture using this spark. A very high voltage is needed to jump the gap at the tip of the spark plug. It is made of a metal threaded shell that is electrically isolated from a central electrode by a ceramic insulator. The central electrode, which may contain a resistor, is connected by a heavily insulated wire to the output terminal of an ignition coil, which is a step-up transformer. The spark plug screws into the engine cylinder head and is electrically grounded. The central electrode protrudes through the porcelain insulator into the combustion chamber forming one or more spark gaps between the inner end of the central electrode and the ground electrode(s). For more information on *Ignition System Nomenclature and Terminology*, see **J139_202002** [6].

Spark Scatter

A timing variation that can occur between cylinders. This timing variation is called spark scatter and is caused by the slack and clearance between the camshaft, distributor gear, and timing chain.

Spark Tester

A special spark plug that measures the voltage to jump the spark gap. For more information on *Ignition System Nomenclature and Terminology*, see **J139_202002** [6].

Spark Voltage

The inductive portion of a spark that maintains the spark in the air gap between the electrodes of a spark plug, usually about one-quarter of the firing voltage level. For more information on *Ignition System Nomenclature and Terminology*, see **J139_202002** [6].

Specific Fuel Consumption

See **Brake Specific Fuel Consumption**.

Specific Gravity

Weight of a given volume of a liquid divided by the burden of an equal volume of water.

Specific Gravity Battery Test

On unsealed battery designs, the precise gravity of the electrolyte is often measured to relinquish a reasonably good idea of the battery state of charge(. A hydrometer is employed to perform this test. A basic battery hydrometer consists of a glass tube or barrel, rubber bulb, rubber tube, and glass float or hydrometer with a scale built into its upper stem. The glass tube encases the float and forms a reservoir for testing the electrolyte. Squeezing the bulb pulls the electrolyte into the reservoir. When full of test electrolyte, the hydrometer float bobs within the electrolyte. The depth at which the glass float sinks within the test electrolyte indicates its relative density. The reading is kicked off the size by sighting along the amount of the electrolyte. If the hydrometer floats deep within the electrolyte, the precise gravity is low. If the hydrometer floats shallow within the electrolyte, the particular gravity is high.

Speed Density

Measures intake air flow by sensing changes in intake manifold pressures resulting from engine load and speed changes. The ICE management computer uses a manifold absolute pressure (MAP) sensor to read these pressures. The computer combines MAP along with temperature, RPM, estimates of volumetric efficiency and volume, and exhaust gas recirculation to calculate mass air flow.

Speedometer

"A device used to measure the instantaneous vehicle speed. SAE J678_201108 applies to speedometers, odometers, and speedometer drives typical of passenger vehicles, buses, and trucks used for personal or commercial purposes. The method of determining wheel revolutions per unit distance (3.1) and overall system design variation (3.3.3) are applicable to passenger cars only. Comparable recommendations for trucks and buses are under development. The data of tachometers is applicable to vehicular use, as previously described, and also to stationary and marine engines and special vehicles. Most modern speedometers turn 1,000 RPM per mile. Older mechanical speedometers turned at 1,440 RPM per mile." For more information on this, see **J678_201108** [11].

Spindle

This device holds the wheel hub and bearings and offers vehicle stability. When the wheel is turned in or out, the spindle lowers and raises the vehicle. When the wheels are released from a turn, the load of the vehicle helps move each spindle back toward the midpoint until the load is equal on both front wheels.

Spiral Bevel Gearset

Spiral bevel gears have curved teeth, similar to those on a helical gear, that increase the loads they can carry and decrease the noise they produce. Manufacturers usually supply spiral bevel gears in matched sets. In this combination, the larger gear is a ring gear, and the smaller gear is the pinion gear.

FIGURE S.18 Spiral bevel ring and pinion gearset.

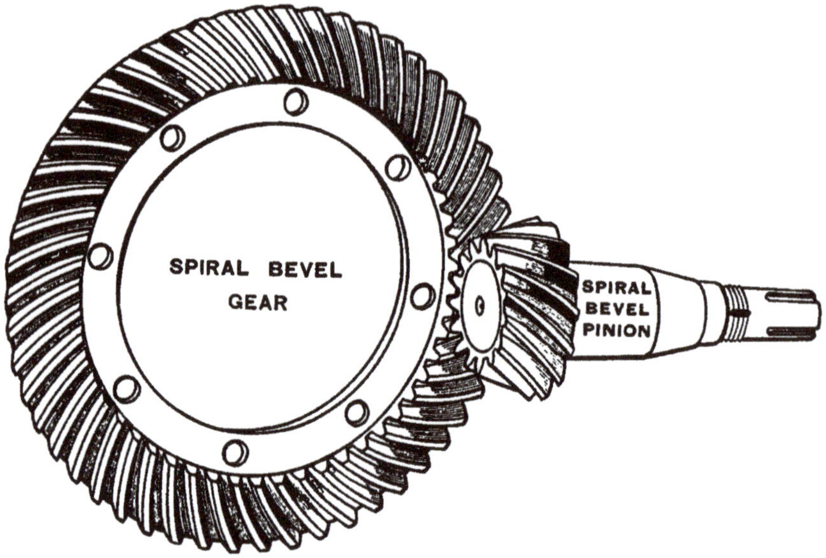

Split Safety Lock Wheel Rim

A safety-rim wheel has raised ridges between the wheel drop center (well) and the flange. The ridge holds the tire bead in position so the vehicle can be controlled for a safe stop if a flat occurs. The ridges also hold the tire bead in position during hard cornering. Most wheels also contain a drop center; this allows the technician to place a bead into the dropped center providing clearance to the outer bead so it can be slipped over the rim flange. For information on this, see **J175_202107** and **J393_201710** [12, 13].

Spring Brake Chamber (Air Brakes)

The spring brake chamber is a combination service and spring brake chamber assembly for truck, tractor, trailer, and bus applications. The design basically provides two individual chambers in one housing with their own diaphragms, springs, and separate air supplies. Spring brake chambers provide a safe parking brake arrangement when system air pressure is exhausted from the spring brake chamber. A large coil spring then forces the pushrod forward to apply the brakes mechanically. For the combination service and spring parking brake systems to operate independently of each other, both service and parking brake air lines must be used. The service line transfers air from the brake pedal only to the service chamber portion of the assembly. The parking air line is connected to the dash-mounted spring brake control valve button and can direct air only into the spring brake chamber portion of the assembly. Spring brakes are either sealed or clamp-type designs. Those that are attached to the rear of the service brake chamber are referred to as piggyback spring brakes. For more information on *Braking System Definitions—Truck and Bus*, see **J2627_200908** [8].

FIGURE S.19 Truck air brake spring brake chamber.

Spring Brake Control Valve (Inversion Valve Air Brakes)

The spring brake control valve supplies air to the hold-off chambers to release the spring brakes and enables the vehicle to maneuver. During normal operation, it limits the hold-off pressure to the spring brake chambers to worth around 90 psi. This action speeds application times within the event of an emergency and permits a uniform hold-off pressure (system pressure fluctuates between cut-in/cut-out pressure values). Some systems use an inversion valve. In the event of a primary air system failure, the inversion valve allows for a modulated application of the emergency air brake chambers by relieving air from the hold-off chamber in direct proportion to the applying pressure at the treadle valve. Spring brake control valves are often incorporated in multifunction valves that contain parking, service, and inversion functions. The trailer air system has its own spring brake control valve, which is intended to:

- Automatically apply trailer spring brakes in event of a breakaway or trailer itinerary failure
- Protect trailer reservoir(s) gas pressure in event of a breakaway or trailer supply route failure
- Prevent an automatic trailer spring brake application with atmospheric pressure loss within the trailer reservoir(s)
- Offer an optional anti-compounding feature and stop overriding the service brake signal while the trailer supply route is at gas pressure (no pressurized air in line)
- Prevent charging of the trailer system should a failure occur within the emergency brake circuit

For more information on *Air Brake Valve - Performance Requirements*, see **J1410_202012** [15].

FIGURE S.20 Spring brake control valve.

Springs (Automotive Suspension)

Springs are located between the frame and so the axle assemblies and are shaped to suit specific applications. Leaf springs are normally semi-elliptical. They absorb the applied force (pressure of the load) by flattening out under load. Springs are used on the underside of a car or truck to help the vehicle carry larger loads. Coil springs are formed in an exceedingly spiral from one steel rod. They absorb the force of impact by twisting and compressing. They are used on smaller vehicles to smooth the ride and improve handling.

Torsion bars are held rigid at one end and denote their center because the control arm is deflected. Control arms, also called A-arms or wishbones, attach to the chassis with rubber bushings and permit for the movement of the tire and wheel assembly. Torsion bars all return to their original shape when the deflecting force (a force that moves an object in a very different direction or into a special shape) is removed. They are typically employed within the front of most pick-up trucks because they handle better than leaf springs. For information on *Helical Compression and Extension Spring Terminology*, see **J1121_201608** [14].

FIGURE S.21 Suspension springs.

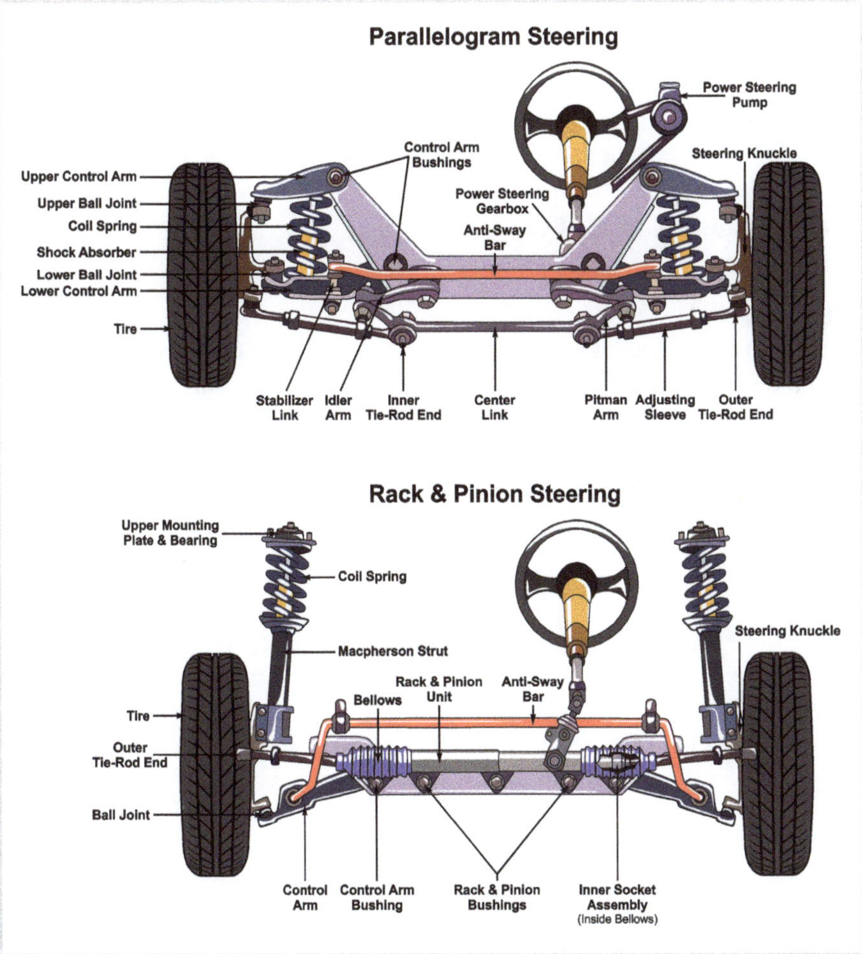

Sprung/Unsprung Weight Ratio

Some suspension parts are attached to the frame on one end and to the wheel on the other, and their weight ratio is calculated as an average between sprung/ unsprung weight. The axle shaft and lower suspension arm are both 50% unsprung weight and are an average between the two ends. The ratio of sprung/ unsprung weight has an effect on the handling of a vehicle on rough roads. It can be calculated by dividing the total sprung weight of the car by the total unsprung weight of all four wheels. Thus a ratio of 5 is good while a ratio of 2 is poor. A vehicle that has a high sprung/unsprung weight ratio will stick better on a rough road than one with a lower ratio.

To get better handling on bumps, it is much better to reduce the unsprung weight, while keeping the sprung weight as light as possible. For example, switching from steel wheels to aluminum ones has a noticeable effect on how a car takes bumps and how it rides. The reason for buying a car with an independent rear suspension is to reduce unsprung weight and thus improve handling. For more information on this, see **J670_200801** [10].

Sprung Weight

The weight of the chassis and all parts mounted on the chassis. This also includes one-half the weight of the springs, shocks, and control arms. For more information on this, see **J670_200801** [10].

Spur Gear

Gears with straight teeth that are mounted on parallel shafts. You will generally only find spur gears on heavy transmissions used in medium- and heavy-duty trucks and construction equipment. Each time a gear tooth engages a tooth on the other gear, the teeth collide, and this makes noise. It also increases stress on the gear teeth. To reduce the noise and stress in the gears, most automotive gears are helical.

Square Engine

A term used to define an engine where the bore diameter is equal to the piston stroke or travel.

Stabilizer Bar (Sway Bar)

Help limit the consequences of body roll while cornering. The bar acts very like a torsion bar. The bar is U-shaped and is typically connected to the two lower control arms and frame through bushings. It will be mounted ahead of or behind the suspension. Because the vehicle enters a turn, the vehicle rolls and causes the surface wheel to compress upward (jounce). This forces the stabilizer end upward. The bar then rotates in its frame attachment bushings and tries to force the alternative side down. Since it cannot force the opposite side down, it applies pressure to the jounced wheel control arm, which counteracts the spring tension and vehicle weight. This helps to keep up the vehicle in a very level position. Stabilizer or sway bar diameters can affect oversteer and understeer. Stabilizer bars are used rather than struts on many vehicles because they hook up with control arms within the same fore and aft direction. Some vehicles use a control bar with a vertical link. The vertical link attaches to a ball joint.

$$\text{Cornering equation:} \Sigma F_y = F_{yf} + F_{yr} = MV2/R \quad \text{(S.4)}$$

where

F_{yf} is the lateral cornering force at the front axle
F_{yr} is the lateral cornering force at the rear axle
M is the weight of the vehicle
V is the forward velocity
R is the radius of the turn

FIGURE S.22 Sway or stabilizer bar.

Stall

The unexpected stopping of an ICE for various reasons, such as an increase in load, loss of spark, or loss of fueling.

Star Configuration

In a vehicle, it is a computer network configuration in which one main computer or master electronic control module is connected to several less complex controller. All signals from the less complex modules must go through the master engine management computer.

Star Wheel

A toothed wheel adjusting screw used to adjust drum brakes or a truck clutch.

Starter Brushes

The brushes of the starter are mounted around the commutator (see Figure S.21). They are spring loaded so they are always touching the commutator surface. Starter brushes allow current to pass from the field coils to the armature windings.

Starter Generator

It could be a combination of two different engine components. The component will be found in hybrid electric vehicles (HEVs), lawn tractors, garden tractors, and aircraft. The starter component generates the initial ignition, starting the vehicle, and also the generator maintains the voltage while the vehicle is being operated. A HEV saves fuel in four aspects: engine idle-stop, regenerative braking, engine downsizing, and engine load change. Saving fuel by 5–10% is achievable with an optimized fuel cut-off strategy. However, the emission and drivability trade-off must be carefully treated. The conventional automatic drive requiring a mechanical pump driven by the engine crankshaft offers hydraulic function, lubrication, and cooling.

The transmission fluid pressure would not be sufficiently maintained to allow clutch engagement. Once the ICE restarts, the engine speed profile can cause the transmission fluid pressure to be too high, which creates a bump during vehicle takeoff. In most vehicles equipped with a traditional transmission (AT) moreover as a start-stop function, an electrical pump is installed to control the fluid pressure. However, cost and complexity increase, and therefore the electric pump is redundant to the mechanical pump during normal driving. SAE Technical Paper 2013-32-9071 on *Development of Belt-Driven Starter-Generator Control Strategy for Hybrid Electric Vehicle* presents a technique for smoothing the change of transmission system fluid (ATF) pressure by utilizing the belt-driven starter generator (BSG) using field oriented control (FOC) method and therefore the correlation curve between AT fluid pressure and rotational speed of BSG. The proposed system reduced the number of changes and marginal cost because of additional start/stop technology applied to the present vehicle design. For more information on *Hybrid Electric Vehicle (HEV) and Electric Vehicle (EV) Terminology*, see **J1715/2_202108** [1].

Starter Motor (Cranking Motor)

The starter motor is an AC or DC motor that includes the following components:

- Housing
- Starter drive
- North (N) pole and south (S) pole of a magnet (field coils)
- Two halves of a split copper ring
- U-shaped conductor with its two ends connected to the two ring halves
- Stationary brushes connected via cable to a battery

The combination of the U-shaped conductor loop and the split copper ring is called the commutator because they rotate together. Current flows from the positive (+) battery terminal, through the brush and copper ring near the north (N) pole, and through the conductor, then to the copper ring and brush nearer the south (S) pole and back to the negative (–) battery terminal. The electromagnet is created in a loop of wire that is placed between two electromagnetic poles.

Motors work on the principle of magnetic repulsion. This magnetic repulsion takes place when a straight loop wire conductor made up of an armature, commutator, and brushes is located within a magnetic field and current flows through that wire loop. This situation creates two separate magnetic fields. One is produced by the magnet (poles of the magnetic field winding) and the other produced by the current flowing through the conductor (armature/commutator/brushes). The magnetic field of the magnet moves from the S pole to N pole, and the conductor magnetic field flows around the conductor.

The magnetic lines of force have a rubber band characteristic. The magnetic flux stretches and tries to shorten itself. There is a stronger magnetic field on one side of the looped conductor (armature-commutator-brushes) with a very weak magnetic field on the other side. Then the armature, which is the conductor, is repulsed by a strong magnetic field and turns toward the weaker magnetic field.

As the current in the conductor, which is the motor armature and the strength of the magnetic field windings, increases, the armature then rotates, and as it turns, the two sides of the conductor loop reverse positions and two halves of the split copper ring alternately make contact with the opposite stationary brushes. This causes the flow direction of electrical current to reverse (alternating current) through the commutator and the commutator to continue to rotate in the same direction.

FIGURE S.23 Starter or cranking motor.

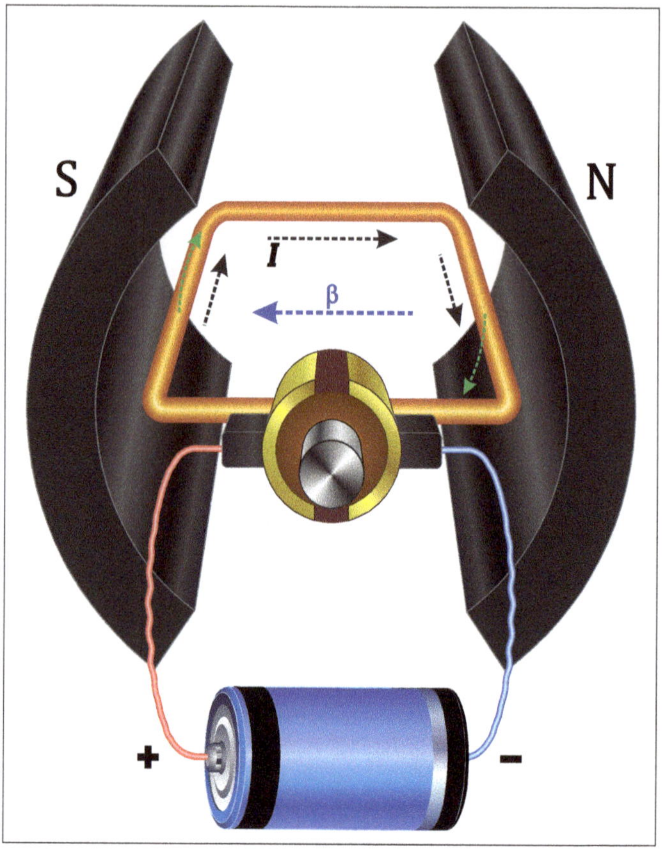

Starter Motor 12/24 V Protocol

"The scope of SAE J3053_201901 is to describe a design standard to define the maximum recommended voltage drop for starting motor main circuits, as well as control system circuits, for 12- through 24-V starter systems. This document is intended to give the industry a design standard to define the design of the electrical circuits used in conjunction with 12/24-V electric starting motors. The document will include all elements of the electric starter system main cranking and control circuits." For more information on *Heavy Duty Truck and Bus Electrical Circuit Performance Requirement for 12/24-Volt Electric Starter Motors*, see **J3053_201901** [16].

Starter Relay

Some starting systems use a relay for motor activation. A relay is a type of magnetic switch that basically allows a low amount of current to control a high amount of current. On starting systems with relays, when the ignition key is turned to the start position, a low current is sent to the relay instead of directly to the starter solenoid to operate the starter motor (cranking motor).

Starter Solenoid

An electromagnetic coil wound in the form of a cylinder. It is mounted on top of the starting motor. When energized, the solenoid pulls a plunger into the coil and holds it there. The plunger connects to a shift lever that connects to the pinion gear. When current is sent to the solenoid, the plunger pulls the shift lever, which moves the pinion gear into mesh with the flywheel ring gear. The solenoid contains two different sets of windings: the hold-in winding and the pull-in winding. A contact disc is also pushed into contact with the solenoid battery and starter motor terminals.

FIGURE S.24 Starter solenoid in an exploded view of a starter motor.

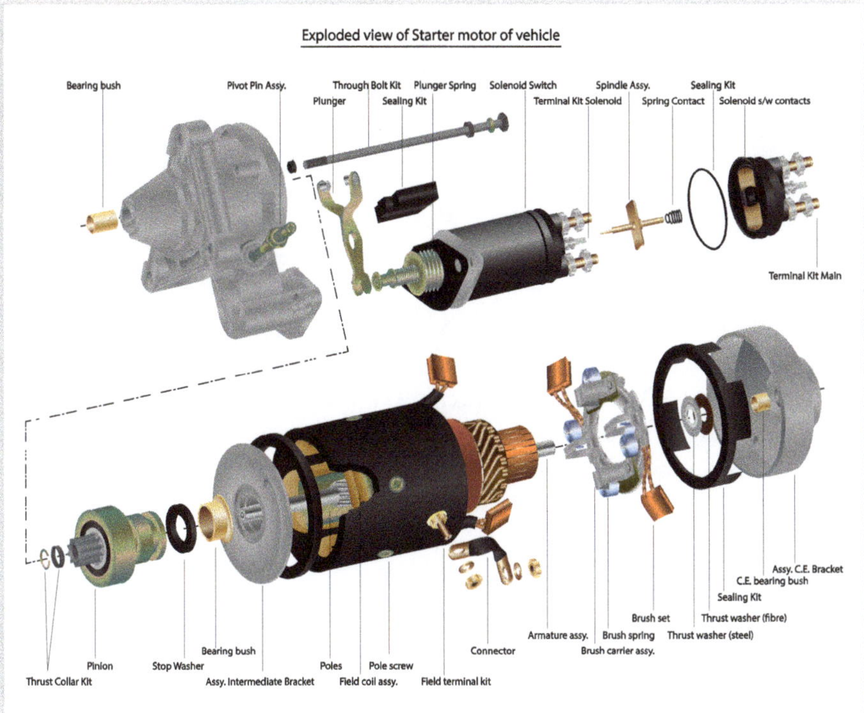

Starting System

The starting system coverts the electrical energy from the battery into mechanical energy to crank an ICE. This is done by creating torque, or twisting force. A small electrical motor, the starter, is able to turn with a lot of torque because it operates under a very high current. Reduction gear cranking motors contain a small transmission that creates additional torque to start the ICE.

FIGURE S.25 Starting system.

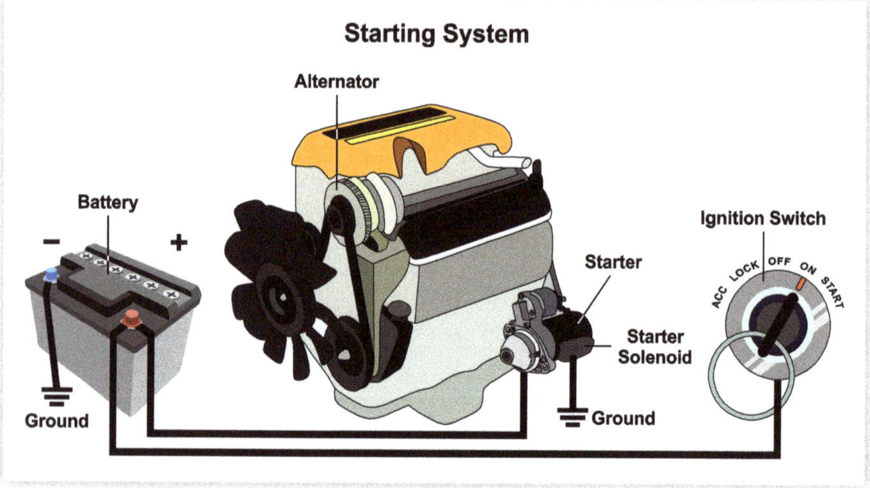

Static Electricity

Electricity is electricity at rest or without any motion. When someone walks across a nylon-based carpet, electrons are taken from the carpet surface and have an electrifying effect on both the substance where electrons are taken (carpet) and the moving body that performs the removal. When this person has obtained a sufficient charge differential that is measured in voltage, the excess electrons will be discharged when they touch a metal object such as a doorknob. An arc will occur dispersing the electrons and balancing the charge. This electrifying process results in both attractive and repulsive charges. Like charges repel and unlike charges attract in electricity.

FIGURE S.26 Static electricity.

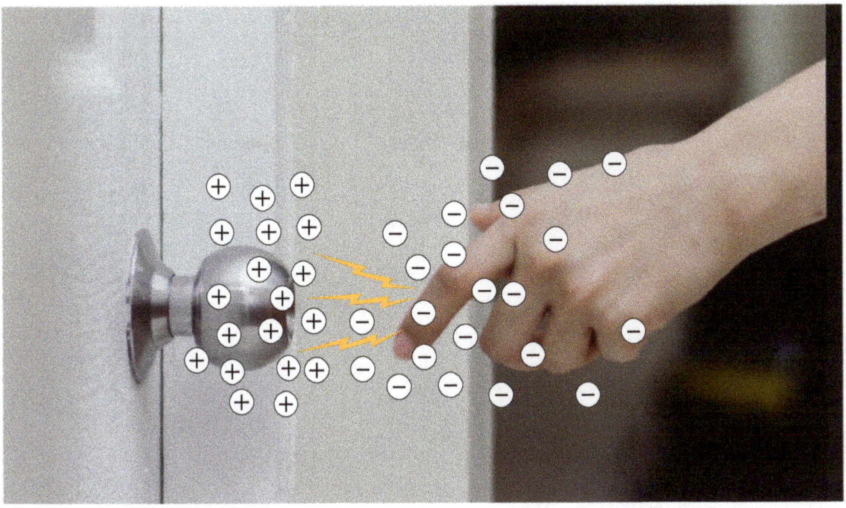

Steering Geometry

A geometric arrangement of linkages within the steering of a vehicle designed to undo the matter of keeping the wheels properly oriented through various positions of the steering and suspension systems. Due to the fact that the wheels move up and down relative to the body, the steering swings vertically through an arc. The wheel would turn in and out because the vehicle goes over bumps if it were not for steering geometry. The pivots for the suspension components cause the wheel to travel through the same arc because the steering components, allowing the wheel to trace straight ahead, or in an exceedingly consistent direction if it is during a turn.

Steering Knuckle

A suspension component that attaches the steering and suspension system to the wheel assembly. Knuckles are forged as heavy, one-piece constructions and provide attaching points for control arms, steering arms, tie rods, struts, and brake calipers, depending on the system used.

FIGURE S.27 Steering knuckle.

Steering System (Parallelogram Type)

A system that allows the driver of an automobile to direct its movement. It provides a means for proper handling, good directional control, and stability. The parallelogram and rack-and-pinion steering systems are the most common. In a parallelogram steering system, all the steering linkage connecting points on the right side of the vehicle are parallel to those on the left side. It allows the motion on one side of the vehicle to be duplicated on the other. The pivot points can be aligned with pivot points in the suspension system. As a result, the steering motions are transmitted to the front wheels independent of suspension status (jounce or rebound). Additionally, the turning angles are the same for both front wheels, allowing for toe-out on turns, regardless of the wheel vertical position. Both recirculating ball or integral non-rack-and-pinion and rack-and-pinion steering gears are designed to operate in a parallelogram.

FIGURE S.28 Parallelogram steering system.

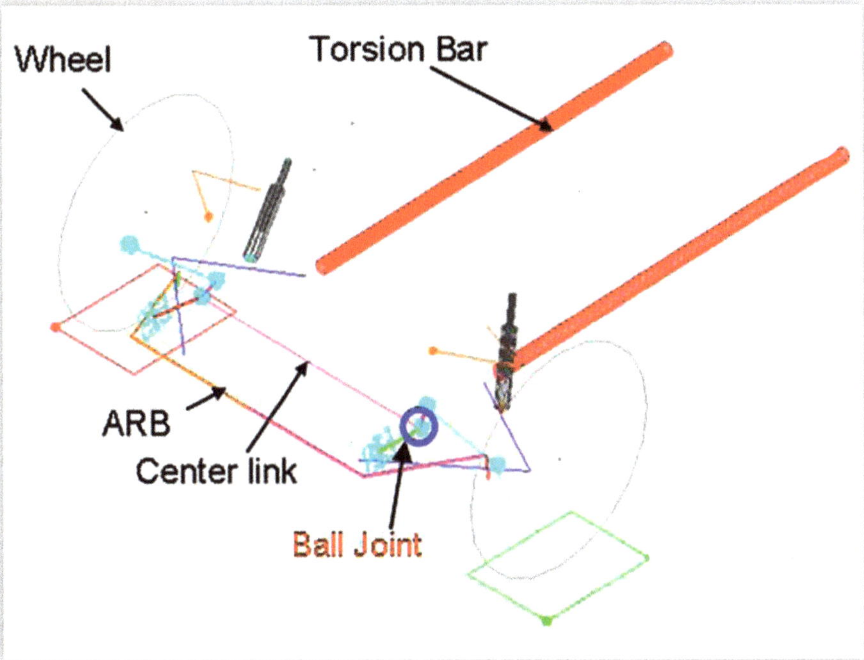

Steering System: Parallelogram Integral Steering Gear

The pitman arm is attached at one end to the sector shaft of the steering gear. The alternative end is connected to the centerlink (relay rod or draglink). The pitman arm is the lever that converts the rotation of the globe shaft into side-to-side (lateral) motion. The pitman arm is securely attached to the globe shaft by splines so any movement of the globe shaft is transmitted to the pitman arm and centerlink. The idler arm is attached to the choice end of the centerlink from the pitman arm. It moves within the identical plane because of the pitman arm, which keeps the linkage parallel. It always bolts to the vehicle frame. It supports and guides the outer end of the centerlink through the identical path (arc) because of the pitman arm.

The centerlink moves to the left and right under the vehicle because the wheel turns. The centerlink is connected between the pitman arm and also the idler arm. The connections between the pitman arm and idler arm are usually a ball-and-socket type; therefore, the centerlink can swivel and move through various angles. Tie rods are attached to the centerlink with similar ball-and-socket-type joints. Thanks to the actual fact that the centerlink moves, it causes the tie rods to maneuver the wheel assembly.

Draglinks are a variation of the centerlink and are used on many modern-day trucks and automobile vehicles. This design places the pitman arm in proper alignment with the wheel assembly and sometimes eliminates the need for an idler arm. The rod connects the centerlink or draglink to the wheel assembly. This causes any centerlink movement to be transferred to the wheels. Tie rods have ball studs (ball and sockets) on both ends to allow a full range of motion due to the vertical and turning movements of the wheels.

FIGURE S.29 Integral steering gear steering linkage.

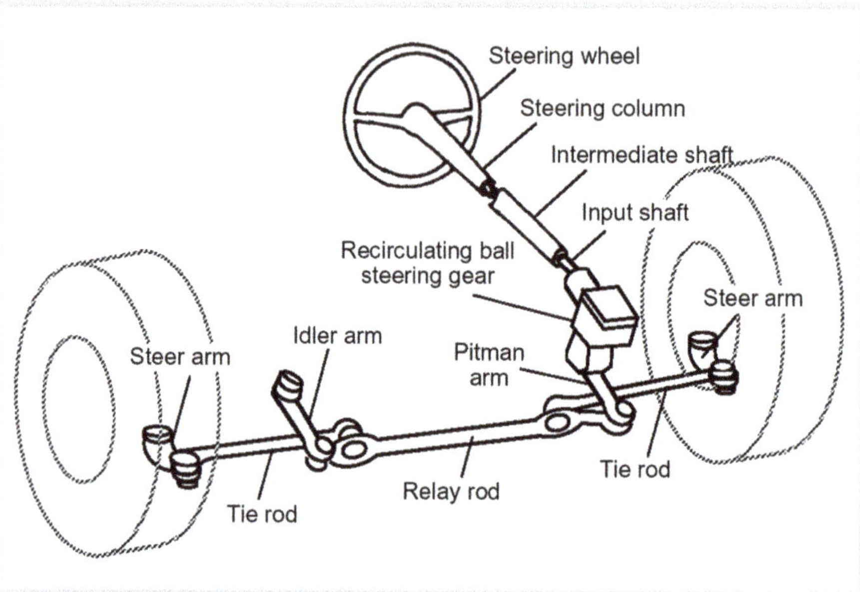

Steering System (Rack and Pinion)

The rack-and-pinion gearbox contains a pinion connected to the underside of the steering column. This pinion runs in mesh with a rack that is connected to the steering tie rods. This connection gives more direct operation, which enables the driving force to feel the road better. Turning the handwheel rotates the pinion and moves the rack from side to side. On end takeoff racks, ball sockets at the tip of the rack locate the tie rods and permit movement within the steering and suspension. On center takeoff racks, the tie rods connect with the middle of the rack and pinion. The middle of the rack is what moves on this variety of rack and pinion.

FIGURE S.30 Steering system comparison: Parallelogram to rack and pinion.

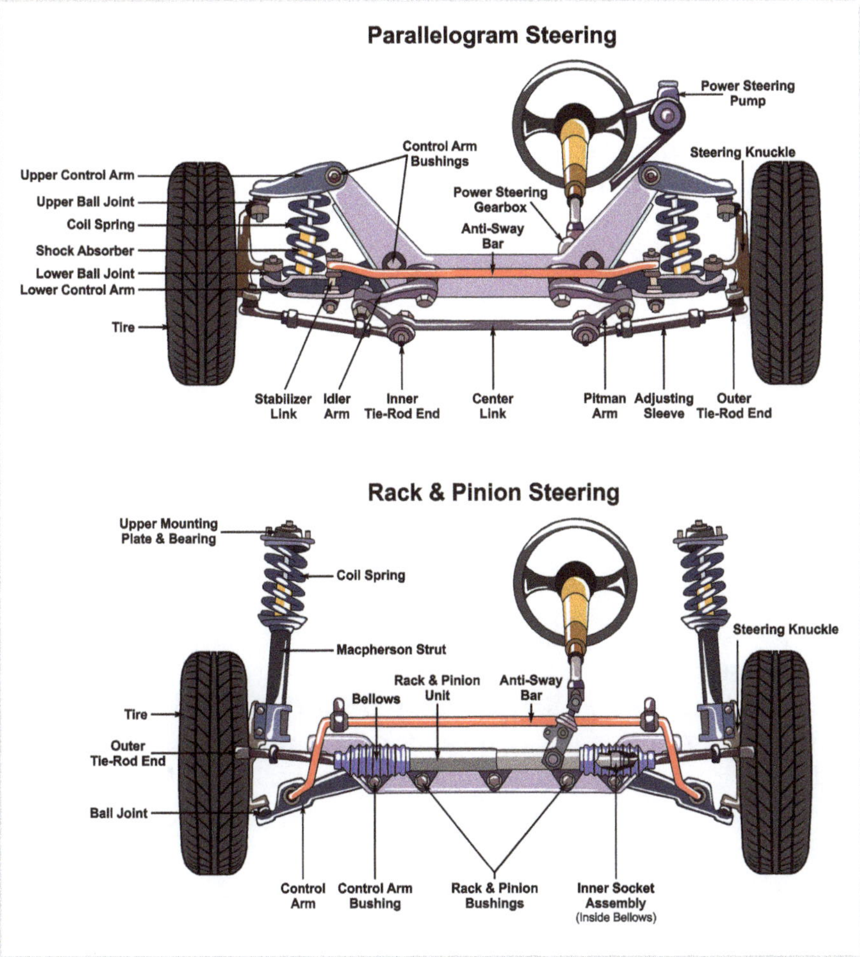

Stepper Motor

A motor in which the armature can be stopped at specific points in its rotation; also called a servo motor.

Stoichiometric Air-Fuel Mixture

An air-fuel mixture that is chemically correct at 14.7 parts air to 1 part fuel.

Stopping Distance (Brakes)

Braking stopping distance refers to the space a vehicle will travel from the purpose when its brakes are fully applied to when it involves a whole stop. It is primarily plagued by the first speed of the vehicle and therefore the coefficient of friction between the tires and therefore the paved surface, and negligibly by the rolling resistance of the tires and vehicle air drag. The sort of brakes in use only affects trucks and huge mass vehicles, which cannot supply enough force to match the static frictional force.

$$SD = \frac{Vo^2}{2\frac{V_o}{F_{xt}}} = \frac{Vo^2}{2D_x} \quad \textbf{(S.5)}$$

where

SD is the stopping distance
V_o is the forward velocity
F_{xt} is the total of all longitudinal deceleration forces on the vehicle
D_x is linear deceleration

Strain-Gauge MAP Sensor

MAP sensor that uses a piezo-resistor chip (a chip that changes internal resistance in response to pressure). The manifold pressure flexes the piezo-resistor, producing variable resistance and thus changes in current that represent pressure.

Strategy-Based Diagnostics

A scientific process of elimination to unravel an automotive diagnostic concern. It provides a way to logically narrow down possible problem sources.

Stroboscopic

A very bright, pulsing light triggered from the firing of one plug accustomed check and adjust ignition timing.

Stroke

The distance the piston travels from top dead center (TDC) when the piston is at the highest of its stroke to bottom dead center (BDC), when the piston is at the underside of its stroke. At TDC the crank is at some extent nearest the combustion chamber. At BDC the piston is farthest from the combustion chamber. The space between the piston and the head when the piston is at TDC is the clearance space or clearance volume.

Surface Finish

The smooth finish on a turning shaft against bearings. It's measured in micro-inches and also the smaller the quantity, the smoother the surface. The overall specification for main and rod crankshaft journals is between 10 and 20 roughness average (Ra). This very smooth surface finish is achieved by polishing the crank journals after the grinding operation.

Supercharger

Engine belt-driven vacuum pump that supplies gas pressure greater than 14.7 psi into the manifold to spice up engine torque and power by compressing the charge. It provides a direct increase in power without the turbo lag related to turbochargers. It takes horsepower far away from the engine to work and, thus, does not seem to be as efficient as a turbocharger. They operate on the positive-displacement principle, which suggests they work at a relentless rate of flow irrespective of the exit pressure and might operate with or without internal compression. The Roots-type supercharger does not have internal compression.

SuperCruise

The GM SuperCruise is a combination of many systems and modules that acts as a form of self-driving. It uses the following driver assist systems running simultaneously, which are also defined in this text:

- Active safety system
- Lane centering control
- Adaptive (radar) cruise control

Suspension System

Separates the vehicle body from road shocks and vibrations that might rather be transferred to the driving force and passengers. It also must keep the tires in touch with the road. A mechanical system must be strong enough to resist loads imposed by the vehicle mass during cornering, accelerating, braking, and uneven road surfaces. When a tire hits an obstruction, there is a reaction force, meaning the tire will move in response to the force applied by the obstruction. The dimensions of this reaction force depend on the unsprung weight (mass) of every wheel assembly. The sprung mass is a part of the vehicle supported by the springs, like the body, the frame, the engine, and associated parts. The unsprung mass is all the parts of the vehicle that are not supported by springs, including the components that follow the road contours, like wheels, tires, axles, and outboard brake assemblies.

FIGURE S.31 Suspension system isolates the vehicle body from road shocks and vibrations that might rather be transferred to the driving force and passengers.

Sweep Rate

The frequency or time division shown on the screen of an oscilloscope during each test.

Synchronizer

A connection or clutching device that is used to sync or match the speed of the engine-driven countershaft to one of the speed gears, so the shift will occur without gear noise. Most synchronizer assemblies ride on the output shaft between two gears.

A synchronizer assembly is named for the gears on either side of it, which are the two speeds that it engages. There are design variations, but most include a hub, a sliding sleeve, blocker rings, keys, and springs. In addition, the tapered cone and coupling teeth machined on the speed gear are part of the synchronizer assembly. Splines are attached to the center hub of the synchronizer to the output shaft, so the hub and output shaft rotate together. There are also splines machined on the outer circumference of the hub. An outer sleeve rides on the external hub spines with enough clearance so that it slides freely. The splines on the sleeve also match the small coupling teeth of the blocker ring and speed gear.

Coupling teeth are also called engagement or clutch teeth. The sleeve is splined to the hub, so it rotates with the output shaft. A blocker ring sits between the speed gear and the sleeve. The coupling teeth on the blocker (synchronizer) ring match those on both the sleeve and speed gear. The blocker or clutch ring also has a tapered cone to match the cone machined on the speed gear to synchronize the engagement.

FIGURE S.32 Manual transmission synchronizer.

Synchronous (AC Motor)

Refers to the rotation of the armature in a motor being timed with the three-phase alternating current (AC) entering the stator or coil windings. The three-phase AC flow sets up a three-wave magnetic field that moves around the stator to push and pull the armature around with it.

Synthetic Oil

Oil created by causing a chemical reaction in various complex molecules to form a new molecular structure. Petroleum, along with several types of acids, and alcohols can be used in fluctuating proportions to make synthetic base oils. They contain less wax than petroleum oils, so they tend to remain liquid when cold. This permits faster lubrication in very cold temperatures.

System Wiring Diagram

"A diagram that contains only the components and wires for a particular circuit. SAE J2202_201912 provides general guidelines on the material selection, construction, and qualification of components and wiring systems used to construct nominal 12 VDC and/or 24 VDC electrical wiring systems for heavy-duty vehicles The guidelines are limited to nominal 12 VDC and/or 24 VDC primary wiring systems and includes cable sizes American Wire Gage 20 to AWG 4 on heavy-duty on-highway trucks. The future of the heavy truck electrical systems will offer a multi-voltage electrical system, which will include voltages above a nominal 14 VDC electrical wiring system, along with new technologies and requirements not included in the current SAE J2202. With the need to include new technologies addressing higher voltage, it was decided to limit the scope of the SAE J2202 to a nominal 12 VDC and/or 24 VDC electrical system." For more information on this, see **J2202_201912** [17].

REFERENCES

1. Society of Automotive Engineers, Standard J1715/2_202108: Hybrid Electric Vehicle (HEV) and Electric Vehicle (EV) Terminology, SAE Publishing, Warrendale, PA, 2021.
2. Society of Automotive Engineers, Standard J1587_200202: Electronic Data Interchange between Microcomputer Systems in Heavy-Duty Vehicle Applications, SAE Publishing, Warrendale, PA, 2002.
3. Society of Automotive Engineers, Standard J1939/21_2021: Data Link Layer, Warrendale, PA, SAE Publishing, 2021.
4. Society of Automotive Engineers, Standard J1708_201609: Serial Data Communications between Microcomputer Systems in Heavy-Duty Vehicle Applications, SAE Publishing, Warrendale, PA, 2016.
5. Society of Automotive Engineers, Standard J2047_201911: Tire Performance Terminology, SAE Publishing, Warrendale, PA, 2019.
6. Society of Automotive Engineers, Standard J139_202002: Ignition System Nomenclature and Terminology, SAE Publishing, Warrendale, PA, 2020.
7. Society of Automotive Engineers, Standard J3085_201701: Terminology for Light and Medium Duty Trailers and Components, SAE Publishing, Warrendale, PA, 2017.
8. Society of Automotive Engineers, Standard J2627_200908: Braking System Definitions—Truck and Bus, SAE Publishing, Warrendale, PA, 2009.
9. Society of Automotive Engineers, Standard J3063_202103: Active Safety Systems Terms and Definitions, SAE Publishing, Warrendale, PA, 2021.

10. Society of Automotive Engineers, Standard J670_200801 Vehicle Dynamics Terminology, SAE Publishing, Warrendale, PA, 2008.

11. Society of Automotive Engineers, Standard J678_201108: Speedometers and Tachometers – Automotive, SAE Publishing, Warrendale, PA, 2018.

12. Society of Automotive Engineers, Standard J175_202107: Wheels - Lateral Impact Test Procedure - Road Vehicles, SAE Publishing, Warrendale, PA, 2021.

13. Society of Automotive Engineers, Standard J393_201710: Nomenclature - Wheels, Hubs, and Rims for Commercial Vehicles, SAE Publishing, Warrendale, PA, 2017.

14. Society of Automotive Engineers, Standard J1121_201608: Helical Compression and Extension Spring Terminology, SAE Publishing, Warrendale, PA, 2016.

15. Society of Automotive Engineers, Standard J1410_202012: Air Brake Valve - Performance Requirements, Warrendale, PA, SAE Publishing, 2020.

16. Society of Automotive Engineers, Standard J3053_201901: Heavy Duty Truck and Bus Electrical Circuit Performance Requirement for 12/24-Volt Electric Starter Motors, SAE Publishing, Warrendale, PA, 2019.

17. Society of Automotive Engineers, Standard J2202_201912: Heavy-Duty Wiring Systems for On-Highway Trucks, SAE Publishing, Warrendale, PA, 2019.

T

TAC (Thermostatic Air Cleaner)

A specific type of ICE air cleaner where heat is picked up from around the exhaust manifold and routed to the air cleaner inlet. A thermostatically controlled bimetallic switch adjusts a vacuum motor. It controls the amount of heated air used.

Tachometer

A device used to measure ICE speed in rpm (revolutions per minute). SAE J678_201108 applies to speedometers, odometers, and speedometer drives typical of passenger vehicles, buses, and trucks used for personal or commercial purposes. The method of determining wheel revolutions per unit distance (3.1) and overall system design variation (3.3.3) are applicable to passenger cars only. Comparable recommendations for trucks and buses are under development. The data of tachometers are applicable to vehicular use, as previously described, and also to stationary and marine engines and special vehicles. For more information on this, see **J678_201108** [1].

Tail Lamp

"Lights that are located in the rear of a vehicle to alert the presence of a vehicle to the motorist. SAE J2040_201601 provides test procedures, requirements, and guidelines for tail lamps intended for use on vehicles 2,032 mm or more in overall width. Tail lamps conforming to the requirements of this document may also be used on vehicles less than 2,032 mm in overall width. SAE J585_201406 provides test procedures, requirements, and guidelines for tail lamps (rear position lamps) intended for use on vehicles of less than 2,032 mm in overall width." For more information on this, see **J2040_201601** and **J585_201406** [2, 3].

Tail Shaft

The output shaft of a transmission or engine.

Tailpipe

The last exhaust pipe in the exhaust stream that carries the exhaust fumes out from under the vehicle and into the atmosphere.

Tall Firing Line

A tall spike or line on a parade pattern, which indicates high resistance in the ignition secondary caused by an open spark plug wire, a wide spark plug gap, a burned coil pack connection, or a burned secondary connection in a distributorless ignition.

TAME

Tertiary-amyl methyl ether (TAME) is a gasoline additive that aids in complete combustion. It is used to improve the octane number of the fuel. It is an oxygenated additive. For more information on this, see **J312_201902** [4].

Tandem Drive Axle

A heavy-truck rear final drive assembly that combines two single-axle assemblies with a power divider, connected via a short driveshaft.

FIGURE T.1 Class 8 truck tandem drive axle.

Ake Anichai Chumsri/Shutterstock.com

Tandem Master Cylinder

See **Dual Master Cylinder**.

Tank Sending Unit

A variable resistor used to indicate fuel level. It contains a float and an arm that operates a 90-ohm variable resistor. The float may be on an arm or on a shaft with a float that rides up and down the shaft depending upon the fuel level. Inside the float is a magnet, that closes reed switches to connecting a resistive load to the output presenting a voltage divider signal output for interpretation in the instrument cluster.

Tape Player

An electronic device that has tape heads that convert the magnetic charges on a tape into voltage signals as the tape slides over the heads. This device may be incorporated into a vehicle radio.

Tapered Leaf Spring

A single mono-leaf spring that uses a single strip of steel or fiber-reinforced plastic that is thicker at the center and tapers out along the ends of the spring. The leaf spring carries a proportional amount of load of the vehicle while balancing the stress load.

Tapered Roller Bearings

Adjustable tapered roller bearings are found mainly on the front spindles of rear-wheel-drive vehicles such as light trucks, vans, and older cars. They are also found on the rear spindles of some front-wheel-drive cars. This bearing consists of a cone assembly (inner race, tapered rollers, and a cage) which is mounted on the stationary spindle. A cup, or outer race, is press fit into the rotating hub and rotor assembly. The races and rollers carry the load while the cage keeps the rollers evenly spaced. Two bearings, an "inner" and an "outer," are used on each wheel. Both bearings require periodic lubrication and adjustment (preload) at specified service intervals.

FIGURE T.2 Tapered roller bearing.

Pongchart B/Shutterstock.com

Tappet

See **Cam Follower** or **Lifter**.

Task Manager

On OBD II systems, the PCM incorporates a special segment of software. Ford and GM systems call this software the diagnostic executive. On Fiat Chrysler systems, it is called the task manager. For more information on this, see **J1930DA_202105** [5].

TBI (Throttle Body Injection)

This form of electronic fuel injection was introduced by GM through Rochester Carburetors and Delco Electronics in around 1981. It was used for several decades and was a throttle body with one or two electronic fuel injectors positioned over the throttle plates. This was a full authority system using a number of input sensors including a closed-loop oxygen sensor.

TCC (Torque Converter Clutch)

A TCC or lock-up converter pressure plate connects the turbine to the converter housing to transfer torque from the crankshaft through the torque converter pump housing to the converter turbine to the transmission/transaxle. The torque converter clutch applies during third or fourth gear operation after the transmission control module receives the appropriate input signals to energize the TCC solenoid, and if equipped, the PWM solenoid is ramped from 0 to 100% duty cycle. For more information on automatic transmissions, see **J649_198807** and **J648_201106** [9, 10].

TCM (Transmission Control Module)

Transmission management computer that controls the transmission by receiving and processing signals from various switches and sensors. This computer controls shift sequences, shift timing, and clutch apply and release characteristics. It is an independent controller. The transmission fluid pressure switch and also the internal mode switch provide driver input to the TCM. Other data sent to the TCM include throttle position; engine, turbine, and output speeds; and sump temperature. Any active special function, like ABS (anti-lock brakes) or PTO (power takeoff), is additionally an input to the TCM.

The TCM processes this data to work out proper shift points, to look at this range, to perform ratio tests, and to collect diagnostic data. The TCM is programmed to protect the transmission and other vehicle driveline components by inhibiting actions, like full-throttle neutral-to-range shifts and high-speed direction changes. The TCM determines if a system malfunction exists and stores diagnostic codes related to the malfunction. The codes, accessed by the service technician, are utilized in diagnosing persistent or intermittent trouble within the system using an electronic service tool. The TCM receives input on throttle position/torque management from an indication transmitted by the engine control module (ECM). The ECM communicates to the transmission electronic controls over an SAE J1850 or J1939 serial communication interface electrical circuit. The TCM must be calibrated to receive these signals. The TCM must be calibrated to receive these signals. For more information on this, see **J1930DA_202105** [5].

TCS (Traction Control System) Car and Light Truck

A secondary function of the electronic stability control on production vehicles, designed to prevent loss of traction (wheel spin) of the driven road wheels. TCS is activated when throttle input and engine power and torque transfer are mismatched to the road surface conditions.

TDC (Top Dead Center)

TDC is a term identifying the position of an ICE piston at the top of the cylinder compressing the air-fuel mixture into the combustion chamber quench area.

It is the highest point in the cylinder that the piston can travel. The measurement from bottom dead center (BDC) to TDC determines the stroke length of the crankshaft. The piston is at the top of its travel in the center of the ICE.

Technical Assistance Hotlines

A telephone or Internet support line that allows technicians in the field to communicate directly to specially trained technical support personnel at the OEM or aftermarket level.

Telematic Satellite Navigation

Using a GPS and electronic mapping tool to enable the driver of a vehicle to locate a position, plan a route, and navigate a journey.

Telematics

Field of telecommunications, like OnStar vehicular technologies, road transportation, road safety, applied science (sensors, instrumentation, wireless communications, etc.), and computing (multimedia, Internet, etc.):

- The technology of sending, receiving, and storing information via telecommunication devices in conjunction with effecting control on remote objects
- Integrated use of telecommunications and informatics for application in vehicles and with control of vehicles on the move
- Global navigation satellite system technology integrated with computers and mobile technology in automotive navigation systems

The word telematics comes from the French word télématique which refers to the transfer of data over telecommunications.

Telltale Light

Dash warning light that is referred to as an idiot light.

Temperature

Coldness or hotness of something solid, liquid, or gas. It is the measure of the speed of the molecule vibration of that thing. An increase in temperature indicates that the speed of the molecules has increased. When you remove heat from a body, it becomes cold with a lower temperature: heat always flows from the hotter (warmer) object to the colder object. A drop in temperature means a decrease in molecular speed. Temperature is not a measure of the total quantity of heat; instead, it is a measure of the degree of heat that something possesses and tells you whether something has gained or lost heat. A thermometer is used to measure temperature such as an infrared digital thermometer is used [6].

Temperature Gauge Circuit

A circuit that uses a sending unit located in an engine coolant passage to operate the gauge in the dash to indicate engine temperature.

Terminal

A connection point for two or more wires.

Tesla Autopilot

Advanced driver-assistance system feature offered by Tesla that has lane centering, adaptive controller, self-parking, and ability to automatically change lanes without requiring driver steering, and enables the car to be summoned to and from a garage or parking spot. Planned improvements to Enhanced Autopilot include transitioning from one freeway to a different and exiting the freeway when your destination is near. As an upgrade above and beyond the capabilities of Enhanced Autopilot, the company's stated intent is to supply full self-driving at a future time, acknowledging that legal, regulatory, and technical hurdles must be overcome to attain this goal.

Test Light (12 V)

A 12-V test light looks like a screwdriver and is used to check a circuit for power or continuity. It uses an alligator clip to connect to the ground and at the other end is a scratch awl probe with a 12-V light bulb that is used to tough the positive circuit. Also called a continuity tester.

FIGURE T.3 A 12-V test light used on 12-V non-electronic circuits to test for continuity.

Tetraethyl Lead

A lead additive used in gasoline to reduce detonation or spark knock and to boost octane. It was prohibited by the EPA. For more information on this, see **J312_201902** [4].

Thermal Efficiency

The thermal efficiency of an ICE is the proportion of the heat energy furnished by the fuel that acts at the crankshaft. If two engines produce the same horsepower at the crankshaft, and one burns more fuel than the other, the thermal efficiency is inferior for that engine. The heat energy supplied by the fuel is lost in the hot exhaust gases because the coolant surrounding the cylinders carries away some of the heat of combustion to cool the engine and some are used in overcoming friction. What heat is left is used for engine output.

FIGURE T.4 Thermal efficiency.

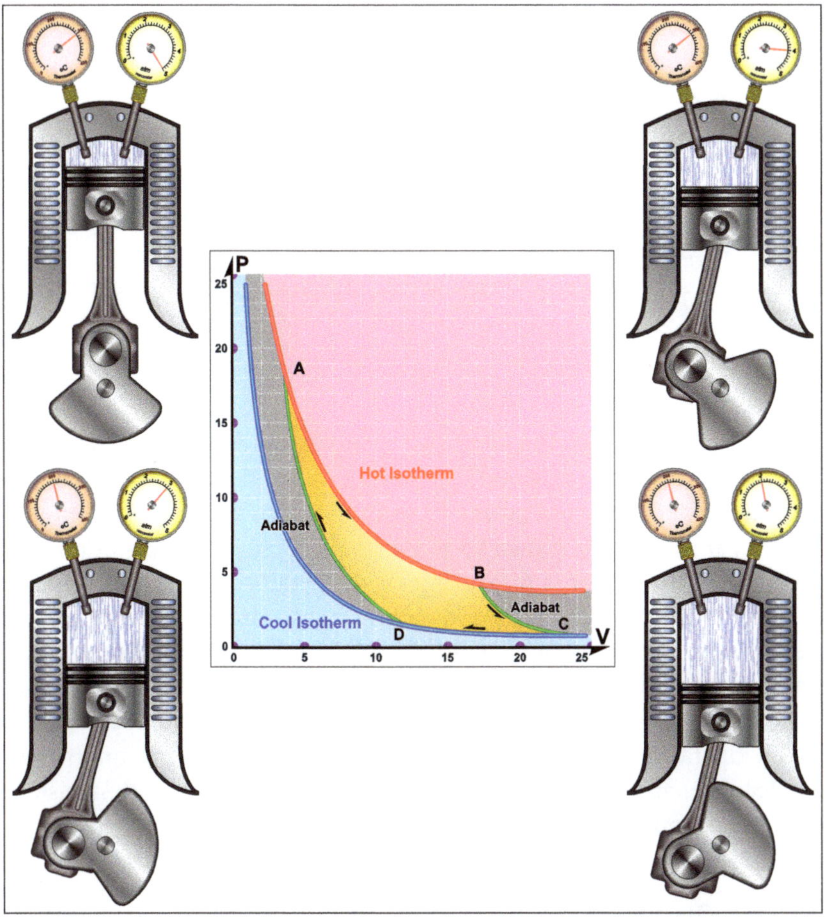

Thermal Expansion

This term describes the dimensional changes exhibited by solids, liquids, and gases during changes in temperature at constant pressure. The solid engine block must constantly and efficiently deal with different types of liquids and gases. Engine coolant has to be able to withstand drastic temperature changes. This liquid tends to expand and contract based on these differences. Throughout the expanding and contracting events, the coolant must withstand the pressures in the engine while maintaining integrity. You multiply the thermal constant of expansion by the temperature change.

$$\Delta L = \alpha_L L_o \Delta T \quad \text{(T.1)}$$

where

L_0 is the original length
α_L is the length expansion coefficient
ΔT is the temperature difference
ΔL is the change in length

Thermal Loading

The heat and temperature effect on the mechanical and structural components of an ICE.

Thermistor

A thermistor is a temperature-sensitive rheostat commonly used to measure the temperature of the coolant, oil fuel, and intake air. The name itself combines the terms resister and temperature. Its resistance is in proportion to temperature. They provide analog data to ICE management computers as an input sensor. When the sensor is measuring air temperature, the sensor is usually constructed with a plastic body to reduce heat transfer from surrounding metal. When measuring coolant or oil temperature, the sensor element is enclosed in a brass case to make it more awake to natural processes. Thermistors are solid-state components with no moving parts. There are negative (NTC) and positive (PTC) temperature coefficient thermistors. In the NTC thermistor, the resistance decreases as temperature increases. In the PTC thermistor, the resistance increases as temperature increases.

Thermocouple

When two dissimilar metals are heated and electrical connections take place.

Thermodynamic Cycle

The sequence of the operating stages of an engine. See **4-Stroke Cycle**.

Thermodynamics

Deals with energy and the various concepts and laws that describe the conversion of one form of energy to another along with the systems used to affect these conversions. These systems are in balance, and the quantity of matter or stuff under consideration is generally called the system and everything else is called the surroundings. There are open and closed systems. In a closed system there is no interchange of matter between systems and surroundings; in an open system, there is an interchange in what is called a process. Any process or series of processes where the system returns to its original condition is called a cycle. An ICE cooling system can be closed or open, but they are actually a combination of the two.

Thermoelectric Principle

The production of current flow created by heating the connection of two dissimilar metals.

Thermoelectricity

Voltage resulting from an unequal transfer of electrons from one metal to another when one of the metals is heated. Applying heat to the connection point of two dissimilar metals can create electron flow, which is thermoelectricity. It is known as the Seebeck Effect. It is called a thermocouple, which is a small device that gives off a low voltage when two dissimilar metals are heated. The pyrometer, which is a thermocouple, is used to measure exhaust gas temperature and consists of two dissimilar metals (iron and constantan, a copper-tin alloy) joined at the positive end and connected to a voltmeter at the gauge end. As temperature increases at the hot end, the reading will increase at the millivolt meter on the display gauge.

Thermosiphon Coolant Circulation

ICE coolant circulation from the radiator to the engine block, which relied on the forward movement of the vehicle. The fan helped to move enough air through the radiator to cause a difference in temperature between the radiator and the engine block to cause the thermosiphon effect.

Thermostat

A device that controls the coolant flow in an ICE cooling system. It is also used to reduce engine warm-up time, maintain engine operating temperature, and heat the inside of the vehicle. The thermostat consists of a valve that is opened by a thermal expansion device that is generally a wax pellet and then it is closed by the return spring. The valve opening is directly related to the coolant temperature around the thermostat. The thermostat is usually installed between the cylinder head coolant outlet and the inlet to the radiator, but there are other locations, depending on engine design. There are also computer-controlled electric thermostats.

FIGURE T.5 A thermostat is used to regulate the temperature in an ICE: thermostat and thermostat flow.

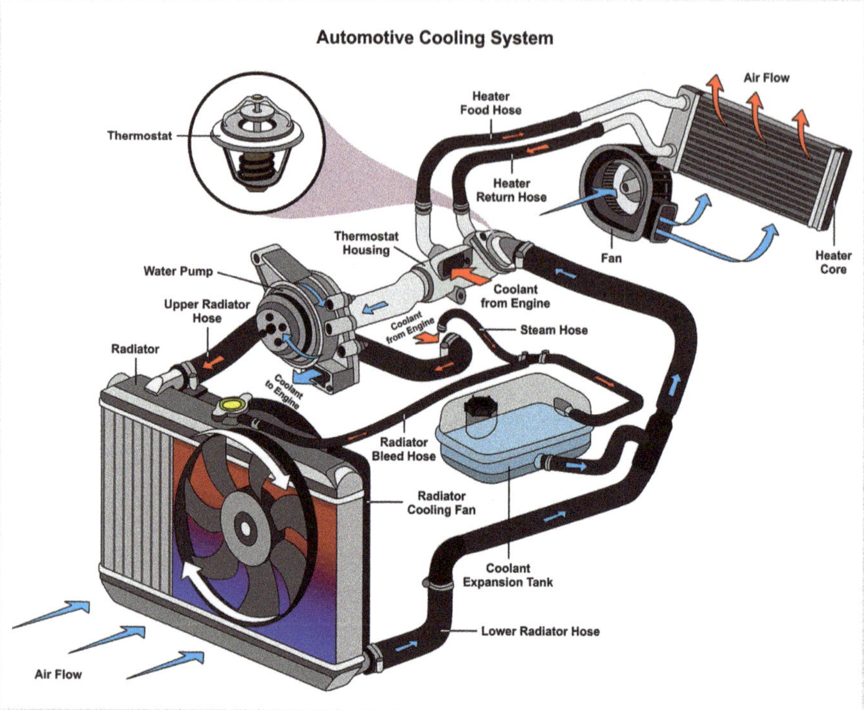

Thixotropic

The thickness of a material like paint, which is solid in its undisturbed state but fluid when stirred.

Three-Pedal System

Automated manual truck transmission that uses a driver-actuated clutch pedal. The driver is required to use the clutch pedal to break torque when stopping or starting or as an option. So the truck has three pedals: accelerator, brake, and clutch pedals.

FIGURE T.6 Three-pedal truck system brake clutch-accelerator pedal-manual transmission.

Three-Quarter-Floating Axle

In this configuration, the bearings are mounted in the brake drum or rotor hub, which is mounted on the axle housing. The outer extension of the hub fits onto the end of the axle, which is usually splined and tapered, and the axle shaft splines to the side gears inside the differential. The wheels are mounted on the hub with lug bolts or nuts, and like the full-floating axle, the axle housing and bearings in the hub support the weight.

Three-Wire Oxygen Sensor

The three-wire sensor design uses an electric resistance heater to help get the O_2S up to temperature more quickly and to help keep the sensor at operating temperature even at idle speeds. For more information on this, see **J1930DA_202105** [5].

Threshold Monitoring

Measures the level of exhaust emissions using one of the input sensors to the engine management computer used to identify a noxious emission. Diesel engines use ammonia, PM (particulate matter), or NOx (oxides of nitrogen) sensors as input sensors for threshold monitors. Gas engine cars under OBD II use an oxygen sensor for this function, and also some diesel engines use an oxygen sensor for the diesel oxidation catalyst. For more information on this, see **J1930DA_202105** [5].

Throating

The valve seat process of raising the valve seat by using a 60° or 70° cut stone or valve seat cutter.

Throttle

The narrow end of an opening like the Venturi of a carburetor. It is the common name for the butterfly valve that controls the inlet of air to an ICE.

Throttle Body

The housing that contains the ICE throttle valve to regulate the airflow into the intake manifold. In an electronically fuel-injected car, a throttle position sensor and air flow sensor communicate with the engine management computer, which supplies the corresponding amount of fuel needed at the injectors.

Throttle Body Electronic Fuel Injection

The use of a pulse width modulated electronic fuel injector placed over the Venturi of a throttle body similar to a carburetor.

Throttle Cable

A Bowden-type cable that replaces the mechanical linkage that connects the vehicle accelerator pedal to the throttle valve.

Throttle Relaxer

Some traction control systems use a throttle close-down strategy that closes the throttle when commanded by the ABS/TCS controller. This may be done by a throttle relaxer or throttle adjuster assembly.

Throttle Valve

Valve mechanism or butterfly where fluid flow is reduced or increased depending on the opening. An ICE power can be increased or decreased by the restriction of inlet gases using a throttle valve. The term throttle means any mechanism where the power or speed of an engine is regulated.

FIGURE T.7 Throttle valve in a throttle body.

Throttle Valve Detent Control

An automatic transmission/transaxle range downshift device. There is a detent or downshift valve in the unit that is operated by a detent cable or mechanical linkage. When the accelerator pedal is pushed to the floor, it moves the downshift/detent valve open to a fluid passage. The fluid pushes back on a shift valve and pushes it to downshift to a lower gear. In the past it was called passing gear.

Through-Drive Differential Power Flow

The most common type of through-drive differential is the single-reduction drive unit, which is used on most on-highway trucks. The single-reduction rear axles are made up of the following two main components: rear axle housing and differential carrier assembly.

Thrust Angle

Direction the rear wheels are pointing. If the rear suspension is not damaged and the rear toe is properly adjusted, the thrust line and the geometric centerline of the vehicle are the same. The thrust angle is the difference between the thrust line and the geometric centerline. A thrust angle to the right is positive; a thrust angle to the left is negative. The thrust angle is measured in degrees. When the thrust angle is too large, it may create problems such as dog tracking and an off-center steering wheel. Dog tracking, or a slight sideways movement, occurs when the thrust angle is excessive and the driver must turn the steering wheel to one side to keep the vehicle moving straight ahead. Even so, the vehicle travels slightly sideways, or dog tracks. In addition, excessive thrust angle is one of the primary causes of a steering wheel that is not centered. Thrust angle measurements provide important diagnostic information. For example, if the left rear is toe-in and the right rear toe-out on a vehicle with a fixed rear axle, it is possible that the axle has shifted or the vehicle suffers from a damaged frame rail.

FIGURE T.8 Steering angles.

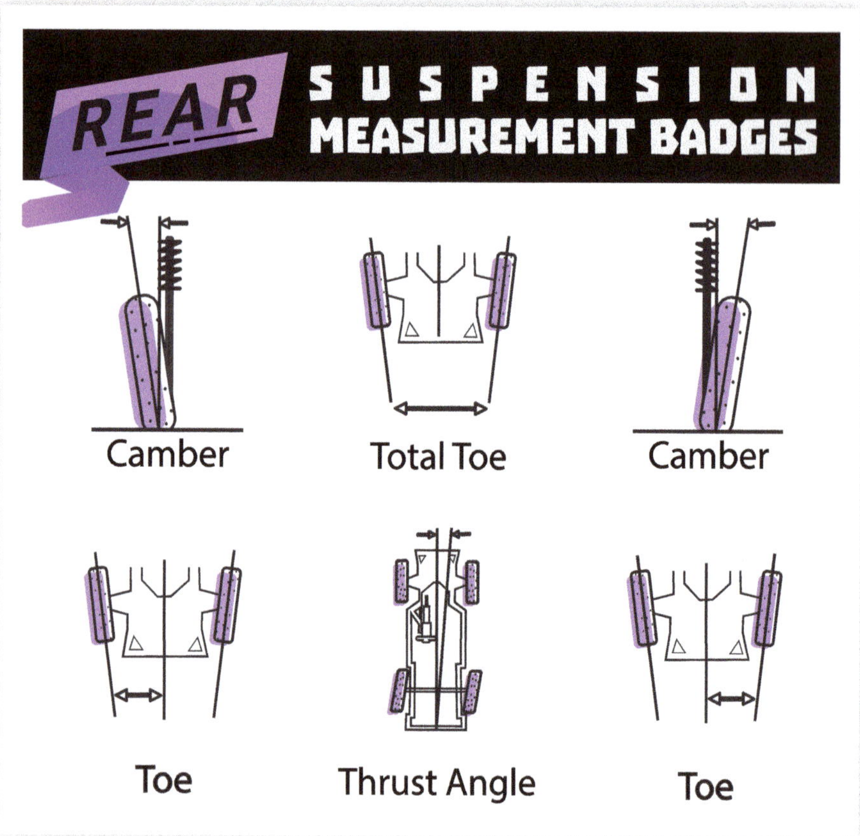

Chimera_art/Shutterstock.com

Thrust Bearing

An ICE or transmission bearing that can be split or solid that controls the axial thrust of a crankshaft or main shaft.

Tie Rod

Connects the steering knuckle to the rack on rack-and-pinion steering and to the drag link or relay rod on parallelogram steering. The inner tie rod is attached to the top of the rack and allows for suspension movement and slight changes in steering angles. The outer tie-rod end is attached between the tie-rod shaft, and therefore the steering arm. It transfers the movement of the rack. The outer tie rod pivots because the rack is extended or retracted when the vehicle is negotiating turns. Some tie rods and tie-rod ends are left or right threaded, which permits toe-in or toe-out to be adjusted to the OEM specifications. Toe-in may be a condition where the fronts of the wheels, as seen from above, are closer together than the rears of the wheels. Toe-out is the opposite condition where the rears of the wheels are closer together than the fronts. A tie-rod end connects the middle link to the steering knuckle on vehicles with parallelogram steering systems.

FIGURE T.9 Tie rod on a rack-and-pinion steering.

Timing

The arrangement of events in an ICE, such as the opening and closing of the poppet valves or at what crankshaft position should the spark plug should fire.

Timing Belt

A toothed cog belt that connects the ICE crankshaft sprocket to the camshaft(s) sprockets and must be timed to the correct position by design.

FIGURE T.10 Timing belt.

Timing Chain

A Morse single or double roller chain that connects the ICE crankshaft sprocket to the camshaft(s) sprockets and must be timed to the correct position by design.

FIGURE T.11 Timing chain.

Timing Light

An oscillating strobe light that is triggered by the activation of an ignition spark through an inductive pickup used to set and adjust the ICE ignition timing.

Tinning

The process of cleaning the tip of a soldering iron or gun to solder electrical wiring to components. You place a thin coat of solder on the iron or gun tip after cleaning.

Tire

Tires are hollow, donut-shaped structures that are mounted onto the wheels. They produce traction to the wheel assembly and act as a cushion to absorb shock from road surfaces. Tires must perform effectively and consistently under different and constantly changing driving conditions. These include variations within the horizontal surface (gravel, asphalt, dirt, concrete), temperature fluctuations (freezing, thawing, heat), and wet and dry conditions.

While originally made up of solid rubber, tire design evolved within the late 1800s to introduce pneumatic, or air-filled, models. Early pneumatic tires for automobiles were first introduced in France by Andre and Edouard Michelin, consisted of an inner, air-filled tube and an outer casing to safeguard the tube and provide traction. Pneumatic tires design evolved again as tubeless tires began replacing those with inner tubes and eventually became the norm on cars and most other vehicles.

The inner liner of the tire is formed from a synthetic product that is somewhat softer than the rubber used on the sidewalls and treads. The inner liner must be flexible and airtight. It helps protect the plies and prevents air loss from small punctures. The layer immediately above the inner liner consists of body plies. Plies are a form of reinforcing material, or the fabric of the tire, that gives the tire the ability to hold its shape. The plies are made of many thin, strong cords that are coated with a rubber compound.

Until about the late 1950s, cords were fabricated from cotton; modern fiber cords are now made from nylon, aramid, rayon, and polyester. Steel cords, called cables, could also be utilized in ply construction. The cables are made of brass-coated strands of steel wires that are twisted together. Plies give the tire its shape or structure, so they are called the tire carcass. The bead area is the portion of the tire that grips the wheel rim, connecting the tire with the wheel. The beads are bands of bronze-coated high-tensile steel wire located at the underside edges of the tire carcass. The cord plies wrap around the bead to form a section of the stiffness to hold the bead against the bead seat and to seal to the wheel rim. The length of the wire used for the bead is determined by the rim diameter of the tire. The bead fits inside the rim of the wheel and is held there in a channel when the tire is inflated.

Tires are going to be inflated through a valve assembly located within the wheel. Removing the tire from the rim might be a challenging job that needs a special tool called a bead breaker to separate the tire from the wheel. For more information on this, see **J2047_201303** [7].

FIGURE T.12 Tire construction.

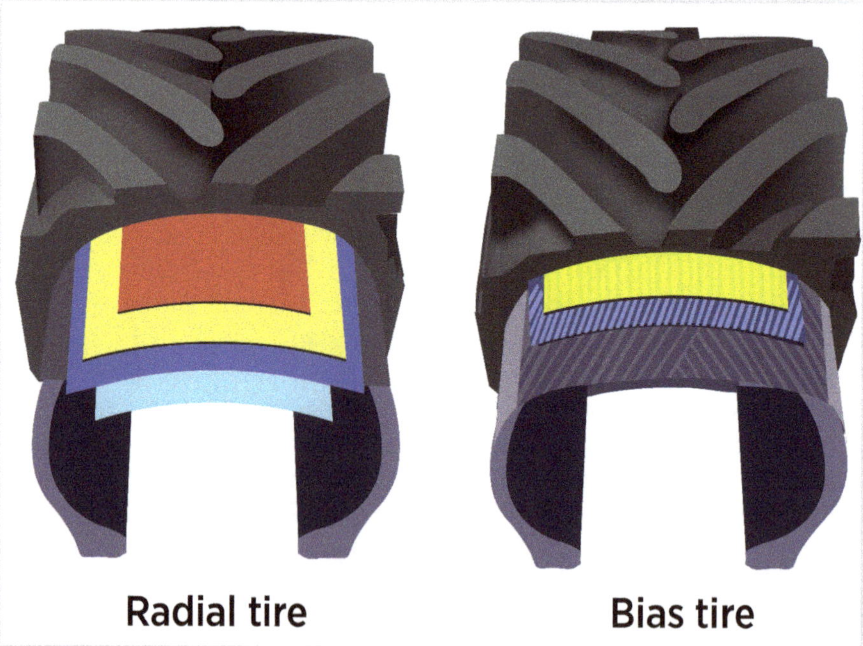

Tire Bead

The bead is located where the tire grips the inside of the wheel rim. It is constructed of many turns of copper or bronze-coated steel wire. The layers of tire body material are wrapped around the bead. For more information on this, see **J2047_201303** [7].

Tire Belt

The tire belt is two or more layers of material applied over the body plies and under the tread area only to stabilize the tread and increase tread life and handling. Belt material can use the following materials:

- Steel mesh
- Nylon
- Rayon
- Fiberglass
- Aramid

For more information on this, see **J2047_201303** [7].

Tire Blowout

An unexpected loss of air pressure in an inflatable tire with a volatile sound. It is called a blowout because of the sudden loss of air pressure with a loud pop sound where the air blows out of the tire. Low air pressure, extreme heat, impact damage, overloading, or a combination of these factors can contribute to over-deflection of the tire, which is a combination of vehicle load versus tire pressure.

Tire blowouts were common during the tubed tire period. In the use of inner tube tires, a puncture in the tube was considered a blowout. Tubed tires were very vulnerable to blowout failures. Blowouts are less common with steel-belted radial tires. The design of the steel-belted radial tire is such that a puncture is more likely to cause slow air loss. Most of the time what is called a blowout is actually tread separation. Most of the time it is due to a defect in the tire as in belt adhesion or underinflation. Underinflation causes the sidewalls to flex and generate heat. Low tire pressure causes a large amount of the tire surface area to be in contact with the road, which increases friction. Increased friction will cause the tires to overheat, which will lead to premature wear, tread separation, and the tire blowing out. For more information on this, see **J2047_201303** [7].

FIGURE T.13 Tire with tread separation is often referred to as a blowout.

Tire Body Ply

This is where a tire gets its strength. Body plies are layers of fabric wrapped around both beads under the tread and sidewall rubber. The body ply creates the tire carcass and these body plies may also be called carcass plies. A four-ply tire has four separate layers of fabric. If the body plies overlap at an angle (bias), the tire is termed a bias-ply tire. If just one or two body plies are used and that they do not cross at an angle, but lie directly from bead to bead, then the tire is named radial ply. For more information on this, see **J2047_201303** [7].

Tire Cap Ply

An extra layer or two of polyester fabric or nylon to help hold everything in place. These cap plies are not found on all tires; they are mostly used on tires with higher speed ratings to help all the components stay in place at high speeds. For more information on this, see **J2047_201303** [7].

Tire Inner Liner

The soft rubber lining on the within of the tire that protects the body plies and helps provide for self-sealing of small punctures. When a tire is assembled on a tire-building machine, the body plies, belts, and tread rubber are spliced together. The material is overlapped approximately five threads. The purpose where the bulk of those overlaps occur is named the foremost splice, which represents the stiffest part of the tire. For more information on this, see **J2047_201303** [7].

Tire Ply Rating

Expression of the tire strength and capacity. The ply rating translates in an exceedingly straightforward thanks to tire capabilities and limitations, especially when it involves inflation and cargo capacity. Tire ply rating is more commonly expressed as load range. The times of cotton layers or plies forming the inner construction of tires are long gone; therefore, the ply rating does not contain a direct connection to the tire construction and number of layers. With modern tire engineering and materials, it is possible to form greater tire strength and capacity without adding layers. For more information on this, see **J2047_201303** [7].

Tire Pressure Sensor

A tire pressure sensor used for the tire pressure monitoring system (TPMS) that is integrated into the air valve in the tire rim and operates with a lithium battery, which lasts about 10 years. If the battery voltage drops, the tire pressure monitor receiver stores a diagnostic trouble code (DTC) into its memory. If the voltage drops even further and the sensor stops working, the tire pressure monitor receiver stores the DTC in memory and illuminates the tire pressure warning light to alert the driver. For more information on this, see **J2657_201912** [12].

Tire Regrooving

Tire regrooving involves recutting the grooves of a worn-out tire. A handheld tool is usually accustomed to perform the method, and one wrong move can render the tire unusable. The grooves are recut to confirm that the tires have a tread depth of 4 mm. The new tread depth may be the maximum amount of 25% of the initial tread depth. It is done to extend the sturdiness of the tires. The tire does not have to be dismounted and might be performed on mounted tires. For more information on this, see **J2047_201303** [7].

Tire Retread or Recap

A process of placing a tread strip around the circumference of a second used tire carcass or casing with a bonding media interposed between the strip and casing. Cover the tread strip with a high temperature nylon film having perforations uniformly distributed across the surface of the said film. Cover the tread strip and film with an envelope having a negative embossment of the tread design aligned with the positive tread design on the tread strip, and then cure the said tread strip and bonding media at a minimum temperature of substantially 300°F. The high temperature nylon film incorporates a thickness within the order of 0.068–0.085 in. The perforations are spread across the surface of the film. These perforations are within the order of 1/16–1/8 in. in diameter and are separated from one another by a distance within the order of roughly 3/4 to 11/4 in. For more information on this, see **J2047_201303** [7].

Tire Sidewall

The tire part between the tread and also the wheel. The sidewall contains all the scale and construction details of the tire. Some tires turn brown on the sidewalls after a brief time because of atmospheric damage that causes the rubber to oxidize. For more information on this, see **J2047_201303** [7].

Tire Sipes and Grooves

Most radial tire treads are designed for year-round operation. Grooves are provided to disperse water that could cause hydroplaning. The grooves also allow space for the rubber to shift as the tire goes through its stress cycle. If the space were not available, the rubber in the tread would pile up in the front of the contact patch. Small grooves (also called sipes) are also added inside the tread which open under tire flexing, increasing the traction edges of the tire. For more information on this, see **J2047_201303** [7].

Tire Stress Cycle

Tires go through stress cycles as they rotate. The tire contracts as it enters the road contact patch area, squeezing the tread. The tire expands as it leaves the patch area. This expansion and contraction causes tire wear as the tread slides across the road surface. Radial tires eliminate most of this stress. The slip angle is much lower in radial than bias-belted tires, improving traction. Radial tires have the same load rating as bias types. Radials improve traction and gas mileage because the tread is more rigid and the sidewall more flexible. Therefore, there is less rolling resistance, improving gas mileage. For more information on this, see **J2047_201303** [7].

Tire/Wheel Runout

Runout measurements are taken with a runout gauge. A runout gauge is a dial indicator with a roller on the plunger with the gauge mounted on a stand. For a radial runout check, place the plunger wheel against the tire tread (in an area of least tread interference) and rotate the tire. Mark the tire in a number of locations and record the readings. Average the readings and compare each reading (area of the tire) to this average to see how much it varies from the true center.

Measurements are compared to the specifications found in the tire section of most service manuals. Tires that are out of specification should be match mounted to the wheel to reduce runout. Rotate the tire for almost half a turn on the wheel and recheck. Mark the high spot of a tire and align it with a low spot on the wheel. If the wheel does not have a high or low spot, check for runout in the hub (wheel studs in drum or rotor). Relocation of the tire/wheel assembly to the hub may eliminate the excess runout. If the tire cannot be brought to specification, replace it.

Lateral runout is measured on the sidewall of the tire and side of the wheel. This is checked the same as radial runout: by marking the tire and comparing readings to an average. Always check wheel runout first; this will eliminate the possibility of a bent rim/axle. If lateral runout is excessive, it is match mounted to the wheel if the wheel has some lateral runout. If match mounting cannot correct the excess runout, replace the component causing the runout. For more information on this, see **J2047_201303** [7].

Titania Oxygen Sensor

An oxygen sensor that uses a thick film of titanium to detect the amount of oxygen present in the exhaust gases. It provides a correct oxygen content signal almost instantly on cold start-up.

TMC

The Technology and Maintenance Council, or TMC, is the technical information division of the American Trucking Associations, or ATA. They provide the Recommended Practices manuals for the trucking industry.

Toe

The angle of the tires comparative to every other when viewed from above the suspension. The condition during which the fronts of the wheels are closer together than the rears of the wheels is termed toe-in. The condition during which the fronts of the wheels are farther apart than the rears is termed toe-out. Some manufacturers use the terms positive toe for toe-in and negative toe for toe-out. Toe is measured in inches or degrees. The static toe setting is meant to compensate for the slight wear in steering joints and components, which can cause the wheels to splay outward or inward while the vehicle is being driven. This effect is intended for vehicle support in several factors, like front-wheel drive versus rear-wheel drive, front/rear brake split versus diagonal brake split, and specific desired handling characteristics. OEMs specify the right static toe for the look of the vehicle such the wheels are parallel when the vehicle is in motion, which avoids scrubbing of the tires. For more information on this, see **J670_200801** [8].

TOOT (Toe-Out on Turns, Turning Radius)

A relative toe setting of the front wheels on turns. When a vehicle makes a turn, each wheel should rotate with true rolling motion, that is, free from tire scrub. True rolling motion is obtained only if each wheel is at 90° to a line drawn between the steering axis and therefore the center point of the turn. The steering axis is the point where the tire turns at the knuckle. Because the rear wheels are fixed, the midpoint of the turn will lie somewhere along the centerline of the rear axle, betting on how far the handwheel is turned from the straight-ahead position. To produce true rolling motion, the inner wheel must be turned at a greater angle than the outer wheel. This permits the inner wheel to show through a smaller turning radius than the outer wheel. For more information on this, see **J670_200801** [8].

FIGURE T.14 Ackerman principle of TOOT.

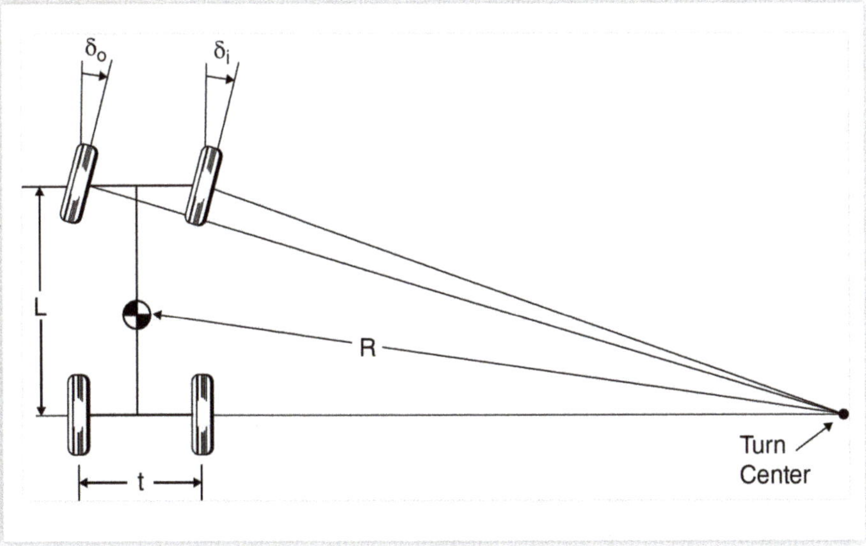

Torque

Torque is a force that tries to twist something around. It is a form of work, which is force times distance. It is defined as the creation of a force and the perpendicular distance between the force line of action and the axis of rotation. Torque is force multiplied by length. The length of the lever is the length of the throw of the crankshaft journal. It is a twisting effort expressed because of the ICE capacity to try to work, where horsepower is defined as the rate at which the engine can work. Engines produce torque by combustion force which pushes down the piston and moves a lever, that is, the stroke of the crankshaft. Torque is mostly expressed in foot-pounds or Newton-meters (1 ft-lb = 1.355 N-m and 0.737 ft-lb = 1 N-m). Torque is expressed mathematically in the following formula:

$$T = FR \tag{T.2}$$

where

T is the torque, in ft-lb or N-m
F is force, in lb or N
R is the radius, or torque-arm distance, in ft or m

The below formula is one used by almost all engine dynamometers. The number 5252 is a mathematical constant derived from the horsepower formula. One can measure engine torque using either a cradle absorption dynamometer, transmission dynamometer, or eddy current dynamometer. The cradle absorption dynamometer is capable of absorbing the power output of the engine. The transmission dynamometer measures power transmitted from an engine using a torque converter, but it does not absorb any of the power. The eddy current dynamometer uses electrical resistance to measure the torque of an engine. The rotor is rotated by the engine and the voltage is applied to the coil or stator housing. A magnetic flux is generated and the rotor cuts off these magnetic fluxes and becomes an electromagnetic brake measuring the power.

$$\text{Torque} = \frac{\text{hp} \times 5252}{\text{rpm}} \tag{T.3}$$

where

5252 is a constant
hp is horsepower

FIGURE T.15 Torque = Force times length or distance.

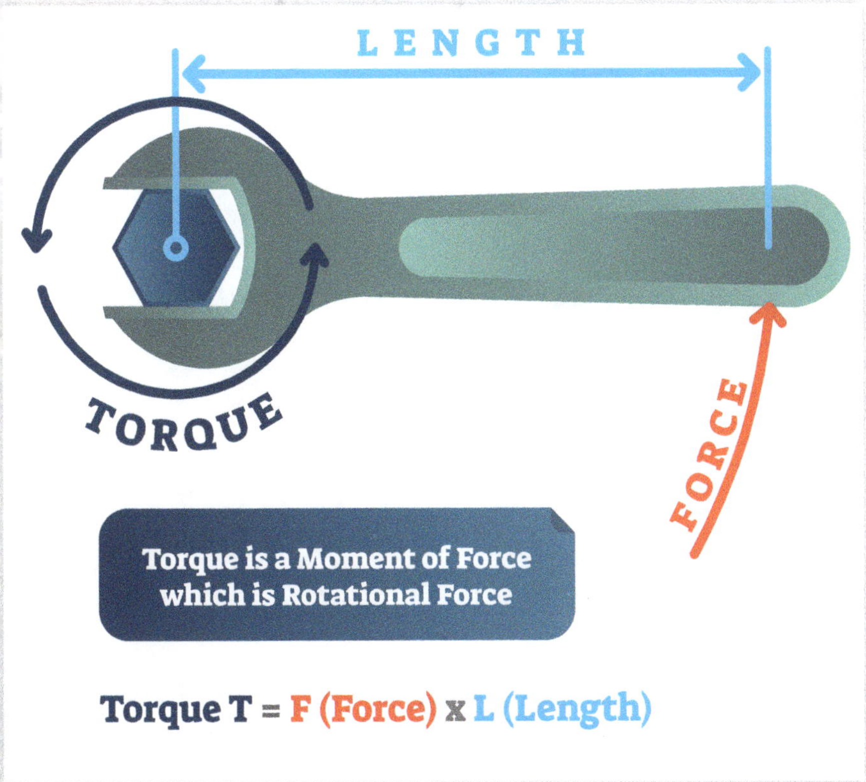

Torque Converter

The torque converter is a single-stage, polyphase, three-element unit, consisting of a pump (impeller), stator, and turbine. At lower output speeds, the torque converter multiplies torque and provides fluid coupling to the engine. At higher speeds, the torque converter clutch (TCC) is automatically engaged to provide direct drive from the engine to the transmission. Hydraulic fluid for converter charging pressure comes from the sump and is supplied by the input pump. The TCC is applied or released by changing the direction of the fluid in the torque converter. An integral converter damper minimizes the need for additional engine vibration control. For more information on this, see **J649_198807** and **J648_201106** [9, 10].

FIGURE T.16 Torque converter showing stator, turbine, converter clutch, and pump.

Torque Converter Formula

The ability of the torque converter to multiply torque can be approached by applying Newton's Third Law of Motion "Action-Reaction: For every action there is an equal and opposite reaction." In the converter, the impeller, the turbine, and the reactor are points of action and reaction with respect to oil flow. During the period of torque multiplication, the reaction of the stationary stator (reactor) blades to the oil is in the same direction as the impeller rotation. In accordance with Newton's law, the reaction of the turbine blades on the oil must be equal to the combined reactor and impeller torque (A + B = C). Turbine torque is, therefore, greater than impeller torque by the amount of the reactor (stator) reaction torque. Therefore, C – B = A.

Using Newton's law again, it is obvious why a coupling cannot increase torque. Without a stator or reactor, the fluid coupling has only two points of action and reaction: the impeller and the turbine. The impeller action on the fluid is opposed only by the reaction of the turbine, so the action-reaction between the two must always be equal. The torque ratio of a torque converter is usually around 2:1, similar to a first gear ratio and determined by several factors. These factors are N speed (RPM), size of the fluid flow path (D), and capacity coefficient (C). Torque ratio determines the stall speed. The fluid flow design path is also known as the torus. The torque (speed) ratio represents a particular set of vector flow relationships inside the torus circulation system. The flow path determines the lever length in the torque equation.

$$T = CN2D5 \tag{T.4}$$

The total torque developed in a torque converter is the sum of the torques of all three elements given in the formula:

$$\Sigma T = 0 = +T_P + T_T + T_R \tag{T.5}$$

For more information on this, see **J649_198807** and **J648_201106** [9, 10].

Torque Rise

Increase in engine torque as the engine speed is reduced from its maximum rated speed to its operational speed. A turbocharged diesel engine can develop 1,000 ft-lb of torque at the speed of 2,100 RPM, and the torque rises to 1,500 lb-ft peak torque at 1,200 RPM. When the RPM is reduced to 1,200 RPM, which is the peak torque, the rate of the rise is equal to 50%. If the 900-RPM drop from the rated divides this 50% torque rise to the peak torque RPM, this engine has established a 5.55% rise from every 100-RPM decrease.

Torque rise formula:

$$\text{Torque rise} = \frac{\text{Peak torque} - \text{Rated speed torque}}{\text{Rated speed torque}} \tag{T.6}$$

$$\text{Torque rise} = \frac{1{,}500\ \text{ft–lb} - 1{,}000\ \text{ft–lb}}{1{,}000\ \text{ft–lb}} = 50\% \tag{T.7}$$

Torque Steer

When the vehicle pulls to one side during hard acceleration in a front-wheel-drive or all-wheel-drive vehicle. When half-shafts are not equal lengths, more torque is applied to the side with the short half-shaft, creating the pulling condition.

Torque Tube

A hollow tube that contains the driveline driveshaft that forms a torsionally rigid structure that connects the transmission to the final drive or rear axle.

Torque Wrench

A wrench that is used to apply the applied torque or turning force to a fastener. It can use a moving beam, spring setting in a click type, or digital electronics to indicate the measured torque.

Torsen Differential

A type of locking differential that applies torque to both drive wheels through the use of interlocking gears. The name is a combination of torque and sensing. SAE J2817_201908 covers passive torque biasing axle and center differentials used in passenger car and light-truck applications. Differentials are of the bevel gear, helical gear, and planetary types, although other configurations are possible. For more information on this, see **J2817_201908** [11].

Torsion Bars

Straight bars of round spring steel. Used on an independent front or rear suspension system, an inline torsion bar can apply pressure near the lower control arm pivot. If the vehicle hits a bump, the wheel moves upward (compression). This causes the lower control arm to twist the torsion bar. When the wheel moves downward, the bar untwists. The twisting action provides the identical result as a coil or spring. Each front wheel has its own torsion bar.

FIGURE T.17 Torsion bar.

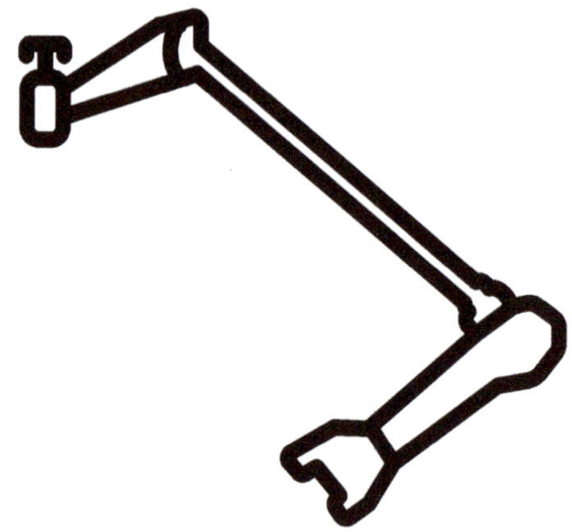

Torsional (Vibration) Damper

See **Harmonic Balancer**.

TP (Throttle Position) Switch

A three-wire variable resistor (potentiometer) mounted to the throttle body. The throttle body is opened and closed by way of a cable or DC motor that is connected to the gas pedal. When the throttle is closed, the computer reads a low voltage signal. When the throttle is wide open, the computer reads a high-voltage signal. Voltage signal changes relative to throttle position about 0.5 V at idle and about 4.5 to 5 V at wide-open throttle. It tells the engine management computer the position of the throttle open or closed. This switch is used to control fuel delivery as well as spark advance and the shift points of the transmission/transaxle.

FIGURE T.18 Throttle position sensor or switch.

TPMS (Tire Pressure Monitoring System)

Beginning September 1, 2007, the US National Highway Traffic Safety Administration requires the use of a tire pressure monitoring system (TPMS) in all passenger vehicles and light trucks. This new Federal Motor Vehicle Safety Standard requires a TPMS to be installed in passenger vehicles and light trucks to inform the driver when a tire falls below specified pressure levels. Two basic

TPMS designs are used. The first system uses a sensor mounted inside the tire/wheel assembly that directly measures tire pressure. This data transmits to a receiver inside the vehicle through a wireless radio frequency device built into the sensor. The other TPMS (indirect measurement) calculates tire pressure using ABS wheel speed sensor information and software in the electronic brake control module. Both system types will illuminate a warning light on the dash, and some may be reset via a TPM dash switch. For more information on this, see **J2657_201912** [12].

TPS (Throttle Position Sensor)

A position sensor that is a common application of a potentiometer. Several problems can occur with a TPS producing incorrect signal voltage. Return ability is the consistency of the voltage measurement at idle. A throttle pedal should have approximately the same voltage at idle every time the pedal returns to idle. Without it, the engine may not idle at the correct rpm and some powertrain operational strategies may not work. The TPS should reach a maximum value at wide-open throttle. If the TPS is misadjusted or has some other problem, a low-power complaint can occur.

Trace

A pattern of circuit voltages in relation to very short time spans drawn on the display of an oscilloscope.

Tracing Powder/Chalk Diagnostic Test

A test or diagnostic technique used to locate automotive component fluid leaks. Thoroughly clean the suspended area, then spray tracing power like a foot spray, such as Desenex or its equivalent, on that area. It puts a white frosting over the area that will show the presence of a fluid leak.

Track Rod

See **Tie Rod**.

Traction

The drive force of a vehicle or its pulling power.

Tractor

Vehicle specifically designed to deliver a high tractive effort at slow speeds for the purpose of hauling a trailer or machinery such as that used in agriculture or mining. For more information on *Terminology for Agricultural Equipment*, see **J1150_200305** [13].

Tractor Protection Valve (Air Brakes)

This air brake air control valve protects the tractor air supply in the event of a trailer breakaway or total air loss. Two airlines connect the tractor brakes to the trailer air circuit. The trailer path supplies the trailer with air for braking and the other pneumatic systems like its suspension. The trailer line is the service brake signal line. The trailer lines are connected to the tractor protection valve. When the trailer supply dash valve is in the off position, the tractor protection valve is off. When the drivers make a service brake application, no air will exit the tractor protection valve service signal line to the trailer. When the trailer air supply valve is open, this air will open the tractor protection valve and provide the trailer with air. When a service brake application is made, the service brake signal air is transmitted through the tractor protection valve to use its service brakes. For more information on this, see **J1410_202012** [14].

Traffic-Sign Recognition

A technology by which a vehicle is ready to acknowledge the traffic signs placed on the road (e.g., regulation or children). It is a part of the ADAS, which help in the driving process. When designed with a secure human-machine interface, they must increase car safety and, more generally, road safety. It uses image processing techniques to detect the traffic signs. The detection methods may be generally divided into color-based, shape-based, and learning-based methods.

Trailer

A non-powered vehicle that is towed by a truck or a tractor.

Trailer Hand Brake (Air Brakes)

The trailer control valve is employed to actuate the trailer service brakes on a trailer independently of the tractor service brakes. This valve is additionally known by the terms trolley valve, broker brake, and spike. The source of air supplied to the trailer control valve is typically the secondary circuit. The pressure air is modulated proportionally with control valve travel. Some drivers are known to use this valve to incorrectly stop a semi using just the trailer brakes. For more information on this, see **J1410_202012** [14].

FIGURE T.19 Truck air brake trailer hand brake.

Trailer Tracking

Tracking the movements and position of an articulated tractor-trailer unit through the use of a location unit fitted to the trailer and a method of returning the position data via a mobile communication network or geostationary satellite communications for use through either a personal computer or cloud-based software. Freight containers can also be tracked by global positioning system (GPS) using a similar approach to that used for trailer tracking like a battery-powered GPS device communicating its position using a cell phone. This use provides increased security and the possibility to reschedule the container transport movements based on accurate information about its location.

Trailing-Arm Suspension

Referred to as trailing link, this may be a vehicle suspension design during which one or more arms (or links) are connected between (and perpendicular to and forward of) the axle and a pivot point (located on the chassis of a motor vehicle). It is typically used on the rear axle.

Transaxle

A transaxle is a combination of the transmission and final drive or rear axle. It uses gear combinations, with lower gear ratios (higher speeds), as the vehicle speed increases. You would use a 4:1 gear reduction to start the vehicle, then upshift to a ratio of 3:1, then 2:1, then 1:1, and finally shift into overdrive at 0.6:1. The final drive ratio would be in addition to the final transmission ratio.

Transducer

A device that changes one form of energy into another. An electrical and mechanical speed sensing and control unit used on cruise control systems.

Transfer Case

A two-speed transmission that provides a connection between the vehicle transmission and the front axle for four-wheel-drive (4WD) operation. Gear-to-gear transfer cases have been used. Three gear shafts are in mesh within the transfer case, one gear shaft is attached to the transmission output shaft. The second shaft acts as an idler, and the third shaft is the output to the front axle. Gear-to-gear transfer cases, in most cases, have two speeds. The first, 4WD Low, is a gear reduction that is usually around two to one. The second gear in the transfer case is a direct drive. The gears are engaged by sliding collars or synchronizers to lock the gears to the shaft. Neutral takes place when neither gear collar is locking a gear in place. Chain-driven transfer case uses a Morse chain to transfer power to the front axle with the use of different types of clutches and a planetary gearset for gear reduction and/or differential action.

FIGURE T.20 Morse chain-driven transfer case.

Transformer

When two coils are approximate, energy could also be transferred from one to the opposite by a magnetic coupling called mutual induction. Mutual induction implies that the expansion or collapse of the force field around one coil induces a voltage within the second coil. There are step-up transformers that increase voltage and step-down transformers that decrease voltage.

FIGURE T.21 Step-up transformer used to increase voltage.

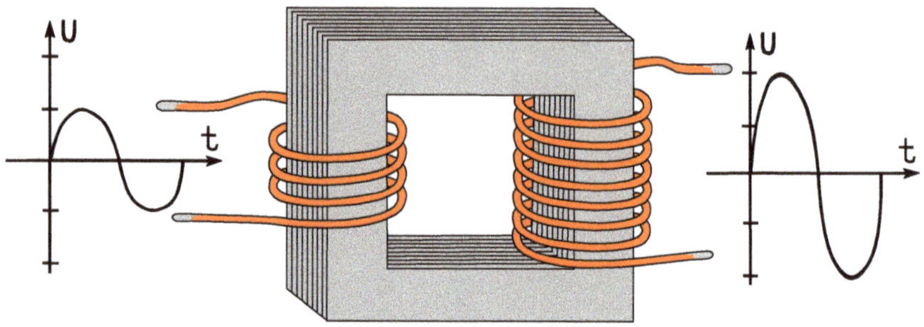

Transistor

A solid-state semiconductor device that can operate as an amplifier or an electrical switch to function as a relay in an electronic circuit.

Transmission

A transmission uses gear combinations, with lower gear ratios (higher speeds), as the vehicle speed increases. You would use a 4:1 gear reduction to start the vehicle, then upshift to a ratio of 3:1, then 2:1, then 1:1, and finally shift into overdrive at 0.6:1.

Transponder

A device that receives a signal, alters it, and then retransmits it in an altered state to the engine management computer.

Transverse Engine

The mounting of the ICE at a right angle to the direction of the vehicle. Some refer to this mounting as an east/west mounting. This design is used on front-wheel-drive or all-wheel-drive vehicles.

Tread Depth

The height of the tire tread groove from its bottom to the top road contact surface of the tire. For more information on this, see **J2047_201303** [7].

Tread Separation

Tire tread and top belts separate from the main tire casing. Most of the time it is due to a defect in the tire as in belt adhesion or underinflation. Underinflation causes the sidewalls to flex and generate heat. Low tire pressure causes a large amount of the tire surface area to be in contact with the road, which increases friction. Increased friction will cause the tires to overheat, which will lead to premature wear and tread separation. For more information on this, see **J2047_201303** [7].

FIGURE T.22 An example of tread separation on a heavy-duty tire.

Trigger

A current level that determines the start point of a waveform. It is also any electrical switch that initiates an ignition event in an electronic circuit.

Trimetal Engine Bearing Material

A copper-lead alloy that is a stronger and more expensive bearing material than Babbitt. It is used for intermediate- and high-speed ICE applications. Tin, in small quantities, is alloyed with copper-lead bearings. This bearing material is most easily damaged by corrosion from acid accumulation. Bearing material can be eroded by acids. Many of the copper-lead bearings have an overlay, or third layer, of metal. This overlay is usually of Babbitt. Babbitt is overlayed for high fatigue strength, conformity, embedability, and corrosion resistance. The overplated bearing is a premium bearing.

Triple Motor-Generator Hybrid

A hybrid vehicle that uses three motor generators in one vehicle drivetrain. Two motor generators are located within the transmission or transaxle, and one is also located within the rear drive-axle assembly. A triple motor-generator arrangement is sometimes employed in four-wheel-drive vehicles to help electrically drive all four wheels and tires. For more information on this, see **J1715/2_202108** [15].

Truck Frame

Two large pieces of longitudinal angle or channel iron, with a number of cross-member steel rails that are welded, riveted, or bolted together to form the vehicle component support system, such as axles and suspension systems. On a tractor, the trailer weight is added to the fifth-wheel assembly. On a straight-body truck, the gross vehicle weight is distributed across the total length and width of the frame. Just how much force the frame absorbs depends on where the front and rear axle(s) are located and how the load is distributed.

TSBs (Technical Service Bulletins)

Bulletins published by vehicle manufacturers that post recent technical changes, recalls, and other service-related information.

Tuftriding

A variation of the nitrating crankshaft hardening process that involves heating the crankshaft in a molten cyanide salt bath. This is a GM trade name.

Turbine

A wheel that uses blades or vanes that is driven by a flow of fluid or gas. These are found in torque converters, fluid couplings, turbochargers, and turbine engines.

Turbocharger

Turbocharging is used to produce higher power outputs, lower emissions levels, and improved efficiency from a diesel- or gasoline-fueled engine. The turbocharger charges the intake manifold at pressures above atmospheric to make the charge denser, thereby packing more molecules in the air-fuel charge and creating greater combustion pressure when the mixture explodes and pushes the piston. This higher pressure is called boost. A turbine wheel is driven by the exhaust, which, in turn, drives a compressor turbine wheel which compresses the intake air above 14.7 psi. A wastegate device like an open door in the exhaust stream is used to control the boost pressure. You can also use a charge air cooler to lower the air temperature to make it cooler and therefore denser for more power. You either cool the air before it enters the turbo or after.

"The supercharging of small-displacement gasoline engines requires high pressure ratios combined with a wide range of air flow rate. To resolve this conflict, two-stage turbo charging with two turbochargers or the combination of a turbocharger and a mechanical compressor is used. But this is associated with an increase in complexity. The highest potential for avoiding a multi-stage system is provided by the systematic modification of the turbo-machinery operating maps, e.g., on the turbine side by using variable turbine geometry. The reduction of particulate emissions of turbodiesel powered vehicles has limited the amount of fuel that can be injected at low engine speeds. Supercharging under these conditions provides increased air flow, increased BMEP, and reduction of smoke. A Sequential Hydro-Supercharging System (SHS) has been developed to provide increased air flow to the engine in a controllable fashion over the engine's entire speed and load ranges. The SHS utilizes two standard turbocharger compressor impellers driven by well proven, miniature, high speed, high efficiency hydraulic turbines.

Each of these turbines is capable of producing 10 HP at 70,000 rpm. A complete SHS package suitable for a 230 hp turbodiesel is comprised of two TO4B size hydro-superchargers, two sequential gear pumps, solenoid valves, oil cooler and a 2 liter oil tank. The entire SHS weighs less than 16 kilograms" [17]. For more information on *Turbocharger Nomenclature and Terminology*, see **J922_201106** [16].

FIGURE T.23 Turbocharger.

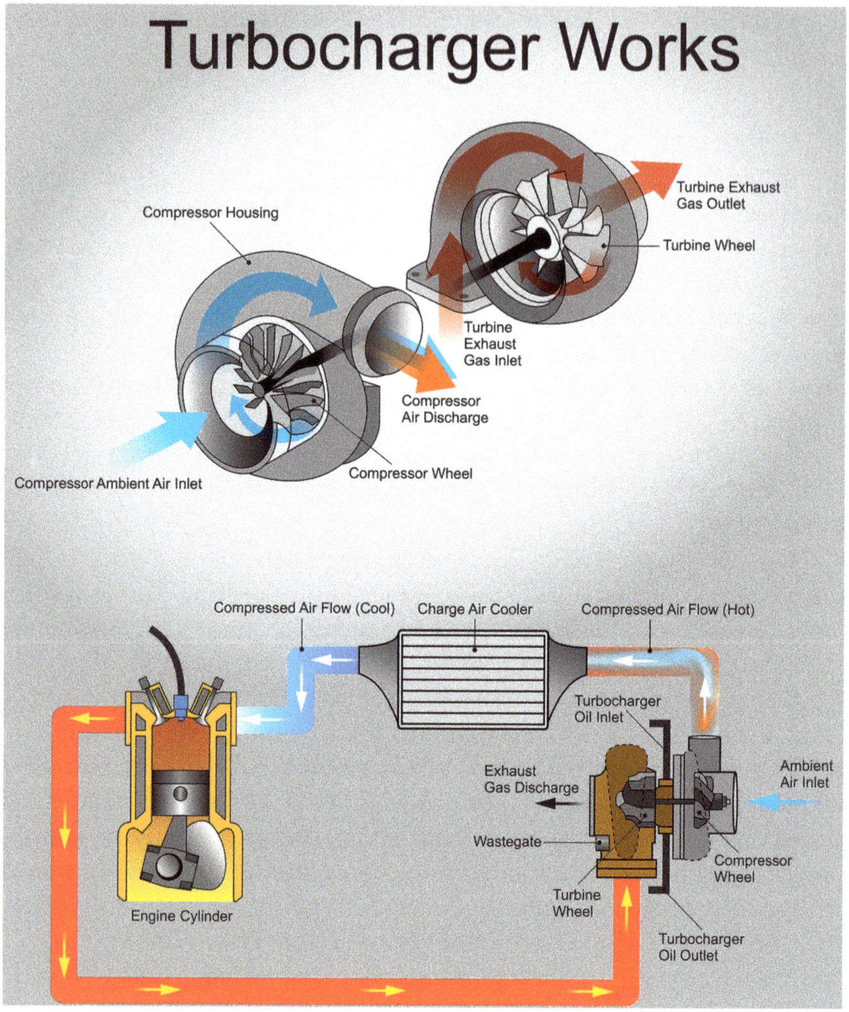

Turning Radius

Measured at the center of the arc made by the outside wheel during its tightest turn. Whenever the vehicle is moving straight ahead, the front wheels should be parallel. In a turn, however, parallel operation of both wheels will result in scrubbing since the inside wheel is traveling a shorter distance through the turn than the outside wheel. See **TOOT**.

Turntable

The flat bearing surface of the fifth wheel of a truck.

Twin Countershafts

A floating main shaft situated between two countershafts, which are commonly known as the left and right countershaft. The number of gears and the tooth count on the two countershafts are the same, with the exception of the PTO (power takeoff) gears added to provide a different drive ratio. The PTO gears also allow the technician to identify the right from the left countershaft. For example, some models of Roadranger twin countershaft transmissions use 45 teeth on the right countershaft and 47 teeth on the left countershaft. The majority of transmissions, however, use five gear ranges in the main gearbox housing. The main housing is directly bolted to an auxiliary housing at the rear.

FIGURE T.24 Twin countershaft autoshift manual gear transmission.

Twist

Twist occurs when the end of one frame rail is bent upward or downward in relation to the opposite frame rail. The following causes can result in a twisted frame: collision damage, rollover, rough terrain operation, and uneven loading.

Two-Gear Reduction Axle

A final drive rear axle that has two separate output ratios available from the same differential assembly. Much like a 4WD system found in an SUV, the driver is able to switch between each output mode using a switch in the truck.

Two-Pedal System

An automated manual truck transmission system that does not use a clutch pedal. The two pedals are the accelerator and brake pedal, e.g., Eaton CEEMAT (Converter Enhanced, Electronically Managed, Automatic Transmission) or Eaton Ultrashift and Autoshift.

Two Speed Axle

The two-speed rear axle contains two different gear ratios or ranges. The two-speed rear axles operate as single-speed rear axles. The driver selects the range using a control switch on the instrument panel. A shift motor assembly mounted on the rear carrier engages a shift fork which selects the range chosen by the driver. The shift motor assembly is electrically operated. The shift fork will perform the following functions: slide a collar onto the sun gear of a planetary gearset and slide the entire sun gear assembly in or out of the gearset.

Two-Stroke Engine

A two-stroke engine also goes through intake compression, power, and exhaust actions to finish one operating cycle. However, the intake and compression actions are combined in one stroke, and the power and exhaust actions are combined within the other stroke. Two-stroke engines are utilized in lawnmowers, heavy trucks, construction equipment, and ships.

When compared to the four-stroke cycle, the two-stoke cycle completes two strokes or actions in one turn of the engine (i.e., compression and power). The four-stroke cycle requires two turns of the engine to complete the cycle. The following items apply to the two-stroke cycle:

- The intake valve has been eliminated and replaced by a row of ports around the cylinder liner.
- A Roots-type blower is placed on the engine to blow air into the cylinder whenever the piston uncovers the ports. These ports have an angular inlet called the Hesselman helical loop to swirl the air creating turbulence with a resultant improvement in the air-fuel mixing. The exhaust valve or valves open to allow the blower to push the exhaust gases out of the cylinder. The blower blows intake air alone into the cylinder when the ports are uncovered. The piston and the exhaust valve close and then cover the ports. The compression of the air takes place as the piston approaches TDC. The injector is timed to inject fuel, finely atomized, into the hot compressed air in the cylinder. The rapid expansion of the burning gases forces the piston down toward BDC and the exhaust valve opens. The piston uncovers the inlet ports and the cylinders are swept clean with scavenging air.

FIGURE T.25 Two-stroke cycle.

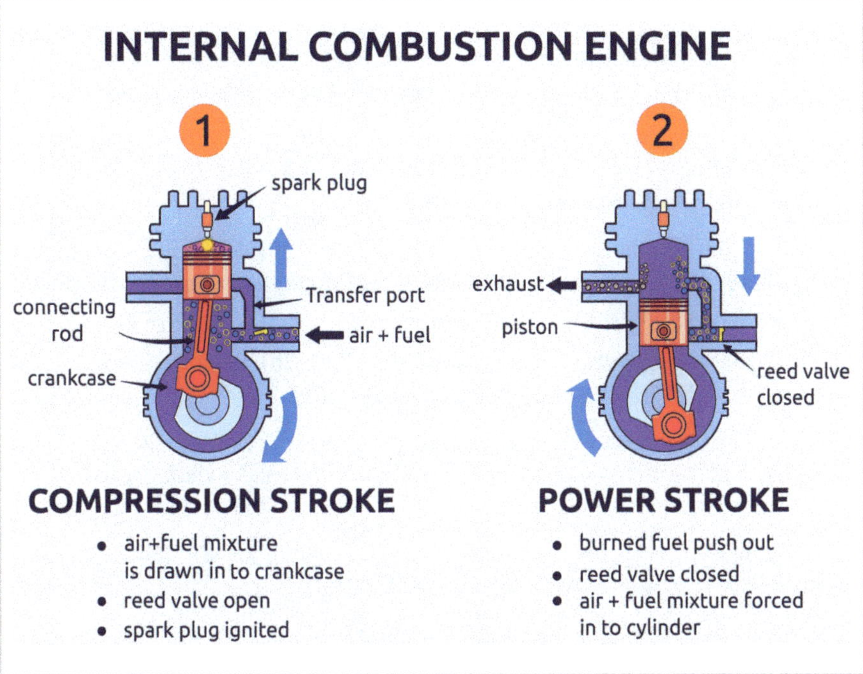

Two-Wire Oxygen Sensor

A two-wire oxygen sensor has a signal wire and a ground wire for the O_2S. For more information on this, see **J1930DA_202105** [5].

Type A Misfires

Ignition misfire that is severe enough to cause catalytic converter damage. For more information on this, see **J1930DA_202105** [5].

Type B Misfires

Ignition misfire that causes an increase in emissions, but not catalytic converter damage. For more information on this, see **J1930DA_202105** [5].

REFERENCES

1. Society of Automotive Engineers, Standard J678_201108: Speedometers and Tachometers—Automotive, SAE Publishing, Warrendale, PA, 2018.
2. Society of Automotive Engineers, Standard J2040_201601: Tail Lamps (Rear Position Lamps) for Use on Vehicles 2032 mm or More in Overall Width, SAE Publishing, Warrendale, PA, 2018.
3. Society of Automotive Engineers, Standard J585_201406: Tail Lamps (Rear Position Lamps) for Use on Motor Vehicles Less than 2032 mm in Overall Width, SAE Publishing, Warrendale, PA, 2018.
4. Society of Automotive Engineers, Standard J312_201902: Automotive Gasolines, SAE Publishing, Warrendale, PA, 2019.
5. Society of Automotive Engineers, Standard J1930DA_202105: Electrical/Electronic Systems Diagnostic Terms, Definitions, Abbreviations, and Acronyms Web Tool Spreadsheet, SAE Publishing, Warrendale, PA, 2021.
6. Kershaw, J.F., *High-Performance Automotive Cooling Systems* (Forest Lake, MN: CarTech, Inc., 2019), ISBN:978-1-61325-504-9
7. Society of Automotive Engineers, Standard J2047_201911: Tire Performance Terminology, SAE Publishing, Warrendale, PA, 2019.
8. Society of Automotive Engineers, Standard J670_200801: Vehicle Dynamics Terminology, SAE Publishing, Warrendale, PA, 2008.
9. Society of Automotive Engineers, Standard J649_198807: Automatic Transmission Functions—Terminology, SAE Publishing, Warrendale, PA, 1988.
10. Society of Automotive Engineers, Standard J648_201106: Automatic Transmission Hydraulic Control Systems—Terminology, SAE Publishing, Warrendale, PA, 2011.
11. Society of Automotive Engineers, Standard J2817_201908: Definition and Measurement of Torque Biasing Differentials, SAE Publishing, Warrendale, PA, 2011.

12. Society of Automotive Engineers, Standard J2657_201912: Tire Pressure Monitoring Systems for Light Duty Highway Vehicles, SAE Publishing, Warrendale, PA, 2011.
13. Society of Automotive Engineers, Standard J1150_200305: Terminology for Agricultural Equipment, SAE Publishing, Warrendale, PA, 2003.
14. Society of Automotive Engineers, Standard J1410_202012: Air Brake Valve - Performance Requirements, SAE Publishing, Warrendale, PA, 2020.
15. Society of Automotive Engineers, Standard J1715/2_202108: Hybrid Electric Vehicle (HEV) and Electric Vehicle (EV) Terminology, SAE Publishing, Warrendale, PA, 2021.
16. Society of Automotive Engineers, Standard J922_201106: Turbocharger Nomenclature and Terminology, SAE Publishing, Warrendale, PA, 2011.
17. Herbst, F., Stöber-Schmidt, C., Eilts, P., Sextro, T. et al., "The Potential of Variable Compressor Geometry for Highly Boosted Gasoline Engines," SAE Technical Paper 2011-01-0376, 2011, https://doi.org/10.4271/2011-01-0376.

U

Ultra-capacitor Module

A control module that supplements the cranking current for a 24-V battery system on a heavy-duty vehicle, which reduces the starting time. Also known as super capacitor.

FIGURE U.1 Ultra-capacitor module.

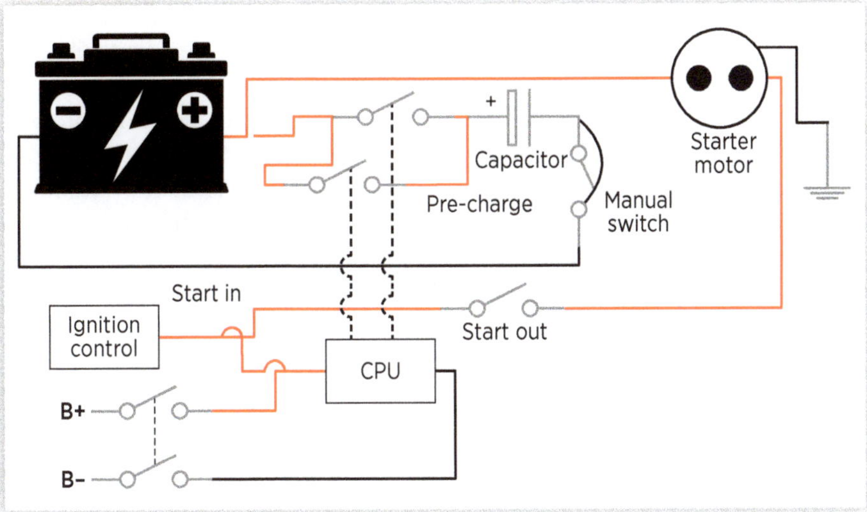

UltraShift Truck Transmission

The UltraShift electrically controlled manual transmission is fully automatic. Although they use a clutch, a no-clutch pedal is required; hence, they are referred to as two-pedal units because they use a clutch that is electro-pneumatically actuated to link the clutch electronically to the management computer and pushbutton or tower-type shift controls. Both the semi- and fully automated transmission types must be mated to an electronically managed engine using the SAE J-1939 CAN standard between the shift control and the engine management computer.

The Eaton fully automatic UltraShift transmission evolved from the semi-automated AutoSelect and AutoShift models. UltraShift models do not require a clutch pedal because they use a clutch that is electro-pneumatically actuated to link the clutch electronically to the transmission management computer. All semi- and fully automated or tower-type models offer shift-by-wire controls. The ten-speed UltraShift model uses an Eaton Fuller 15 1/2 in. two-plate datalink mechanical clutch that is computer controlled and a high-capacity inertia brake to speed automated upshifts. The inertia brake also serves as a clutch brake. The UltraShift models are also designed with advanced Generation II electronics. The UltraShift can be used in full-automatic mode or with manual intervention, according to the driver's preference.

The UltraShift ten-speed provides automatic shifting as operating and load conditions allow, plus float shifting between gears to extend clutch life. The UltraShift transmission uses a three-position pushbutton selector, which offers a manual mode to hold the current gear and permits driver-selected upshifts or downshifts. The low mode feature can be used in any gear to prevent the transmission from upshifting and initiates downshifts to optimize engine braking performance. The low mode can also be used to select the first gear at a standstill and will hold in low until drive mode is selected. The low mode also offers maximum engine braking.

Ultrasonic (Sensor)

This type of sensor uses ultrasonic waves to detect the vortexes that are produced and generates a digital (on and off) signal where the frequency is proportional to the amount of air passing through the sensor. An example is a fuel level sensor. For more information on *Active Safety Systems Terms and Definitions*, see **J3063_202103** [4].

Ultrasonic Sensors Parking Assist System

The parking assist system identifies and notifies the motive force of an object within the vehicle path when reversing at speeds of (5 mph). The space and site of the item is set by four ultrasonic object sensors located within the rear bumper. The system will notify the driving force using audible beeps through the radio or haptic pulses through the driver's seat. The driving force can select between audible or haptic alerts within the vehicle personalization menus. See the user's manual for more detailed information on vehicle personalization. The parking assist system is made up of the following components:

- Parking assist control module
- Rear object sensors
- Parking assist switch
- Parking assist indicator

For more information on *Active Safety Systems Terms and Definitions*, see **J3063_202103** [2].

FIGURE U.2 Ultrasonic sensors for parking assist.

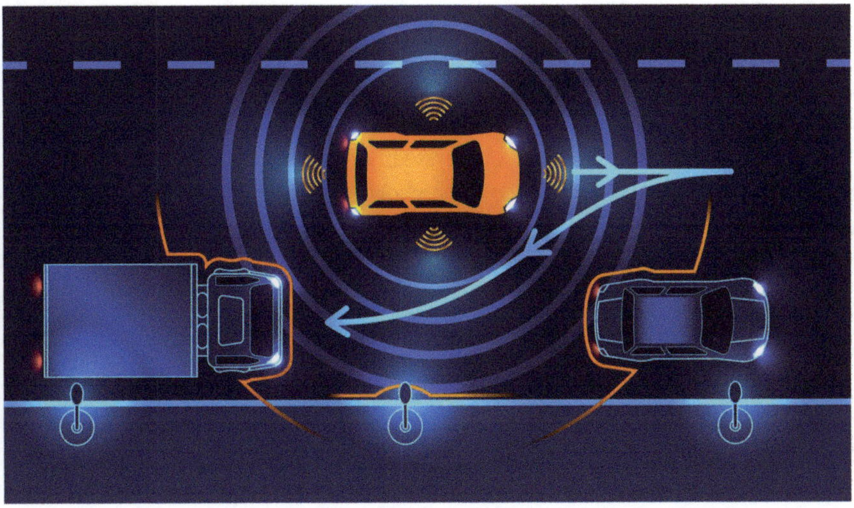

Ultrasound

Sound waves with frequencies on top of the upper audible limit of human hearing. Ultrasound is not any different from "normal" (audible) sound in its physical properties, except that humans cannot hear it. This limit varies from person to person and is approximately 20 kHz (20,000 Hz) in healthy young adults. Ultrasound devices operate with frequencies from 20 kHz up to several gigahertz using ultrasonic sensors. Ultrasonic sensors are accustomed detect objects and measure distances. Ultrasound imaging or sonography is typically employed in medicine. Ultrasound is utilized to detect invisible flaws. Industrially, ultrasound is utilized for cleaning and mixing and to accelerate chemical processes. Animals like bats and porpoises use ultrasound for locating prey and obstacles.

Undercarriage

The wheels, tires, frame, and suspension of a motor vehicle.

Undersquare Engine

The term used to describe an engine in which the cylinder bore diameter is smaller than the stroke dimension. Most truck and bus diesel engines are undersquare.

Understeer

Understeer occurs when the front end moves less than desired in relation to the actual turning position of the front tires. This occurs when the front wheels loose traction in a turn (plowing). It occurs when the driver turns the wheel farther and the vehicle turns less sharply. Understeer results in the front of the vehicle being pushed toward the outside of the corner. The vehicle wants to push instead of turning. Understeer takes place when the front slip angle is larger than the rear slip angle. Understeer results in the front of the vehicle being pushed toward the outside of the corner. Understeer is when the front slip angle is larger than the rear slip angle. For more information on this, see **J670_200801** [3].

Unit Fuel Injector

A high-pressure integrated direct fuel injection system used in diesel engines, combining the injection nozzle and the injection pump into a single component. The scroll plunger pump is driven by a shared camshaft. A high-pressure injection delivers fuel as a larger number of smaller droplets, giving a much higher ratio of surface area to volume. This provides improved vaporization from the surface of the fuel droplets and so more efficient combining of atmospheric oxygen with vaporized fuel, delivering more complete and cleaner combustion. Also called a mechanical unit injector, or MUI (see **EUI**).

Unitized Hub

To reduce the service requirements of wheel hubs, a unitized wheel end is a pre-packaged system in which the bearings, seals, and lubricant are pre-installed into the hub. Unitized hubs can be filled with synthetic grease and sealed to avoid lubrication intervals, bearing adjustment, and seal replacement, thereby eliminating hours of typical wheel-hub maintenance. The bearings can either be integrated into the hub or pressed in as a separate cartridge. The bearing adjustment is set during the manufacturing process, and the hub nut acts to retain the assembled hub on the axle spindle. This hub cannot be serviced in the field but must be returned to the manufacturer for rebuilding or exchange.

Unitized Seal

One-piece unitized seals are the most common found in today's heavy-duty truck applications. This design incorporates a wear sleeve and seal into a single unit, simplifying installation. The seal is pressed into the hub and rotates with the hub.

Universal Joint Operation

The universal joint, or U-joint, allows movement and flexibility in the driveline between the transmission and axles. As vehicles travel on the road, over bumps and irregular surfaces, it is important that the axles are able to "float" up and down, and that means that the driveshaft must also be able to accommodate this movement, essentially changing the angle of operation between the transmission or transfer case and the final drive axles. This accommodation of up-and-down motion is made possible with the use of U-joints at each end of the driveline.

FIGURE U.3 A single cardan is a propeller shaft with one U-joint at each end of the assembly. Technically speaking, even though it is called a "single" cardan shaft, there are actually two U-joints on a tractor (left) and Class 2 truck (right).

A double-cardan joint is a one-piece driveshaft with three (or more) U-joints. A double-cardan driveshaft can split the operating angle of the U-joint into two separate angles that are each exactly one-half of the original angle. Double-cardan driveshafts are used in applications where it would not be possible to align a single-cardan driveshaft properly. The double cardan allows the operating angle to be split across the two halves of the joint. It is also possible to accomplish this with two constant velocity joints on a driveshaft, but U-joints tend to deliver better reliability. Torque from the engine is transferred through the U-joint to the final drive axle. However, while the engine drives or rotates the output shaft and U-joint at a constant speed, the output speed changes anytime an angle is introduced at the U-joint, which is also referred to as the joint working angle.

FIGURE U.4 Double-cardan joint acts as a constant velocity joint.

Nordroden/Shutterstock.com.

Unsprung Weight

The tires, wheels, and suspension parts that move along with the wheels. This also includes half the weight of the springs, shocks, and control arms.

Urea

See **Diesel Exhaust Fluid**.

UTQGS (Uniform Tire Quality Grading System)

The US Department of Transportation requires all passenger car tires (excluding snow tires, temporary spare tires, and deep) to be rated and labeled for tread wear, traction, and temperature. This is known as the Uniform Tire Quality Grading System (UTQGS). The UTQGS is not a guarantee of tire life or performance; it serves only for consumer reference. For more information on *Tire Performance Terminology*, see **J2047_201911** [1].

REFERENCES

1. Society of Automotive Engineers, J2047_201911: Tire Performance Terminology, SAE Publishing, Warrendale, PA, 2019.
2. Society of Automotive Engineers, J3063_202103: Active Safety Systems Terms and Definitions, SAE Publishing, Warrendale, PA, 2021.
3. Society of Automotive Engineers, J670_200801: Vehicle Dynamics Terminology, SAE Publishing, Warrendale, PA, 2008.
4. Society of Automotive Engineers, J3063_202103: Active Safety Systems Terms and Definitions, SAE Publishing, Warrendale, PA, 2021.

V

V2I (Vehicle-to-Infrastructure)

A communication system that permits motorized vehicles to share information with the components that support the road and highway systems. These include overhead readers and cameras, traffic lights, lane markers, streetlights, signage, and parking meters. V2I communication is wireless and bidirectional; that is, data from infrastructure components are often delivered to the vehicle over a VANET (vehicular ad hoc network) or MANET (mobile ad hoc network). Almost like V2V (vehicle-to-vehicle) communication, V2I uses DSRC (dedicated short-range communication) frequencies to transfer data.

V2V Communication (Vehicle-to-Vehicle Communication)

Wireless transmission of information between vehicles. V2V communication is meant to forestall accidents by allowing vehicles in transit to send position and speed data to at least one another over a commercial ad hoc network, which is defined as a network formed or arranged for a special purpose. V2V is an automobile technology designed to permit automobiles to speak with one another. V2V systems will use a 5.9 GHz band which was set aside by the UN General Assembly in 1999, the unlicensed frequency is also used for Wi-Fi, which is a type of WLAN (wireless local area network), to connect the vehicle to the environment outside of the vehicle. V2V is additionally called VANETs (vehicular ad hoc networks). It is a variation of MANETs (mobile ad hoc networks), with the stress being now that the mode is for vehicles.

Vacuum

When a space contains a vacuum, it contains a smaller quantity of air than that same space is capable of containing (determined by atmospheric pressure). Vacuum could be measured in pounds per square inch (psi), but inches of mercury (Hg) is also used. Atmospheric pressure is available outside the engine air intake. When a piston moves downward with an intake valve open, a vacuum is created in the cylinder above the piston. The air moves rapidly from the high pressure outside the air intake to the lower pressure in the cylinder. Vacuum is the absence of pressure or just low pressure.

Vacuum is measured in inches of mercury (Hg) or water. A complete vacuum or absence of pressure is equal to 29.995 in. Hg (101.06 kPa). Atmospheric pressure at sea level is 14.7 psi and decreases as elevation increases. At 5,000 ft (1,524 m) above sea level, a 1 in.^{2} column of air from Earth's surface to the outer edge of the atmosphere is 5,000 ft (1,524 m) shorter than the same column at sea level. Therefore, the weight of this column of air is less at 5,000 ft (1,524 m) elevation than at sea level. As altitude continues to increase, atmospheric pressure continues to decrease [2].

FIGURE V.1 Atmospheric pressure.

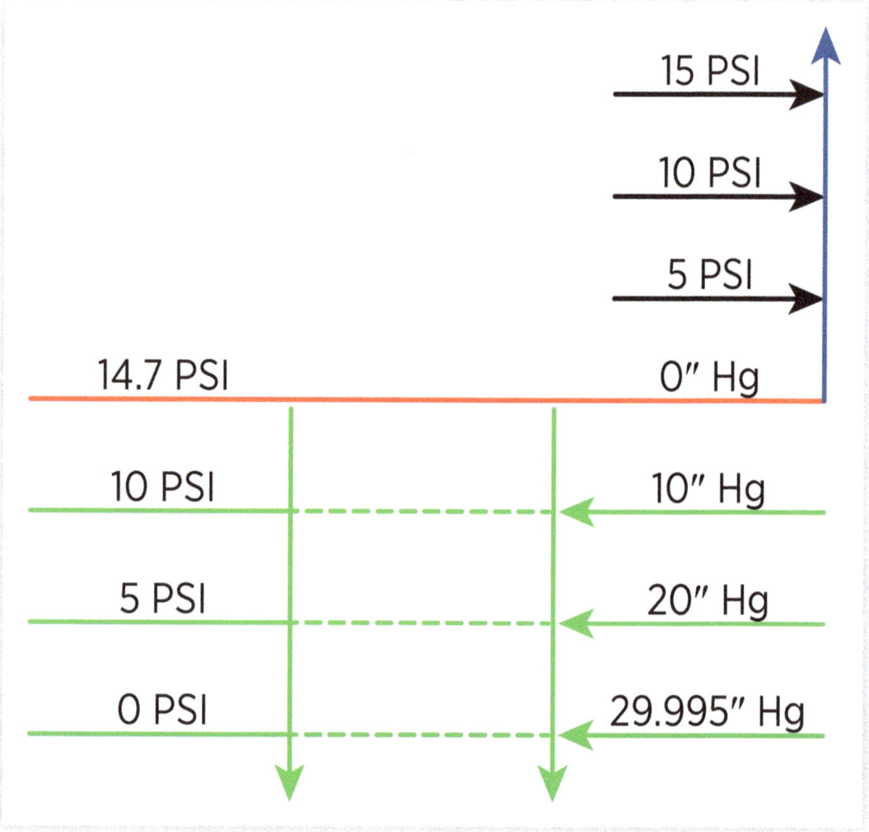

Vacuum Gauge

Measures vacuum or low pressure.

Vacuum Pump

Mechanically or electrically driven pump that provides a source of vacuum or low pressure.

Valence

Electrons orbit around the nucleus in paths. These paths form shells, like concentric rings circling the nucleus. Only a particular number of electrons can orbit within each shell. If there are too many electrons for the primary and closest shell to the nucleus, the others will orbit in additional shells until all electrons have an orbit within a shell. There are often as many as seven shells around one nucleus. The outermost group or ring is called the valence ring. The quantity of electrons during this ring determines the valence of the atom and indicates its capacity to mix with other atoms. If the valence ring of an atom has three or fewer electrons in it, the ring has room for more. The electrons there are held very loosely, and it is easy for a drifting electron to hitch the valence ring and push another electron away. These loosely held electrons are called free electrons. When the valence ring has five or more electrons in it, it is full. The electrons are held tightly, and it is hard for a drifting electron to push its way into the valence ring. These tightly held electrons are called bound electrons. The movement of those drifting electrons is named current.

Valve

A device that controls, restricts, or interrupts the flow of a gas or liquid.

Valve Face

The sealing area of an ICE poppet valve head with a precision ground valve face, which seals against the seat. The face is circular and on most valves, it is at a 45° angle to the highest of the pinnacle. The valve face seals against a seat within the plate on overhead-valve and overhead-camshaft engines.

FIGURE V.2 Valve face.

Valve Float

Valve float occurs when the valve spring cannot keep the valve in contact with the camshaft lobe. As the camshaft lobe starts to open the valve, it sets the valve train in motion. When the valve reaches the fully open position, the inertia of the valve train tends to continue to open the valve. If the inertia is big enough, it will continue opening the valve until the spring is completely compressed. Then the spring will attempt to close the valve with force. As the camshaft lobe has moved out from under the lifter, there is no resistance against the spring, and the parts of the valve train will come colliding together. This inertial opening and closing is called valve float, and it will quickly damage the valves, seats, and springs.

Valve Guide

A tubular guide installed in the cylinder head of an ICE that the poppet valve rides in and provide a sealing of the engine oil from the combustion chamber.

Valve Keeper

A small cone-shaped lock that is in two halves which snap into a groove at the end of the valve stem. It is designed to secure the valve spring, valve spring retaining washer, and valve stem all together.

FIGURE V.3 Valve components.

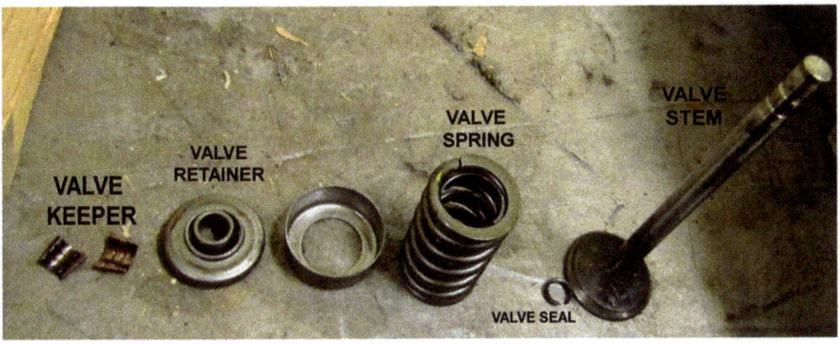

Valve Lift

Valve lift is the amount the valve is open or the distance that it is the valve is lifted off the valve seat. It is measured in thousandths of an inch or hundredths of a millimeter. Each cam lobe has a specific amount of lift, called the lobe lift. There are two types of lift you can achieve. The first is a reliable amount, which is the conservative mechanical lift that will affect the least wear yet and work for long-term use. This is what can be found in the stock camshaft. There are also high lift cam designs that give maximum power, but with greater wear to the valve train. Normal high-performance camshafts have valve lifts in the 0.400 to 0.432 range, and high lift cams have greater lifts in the 0.450 to 0.600 in. range.

Valve Lift Measurement

The most accurate way to measure the camshaft lobe lift is to rotate the camshaft between centers and measure the rise and fall with a dial indicator on a mockup of your engine. Valve lift is significant to the breathing of an engine because, up to a certain point, the valve head is a restriction to airflow. Once a valve in a street engine reaches about 0.300 in. lift, the port flows as if the valve were not there. However, it is helpful to lift the valve higher than this point because the valve then spends more cam duration time at or above 0.300 in. For best airflow, it is important to keep the valves open as long as possible practical. High lift places greater loads on the valve train. High valve lifts may also cause valve-to-piston interference.

Valve Lifter

It rides on the ICE camshaft and transfers the motions of the camshaft lobe through the pushrods and rockers to open and close the poppet valves.

The size and shape of the cam lobe under the lifter (multiplied by the ratio of the rocker arms) determines valve lift and duration. Also see **Cam Follower, Hydraulic Valve Lifter**, and **Lifter**.

Valve Overlap

The time the intake and exhaust valves of an internal combustion engine (ICE) is both open. When the intake valve opens before the exhaust valve has closed, many four-stroke ICEs use a short overlap period. A long-duration camshaft usually extends the overlap period, which helps make large amounts of power available at high RPM. The overlap period packs more air and fuel mixture into the cylinder by opening the intake valve near the end of the exhaust stroke while the exhaust valve is still open. The burst of exhaust gas rushing out of the cylinder helps to pull in the intake mixture, even though the piston has not yet reached TDC. The exhaust valve is still open when the piston starts going down on the intake stroke. This prevents exhaust gases from being sucked back into the cylinder at high RPM, as the air velocity is high and the outward movement of the exhaust gases is still pulling the air-fuel mixture into the cylinder.

Valve Retainer

A flat washer that sits on top of the valve spring and is the retaining seat for the valve keeper. See Figure V.4.

Valve Seat

The angled and beveled surface that the poppet valve seals against in the cylinder head.

Valve Spring

The return spring that is mounted on top of the valve stem and held by the valve retainer and valve keepers. See Figure V.4.

Valve Spring Tester

A mechanical tester that measures the tension of an ICE poppet valve spring in pounds.

FIGURE V.4 Valve spring tester.

Valve Stem

The long neck of the poppet valve that rides in the valve guide and contains grooves for the valve keepers that retain the valve in the cylinder head. See Figure V.4.

Valve Stem Seal

A rubber O-ring or lip seal that fits around the valve stem that prevents the engine oil from entering the combustion chamber.

Valve Timing

Camshafts are rated by their valve timing which determines how much, how long, and when the valves open. Since the valves allow the air-fuel mixture into the engine, these factors affect horsepower and torque. The valve timing determines the engine operating range. The valve timing optimizes the airflow and cylinder pressure in a definite range. No one camshaft works best at all engine speeds, so valve timing is a compromise. The important terms used to rate or measure valve-timing events are as follows: valve lift, duration, and overlap.

Valve action is stated with degrees of crankshaft rotation about the TDC or BDC positions of the piston in the cylinder.

In the four-stroke cycle, the intake stroke occurs only during the piston downstroke and the exhaust stroke only during the piston upstroke. Moving air has inertia or the resistance to a change in motion and requires opening the valves longer than the piston stroke in an operational engine. The piston moves at varying speeds during its stroke because it must stop at each end of the stroke, gain speed toward midstroke and slow down to a stop at the opposite end. The inertia of the air is used to keep the exhaust gas or intake air moving even though the piston may be stopped. If the valve is held open, the intake stroke now lasts longer than the down travel of the piston. The exhaust stroke also lasts longer than the upstroke of the piston.

- *Exhaust Valve Timing*: The exhaust valve is timed to open before BDC in the power stroke and to close after TDC in the exhaust stroke. The reason for this timing is to get rid of all of the exhaust gases. The valve is opened before BDC in the power stroke to release the low pressure remaining from combustion and to make sure the exhaust valve is fully open when the exhaust stroke begins. This avoids high backpressure against the piston during the start of the exhaust stroke and reduces withdrawing of the gases through the valve and seat. There is a small power loss in doing this, but is canceled out by a gain through the exhaust gases released. The exhaust valve is open past the end of the exhaust stroke to provide an overlap of the exhaust valve and intake valve opening. This permits a thorough clearing of exhaust gases from the clearance space and the creating of a partial vacuum. The inertia of this flow of exhaust gases creates this suction or partial vacuum in the clearance volume. All the burned gases are expelled and the charge of intake air is fresh air.
- *Intake Valve Timing*: In a four-stroke cycle diesel engine, the intake valve is timed to open before TDC and close after BDC of the intake stroke. This is done to achieve the highest volumetric efficiency. The opening of the intake valve close to the end of the exhaust stroke and before the closing of the exhaust valve allows the removal of burned gases before the beginning of the intake stroke. The intake air coming in partially displaces the small amount of exhaust that remains and reduces the amount of intake air contamination. This is sometimes referred to as the scavenging period of the four-stroke cycle. During the period that both valves are open, near the end of the exhaust stroke and the start of the intake stroke, there is a slight movement in the piston. Yet the volume of exhaust gas that has been forced out into the exhaust path by the piston continues outward through the exhaust manifold. The inertia of this flow

of exhaust gases creates a suction or partial vacuum in the clearance volume. So all the burned gases are expelled and the charge of intake air is fresh air and the quantity of oxygen is increased. The intake valve is held open after the piston has reached the end of the intake stroke. The tendency is for air to continue flowing into the cylinder due to the partial vacuum created by the piston during suction and the inertia imparted to the column of incoming air.

FIGURE V.5 ICE valve timing.

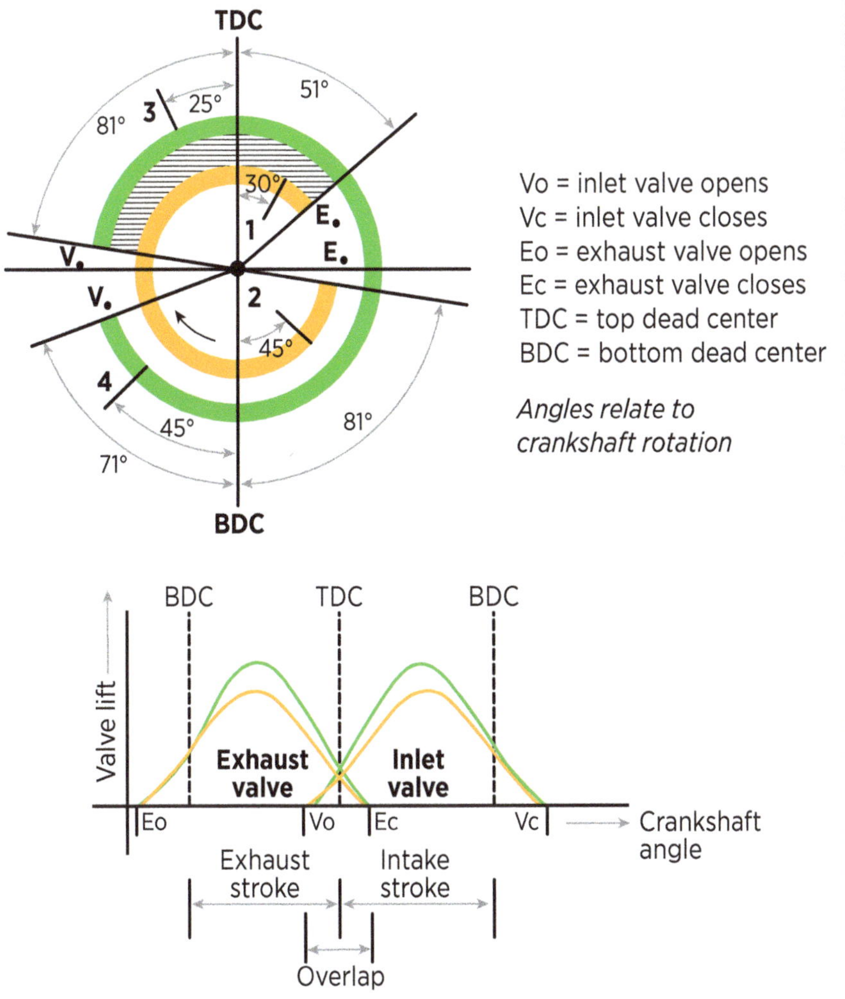

Valve Train

The total mechanism from the internal combustion engine (ICE) camshaft that opens and closes the intake and exhaust poppet valves. The parts of an ICE that make the valves operate. The valve train includes the camshaft(s), related drive components, and the various parts that convert the camshaft rotary motion into reciprocating motion, along with the valves and their associated components.

VANET (Vehicular Ad Hoc Network)

A network created by applying the principles of mobile ad hoc networks (MANETs). They allow more precise vehicle-to-vehicle (V2V) and vehicle-to-infrastructure (V2I) architectures to coexist and provide road safety, navigation, and other roadside services and are a key part of the intelligent transportation systems (ITS) framework, often referred to as ITS. VANETs were seen as a one-to-one application of MANET principles, but they have since then developed into a field of research in their own right. VANETs have become synonymous with the more generic term IVC (inter-vehicle communication) although the focus remains on the aspect of spontaneous networking, much less on the use of infrastructure like RSUs (roadside units) or cellular networks.

Vapor

A substance that can be easily compressed, has no resistance to flow, and no fixed volume.

Vapor Lock

Vaporized fuel, usually in the fuel line, that prevents or retards the necessary fuel delivery to the cylinders. If the brake fluid absorbs too much heat, it will boil into a compressible gas. This causes an increase in brake pedal travel because the vapor must be compressed before the fluid can transmit force. If enough fluid boils, the brake pedal will go all the way to the floor and braking power will be lost.

Variable Damping Shocks

A shock absorber or strut that has multiple orifices to increase or decrease damping rates.

V-Belt

SAE J636_201912 specification covers standard dimensions, tolerances, and methods of measurement of V-belts and pulleys for automotive V-belt drives. The drive belt system may use one V-belt or two V-belts. The drive belt is thin so it can bend backward and has several ribs to match the grooves within the pulleys. There could also be a V-belt-style belt to drive just certain accessory drive components. The drive belts are fabricated from differing types of rubbers: such as chloroprene or EPDM rubber (ethylene propylene diene monomer rubber). These belts have different layers or plies containing either fiber cloth or cords for reinforcement. Either side of the drive belt could also be accustomed to drive the various accessory drive components. When the rear side of the drive belt is employed to drive a pulley, the pulley is smooth. The crankshaft balancer pulley pulls the drive belt across the accessory drive component pulleys. The spring-loaded drive belt tensioner keeps constant tension on the drive belt to forestall the drive belt from slipping. The drive belt tensioner arm will move when loads are applied to the drive belt by the accessory drive components and also the crankshaft. The drive belt system may have a pulley block, which is used to cause the adjacent pulleys to wrap. Some systems use a pulley block in situ of an adjunct drive component when the vehicle is not equipped with the accessory. For more information on this, see **J636_201912** [1].

VCO (Valve Covering Orifice) Diesel Fuel Injector

This type of diesel fuel injector uses a moving needle inside the injector. This needle covers the multiple holes at the end of the injector and causes a very clean end to the injection event. The VCO nozzles have much more material to work with than the SAC style, so a SAC injector only has an orifice of up to 0.020 in. in size. The VCO can go all the way up to 0.028-in. orifice size. One of the downsides of a VCO nozzle is when it starts to open, the needle is mostly blocking the passage that feeds out of the nozzle, so the fuel flow is weak and the spray pattern is not that clean at the start of injection.

Vehicle Re-identification Methods

Sets of detectors mounted along the road employing a singular serial number for a tool within the vehicle are detected at one location and then detected again (re-identified) further down the road. Travel times and speed are calculated by comparing the time at which a selected device is detected by pairs of sensors. This might be done using the MAC (media access control) addresses from Bluetooth devices or using the RFID (radio frequency identification) serial numbers from ETC (electronic toll collection) transponders (toll tags) better cited as EZ Pass.

Vehicle Telematics

Junction of telecommunications (communication over a distance by cable, telegraph, telephone, or broadcasting) and data processing. This term grew to refer to automation in automobiles, such as the invention of the emergency warning system for vehicles. GPS navigation, integrated hands-free cell phones, wireless safety communications, and automatic driving assistance systems all are covered under vehicle telematics.

Vehicle Tracking

Monitoring the movements, status, and behavior of a vehicle. This could be achieved through a combination of a global positioning system (GPS) receiver and a device (usually comprising a GSM [Global System for Mobile Communications] modem or sender) installed in each vehicle, communicating with the user (dispatching, emergency, or coordinating unit) and PC-based or Web-based software. The data became information by management reporting tools in conjunction with a visual display on computerized mapping software. Vehicle tracking systems may use odometry or dead reckoning as another, or complementary, means of navigation. GPS tracking is usually accurate to around 32 to 65 ft or 10–20 m. For more information on *Taxonomy and Definitions for Terms Related to Driving Automation Systems for On-Road Motor Vehicles*, see **J3016_202104** [7].

Vehicle (Truck) Weight Distribution Equations

The following weight distribution equations are derived from basic mechanical engineering and will use the following definitions:

- L is the payload (P/L) plus the body or trailer weight for an evenly distributed load or for tractor service
- F/X is the front axle
- C/L is the centerline
- R/X is the rear axle
- CG is the load center
- WB is the wheelbase
- KP is the distance of the CL of the rear axle to the CG
- T is the distance of the CL of the front axle to the CG

$$\text{Load transfer to the front axle: } F/X = \frac{KP \times L}{WB} \tag{V.1}$$

$$\text{Load transfer to the rear axle: } R/X = \frac{T \times L}{WB} \tag{V.2}$$

$$\text{Location of CG}\left(\text{load center}\right)\text{:KP} = \frac{\text{WB} \times \text{L on F/X}}{\text{Total load “L”}} \tag{V.3}$$

$$\text{Chassis wheelbase: WB} = \frac{\text{T} \times \text{L}}{\text{L on R/X}} \quad \text{or} \quad \frac{\text{KP} \times \text{L}}{\text{L on F/X}} \tag{V.4}$$

Velocity

The rate of change of position with respect to a frame of reference, and is a function of time. It is a vector quantity having both direction and magnitude as opposed to speed, which is a scalar quality having only direction. The velocity of an object is the rate of change of its position. Velocity is the vector quantity of a vehicle speed and direction of motion, for example, 100 km/h to the south.

Venturi

A tube or passage with a tapered or narrow area in the middle of this tube which is, for this definition, the throat of an ICE carburetor. It causes an increase in the speed of the air flowing through that makes suction or negative pressure from the pistons sucking the fuel air into a carburetor. When the air molecules move through the narrow passage of the boost or auxiliary venturi (small hole venturi inside a big hole venturi) orifice, they must separate and the stream of air gets thinner with fewer molecules. This causes the pressure in the smaller hole or boost (auxiliary) venturi to be less than the pressure above it. These designs are also called a multi-venturi carburetor. To keep it simple, wewill refer to the boost or uxiliary venturi as the boost venturi.

VFD (Vacuum Fluorescent Display)

A display where electrons from a heated filament strike a phosphor material that emits light.

Vibrations

The oscillation of the parts of a fluid or solid component whose equilibrium has been disturbed, causing it to maneuver back and forth or up and down. Rotating automotive components will cause vibrations when excessive imbalance or runout is present. During vibration diagnosis, the amount of imbalance or runout should be considered a tolerance. The less imbalance or runout the less vibration.

Rotating components will cause a vibration concern once they are not properly isolated from the passenger compartment: Engine firing pulses are detected as a vibration if a motor mount is collapsed. A vibrating component operates at an identical rate (km/h, mph, or RPM). Measure the speed of vibration in question. When the rate/speed is decided, relate the vibration to a component that operates at an equal rate/speed to pinpoint the source. Vibrations also tend to transmit through the structure to other components. Therefore, simply because the seat vibrates, it does not mean the source of vibration is within the seat. Vibrations incorporate the subsequent three elements:

- Source—explanation for vibration
- Transfer path—trail the vibration travels through the vehicle
- Responder—component where vibration is sensed

In the world of vibrations, there are first-, second-, third-order vibrations. Order refers to the number of times the vibration occurs during one revolution of a rotating component. A tire with one item would create a disturbance once for each revolution of the tire. An oval-shaped tire with two high spots would create a disturbance twice for each revolution. This can be called second-order vibration. Three high spots would be third order. Two first-order vibrations may add or subtract from the amplitude of the disturbance, but that is all. Two first-order vibrations do not equal a second order. Thanks to the force, an unbalanced component will always create a minimum of a first-order vibration.

Video Vehicle Detection Systems

A system that uses video cameras to live traffic-flow measurement and automatic incident detection as a method of car detection. Since video detection systems like those employed in automatic number plate recognition do not involve installing any components directly into the horizontal surface or roadbed, this will be a non-intrusive method of traffic detection. Video from cameras is fed into processors that analyze the changing characteristics of the video image as vehicles pass. The cameras are generally mounted on poles or structures above or adjacent to the roadway. Most video detection systems require some initial configuration to instruct the processor the baseline background image. This usually involves inputting known measurements just like the gap between lane lines or the height of the camera above the roadway. One video detection processor can detect traffic simultaneously from one to eight cameras, depending on the brand and model. The everyday output from a video detection system is lane-by-lane vehicle speeds, counts, and lane occupancy readings. Some systems provide additional outputs including gap, headway, stopped-vehicle detection, and wrong-way vehicle alarms.

VIN (Vehicle Identification Number)

"A vehicle identification number (VIN) is a unique code, including a serial number, used by the automotive industry to identify specific motor vehicles, towed vehicles, motorcycles, scooters, and mopeds, as defined by the ISO (International Organization for Standardization) in ISO 3779 and ISO 4030 (location and attachment). VINs were first used in 1954 in the United States. From 1954 to 1981, there was no accepted standard for these numbers, so different manufacturers used different formats. In 1954, US car OEMs and the Automobile Manufacturers Association (AMA) created a new, standardized vehicle identification numbering system named the Vehicle Identification Number (VIN) with an agreed upon digit sequence and concealed chassis markings of this VIN. In 1981, the National Highway Traffic Safety Administration standardized the format. It required all on-road vehicles sold to contain a 17-character VIN. The VIN does not use the letters O, I, and Q (to avoid confusion with numerals 0, 1, and 9). SAE J1877_201207 describes the basic content requirements, bar code specifications, and functional test specifications of the vehicle identification number (VIN) label. On the vehicle, the VIN label is to be mounted in a readily accessible location for use of a bar-code scanning device." For more information on this, see **J1877_201207**, **J218_200910**, and **J1930DA_202105** [3, 4, 6].

FIGURE V.6 VIN (left) and VIN plate (right).

Viscosity

Tendency of a liquid to resist flowing. A quantity of internal friction between the substance layers as measured by the force per unit area resisting the flow of a liquid in parallel layers with speed relative to one another. Viscosity is an important fluid property when analyzing liquid behavior and fluid motion near solid boundaries. The viscosity of a fluid is a measure of its resistance to gradual deformation by shear stress or tensile stress. The shear resistance in a fluid is caused by intermolecular friction exerted when layers of fluid attempt to slide by one another.

Viscosity Index (VI) Improvers

Helps the oil resist thinning at high temperatures. Since VI improvers actually increase viscosity at low as well as high temperatures, a 5W-30 multi-grade motor oil starts out as slightly less than a 5 weight base stock oil. Oil companies then add sufficient VI improvers so that the oil is able to perform like a 30 weight base oil at high temperatures.

Viscous Coupling

A coupling device that uses a selection of steel plates housed in a sealed steel drum. A viscous fluid such as silicon-based oil fills the areas between the plates. The viscous coupling is not active during equal traction conditions. However, the viscous coupling is actively transferring torque during light to moderate cornering, but there is a certain amount of slippage under these conditions to prevent driveline windup. If there is a loss of traction, the speed differences between the front and rear axles increase and heat the silicon fluid, causing it to thicken to the point that it transfers more torque to the axle losing traction. During a severe loss of traction, the viscous fluid thickens enough to lock the plates together, dividing the engine torque equally 50/50 between the front and rear axles.

FIGURE V.7 Driveline viscous coupling.

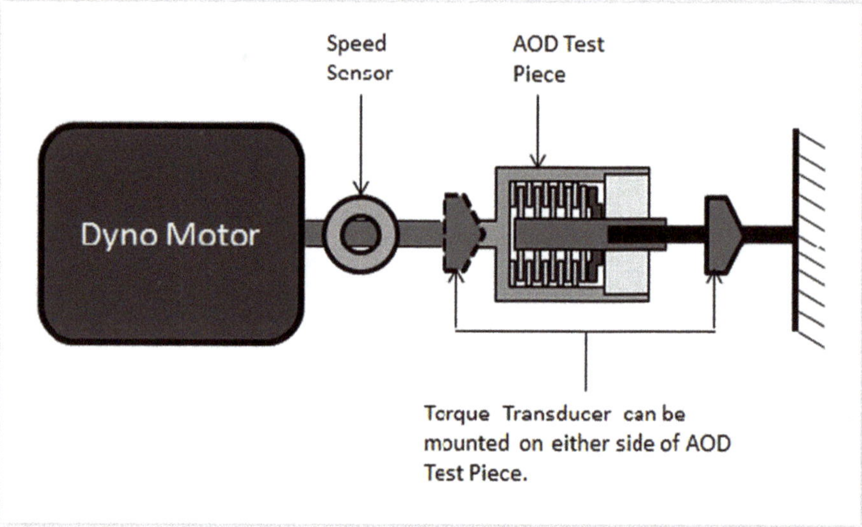

Viscous Fan Clutch

The fan clutch engages the fan blades to provide maximum cooling. When maximum cooling is not needed, the fan "slips" to prevent wasting engine power. The viscous fan clutch uses a silicone-fluid coupling, somewhat like an automatic transmission torque converter to connect the cooling fan to the engine drive. There are two viscous members, a drive member and a driven member. The clutch plate is splined to the pulley shaft and is the drive member. The driven member is integral to the drive housing, which is bolted to the fan blades. Both of these members have circular grooves that are closely mated to each other. The viscous coupling fluid is silicone that is stored in a reservoir chamber in front of the pump plate. A bi-metallic spring controls the flow of fluid using a valve attached to the spring. This bi-metallic spring senses the air temperature directly behind the radiator to engage or disengage the viscous clutch as needed.

FIGURE V.8 Viscous fan clutch.

Volatile Memory

Automotive computers use two types of RAM: volatile and nonvolatile. Volatile memory erases when power is removed from the computer. Nonvolatile is removed by some external action.

Volatility (Volatile)

A measurement of the tendency of a liquid to change to vapor.

Volt

A unit of measurement for the amount of electromotive force (EMF) or electrical pressure. It is shown in formulas and diagrams by the letters *V* or *E*.

Voltage

The electrical force, or pressure, that pushes free electrons and causes current. It is also referred to as electromotive force (EMF) or potential difference. A number of electrons gathered in one place affect an electrical charge. We call this charge a potential difference or voltage. Voltage is measured in volts (V). Since it is used to "move electrons," an externally applied potential difference is sometimes called an "electromotive force," or EMF. Potential, voltage, and EMF all mean the same thing. You can think of voltage as the electrical "pressure."

Voltage Drop Measurement/Voltage Drop Tests

Voltage drop by a circuit is dynamic resistance and is like a kink during a water hose. In a water hose with too much water, the pressure upstream of the kink is higher than downstream of the kink. Voltage is higher before the resistance than it is after the resistance. The voltage change across the resistance is called voltage drop and is the result of a fully applied voltage that is not equal at either end of the load circuit. Whenever the voltage available to a tool (load) is a smaller amount than the source voltage, there is a resistance between the two components.

The resistance in exactly one circuit opposes the electron flow, resulting in voltage loss across the load. We call this loss a "voltage drop." Kirchhoff's voltage law states that the sum of the voltage drops in any circuit will equal the source voltage. Drop is the voltage lost or consumed as current moves through resistance.

The voltage is highest where the conductor connects to the voltage source, but decreases slightly because it moves through the conductor. If you measure voltage before it goes into a conductor and measure it again on the opposite side of a conductor, you may find that the voltage has decreased. This can be voltage dip, and depending on the circuit, this might be too small of a difference to dwell on. Once you connect several conductors to each other, the voltage will drop every time the current passes through another conductor.

Voltage Regulator

A conductor that maintains a controlled voltage output in an alternator. An electronic or mechanical unit that controls the output voltage of an electrical generator or alternator by controlling the sphere current of the generator by adding resistance serial.

FIGURE V.9 Voltage regulator.

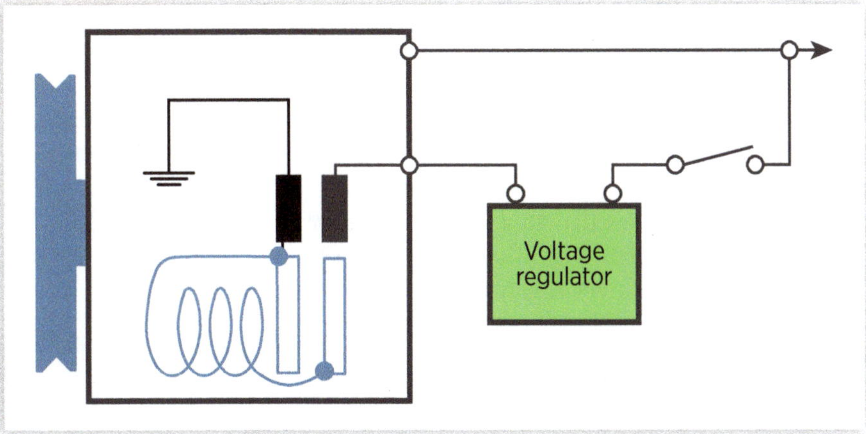

Voltmeter

A test instrument used to measure volts or units of electrical pressure as a stand-alone device or part of a digital multimeter. The voltmeter is connected in parallel with the unit or circuit being tested.

Volts per Division

The vertical scale of an oscilloscope that has eight divisions or graticules. If each division is set to equal 1 V, the display will show 0 to 8 V. The volts per division (V/div) should be set so that the entire anticipated waveform can be viewed.

V-ref

Reference voltage from the engine management computer.

VRLA (Valve Regulated Lead-Acid) Battery

VRLA batteries provide improved grid materials to allow some sealed, maintenance-free batteries to develop fully charged, open-circuit voltage of 2.1 V per cell, or 12.6 V total for a 6-cell 12-V battery. Separators made from a porous material reduce the internal resistance and generate higher voltage and current. The electrolyte is housed inside plastic envelope-type separators around the plates. The entire case is not flooded with electrolyte, which eliminates the sloshing of the electrolyte or leaks from a cracked battery. Because the battery is not vented, terminal corrosion from battery gassing and electrolyte spills or spray is also eliminated. There are two types of AGM (Absorbent Glass Mat) or gel. For more information on this, see **J1715/2_202108** [5].

VSM (Value Stream Map)

All the steps necessary (activities/tasks) to deliver a product or service from supplier to customer (cradle to grave). It represents the input of raw material through the delivery of a finished product or service. The total value stream may include multiple companies and groups within the business. It is a big-picture perspective needed to avoid selective implementation resulting in isolated islands of lean within an overall non-lean process. The VSM is a snapshot in time of the process with metrics that flows from left to right.

VSS (Vehicle Speed Sensor)

This sensor measures the vehicle speed using a sensor located at the output of the transmission/transaxle or by monitoring sensors at the wheel speed sensors.

The VSS provides vehicle speed information to the engine management computer. The analog version uses a permanent magnetic generator mounted within the speedometer cable opening within the transmission/transaxle housing. The static magnet generator produces a pulsing (AC) voltage whenever the vehicle speed is over 3 mph. The AC voltage level and therefore the number of pulses increase with vehicle speed. The VSS output is AC voltage, which cannot be directly utilized by digital electronic components, just like the computer. The AC voltage is converted into a digital signal by the VSS buffer. The personal computer (PC) needs information about the vehicle speed to work. Current VSS are Hall-effect sensors (see **Hall-Effect Sensor**). Most up-to-date speedometers turn 1,000 RPM per mile. Older mechanical speedometers turn at 1,440 RPM per mile.

REFERENCES

1. Society of Automotive Engineers, Standard J636_201912: V-Belts and Pulleys, SAE Publishing, Warrendale, PA, 2019.
2. Kershaw, J.F., *High-Performance Automotive Cooling Systems* (Forest Lake, MN: CarTech, Inc., 2019), Chapter 1, ISBN: 978-1-61325-504-9.
3. Society of Automotive Engineers, Standard J218_200910: Passenger Car Identification Terminology, SAE Publishing, Warrendale, PA, 2019.
4. Society of Automotive Engineers, Standard J1877_201207: Recommended Practice for Bar-Coded Vehicle Identification Number Label, SAE Publishing, Warrendale, PA, 2019.
5. Society of Automotive Engineers, Standard J1715/2_202108: Battery Terminology, SAE Publishing, Warrendale, PA, 2021.
6. Society of Automotive Engineers, Standard J1930DA_202105: Electrical/Electronic Systems Diagnostic Terms, Definitions, Abbreviations, and Acronyms Web Tool Spreadsheet, SAE Publishing, Warrendale, PA, 2021.
7. Society of Automotive Engineers, Standard J3016_202104: Taxonomy and Definitions for Terms Related to Driving Automation Systems for On-Road Motor Vehicles, SAE Publishing, Warrendale, PA, 2021.

W

Walking Beam Truck Suspension

See **Hendrickson Equalizing Bar.**

Wander

A wheel alignment condition where the vehicle will wander or pull to one side of the road. It is a tendency to deviate from the direction of the vehicle.

Wankel Engine

ICE that uses an eccentric rotor to convert pressure into rotating motion. This ICE develops more uniform torque with less vibration for a given power than a piston engine. The rotor creates the turning motion without reciprocation and is similar in shape to a Reuleaux triangle, except the sides have less curvature. This ICE delivers three power pulses per revolution of the rotor using the Otto cycle. Gearing is used on the output shaft that allows the rotor to turn three times faster giving one power pulse per revolution. In one revolution, the rotor performs power pulses and exhausts gas simultaneously, while the Otto four-stroke cycle takes place at separate times. The two-stroke piston engine develops one power pulse per revolution like the Wankel engine, but in the four-stroke cycle, it develops one power pulse for every two revolutions. The four-stroke cycle of intake, compression, power, and exhaust in the Wankel takes place at each revolution of the rotor at each of the three rotor faces moving inside the oval-like epitrochoid housing, enabling the three power pulses per rotor revolution. Displacement is for only one face of one rotor or one face multiplied by the number of rotors; however, all three faces of all rotors are working at the same time.

Warm-Up Cycle

A warm-up cycle is defined as achieved when the engine temperature increase of at least 40°F and where the engine temperature reaches at least 160°F (71°C).

Warning Light

A light on the instrument panel that alerts the driver when one-half of a split hydraulic system fails as determined by the pressure differential switch.

Warp

Takes place when two wheels on opposite sides and axles are in jounce, while the remaining wheels are in rebound.

Wastegate

Reduces exhaust gases driving the turbocharger turbine by opening an alternate path for the exhaust gases, reducing the speed of the exhaust turbine. For more information on *Turbocharger Nomenclature and Terminology*, see **J922_201106** [8].

Waste Spark

The spark that occurs on the exhaust stroke of a waste-spark system using one coil for two plugs.

Waste-Spark Ignition

A distributorless ignition system that uses multiple coils where each fires two spark plugs. Each coil is a step-up transformer where the primary winding and secondary coil winding are not electrically connected. Each secondary winding end is connected to a cylinder opposite the other cylinder in firing order, which is called a companion cylinder. Both spark plugs fire at the same time. In a four-cylinder engine, when cylinder #1 is on the compression stroke, its companion cylinder #4 is on the exhaust stroke. The spark that occurs on the exhaust stroke at #4 is called the waste spark because it does no useful work and is the ground path for the secondary ignition coil winding. The voltage required to jump the spark plug gap on the exhaust stroke is 2–3 kV and provides the ground circuit for the secondary coil circuit. The remaining coil energy is used by the cylinder on the compression stroke of cylinder #1. One spark plug of each pair always fires with a straight polarity and the other cylinder always fires reverse with polarity. For more information on this, see **J139_202002** [1].

Water Cooled

A water-cooled system is the most common automotive cooling system. The coolant is circulated outside of the engine and exposed indirectly to the air by a radiator. The air absorbs heat from the coolant so that the coolant can flow back into the engine and absorb more heat. The greater the difference in temperature between the coolant and the air, the more heat will be absorbed by the air. Uniform engine temperatures reduce thermal stress. The consistent temperatures also allow very precise fuel metering, which helps keep exhaust emissions low. See **Cooling System**.

Water Jacket

In a water-cooled ICE, it is the area around the cylinder in a block where the engine coolant flows to remove the heat of combustion.

Water Pump

A centrifugal pump that circulates coolant through an internal cmbustion engine (ICE) cooling system. The water pump got its name from the fact that, in the early ICE engines, water was used as the coolant and pumped through the system before antifreeze was invented, hence the name water pump. Coolant goes into the pump impeller near the rotating axis and is accelerated by the impeller, flowing radially outward into a diffuser or volute chamber from where it exits into the engine water jackets. Water Pump Formula for Water horsepower = minimum power required to run water pump:

$$\text{Water HP} = \frac{\text{TDH} \times \text{Q} \times \text{SG}}{3{,}960} \quad \textbf{(W.1)}$$

where

TDH is the total dynamic head = Vertical distance liquid travels (in ft) + friction loss from plumbing

Q is the flow rate of liquid, in gal/min

SG is the specific gravity of liquid (this equals 1 if you are pumping water)

FIGURE W.1 Water pump used to pump coolant in an ICE.

Watt Rating

Amount of power the resistor can handle without damage.

Watt's Law

Relationships among power, voltage, and current with Watt's Law. The fundamental equation for Watt's Law is $P = I \times E$ or Power = amps × volts. Power is the product of current multiplied by voltage. In a circuit, if voltage or current increases, then power increases. If voltage or current decreases, then the ability will decrease. The unit of measurement for power is watt. One watt is equal to one ampere times one volt. The foremost common application of a rating in watts is maybe the sunlight light bulb. The amount of watts they consume classifies light bulbs. So are resistors, audio speakers, and residential appliances. Here is an example of how we determine power in watts:

- P = IE—I is the current, E is the voltage: If the overall current (I) is up to 10 A, and therefore the voltage (E) is equal to 120 V, then $P = 120 \times 10$ and $P = 1,200$ W.

$$P = 120 \times 10 \quad \text{and} \quad P = 1{,}200 \text{ W} \qquad \textbf{(W.2)}$$

Watt's Linkage

Name for a rigid-axle coil spring suspension that uses two bars similar to a Panhard arm and a pivot point on the axle to keep the axle from moving in turns.

Waveform

The pattern defined by an electrical signal.

Weather Strip

A strip of compressible rubber that is attached around the vehicle doors and windows.

Web

The vertical part of a vehicle frame joining the flanges.

Wedge Brake Design (Truck and Bus Air Brakes)

A wedge-shaped plunger is used to force two rollers against the brake shoes. A wedge-shaped plunger forces two rollers against the brake shoes. Pushrod action forces the shoes against the drum. For more information on *Braking System Definitions—Truck and Bus*, see **J2627_200908** [3].

Wedge-Shaped Combustion Chamber

The wedge combustion chamber provides efficient combustion, but does not breathe as well as the hemispherical chamber. Some OEMs modified the wedge design to give better breathing and produced a chamber with some of the advantages of the hemispherical. The valves remained in line, but manufacturers tilted or canted them away from each other to provide better flow. They opened the combustion chamber, making it closer to be hemispherical in shape. This unshrouds the valves, providing better flow at higher valve lifts. For more information on *Internal Combustion Engines—Piston Vocabulary*, see **J2612_202002** [2].

Weight Transfer

Weight transfer is key to a strong vehicle launch. When the vehicle is at standstill, its weight is spread about in the chassis and body, but upon launch and acceleration, inertia shifts the weight backward, which helps traction (in rear-wheel-drive cars). As the car proceeds and the initial brute acceleration decreases while speed builds up, the weight gradually tries to move forward to regain its original front/rear distribution. This very same weight will shift all the way forward upon braking, which will push the nose of the vehicle down. Also called pitch rotation as viewed from the side of the vehicle. It can be best described as load transfer, or more commonly weight transfer.

Welch Plug

Another name for a core-out or freeze plug used to seal the casting holes in an engine block.

Wet Clutch

A multiple-disc friction clutch that is immersed in an oil bath. They are mostly used in industrial equipment.

Wet Liner

See **Liner**.

Wet Sump

The oil pan that attaches to the bottom of the crankcase that is used as an oil reservoir for the ICE lubricating oil.

Wheatstone Bridge

Electrical bridge circuit used to calculate changes in resistance of an unknown variable resistor. They are used to measure temperature as well as pressure and mechanical strain. The Wheatstone Bridge uses four resistors. Three of the resistors have known values, and the fourth has an unknown resistance value and is the sensing resistor. When all four sensors are the same in value, the bridge is said to be "in balance." In the balanced state, the Wheatstone Bridge voltage sensor will indicate a value of 0 V. In temperature monitoring applications, temperature changes the resistance value of the sensing resistor, which unbalances the circuit. The sensing circuit will receive a voltage reading that is proportional to the change in resistance. When used in a MAP sensor, the Wheatstone Bridge is incorporated into a silicon chip along with a diaphragm. This assembly is placed in a sealed housing. That housing is connected to the intake manifold by a tube. When manifold pressure is applied to the diaphragm through the tube, it causes the diaphragm to deflect. The resistance of the sensing resistors changes in proportion to the amount of deflection, thus indicating the amount of manifold pressure. This resistance change is indicated as a change in voltage by the sensing circuit.

Wheel

"The disc that a tire is mounted on. It typically means the tire and wheel rim. SAE J175_202107 establishes minimum performance requirements and related uniform laboratory test procedures for evaluating lateral (curb) impact collision resistance of all wheels intended for use on passenger cars and light trucks." SAE J393_201710 establishes uniform engineering nomenclature for wheels, hubs, rims, and their components used in truck, bus, and trailer applications. This nomenclature and accompanying drawings are intended to define functional truck wheel, hub, and rim designs. For more information on this, see **J175_202107** and **J393_201710** [4, 5].

Wheel Balancing (ON and OFF Vehicle)

On-vehicle balancing is performed using a spinner, adapters, and a balancing instrument. The spinner is driven by an electric drum capable of spinning a non-driving wheel at different speeds. The adapters are used to mount the balancing instrument to the wheel flange. One type uses a balancing instrument that contains an internal weight that can be rotated by hand for proper placement until the vibration is eliminated. Before balancing, all rocks and wheel weights should be removed from the tread and rim. The tire should be checked for loose wheel bearings. The adapter and instrument are installed on the wheel. Dynamic balancing brings the wheel up to the speed, during which the vibration is most noticeable. Off-vehicle dynamic balancing requires a computerized balancing machine. The tire is mounted on a spindle/hub with adapters or springs with cones. The tire machine spins the tire to determine the correct areas for weight placement. The procedures differ depending on the machine, but most ask for rim and tire measurements and need a reference rim flange measurement for the calculations. The rim flange measurement allows the computer to determine where loads should be placed on the mounting spindle of the machine. Most are capable of static and dynamic balancing. Follow the balancer manufacturer's directions for operation.

FIGURE W.2 Computer wheel balancer.

Wheel Cylinder

A hydraulic cylinder used in drum brake systems to apply the rear or front brake shoes. Pressure from the master cylinder enters the center of the wheel cylinder and pushes two pistons equally outward to push the brake shoes against the drum to stop the wheel.

FIGURE W.3 Drum brake wheel cylinder.

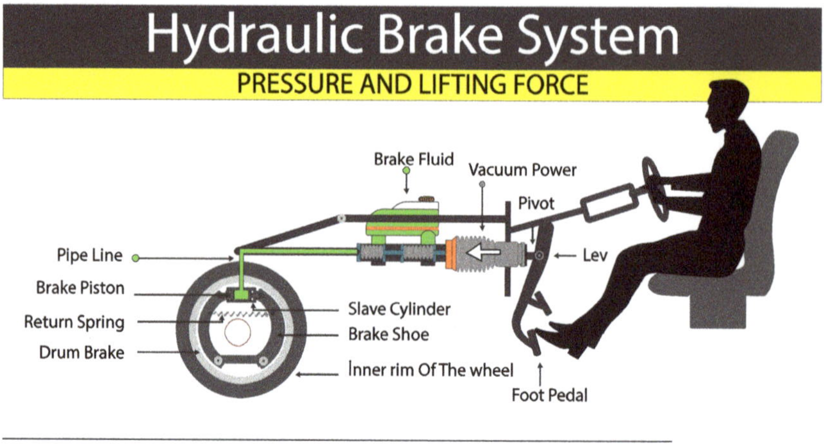

Wheel Disc

That part of a road vehicle wheel between the hub and the rim. For more information on this, see **J175_202107** and **J393_201710** [4, 5].

FIGURE W.4 Wheel assembly.

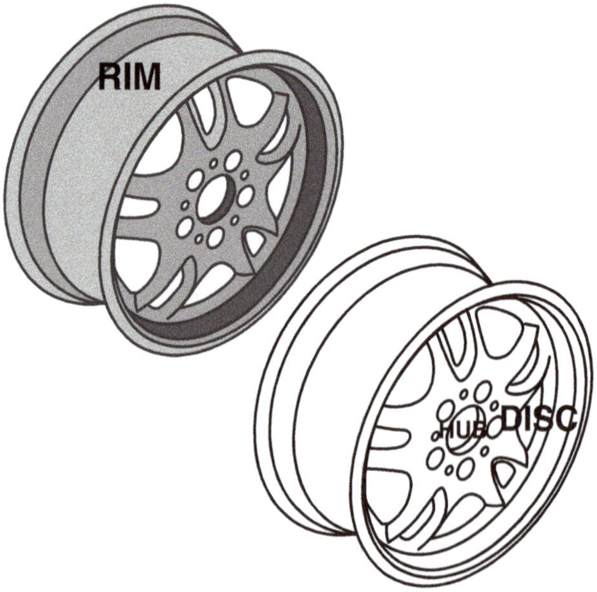

Wheel Hop

Excessive oscillations of the wheel in a radial direction going up and down and losing contact with the road.

Wheel Offset

The distance the mounting face of the tire wheel is located from the actual center of the wheel. Wheels come in a variety of offsets to clear brake components and can alter the vehicle scrub radius. The OEM wheel offset should always be the same for replacement wheels. For more information on this, see **J175_202107** and **J393_201710** [4, 5].

Wheel Rim

The outside of a disc wheel where the tire is mounted. For more information on this, see **J175_202107** and **J393_201710** [4, 5].

Wheel Shimmy

Side-to-side movement of the wheel assembly from a lack of wheel parallelism.

Wheel Speed Sensor (Active)

These wheel sensors require outside electrical power to operate. An active WSS is also a digital wheel speed sensor because it generates a digital on and off output signal and is capable of detecting wheel speeds as low as 0 mph (0 km/h). Two types are used: (1) Hall effect, which uses a permanent magnet in front of an integrated circuit (IC) chip. The magnetic field penetrates the chip and the reluctor. If the reluctor is turning, the magnetic field is changing. The IC measures the change in the flux. The signal is not dependent on the speed. (2) Magneto-resistive, which is the most common type. It uses a magneto-resistive circuit to produce a square waveform where the frequency is proportional to the wheel speed. It can also detect the direction and can be used by the controller for advanced features such as hill holding.

Wheel Speed Sensor (Passive)

"This input sensor provides the ABS with wheel rotational speed information. They generate their own output signal and can operate without vehicle voltage. They are also called variable reluctance (VR)–type sensor. They use a magnetic core surrounded by coil windings. The sensor is mounted at a specific gap from a notch-and-tooth reluctor. The wheel speed sensing system consists of a sensing device and an actuator. The WSS is an inductive pickup sensor: a permanent magnetic core wrapped within a coil of wire. The actuator is a toothed gear tone wheel like a spur gear that rotates with the wheel. It is a reluctor devise that allows magnetic flow and then blocks it to provide a variable voltage signal to the ABS controller. When magnetic flux cut across a conductor, a voltage is generated in that conductor. This variable signal indicates the speed of the wheel. WSS output can vary from wheel to wheel and vehicle to vehicle due to the following variables: Magnetic strength, Air gap. Wheel speed, Winding efficiency (manufacturing of the coil), and Total circuit resistance" [6].

FIGURE W.5 ABS passive wheel speed sensor.

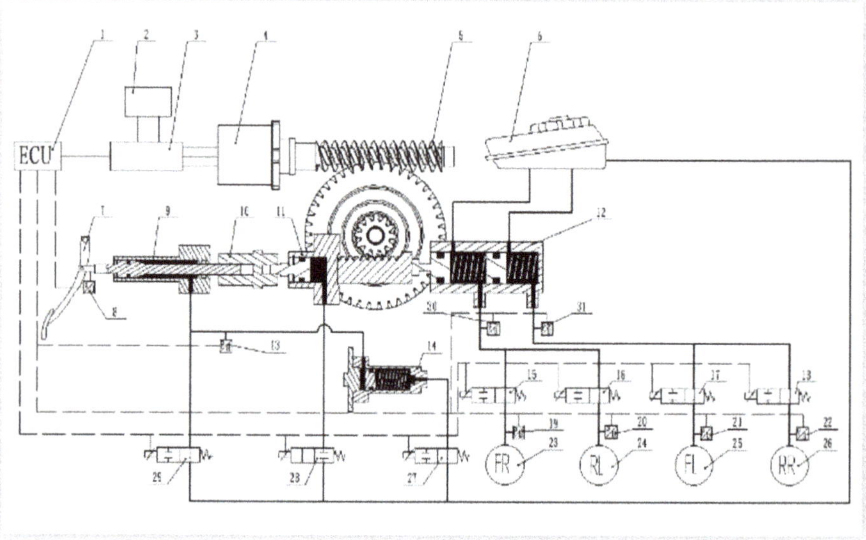

Wheel Spin

The rapid movement of the drive wheels caused by the application of excessive torque.

Wheel Well

The cavity or opening is the structure of a road vehicle to house the wheel. For more information on this, see **J175_202107** and **J393_201710** [4, 5].

Wheelbase

The linear distance between the front and rear wheels of a road vehicle. In dynamic suspension equations, this is the value "L." For more information on this, see **J175_202107** and **J393_201710** [4, 5].

White Smoke
Mixture of water vapor and unburned or partially burnt fuel. As a result of the lower temperature, the fuel ignites so late that combustion is incomplete at the time the exhaust valve opens and fuel is emitted in an unburned or partially burned condition.

Wideband Oxygen Sensor
A sensor that monitors air-fuel ratio over a broader range above and below stoichiometric than a conventional oxygen sensor. It can sense air-fuel mixture ratios between 0.9 and 2.2. Also called an air-fuel ratio sensor.

Wind Tunnel
A testing chamber to measure the aerodynamic air flow over a vehicle. A large high-speed fan directs air and smoke over a full-scale or scale model of a vehicle to determine the aerodynamic drag. The vehicle is fully instrumented and mounted on a tilt table.

FIGURE W.6 Wind tunnel.

Windshield
A window located at the front of the passenger compartment of a motor vehicle. Its British term is windscreen.

Windshield Washers

An electrically operated device accustomed squirt washer fluid onto the windshield where it can help dissolve debris and clean the viewing area of the windshield when the wipers are used. Most vehicle windshield washers use a positive-displacement or centrifugal-type washer pump located within the washer reservoir. A momentary contact switch, which is commonly a part of a steering column–mounted combination switch assembly, energizes the washer pump. Washer pump switches are installed either on the steering column or on the board. The nozzles will be located on the bulkhead, within the hood, or mounted as a part of the wiper arms.

Windshield Wipers

Motor-operated device accustomed to keep the viewing area of the windshield away from the rain. Mechanical device systems and circuits vary greatly between OEMs, furthermore between models. Some vehicles combine the wiper and windshield washer functions into one system. Many SUVs even have a car window wiper and washer system that works independently of the windshield system. All windshield and car window wiper and washer systems operate in an exceedingly similar fashion. Wiper control switches are installed on the steering column.

Windup

See **Axle Windup**.

Wire

A wiring harness is created from a conductor and an insulator. The metal core of the wire, typically made of copper, is the conductor. The plastic or other material jacket coating the core is the insulator. Under normal circumstances, electrons move some inches per second. Yet when an electrical potential is applied to at least one end of a wire, the effect is felt before long at the opposite end of that wire.

Wire Code Identification System

An alphanumeric method of identification that explains the color, location, and gauge size of wires on a wiring diagram.

Wire Gauge

A measurement for a wire determined by the diameter or cross-sectional area of its metal conductor.

Wire Protection Device

A device that protects wires from damage, which includes electrical tape, plastic tubing, retainers, clips, boots, and straps.

Wiring Diagrams

"Drawings that show how the wires connect to each component in an electric or electronic circuit. Symbols represent components, and lines represent wires. Also called wiring schematics. SAE J2202_201912 provides general guidelines on the material selection, construction, and qualification of components and wiring systems used to construct nominal 12 VDC and/or 24 VDC electrical wiring systems for heavy-duty vehicles The guidelines are limited to nominal 12 VDC and/or 24 VDC primary wiring systems and includes cable sizes American Wire Gage 20 to AWG 4 on heavy-duty on-highway trucks. The document identifies appropriate operating performances requirements. This document excludes the male-to-female connection of the SAE J560 connectors. The future of the heavy truck electrical systems will offer a multi-voltage electrical system, which will include voltages above a nominal 14 VDC electrical wiring system, along with new technologies and requirements not included in the current SAE J2202. With the need to include new technologies addressing higher voltage, it was decided to limit the scope of the SAE J2202 to a nominal 12 VDC and/or 24 VDC electrical system." For more information on this, see **SAE J2202_201912** [7].

Wiring Harness

A set of several wires wrapped together with tape or enclosed together in a plastic protection device. Wires and connectors branch out from the harness to connect to various components.

Wiring Schematic

Drawings that show how the wires connect to each component in an electric or electronic circuit. Symbols represent components, and lines represent wires. Also called wiring diagrams.

Wiring Splices

The place where two or more wires are connected.

Work

The distance through which the restraining force acts in one shaft revolution is $2\pi R$, in which R is the radius of the shaft and π is 3.1416. The engine turning the dynamometer and producing a force is comparable to the engine crankshaft standing still and the dynamometer being turned around it with a force, F, acting through a radius arm R. When the speed is not known, work can be found using this formula and produced per revolution as follows:

$$\text{Work} = 2\pi FRN \quad \textbf{(W.3)}$$

where

F is the dynamometer load, in lb
R is the radius arm of the dynamometer, in ft
N is the engine speed, in RPM
$\pi = 3.1416$, a constant

Worm Gear Steering Gears

Uses two gears: a worm and a gear (also called a worm wheel). The worm has teeth cut in an exceedingly helical (spiral) shape and operates identical to a screw. The helix on the worm moves the gear wheel one tooth for every revolution of the worm. The helix provides a smooth and quiet steering operation for the motive force. It converts the movement of the wheel to the linear motion needed to regulate the wheels. This produces an outsized gear reduction. The gear ratio of the gear train in the gearbox may be a comparison of the angles of movement between the wheel, and therefore the wheel assembly, and is expressed as a ratio, for example, 18:1. The gear ratio of a worm gearbox increases the output torque and reduces the hassle the driving force must apply. Within the mechanism, gear reduction transforms an oversized turn of the wheel into a smaller turn of the road wheels. The use of drag reduction makes steering easier for the driving force. Wheels require an outsized force to be steered thanks to the load of the vehicle pushing the tire into contact with the road.

FIGURE W.7 Worm gear steering gear.

WOT (Wide-Open Throttle)

The position of the throttle valve at its maximum opening.

WVSCT (Wireless Vehicle Safety Communications Telematics)

A system that aids in car safety and road safety. It is an electronic subsystem in a car or other vehicle for the purpose of exchanging safety information, about such things as road hazards, and therefore the locations and speeds of vehicles, over a short-range of radio links. This might involve temporary unintended WLANs. Wireless units are being installed in vehicles and, doubtless, in fixed locations such as near traffic signals and emergency call boxes along the road. Sensors in the cars and at fixed locations, additionally as possible connections to wider networks, will provide the knowledge, which can be exhibited to the drivers, on how it will even be possible for traffic lights, which one can expect to become smarter, to use this information to cut back the possibility of collisions.

REFERENCES

1. Society of Automotive Engineers, J139_202002: Ignition System Nomenclature and Terminology, SAE Publishing, Warrendale, PA, 2020.
2. Society of Automotive Engineers, J2612_202002: Internal Combustion Engines—Piston Vocabulary, SAE Publishing, Warrendale, PA, 2020.
3. Society of Automotive Engineers, Standard J2627_200908: Braking System Definitions—Truck and Bus, SAE Publishing, Warrendale, PA, 2009.
4. Society of Automotive Engineers, Standard J175_202107: Wheels—Lateral Impact Test Procedure—Road Vehicles, SAE Publishing, Warrendale, PA, 2021.
5. Society of Automotive Engineers, Standard J393_201710: Nomenclature—Wheels, Hubs, and Rims for Commercial Vehicles, SAE Publishing, Warrendale, PA, 2017
6. Liu, T., Yu, Z., Xiong, L., and Han, W., "Anti-Lock Braking System Control Design on An Integrated-Electro-Hydraulic Braking System," *SAE Int. J. Veh. Dyn., Stab., and NVH* 1, no. 2 (2017): 298-306, doi:https://doi.org/10.4271/2017-01-1578.
7. Society of Automotive Engineers, Standard J2202_201912: Heavy-Duty Wiring Systems for On-Highway Trucks, SAE Publishing, Warrendale, PA, 2019.
8. Society of Automotive Engineers, Standard J922_201106: Turbocharger Nomenclature and Terminology, SAE Publishing, Warrendale, PA, 2011.

Y

Yaw Rate Sensor

A sensor that uses piezoelectric crystals to detect the yaw rate of the vehicle during rapid cornering, deceleration, and acceleration. It can be measured with accelerometers on the vertical axis. Any device intended to measure the yaw rate is called a yaw rate sensor. For more information on *Vehicle Aerodynamics Terminology*, see **J1594_201007** [1].

Yaw Roll Model

A yaw-roll model simulates the turning and rolling behavior of vehicles in constant-speed maneuvers. Turning behavior may be controlled either by defined steering inputs or by a driver model following a prescribed trajectory. For more information on *Vehicle Aerodynamics Terminology*, see **J1594_201007** [1].

Yaw (Rotation)

A yaw rotation is a movement around the yaw axis of a rigid body that changes the direction it is pointing, to the left or right of its direction of motion. The yaw rate or yaw velocity of a car or other rigid body is the angular velocity of this rotation, or rate of change of the vehicle heading. Vehicle movement may be due to yaw, pitch, or roll. For more information on *Vehicle Aerodynamics Terminology*, see **J1594_201007** [1].

Yaw-Weighted Drag Coefficient

Ambient wind magnitude, heading angle, and vehicle path directions have an effect on the overall average aerodynamic drag of a vehicle during a particular duty cycle. "The yaw-weighted drag coefficient is defined as the average drag coefficient during a particular driving schedule and ambient wind input. The wind and driving schedule factors affecting the wind-averaged drag coefficient have not been standardized." For more information on *Vehicle Aerodynamics Terminology*, see **J1594_201007** [1].

Yield Strength

An expression of the strength of the frame material that is used in its construction. It is the maximum load measured in psi (kPa) that can be placed on a material and still have the material return to its original shape once the load that caused the deformation is removed. A typical example is how far a truck frame deflects when the vehicle is loaded versus when it is unloaded.

Yoke

The connection point used on a driveshaft universal joint that connects the slip joint to the driveshaft through the cross shaft or spider.

Y-Pipe

An exhaust pipe in the shape of a Y that is used on "V" engines to connect the right and left exhaust bank to a single exhaust pipe going to the muffler.

FIGURE Y.1 Engine exhaust Y-pipe for V-engine.

Y-Type or Wound Stator

A configuration within which the wire ends from the stator windings is connected to a neutral junction in order that the circuit appears like the letter Y. A Y-type stator provides good current output at lower engine speeds.

FIGURE Y.2 Y-type alternator stator.

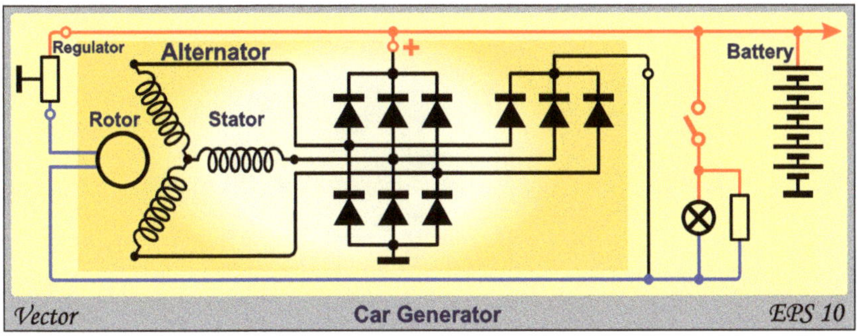

REFERENCE

1. Society of Automotive Engineers, J1594_201007: Vehicle Aerodynamics Terminology, SAE Publishing, Warrendale, PA, 2010.

Z

Zener Diode

A specially constructed (heavily doped) diode designed to control with a reverse-bias current after a specific voltage has been reached. Named after Clarence Melvin Zener.

FIGURE Z.1 Zener diode.

Red sun design/Shutterstock.com.

Zero Resistance

When there is continuity and a resistance of 0 within the tested circuit or component.

Zeroth Law of Thermodynamics

Expounds the conception of temperature or the speed of molecules. It is designed to allow the existence of an empirical parameter, the temperature, as a property of a system specifying that the systems are in equilibrium with each other, that is, have the same temperature. This law is compatible with the utilization of a selected body, for instance, a mass of gas, to match the temperature of other bodies, but does not justify regarding temperature as a quantity which will be measured on a scale of real numbers. If two systems are in thermal balance with a third system, they are in thermal balance with one another. This law helps define the concept of temperature.

Zirconia Oxygen Sensor

An oxygen sensor made up of zirconium oxide with a skinny coat of platinum. It produces a voltage output that represents the oxygen content of the exhaust gases. At an operating temperature of about 600°F (315°C), the element of the oxygen sensor generates a small voltage. For more information on this, see **J1930DA_202105** [1].

REFERENCE

1. Society of Automotive Engineers, J1930DA_202105: Electrical/Electronic Systems Diagnostic Terms, Definitions, Abbreviations, and Acronyms Web Tool Spreadsheet, SAE Publishing, Warrendale, PA, 2021.

Index

B

D

S

T

W

Y

Z

About the Author

)r. **John Kershaw** has over 53 years of experi-
nce in automotive technology. He was an
ngineering Grade member of SAE for many years.
Ie has provided classroom curriculum and training
s an instructor educating students and developing
ersonal employee fulfillment utilizing Passion Test
nd Passion Test for Business. He is an IASSC-
ertified Lean Six-Sigma Black Belt. Dr. Kershaw
pplies his expertise to the forensic investigation of
laims involving automotive failure and determines the cause or causes of
utomotive failures and validates manufacturer defects, or he validates the
ormal operation of vehicles involved in such mechanical claims. He is the
uthor of fifteen General Motors (GM) technical training publications as well
s the published author of six automotive textbooks. In addition he has devel-
ped instructional materials for GM, Nissan, Fiat, Corinthian Colleges,
Iyundai, Mazda, Ohio Technical College, IntelliTec College, Community
College of Philadelphia, NAPA Auto Parts Canada, ConsuLab, General Mills,
nd the University of Missouri in Columbia along with Penn Foster College.

www.ingramcontent.com/pod-product-compliance
Lightning Source LLC
LaVergne TN
LVHW051046210926
848486LV00001B/37

* 9 7 8 1 4 6 8 6 0 4 0 6 1 *